Subcellular Biochemistry

Volume 18

Intracellular Parasites

SUBCELLULAR BIOCHEMISTRY

SERIES EDITOR

J. R. HARRIS, Department of Molecular Structural Biology, Max-Planck-Institute for Biochemistry, Martinsried bei München, Germany

ASSISTANT EDITOR

H. J. HILDERSON, University of Antwerp, Antwerp, Belgium

Recent Volumes in This Series:

Volumes 5–11 Edited by Donald B. Roodyn

Volume 12 **Immunological Aspects**
Edited by J. R. Harris

Volume 13 **Fluorescence Studies on Biological Membranes**
Edited by H. J. Hilderson

Volume 14 **Artificial and Reconstituted Membrane Systems**
Edited by J. R. Harris and A.-H. Etémadi

Volume 15 **Virally Infected Cells**
Edited by J. R. Harris

Volume 16 **Intracellular Transfer of Lipid Molecules**
Edited by H. J. Hilderson

Volume 17 **Plant Genetic Engineering**
Edited by B. B. Biswas and J. R. Harris

Volume 18 **Intracellular Parasites**
Edited by J. L. Avila and J. R. Harris

Volume 19 **Endocytic Components: Identification and Characterization**
Edited by J. J. M. Bergeron and J. R. Harris

A Continuation Order Plan is available for this series. A continuation order will bring delivery of each new volume immediately upon publication. Volumes are billed only upon actual shipment. For further information please contact the publisher.

Subcellular Biochemistry

Volume 18
Intracellular Parasites

Edited by

J. L. Avila
Instituto de Biomedicina
Caracas, Venezuela

and

J. R. Harris
Max-Planck-Institute for Biochemistry
Martinsried bei München, Germany

Springer Science+Business Media, LLC

The Library of Congress cataloged the first volume of this title as follows:

Sub-cellular biochemistry.

London, New York, Plenum Press.
v. illus. 23 cm. quarterly.
Began with Sept. 1971 issue. Cf. New serial titles.
1. Cytochemistry—Periodicals. 2. Cell organelles—Periodicals.
QH611.S84 574.8′76 73-643479

ISBN 978-1-4899-1653-2 ISBN 978-1-4899-1651-8 (eBook)
DOI 10.1007/978-1-4899-1651-8

This series is a continuation of the journal *Sub-Cellular Biochemistry,*
Volumes 1 to 4 of which were published quarterly from 1972 to 1975

Orginally published by Plenum Press, New York in 1992
Softcover reprint of the hardcover 1st edition 1992

Contributors

Marie Laure Ancelin Membrane Interactions, CNRS URA 530, 34095 Montpellier, France

José Luis Avila Instituto of Biomedicina, Apartado 4043, Caracas 1010A, Venezuela

Patrick Bastien Laboratoire d'Ecologie Médicale et Pathologie Parasitaire, Annex de la Faculté de Médecine, 34000 Montpellier, France

Christine Blaineau Laboratoire d'Ecologie Médicale et Pathologie Parasitaire, Annex de la Faculté de Médecine, 34000 Montpellier, France

Clément Bordier Affinity Products, CH-1023 Crissier-sur-Lausanne, Switzerland

Juan José Cazzulo Instituto de Investigaciones Bioquimicas "Luis F. Leloir," Fundación Campomar—CONICET—Facultad de Ciencias Exactas y Naturales, Universidad de Buenos Aires, 1405 Buenos Aires, Argentina

Robert Etges Institute of Biochemistry, University of Lausanne, CH-1066 Epalinges, Switzerland

Brian Fenton Cell Cycle Genetics Group, Department of Biochemistry, Medical Sciences Institute, University of Dundee, Dundee DD1 4HN, Scotland

Laurie Hall Department of Tropical Public Health, Harvard School of Public Health, Boston, Massachusetts 02115

Selma M. B. Jeronimo Department of Biochemistry, Federal University of Rio Grande do Norte, Natal, Rio Grande do Norte 59072, Brazil

Mark A. Jutila Veterinary Molecular Biology, Marsh Laboratory, Montana State University, Bozeman, Montana 59717

Paul B. Kimsey Division of Comparative Medicine, Massachusetts Institute of Technology, Cambridge, Massachusetts 02139

J. Joseph Marr Discovery Research, Searle Pharmaceuticals, Skokie, Illinois 60077, and Department of Internal Medicine, Rush Medical College, Chicago, Illinois 60612

Mary McGurn Department of Tropical Public Health, Harvard School of Public Health, Boston, Massachusetts 02115

Steven R. Meshnick Department of Epidemiology, University of Michigan School of Public Health, Ann Arbor, Michigan 48109

Michel Pagès Centre de Recherche de Biochimie Macromoleculaire du CNRS et Unité 249 de l'INSERM, 34033 Montpellier Cedex, France

Richard D. Pearson Division of Geographic Medicine, Departments of Internal Medicine and Pathology, University of Virginia School of Medicine, Charlottesville, Virginia 22908

Laura Povinelli Department of Tropical Public Health, Harvard School of Public Health, Boston, Massachusetts 02115

Linda D. Saffer Department of Pathology, University of Virginia School of Medicine, Charlottesville, Virginia 22908

Pascal Schneider Institute of Biochemistry, University of Lausanne, CH-1066 Epalinges, Switzerland

Joseph D. Schwartzman Department of Pathology, University of Virginia School of Medicine, Charlottesville, Virginia 22908

Anuraj Shankar Department of Tropical Public Health, Harvard School of Public Health, Boston, Massachusetts 02115

Alwi M. Shatry Veterinary Molecular Biology, Marsh Laboratory, Montana State University, Bozeman, Montana 59717

Clarence A. Speer Veterinary Molecular Biology, Marsh Laboratory, Montana State University, Bozeman, Montana 59717

Cynthia M. Theodos Department of Tropical Public Health, Harvard School of Public Health, Boston, Massachusetts 02115

Richard G. Titus Department of Tropical Public Health, Harvard School of Public Health, Boston, Massachusetts 02115

Salvatore J. Turco Department of Biochemistry, University of Kentucky Medical Center, Lexington, Kentucky 40536

Henri J. Vial Membrane Interactions, CNRS URA 530, 34095 Montpellier, France

David Walliker Institute of Cell, Animal, and Population Biology, University of Edinburgh, Edinburgh EH9 3JN, Scotland

Earl Weidner Department of Zoology and Physiology, Louisiana State University, Baton Rouge, Louisiana 70803, and The Marine Biological Laboratory, Woods Hole, Massachusetts 02543

Michael W. White Veterinary Molecular Biology, Marsh Laboratory, Montana State University, Bozeman, Montana 59717

Preface

Volume 18 of the *Subcellular Biochemistry* series, which specializes in various aspects of the biochemistry of the intracellular parasites, was initially proposed by José Luis Avila and strongly supported by myself, as Series Editor of *Subcellular Biochemistry.* Considerable assistance was received from Professor Frank Wunderlich (University of Dusseldorf) and more particularly from Dr. Michael Miles (London School of Hygiene and Tropical Medicine) during the compilation of the list of possible chapters. Our initial aim was to present a comprehensive survey of this broad field of study. Although some interesting topics have been lost due to authors backing out late in the production schedule of the book, the manuscripts that were submitted have provided a useful overview of the subject, with notable strength within the field of *Leishmania.* The 13 chapters of the book have been grouped according to subject. The first five chapters deal with *Leishmania* and are followed by two chapters on *Tryponosoma cruzi,* two on the malarial parasites, and two on the Coccidia. The final two chapters cover the Microsporidia and chemotherapy, respectively.

The book opens with the contribution of Selma Jeronimo and Richard Pearson, in which they set the scene by discussing the *Leishmania* within the context of adaption to extracellular and intracellular survival. Chapter 2 is a detailed account of the membrane proteins and enzymes of *Leishmania* by Pascal Schneider, Clément Bordier, and Robert Etges. In Chapter 3, Salvatore Turco presents an account of structural and functional aspects of the lipophosphoglycan of *Leishmania.* Immunity to leishmaniasis and the role of T cells and cytokines are dealt with in an extremely competent manner by Richard Titus and his colleagues. The final chapter on the *Leishmania* comes from Patrick Bastien, Christine Blaineau, and Michel Pagès, who discuss at some length the important topic of molecular karyotype analysis. Moving to *Tryponosoma cruzi,* José Luis Avila presents a thorough account of the role of intracellular digestion of endocytosed proteins as a source of amino acids for protein synthesis, and Juan José Cazzulo discusses energy metabolism. A lengthy overview of the malarial lipids comes from Henri Vial and Marie Laure Ancelin and is followed by the contribution of Brian Fenton and David Walliker on the genetic analysis of malaria parasites. The first of the two chapters on the coccidian parasites, from Joseph Schwartzman and Linda Saffer, considers the mechanisms of *Toxoplasma gondii* entry and exit from host cells. Then Michael White and his colleagues present a survey of both biochemical and ultrastructural observations on coccidian host cell interactions. Earl Weidner then covers the microsporidian parasites, with emphasis on the cytoskeletal proteins in the *Thelohania* species and *Spraguea lophii*. In the final chapter of the book, Steven Meshnick and Joseph Marr appropriately present a survey of new developments in chemotherapy; they deal successively with drugs directed at the four main groups of intracellular parasites discussed in the earlier chapters: malaria, leishmaniasis, American trypanosomiasis, and toxoplasmosis.

Thus, despite the unforeseen omissions mentioned above, this volume nevertheless possesses a considerable unity of purpose and content. It is hoped that the book will be of interest and use to all those involved with the investigation of intracellular parasites and treatment of the diseases they cause in humans. Biochemists interested in the cellular metabolism and membrane systems of intracellular parasites should also find much of interest in this book, as will molecular geneticists.

Robin Harris

Martinsried bei München, Germany

Contents

Chapter 1

The *Leishmania:* Protozoans Adapted for Extracellular and Intracellular Survival

Selma M. B. Jeronimo and Richard D. Pearson

Chapter 2

Membrane Proteins and Enzymes of *Leishmania*

Pascal Schneider, Clément Bordier, and Robert Etges

Chapter 3

The Lipophosphoglycan of *Leishmania*

Salvatore J. Turco

Chapter 4

Role of T Cells in Immunity to the Intracellular Pathogen, *Leishmania major*

Richard G. Titus, Cynthia M. Theodos, Paul B. Kimsey, Anuraj Shankar, Laurie Hall, Mary McGurn, and Laura Povinelli

Chapter 5

Molecular Karyotype Analysis in *Leishmania*

Patrick Bastien, Christine Blaineau, and Michel Pagès

Chapter 6

Intracellular Digestion of Endocytosed Proteins as a Source of Amino Acids for Protein Synthesis in *Trypanosoma cruzi*

José Luis Avila

Chapter 9

Genetic Analysis of Malaria Parasites

Brian Fenton and David Walliker

Chapter 10

How *Toxoplasma gondii* Gets In and Out of Host Cells

Joseph D. Schwartzman and Linda D. Saffer

Chapter 11

Biochemical and Ultrastructural Observations of Coccidian Parasite and Host Cell Interactions

Michael W. White, Alwi M. Shatry, Mark A. Jutila, and Clarence A. Speer

Chapter 12

Cytoskeletal Proteins Expressed by Microsporidian Parasites

Earl Weidner

Chapter 13

Intracellular Parasites: New Developments in Chemotherapy

Steven R. Meshnick and J. Joseph Marr

Chapter 1

The *Leishmania*

Protozoans Adapted for Extracellular and Intracellular Survival

Selma M. B. Jeronimo and Richard D. Pearson

1. INTRODUCTION

Leishmania species are among the most fascinating of the protozoal pathogens that infect humans. They are single-celled parasites that are capable of flourishing in widely disparate intracellular and extracellular environments. The *Leishmania* successfully take sanctuary in mononuclear phagocytes, host defense cells responsible for the ingestion and killing of microbes and the processing and presentation of antigens. The *Leishmania* bind to a number of macrophage receptors, are ingested by the phagocytes, avoid death by macrophage oxidative and nonoxidative killing mechanisms, and grow and multiply in the acidic environment of the phagolysosome as amastigotes. Not only are the *Leishmania* able to survive in an intracellular environment that is lethal for other microbes, but in many instances they effectively delay or prevent the development of protective T-

Selma M. B. Jeronimo Department of Biochemistry, Federal University of Rio Grande do Norte, Natal, Rio Grande do Norte 59072, Brazil. **Richard D. Pearson** Division of Geographic Medicine, Departments of Internal Medicine and Pathology, University of Virginia School of Medicine, Charlottesville, Virginia 22908.

Subcellular Biochemistry, Volume 18: Intracellular Parasites, edited by J. L. Avila and J. R. Harris. Plenum Press, New York, 1992.

cell-mediated immune responses. Leishmanial infections have emerged as model systems for the study of cell-mediated immunity (Wilson and Pearson, 1990).

The *Leishmania* are spread among mammals by phlebotomine sandflies. They ingest *Leishmania*-infected macrophages when they take blood meals. The *Leishmania* break out of the macrophages and they develop and multiply as flagellated promastigotes in the sandfly gut where they are exposed to an extracellular environment and ambient temperatures of approximately 25 °C. After maturation, infective metacyclic promastigotes migrate to the proboscis and are inoculated into the next mammalian host when the sandfly attempts to take another blood meal. Studies over the past decade have provided important insights into how the *Leishmania* survive and multiply in these two disparate environments (Pearson and de Sousa, 1990).

2. TAXONOMY OF *LEISHMANIA* SPECIES

The genus *Leishmania* is in the order Kinetoplastida and the family Trypanosomatidae. The *Leishmania* are related to *Trypanosoma cruzi*, the cause of Chagas' disease in Latin America, and to *Trypanosoma brucei* subspecies, which cause sleeping sickness in animals and humans in Africa. The taxonomy of *Leishmania* species is still evolving (Lainson and Shaw, 1987; Grimaldi *et al.* 1989), and new species have recently been identified. There are some morphologic differences among the promastigotes and amastigotes of different *Leishmania* species, but the differences are not sufficient to allow for species identification.

Traditionally, speciation of *Leishmania* was based on the clinical syndromes that they produced in humans, the animal reservoir(s) of infection, and the sandfly species responsible for transmission (Pearson and de Sousa, 1990). More recently, isoenzyme analysis (Ebert, 1973; Gardener *et al.*, 1974; Brazil, 1978; Rassam *et al.*, 1979; Schnur *et al.*, 1981), species-specific monoclonal antibodies (McMahon-Pratt and David, 1981), restriction endonuclease digestion of kinetoplast DNA (Arnot and Barker, 1981; Jackson *et al.*, 1984) and species-specific hybridization of kinetoplast DNA (Wirth and McMahon-Pratt, 1982) have been used for classification, but no method has emerged as the single standard for speciation. Lainson and Shaw (1987) proposed a classification based on all of the above parameters (Table I). Pulse field gradient electrophoresis has been used to separate chromosomes of some *Leishmania* species, and the karyotypes differ (Giannini *et al.*, 1986).

Several *Leishmania* species are responsible for the majority of human cases (Table I). These species produce a spectrum of clinical disease that includes visceral, cutaneous, and mucosal syndromes (Pearson and de Sousa, 1990). Visceral leishmaniasis is usually caused by *L. (Leishmania) donovani* subspecies

Table I
Clinical Manifestations and Geographic Distribution of Leishmaniasis[a]

Syndromes	Parasite	Geographic Areas
Visceral leishmaniasis (kala-azar)		
General involvement of the reticuloendothelial system (spleen, bone marrow, liver etc.)	*Leishmania (L.) donovani*	Indian subcontinent, China
	L. (L.) infantum	Middle East, Mediterranean littoral, Balkans, Western Asiatic area, Northwestern Iberia, China, subsaharan Africa
	L. (L.) chagasi	Latin America
New World cutaneous leishmaniasis		
Single or limited number of skin lesions	*L. (L.) mexicana* (chiclero ulcer)	Mexico, Central America Texas (?)
	L. (L.) amazonensis	Amazon basin and neighboring areas
	L. (L.) pifanoi	Venezuela
	L. (Viannia) braziliensis	Brazil, Peru, Ecuador, Bolivia, Paraguay, Argentina
	L. (V.) guyanensis (pian bois, bush yaws)	Guyana, Surinam, northern Amazon Basin
	L. (V.) peruvian (uta)	Peru, Western Andes
	L. (V.) panamensis	Panama and adjacent areas
Diffuse cutaneous leishmaniasis	*L. (L.) amazonensis*	Amazon basin and neighboring areas
	L. (L.) pifanoi	Venezuela
	L. (L.) mexicana	Mexico, Central America
Mucosal leishmaniasis	*L. (V.) braziliensis* (espundia)	Multiple areas in South America
	L. (V.) panamensis (rare)	Panama and adjacent areas
Old World cutaneous leishmaniasis		
Single or limited number of skin lesions	*L. (L.) major*	Middle East, Central Asia, Africa, Indian subcontinent
	L. (L.) tropica	Mediterranean littoral, Middle East, West Asia, India subcontinent
	L. (L.) aethiopica	Ethiopian highlands
Diffuse cutaneous leishmaniasis	*L. (L.) aethiopica*	Ethiopian highlands, Kenya
Post-kala-azar dermal	*L. (L.) donovani*	India subcontinent
	L. (L.) species	Kenya, possibly Ethiopia and Somalia

[a]Based on Lainson and Shaw (1987).

and by *L. (L.) chagasi,* although on occasion, *Leishmania* species that characteristically cause cutaneous disease have been isolated from animals or humans with visceral syndromes.

Cutaneous leishmaniasis in the Western hemisphere is usually caused by *L. (L.) mexicana* subspecies or *L. (Viannia) braziliensis* subspecies. Cutaneous disease in Europe, Asia, India, and Africa is caused by *L. (L.) major, L. (L.) tropica,* or *L. (L.) aethiopica.* On occasion, *L. (L.) chagasi* is isolated from simple cutaneous lesions. *Leishmania (L.) donovani* occasionally produces nodular or plaquelike skin lesions after resolution of visceral leishmaniasis, a condition known as post-kala-azar dermal leishmaniasis.

Mucosal disease in Latin American is usually due to *L. (V.) braziliensis* and typically follows the resolution of cutaneous lesions by months to years, the median duration being 4.5 years (Zajtchuk *et al.*, 1989). *Leishmania (V.) panamensis* can also cause mucosal lesions.

3. CLINICAL LEISHMANIASIS

Leishmaniasis refers to a spectrum of clinical syndromes produced by members of the genus *Leishmania.* The clinical manifestations of infection depend on complex interactions between genetically determined mammalian immune responses and the genetically defined virulence characteristics of the infecting *Leishmania* species.

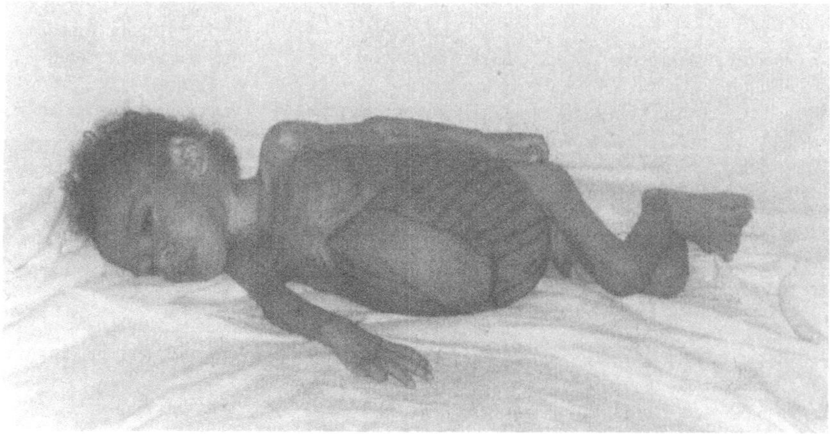

FIGURE 1. Visceral leishmaniasis (kala-azar). The child is cachectic. Abdominal markings indicate hepatosplenomegaly. (Photograph courtesy of Dr. Anastacio de Sousa, Universidade Federal do Ceara, Fortaleza, Brazil.)

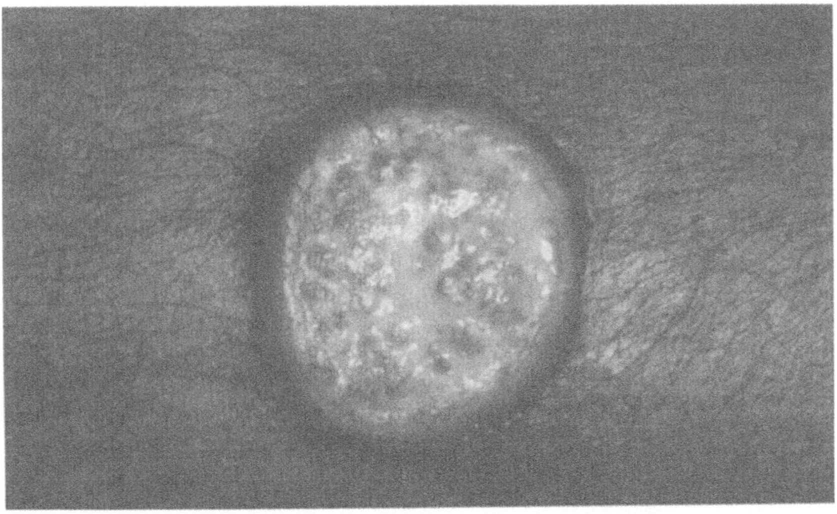

FIGURE 2. American cutaneous leishmaniasis. A single, large lesion on the leg. (Photograph courtesy of Dr. Anastacio de Sousa, Universidade Federal do Ceara, Fortaleza, Brazil.)

Leishmaniasis is divided into visceral, cutaneous, and mucosal syndromes. Visceral leishmaniasis is an important cause of morbidity and mortality in widely scattered areas in India, Africa, Latin America, and the Mediterranean littoral (Figure 1). Cutaneous leishmaniasis is more common. Although rarely associated with mortality, cutaneous leishmanial lesions are usually chronic and may be disfiguring (Figure 2). Mucosal leishmaniasis can result in destructive lesions of the nose, oral pharynx, lips, or face (Figure 3). It is an important cause of disfigurement and morbidity in endemic areas of Latin America.

3.1. Visceral Leishmaniasis

Leishmania (L.) donovani or *L. (L.) chagasi* initially multiply in mononuclear phagocytes at the dermal site where promastigotes are inoculated by sandflies. The majority of infections are asymptomatic, and parasites are eliminated by effective cell-mediated immune responses. In persons with disease, amastigotes disseminate to macrophages throughout the reticuloendothelial system. Some persons have mild symptoms, or a subclinical form of visceral leishmaniasis (Badaró *et al.*, 1986a,b). A subset of those infected develop progressive visceral leishmaniasis, or kala-azar, which is characterized by fever, anorexia, weight loss, hypergammaglobulinemia, hepatomegaly, and massive splenomegaly (Evans *et al.*, 1985).

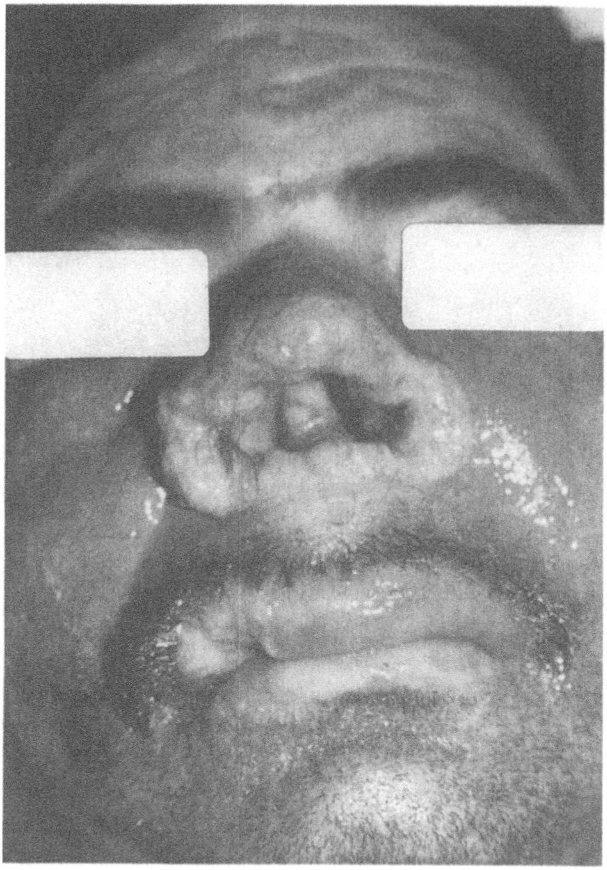

FIGURE 3. American mucosal leishmaniasis. There is extensive involvement of the nose and upper lip with destruction of the nasal septum. From Pearson *et al.* (1983c), with permission.

Patients with progressive visceral leishmaniasis have no evidence of T-cell-mediated immune responses against *Leishmania*. There is no delayed-type hypersensitivity reaction to leishmanial antigens administered intradermally and circulating T lymphocytes neither proliferate nor produce interferon-gamma, a macrophage-activating factor, or interleukin 2 in response to parasite antigens *in vitro* (Carvalho *et al.*, 1985, 1989).

Treatment of progressive infection is usually with a pentavalent antimonial such as stibogluconate sodium of meglumine antimonate (Berman, 1988a; Pearson and de Sousa, 1990). Recently, the combination of pentavalent antimony and recombinant interferon-gamma, which can activate macrophages *in vitro* and *in vivo* to prevent the growth of *Leishmania*, has been shown to be effective in treating patients who did not respond to pentavalent antimony alone (Badaró *et al.*, 1990).

Visceral leishmaniasis is recognized as an opportunistic infection in persons with the acquired immunodeficiency syndrome (Montalban *et al.*, 1989) or other diseases which compromise T-cell-mediated immune responses (Badaró *et al.*, 1986c; Fernandez-Guerrero *et al.*, 1987). Persons with these disorders are more likely to develop progressive visceral disease when infected with *L. (L.) donovani*. They often present with unusual manifestations and respond poorly to antileishmanial chemotherapy.

3.2. Cutaneous Leishmaniasis

Cutaneous leishmaniasis is characterized by the development of a chronic skin ulcer at the site where *Leishmania* are inoculated. The characteristics of the lesions vary depending on the infecting *Leishmania* species and the immune responses of the mammalian host. Typically, a nodule forms at the site where promastigotes were inoculated into the skin. With time, the overlying skin ulcerates. Leishmanial ulcers may persist for months to years but eventually they heal, leaving an atrophic scar as evidence of the disease.

Cutaneous ulceration is probably a consequence of the immune response against *Leishmania*. Ulceration is observed at approximately the time that circulating T cells first demonstrate responsiveness to parasite antigens (Sadick *et al.*, 1984). Despite evidence of cell-mediated immunity, the local immune response is not immediately effective in eradicating organisms from the skin. Once healing has occurred, there is evidence of strong *Leishmania*-specific T-cell-mediated immunity against reinfection.

There are several variants of cutaneous leishmaniasis. In diffuse cutaneous leishmaniasis, an anergic variant, amastigotes initially multiply in macrophages at the site of inoculation and eventually disseminate to macrophages throughout the skin forming nonulcerative metastatic lesions. There is no evidence of T-cell responses to leishmanial antigens. At the other end of the spectrum, patients with leishmaniasis recidiva show *Leishmania*-specific T-cell responses, but they have chronic ulcerative lesions that persist for months to years. These lesions often have slowly advancing margins and tend to heal in the center. A limited number of parasites are found in macrophages at the periphery of the lesions.

3.3. Mucosal Leishmaniasis

In Latin America, a small percentage of persons with cutaneous leishmaniasis due to *L. (V.) braziliensis* develop destructive mucosal lesions, usually years after resolution of their cutaneous lesions (Zajtchuk *et al.*, 1989). Histologically there is a destructive mononuclear cell response. There is evidence of T-cell responses to leishmanial antigens and only a few parasites are found in the lesions. It appears that a hyperergic, cell-mediated immune response causes destruction of host tissue, but it is not effective in eradicating amastigotes.

4. LIFE CYCLE

4.1. Amastigotes

Leishmania reside as amastigotes within mammalian mononuclear phagocytes in parasitophorous vacuoles (Figure 4). They are oval or round in shape and 2 to 3 μm in diameter (Figure 5). The plasma membrane of all stages of *Leishmania* is a typical trilaminar unit. It overlies rows of subpellicular microtubules much like the surface of an umbrella over its metal frame. The nucleus is similar in ultrastructure to nuclei of other eukaryotic cells. It is contained within two nuclear membranes punctuated by pore complexes and it contains a prominent nucleolus (Molyneux and Killick-Kendrick, 1987). The kinetoplast, which is a specialized mitochondrial structure, contains a substantial amount of extranuclear DNA in the form of minicircles and maxicircles. It lies posterior to the basal body.

Amastigotes have a flagellum within a flagellar pocket, but the flagellum does not extend outside the circumference of the cell. Within the cytoplasm there are membrane-associated organelles including the smooth and rough endoplasmic reticulum and Golgi apparatus, which are similar to those of other eukaryotes. A variety of vacuoles including lysosomes are present. Molyneux and Killick-Kendrick (1987) have reviewed the ultrastructure and morphology of the *Leishmania* in detail.

Sandflies of the genera *Lutzomyia* and *Psychodopygus* are responsible for transmission of *Leishmania* in the Western hemisphere and *Phlebotomus* species elsewhere. Sandflies are modified pool feeders; that is, they create a pool of blood by repeatedly proving with their proboscis. Once the pool is formed, they aspirate blood. Sandflies become infected when they ingest mononuclear cells that contain amastigotes.

4.2. Promastigotes

The *Leishmania* undergo differentiation as flagellated extracellular organisms in the gut of the sandfly. As many as ten morphologic forms have been observed with *L. major* as it develops in *Phlebotomus duboscqi* (Lawyer *et al.*, 1990). Some of the intermediate forms adhere to the epithelium of the sandfly gut with their flagella embedded deeply into microvilli. Other forms attach to epi-

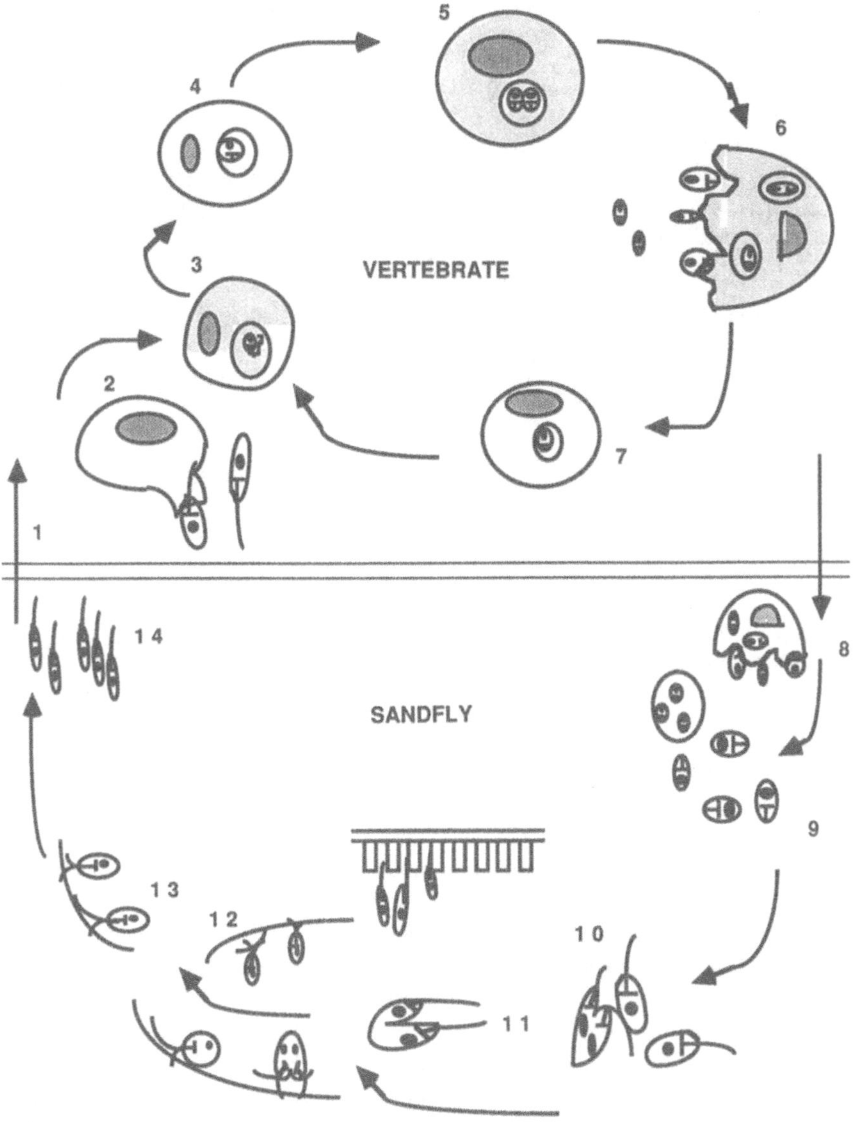

FIGURE 4. Principal events in the life cycle of *Leishmania*: (1) Sandflies inoculate promastigotes into vertebrate; (2) Attachment and engulfment of promastigotes by macrophages; (3) Fusion of phagosome and lysosome; (4) Differentiation of promastigotes into amastigotes; (5) Division of amastigotes; (6) Rupture of parasitized macrophage; (7) Infection of a different macrophage; (8) Sandfly infection after a blood meal; (9–13) Development of promastigotes in the sandfly gut; (14) Metacyclic promastigotes are inoculated into the next vertebrate. Modified from Chang (1990).

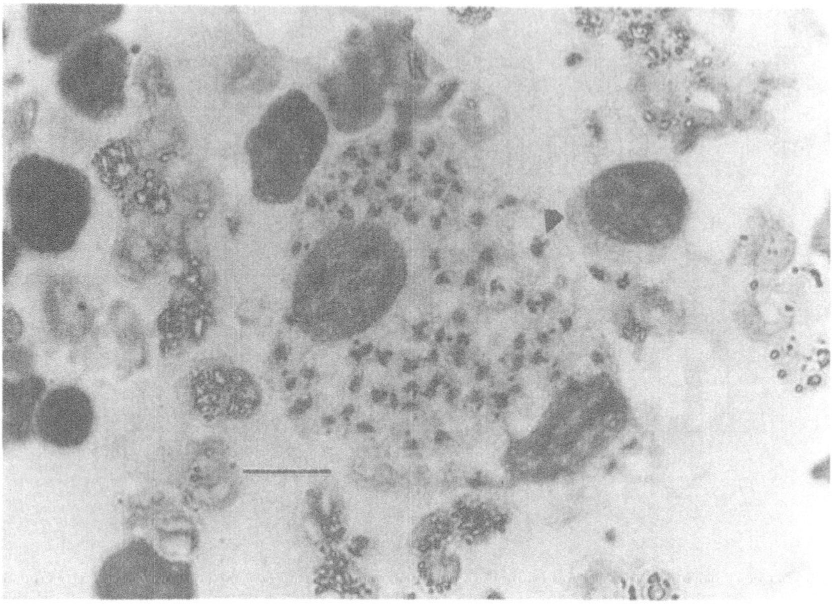

FIGURE 5. Touch preparation of a hamster spleen infected with *L.* (*L.*) *donovani* amastigotes (arrow). The bar equals 10 μm.

thelial cells by hemidesmosomes. The number of morphological forms and the location of their development within the sandfly gut varies among the *Leishmania* species and the sandfly species (Walters *et al.*, 1989; Lawyer *et al.*, 1990).

After development is complete, a process that takes approximately a week, infective, free-swimming "metacyclic" promastigotes migrate to the proboscis. They are spindle-shaped, flagellated, and highly motile. Promastigotes have variable dimensions. The average promastigote is 10–15 μm in length and 2–3 μm in width. The flagellum extends from the basal body out through the flagellar reservoir at the anterior pole. It is composed of nine pairs of peripheral axonemal doublets encircling a central pair. Promastigotes with a long flagellum also have a paracrystalline, latticelike paraxial rod, which runs parallel to the axoneme (Molyneux and Killick-Kendrick, 1987).

The structure of the plasma membrane and the underlying subpellicular microtubules of promastigotes are modified at the point where the flagellum exits from the flagellar reservoir (Molyneux and Killick-Kendrick, 1987). The flagellar reservoir is not lined with microtubules. Desmosomal attachment plaques are formed where the flagellar plasma membrane runs parallel to the reservoir membrane. Uptake of macromolecules is thought to occur through the plasma membrane of the reservoir. Both pinocytosis and omega-type spinycoated vesicles have been observed in the flagellar reservoir of promastigotes (Molyneux *et al.*, 1975).

A number of culture systems are available for *in vitro* cultivation of promastigotes. Most *Leishmania* species grow well *in vitro*. Isolates of *L. (V.) braziliensis* are frequently exceptions. Novy-MacNeal-Nocille (NNN), a biphasic blood-based medium with fetal calf serum, is the oldest. Several liquid media including Schneider's insect media with fetal calf serum (Hockmeyer *et al.*, 1981) are now widely used. Although axenic culture systems have been described, most media require the presence of fetal calf serum for optimal promastigote multiplication.

Cultured promastigotes are a heterogenous population. They are flagellated, but their bodies vary from round to spindle shaped. The morphology and infectivity of promastigotes vary with growth phase and culture conditions. The various forms observed in culture probably reflect to some degree the different developmental forms obtained in the gut of sandflies.

Cultured promastigotes require temperatures of 24–26 °C for optimal growth. This is similar to the ambient temperature of the sandflies' natural environments. In contrast to amastigotes, most promastigotes die at 37 °C. Promastigotes grow well at neutral to relatively acidic pH; they generally grow poorly or die at higher pH.

The infectivity of cultured promastigotes varies with the growth phase of the culture. Stationary-phase promastigotes are more infective than logarithmic-phase organisms. This correlates with changes in surface antigens. Sacks *et al.* (1985) and Sacks and da Silva (1987) observed that *L. (L.) major* promastigotes, which become infective as they enter stationary phase, lose their ability to bind peanut agglutinin as a result of the differential expression of a developmentally regulated surface glycolipid. A similar developmental change is thought to occur in the sandfly gut (Sacks and Perkins, 1984).

The precise sequence of events that follows sandfly inoculation of promastigotes into mammals has not been worked out, but data from hamsters given intradermal inoculations of cultured *L. (L.) donovani* promastigotes suggest that promastigotes quickly gain access to mononuclear phagocytes, convert within them to amastigotes, multiply, and then infect additional mononuclear cells that arrive at the site of infection (Wilson *et al.*, 1987). Control of infection is ultimately dependent on the proliferation of *Leishmania*-specific T cells that are capable of activating macrophages to kill amastigotes (Wilson and Pearson, 1990).

5. METABOLISM

Energy for the *Leishmania* and other higher eukaryotic cells derives from several sources: carbohydrates, which are metabolized in the glycolytic pathway and the pentose shunt; β-oxidation of fatty acids; and catabolism of amino acids. Like other Trypanosomatidae, *Leishmania* have a subcellular organelle called the glycosome that contains enzymes of the glycolytic pathway, as well as enzymes for glycerol metabolism, carbon dioxide fixation, β-oxidation of fatty acids, and synthesis of ether lipids (Hart and Opperdoes, 1984). Mammalian cells lack these structures. Salient aspects of leishmanial biochemistry are reviewed below. Several authors have previously reviewed this subject (Chang *et al.*, 1985; Janovy, 1987; Berman, 1988b; Glew *et al.*, 1988).

5.1. Carbohydrates

Leishmania can metabolize several carbohydrates. Uptake of D-glucose as well as amino acids is driven by a symport type of transport coupled to the proton gradient (Zilberstein and Dwyer, 1985). Enzymes of the glycolytic pathway, the tricarboxylic acid cycle (Krebs cycle), and the pentose shunt have been described for both amastigotes and promastigotes of several *Leishmania* species (Martin *et al.*, 1976; Hart and Coombs, 1982; Meade *et al.*, 1984). In general, *Leishmania* metabolize sugars to carbon dioxide and organic acids (Figure 6), predominantly succinate, which is formed by fixation of carbon dioxide and reverse flux through part of the tricarboxylic acid cycle (Darling *et al.*, 1989). The *Leishmania* do not have lactate dehydrogenase (Martin *et al.*, 1976), the enzyme that normally mediates the reduction of pyruvate to lactate using NADH as a coenzyme and also regulates the cytoplasmic level of NAD^+. Instead, the *Leishmania* have α-glycerophosphate dehydrogenase and glycerophosphate phosphatase, which can regulate the level of NAD^+ (Martin *et al.*, 1976; Meade *et al.*, 1984).

Coombs *et al.* (1982) reported that the activities of glycolytic enzymes were lower in amastigotes than in promastigotes. The production of acetate from different sources indicated that pyruvate dehydrogenase was active (Meade *et al.*, 1984). Acetate was oxidized slowly suggesting that citrate synthase has a low specific activity (Darling *et al.*, 1989).

Promastigotes of *L. (V.) panamensis* grown in media containing glucose as the single source of carbon released succinate, acetate, pyruvate, alanine, and carbon dioxide. Under anaerobic conditions they released glycerol (Darling *et al.*, 1987). Darling *et al.* (1989) subsequently studied the utilization of metabolites of glucose by promastigotes. When promastigotes were incubated with glycerol, they released acetate, succinate, D-lactate, and carbon dioxide. Darling and Blum (1988) demonstrated that promastigotes incubated anaerobically produced D-lactate from glucose, ribose, and methylglyoxal, but not from glycerol, alanine, or pyruvate, suggesting the existence of the enzymes glyoxalases I and II and the absence of D-lactate dehydrogenase. D-Lactate production from glucose was increased over fivefold under anaerobic conditions. D-Lactate is not metabolized by the parasite. Its production is thought to be the mechanism by

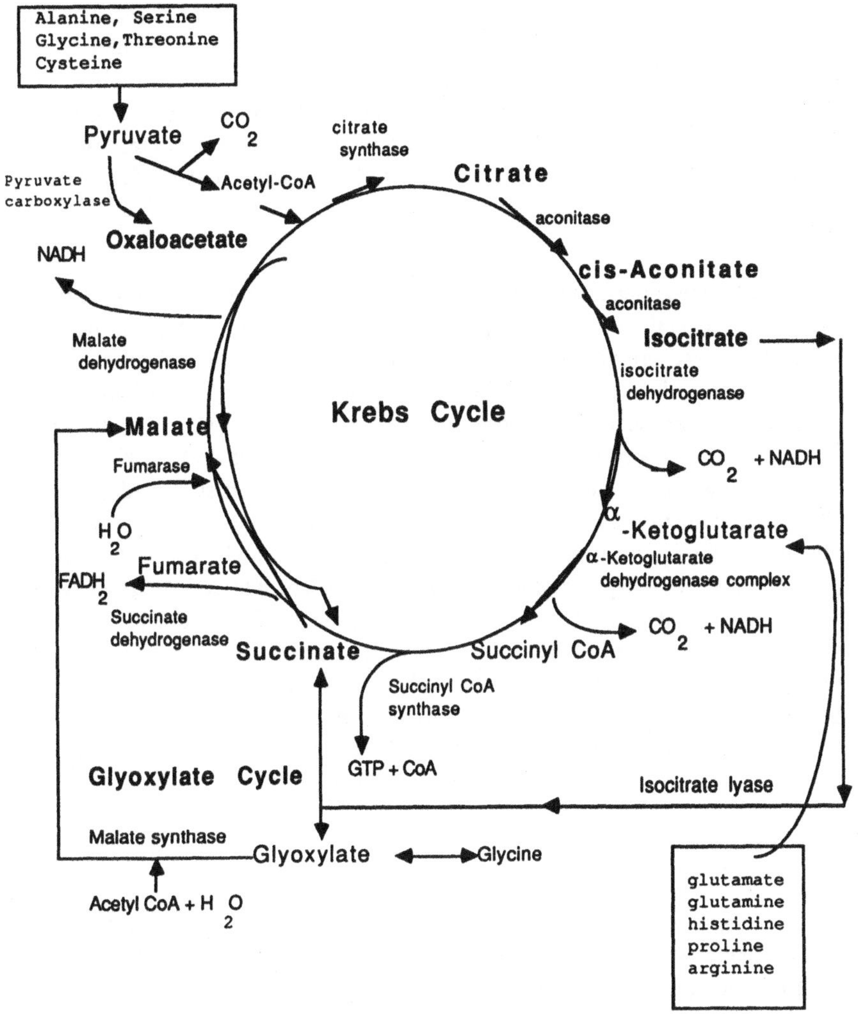

FIGURE 6. Schematic representation of Krebs and glyoxylate cycles in *Leishmania* and entry of some amino acids into the Krebs cycle.

which the parasite detoxifies methylglyoxal produced under anaerobic conditions (Darling and Blum, 1988). Succinate derived from glycerol, like succinate from glucose, originates by fixation of carbon dioxide followed by reverse flux through part of the Krebs cycle. When promastigotes were incubated with glucose and glycerol, the consumption of glucose was unchanged and the consumption of glycerol decreased by one-half. Alanine did not affect the consumption of either glucose or glycerol, but there was a decreased release of carbon dioxide and increased flow of label into alanine. Glutamate inhibited glucose consumption slightly (Darling *et al.*, 1989).

Berman *et al.* (1987a) showed that the rate of utilization of the hexose monophosphate shunt was higher than the glycolytic pathway in amastigotes of *L. (L.) mexicana* by quantifying the metabolism of glucose labeled at 1-^{14}C and 6-^{14}C.

5.2. Amino Acids

Glucose was once thought to be the principle substrate for metabolism, but studies of several *Leishmania* species indicate that they can grow in media without glucose provided that amino acids such as proline and glutamate are present (Mukkada *et al.*, 1974; Steiger and Meshnich, 1977; Mukkada, 1985). Amino acids are deaminated and their carbon skeletons utilized in the tricarboxylic acid pathway (Simon *et al.*, 1983). Figure 6 shows two of the points where amino acids can enter the Krebs cycle once they are deaminated. Leishmanial enzymes that are capable of converting amino acids into intermediates of the Krebs cycle have been described (Simon *et al.*, 1983). *Leishmania (L.) tropica* promastigotes were shown to catabolize intracellular proteins at a relatively high rate during nutritional stress (Simon and Mukkada, 1983). Amino acids that originated from breakdown of proteins were rapidly utilized as carbon and energy sources. Keegan *et al.* (1987) have shown that *L. (V.) panamensis* promastigotes oxidize glutamate at a higher rate than glucose. In spite of this evidence for the utilization of amino acids by leishmania, further studies are needed to determine the relative role of amino acids in the metabolism of both promastigotes and amastigotes.

5.3. Glyoxylate Cycle

The glyoxylate cycle is active in *Leishmania* species (Simon *et al.*, 1978). The cycle is also present among bacteria, fungi, algae, and higher plants, but it is not found in animal cells. The pathway allows cells to convert two carbon units (acetyl) into four carbon units (succinate). The cycle departs from the Krebs cycle at the level of citrate as shown in Figure 6, bypassing the two decarboxylation steps of the cycle. Citrate is cleaved by isocitrate lyase, forming glyoxylate and succinate. Glyoxylate is acetylated to malate by malate synthase. The products formed, succinate and malate, are intermediates of the Krebs cycle and so can proceed to oxaloacetate.

The glyoxylate cycle is active in the presence of glucose, which was shown by Kornberg (1966) to be a repressor of the cycle in some bacteria. As for the Krebs cycle, the activity of the glyoxylate pathway depends on the availability of citrate. In spite of the lower activity of citrate synthase (Darling *et al.*, 1989), the amount of citrate generated by *Leishmania* is adequate to feed both cycles (Simon *et al.*, 1978).

The role of the glyoxylate pathway in *Leishmania* metabolism is not well understood. The glyoxylate cycle could be coupled to gluconeogenesis/glyconeogenesis. This hypothesis is supported by formation of significant amounts of glucose and glycogen simultaneously with an increased level of malate.

The glyoxylate cycle is not the only source of glyoxylate for the parasite. Transaminases such as L-alanine-glyoxylate aminotransferase and glyoxylate-aspartate aminotransferase can also contribute to glyoxylate formation. The combination of alanine-glyoxylate aminotransferase and glyoxylate-aspartate aminotransferase, which are functional enzymes found in *Leishmania,* may generate glyoxylate, which is later acetylated to malate or used as a precursor of glycine (Simon *et al.*, 1978).

5.4. Fatty Acids

Fatty acids are important substrates for *Leishmania*. Amastigotes cleave both short- and long-chain fatty acids (Berman *et al.*, 1987b). Uptake of exogenous palmitic, stearic, and oleic acids is similar by promastigotes and amastigotes. These fatty acids are rapidly esterified in the form of tryglycerides or phospholipids. Studies of leishmanial fatty acyl coenzyme A dehydrogenase kinetics have revealed a difference from classical Michaelis–Menten kinetics. The V_{max} for long-chain fatty acids was similar to the V_{max} for short-chain substrates, and also the V_{max} was higher for the leishmanial enzyme than for the mammalian enzyme (Berman *et al.*, 1987b).

Darling and Blum (1987) showed that a rise in temperature from 26 °C to 34 °C resulted in the formation of ellipsoidal-shaped promastigotes. Blum (1987) studied the oxidation of fatty acids in promastigotes grown at 26 °C and the heat-induced ellipsoidal forms. He found that heat treatment caused an increase in the rate of oxidation of both medium- and long-chain fatty acids, but it decreased the oxidation of glucose through the pentose shunt. Studies of fatty acids labeled at initial and terminal carbons suggested that β-oxidation occurs in two separate compartments: the mitochondrion and the glycosome. Understanding of the lipid metabolism of *Leishmania* may be important in the development of new chemotherapeutic agents (Berman *et al.*, 1987b). Lipid metabolism of *Leishmania* has previously been reviewed (Holz, 1985; Berman, 1988b; Glew *et al.*, 1988).

6. NUCLEOTIDE METABOLISM

6.1. Purines

Mammalian cells obtain purine nucleotides by two means: *de novo* synthesis and a salvage pathway. In synthesizing purines, mammalian cells use amino acids, carbon dioxide, and ammonia as precursors and assemble the purine ring on the carbon-1 of ribose 5-phosphate. This culminates in the formation of inosine 5′-monophosphate (IMP). IMP can be converted to either adenosine 5′-monophosphate (AMP) or to guanosine 5′-monophosphate (GMP), which can be phosphorylated, forming nucleoside di- and triphosphates. During the process of *de novo* synthesis, there is no release of free base or nucleoside. *Leishmania* lack a pathway for the *de novo* synthesis of purines, and therefore they depend on a salvage pathway to obtain these nucleotides (Marr and Berens, 1985).

In mammalian cells, the purine salvage pathway utilizes bases obtained from the digestion of nucleic acids and nucleotides. The ribose phosphate moiety of 5-phosphoribosyl-1-pyrophosphate (PRPP) is transferred to the purine ring by two different enzymes. Adenine phosphoribosyl transferase catalyzes the formation of adenylate and hypoxanthine-guanine phosphoribosyl transferase catalyzes the formation of both inosinate and guanylate.

Other salvage pathways convert free purines to nucleosides and subsequently to nucleotides. These reactions are catalyzed by purine nucleoside phosphorylases using ribose-1-phosphate as the ribose donor. Inorganic phosphate is released with the purine nucleoside. The nucleoside is then converted to nucleotide through the transfer or a phosphate group to position 5′ of the ribose ring. This reaction is mediated by nucleoside kinase or nucleoside phosphotransferase (Marr and Berens, 1985). *Leishmania* have adenine-phosphoribosyl transferase and hypoxanthine-guanine phosphoribosyl transferase. One additional enzyme that transfers xanthine to PRPP was described in *L. (L.) donovani* promastigotes by Tuttle and Krenistsky (1980). They demonstrated differential activity between the hypoxanthine-guanine phosphoribosyl transferase of *Leishmania* and that of human erythrocytes. The difference in activity explains the accumulation in *Leishmania* of nucleotides formed from purine analogues like allopurinol.

Marr *et al.* (1978) showed that the purine requirements of *Leishmania* are better satisfied by adenosine or inosine than guanosine. Adenosine and inosine are converted to a common intermediate, hypoxanthine, before the transformation to nucleotides. Adenine is also transformed to hypoxanthine by adenine aminohydrolase. There is preferential synthesis of adenine nucleotides as indicated by the high A/G ratio.

Purine metabolism in *L. (L.) donovani* amastigotes was shown to be similar to that in promastigotes (Looker *et al.*, 1983), with the exception of adenosine metabolism. Adenosine kinase has a higher activity in amastigotes. Amastigotes deaminate adenosine to inosine through adenosine deaminase, an enzyme not present in promastigotes. Inosine is then transformed to hypoxanthine and phosphoribosylated by hypoxanthine-guanine phosphoribosyl transferase.

An important difference between mammalian cells and *Leishmania* is that the protozoan enzyme hypoxanthine-guanine phosphoribosyl transferase uses xanthine analogues, like allopurinol, as substrates (Hassan and Coombs, 1985). This difference has been used successfully as a chemotherapeutic approach with allopurinol. Looker *et al.* (1983) and Marr and Berens (1985) demonstrated that allopurinol is converted to IMP analogue. Subsequently, adenyl succinate synthase and adenylosuccinate lyase convert it to the ribonucleoside monophosphate, which is phosphorylated to its triphosphate form and incorporated into ribonucleic acids (Spector *et al.*, 1979).

Purine nucleotides are degraded by phosphatases or nucleotidases, yielding nucleosides and inorganic phosphate. The nucleosides are degraded by nucleoside phosphorylase or nucleoside hydrolase. The former yields purine and ribose 1-phosphate; the latter produces the base and free ribose. In mammalian cells the purine ring, if it is not used by the salvage pathway, is catabolized. The base is deaminated and oxidized by xanthine oxidase to uric acid, which is excreted. In *Leishmania* the base is used in the salvage pathway, and there is no formation of uric acid. Promastigotes lack xanthine oxidase (Marr and Berens, 1985).

6.2. Pyrimidines

Mammalian cells and *Leishmania* possess pathways for the *de novo* synthesis and salvage of pyrimidine bases. In the former the pyrimidine ring is synthesized and then transferred to a ribose unit, in contrast to purine synthesis. Hammond and Gutteridge (1982) showed that all six enzymes involved in *de novo* pyrimidine biosynthesis were present in *L. (L.) amazonensis*. LaFon *et al.* (1982) reported that the pyrimidine salvage pathway in *L. (L.) donovani* was much less active than the purine salvage pathway.

Leishmania possess a characteristic thymidylate synthase and dihydrofolate reductase on the same polypeptide (Meek *et al.*, 1985). In this bifunctional protein, the dihydrofolate reductase domain is on the amino terminus and the thymidylate synthase, the enzyme that is responsible for the production of deoxythymidine monophosphate from deoxyuridine monophosphate, is on the carboxyl terminus (Ivanetich and Santi, 1990). Better understanding of this polypeptide may eventually be useful in designing drugs to selectively inhibit its enzymatic activities. Attempts have been made to find selective inhibitors. Sirawaraporn *et al.* (1988) reported that 5-(substituted-benzyl)-2,4-diaminopyrimidines, pyrimidine analogues, are potent inhibitors of the leishmanial dihydrofolate reductase enzyme and selectively inhibit amastigote and promastigote growth. These compounds may eventually be useful as antileishmanial drugs.

7. PROTON PUMP

The *Leishmania* must survive and multiply in two environments: the gut of sandflies, with a relatively neutral pH, and acidic secondary lysosomes, with a pH of approximately 5.0, in macrophages (Rabinovitch *et al.*, 1982). *Leishmania* must regulate the level of their cytoplasmic proton concentration in both environments. Proton pumps are important in regulating cytoplasmic pH in prokaryotes (Padan and Schuldiner, 1987). Zilberstein and Dwyer (1988) and Zilberstein *et al.* (1989) demonstrated an active H^+-ATPase on the surface of *L. (L.) donovani* promastigotes, which is probably the mechanism by which the parasites regulate their intracellular pH. This proton pump is coupled to the transport of glucose and amino acids (Zilberstein and Dwyer, 1985).

8. HEAT SHOCK PROTEINS

There is great interest in identifying the mechanisms involved in conversion of promastigotes to amastigotes and vice versa. Heat shock proteins may play a role. Changes in temperature have been found to induce expression of a small set of heat shock proteins (HSPs) in both eukaryotic and prokaryotic organisms. The genes for HSPs are translated under adverse conditions when translation of many other genes does not occur. The potential roles of HSPs in allowing organisms to react to environmental changes have been extensively discussed elsewhere (Pelham, 1986, 1989; Winfield, 1989; Hurst, 1990), but much is still not known about their general function. In many organisms, HSPs are expressed during certain stages of cell development.

Heat shock proteins are products of highly conserved genes that encode for a limited number of polypeptides with molecular weights in the range of 22 to 110 kDa. The best characterized HSPs from eukaryotic cells are in the 70-kDa and 90-kDa families. The 70-kDa family includes the heat shock protein 70 (hsp 70) and the glucose-regulated protein (grp 78). The former is heat inducible and located in the cell nucleus and nucleolus; the latter is induced by glucose deprivation and is found in the endoplasmic reticulum. The 90-kDa family is induced by heat and, in some cells, by steroids (Hurst, 1990).

In general, the interaction of the HSP 70-kDa family with nuclear and nucleolar proteins requires binding and hydrolysis of ATP. Pelham (1986) suggested that this is associated with the assembly and disassembly of proteins and enables HSPs to bind to hydrophobic domains on denatured proteins thereby preventing the formation of insoluble aggregates that could hamper cell function (Pelham, 1989). The glucose-regulated proteins may serve a similar function, preventing aggregation of hydrophobic proteins prior to glycosylation in the endoplasmic reticulum. The 90-kDa family is less well understood. These proteins complex with receptors for steroids. The interaction with steroids may play a role in immune and inflammatory responses (Hurst, 1990).

Other stress proteins are produced in response to different stimuli. Stress proteins produced in response to heavy metal and thiol-reactive agents have been described (Caltabiano *et al.*, 1986). Keyse and Tyrrell (1987) have also described the production of a 32-kDa stress protein in fibroblasts exposed to ultraviolet radiation and hydrogen peroxide.

Promastigotes reside in the gut of poikilothermic sandflies at temperatures in the range of 25–27 °C and within mammals in mononuclear phagocytes at approximately 37 °C. Heat shock proteins have been demonstrated in *Leishmania* (Lawrence and Robert-Gero, 1985; Van der Ploeg *et al.*, 1985; Lee *at al.*, 1988; Searle *et al.*, 1989). It seems plausible that HSPs or other stress proteins arising in response to changes in pH, exposure to oxidizing agents (e.g., hydrogen peroxide), or exposure to hydrolytic enzymes may contribute to conversion from one form to another.

Van der Ploeg *et al.* (1985) reported differences between promastigotes and amastigotes of *L. (L.) major* in expression of HSPs of 70 kDa. Promastigotes grown at 25 °C were slender while amastigotes and promastigotes at 37 °C were small and round. Stinson *et al.* (1989) reported similar changes in the morphology of *L. (V.) braziliensis;* a shift in temperature was sufficient to induce differentiation from promastigotes to amastigotelike forms. The morphologic changes in *L. (L.) major* were accompanied by expression of four HSP 70-kDa genes (Van der Ploeg *et al.*, 1985). These HSPs were 89% conserved at the amino acid level with the 70-kDa HSPs of *Trypanosoma brucei*. Differentiation from promastigote to amastigote may be under the control of temperature induced expression of HSPs (Van der Ploeg *et al.*, 1985).

Toye and Remold (1989) found that serum deprivation as well as increased temperature increased the synthesis of both 70- and 82-kDa HSPs in *L. (L.) major* promastigotes. There was also uncoordinated regulation of the expression of the α- and β-tubulin genes. As the serum concentration decreased, α-tubulin expression increased while β-tubulin expression decreased.

Amastigotes and promastigotes differ in expression of more than just HSP genes. Hunter *et al.* (1984) reported that *L. (L.) mexicana* promastigotes increased synthesis of proteins of 22, 23, 26, 27, and 83 kDa in addition to 70 kDa when exposed to temperature shifts. Searle *et al.* (1989) characterized two genes of *L. (L.) major* that were related to the 70-kDa HSPs, but these genes were not heat-inducible and they were constitutively expressed in promastigotes.

The role of HSPs and proteins that are developmentally induced in the life cycle of *Leishmania* is still not clear. Cairns *et al.* (1989) have cloned a developmentally regulated gene from *L. (L.) enriettii* that is expressed in higher amounts in promastigotes than amastigotes. The predicted protein encoded by this gene is homologous to the human erythrocyte glucose transporter. This protein might play an important role in the adaption of promastigote in the insect gut.

9. SURFACE MOLECULES

Many of the initial studies of promastigotes focused on the glycosylation of surface antigens. Lectins were used to identify exposed surface carbohydrates.

The pattern of glycosylation varied among the *Leishmania* species. Studies with lectins suggested that *L. (L.) donovani* and *L. (L.) mexicana* had predominantly glycoconjugates with exposed mannose on their surface (Dwyer, 1974, 1977; Wilson and Pearson, 1984); other *Leishmania* species like *L. (L.) major* had primarily exposed galactose (Jacobson *et al.*, 1982; Schottelius, 1982). The lectin-binding characteristics of *L. (L.) donovani* were also observed to change as the parasite differentiated from amastigote to promastigote (Wilson and Pearson, 1984).

Western blot analysis and monoclonal antibody studies subsequently indicated that there were both shared and species-specific leishmanial surface antigens (McMahon-Pratt and David, 1981; Pan and McMahon-Pratt, 1988). Several of the shared leishmanial surface molecules have been characterized. Species-specific monoclonal antibodies have proven useful in identifying isolates.

9.1. Glycoprotein 63 kDa

The first leishmanial surface molecule to be isolated, cloned, and extensively studied was a glycoprotein of 63 kDa (gp63) (Button and McMaster, 1988). Gp63 is a dominate surface glycoprotein on promastigotes of all *Leishmania* species; it is also present on the surface of amastigotes (Frommel *et al.*, 1990). Gp63 is anchored by glycosylphosphatidylinositol in the cell membrane in a manner analogous to the variant surface antigens of *Trypanosoma brucei*. Some other leishmanial surface proteins appear to be anchored in the same manner. Gp63 is a protease and may play a role in parasite metabolism. It appears to be important to parasite survival in macrophages (Kweider *et al.*, 1987).

9.2. Phosphomonoesterases

L. (L.) donovani promastigotes also have phosphomonoesterases with three distinct enzyme activities on their external surface (Dwyer and Gottlieb, 1985). Hassan and Coombs (1987) have shown the presence of these enzymes on both amastigotes and promastigotes of *L. (L.) mexicana*. The enzymes are a nonspecific acid phosphatase, a 3′-nucleotidase, and a 5′-nucleotidase.

Gottlieb and Dwyer (1985) suggested a nutritional role for the acid phosphatase, which is present in most but not all *Leishmania* species (Lovelace and Gottlieb, 1986). The enzyme dephosphorylates serine residues of several phosphoproteins including pyruvate kinase, phosphorylase kinase, and histones. Katakura and Kobayashi (1988) reported that the activity of the acid phosphatase was higher in virulent clones than avirulent clones of *L. (L.) donovani*. Avila *et al.* (1989) found that the acid phosphatase in *Leishmania* exists as isoenzymes and characterized two membrane-bound forms. Lovelace *et al.* (1986) purified and characterized a soluble acid phosphatase of *L. (L.) donovani* promastigotes as well. Bates *et al.* (1987) demonstrated that the soluble and membrane-bound enzymes were different by immunoprecipitation.

By hydrolyzing phosphomonoesters, acid phosphatase could provide the parasite with a source of inorganic phosphate. Alternatively, the acid phosphatase may be involved in the pathophysiological interaction between the parasite and host phagocytes. Remaley *et al.* (1984) found that *L. (L.) donovani* acid phosphatase blocks superoxide anion production by activated human neutrophils. Das *et al.* (1986a) demonstrated that the acid phosphatase could hydrolyze phosphatidylinositol-4,5 diphosphate and inositol triphosphate. Inositol triphosphate and diacylgycerol are important in phagocyte activation. The mechanism by which acid phosphatase blocks the production of reduced oxygen intermediates may involve the degradation of phosphatidylinositol diphosphate or inositol triphosphate. This is discussed in a greater detail in a review by Glew *et al.* (1988).

Nucleotidases are phosphomonesterases that remove the phosphate group from nucleotides (deoxy- and ribonucleotides), yielding nucleosides. Hassan *et al.* (1987) suggested that the 3′-nucleotidases of *L. (L.) mexicana* could cleave exogenous nucleic acids, producing nucleotides that subsequently could be degraded by the 5′-nucleotidase to nucleosides. Nucleosides are easily salvaged by the parasite.

9.3. ATPases

The leishmanial membrane contains a proton-translocating ATPase that presumably helps to maintain pH homeostasis of the parasite and contributes to lysosomal acidification. The proton-translocating gradient, as described previously, drives the transport of sugar and amino acids, which is critical for parasite survival and growth (Zilberstein and Dwyer, 1985). This transport is vital for the parasite as demonstrated by the lethal effects of the antidepressants clomipramine and nitroimipramine, which disturb membrane function. Amastigotes were more sensitive than promastigotes to nitroimipramine (Zilberstein and Dwyer, 1984).

Benain and Romero (1990) demonstrated Ca^{2+}-ATPase activity on plasma membrane vesicles from *L. (V.) braziliensis*. The Ca^{2+}-ATPase had a high affinity for Ca^{2+}, and it was stimulated by calmodulin. The calcium pump in the plasma membrane of *Leishmania* may be responsible for the maintenance of the low cytoplasmic calcium concentration.

9.4. Protein Kinases

Leishmania species contain high levels of protein kinase activity in the outer membrane of the parasite (Das *et al.*, 1986b; Berman, 1988b). Protein kinases catalyze the transference of a γ-phosphate group from nucleoside triphosphates to the hydroxyl group of serine, threonine, or tyrosine residues. Protein kinase activity varies with the parasite growth phase. Kinase activity is low during the exponential growth of promastigotes and increases as the parasites enter stationary phase. In late stationary phase, there is a decline in the activity (Glew *et al.*, 1988).

9.5. Lipophosphoglycan

Studies of the immunology of leishmaniasis indicate that surface glycolipids are important in parasite–phagocyte interactions and in eliciting T-cell-mediated immune responses. A lipophosphoglycan (LPG) was first isolated and characterized from *L. (L.) donovani* (Turco *et al.*, 1984, 1987; Orlandi and Turco, 1987). Turco *et al.* (1987) isolated the carbohydrate fragment of the LPG of *L. (L.) donovani.* There were approximately 16 phosphorylated disaccharides in the fragment. The disaccharides were formed of $PO_4 \rightarrow 6$ Gal(β1 → 4)Man. They were linked by α-glycosidic linkages. The hydrophobic moiety of LPG is a lysoalkylphosphatidylinositol lipid, which anchors the structure in the membrane (Orlandi and Turco, 1987). An antigenically different LPG has been isolated from *L. (L.) major* (Handman *et al.*, 1984). The LPG from *L. (L.) major* seems to be an important antigen in eliciting cellular immune responses (Handman *et al.*, 1986).

10. *LEISHMANIA*–MACROPHAGE INTERACTIONS

10.1. Attachment

Much attention has focused on the interaction of *Leishmania* promastigotes with mononuclear phagocytes (Pearson and Wilson, 1989). *In vitro* cinemicroscopic studies have revealed that attachment of promastigotes to human monocyte-derived macrophages typically occurs at either the flagellar or aflagellar poles of the parasite (Pearson *et al.*, 1983a) (Figure 7). Flagellum-first attachment predominates, probably because the flagellum, which pulls the parasite forward, first makes contact with the macrophage.

The attachment of promastigotes to macrophages is mediated by receptor–ligand interactions. Klempner *et al.* (1983) demonstrated this when they showed that attachment of membrane vesicles made from human mononuclear cells to *L. (L.) tropica* promastigotes exhibited specificity, saturability, and competitive inhibition. Furthermore, attachment of promastigotes to macrophages is calcium dependent (Chang, 1981; Zenian, 1981) and is sensitive to the action of trypsin on the phagocyte but not the parasite (Bray, 1983; Wilson and Pearson, 1986).

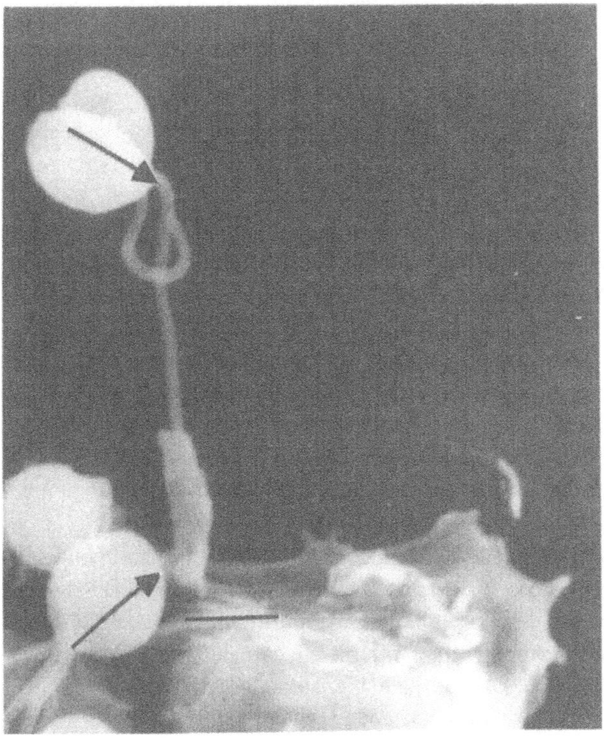

FIGURE 7. Scanning electron micrograph showing attachment of a dividing promastigote to a human monocyte at the flagellar pole of the parasite. The bars equal 10 μm. The arrows indicate the flagellum and pseudopod.

Leishmania promastigotes display an array of glycoconjugates on their surfaces (Dwyer, 1974, 1977; Jacobson *et al.*, 1982; Schottelius, 1982; Wilson and Pearson, 1984). *Leishmania (L.) donovani* and *L. (L.) mexicana* are particularly rich in exposed surface mannose. *In vitro* studies with *L. (L.) donovani* promastigotes indicated that mannan, a mannose polymer derived from yeast cell walls, or the neoglycoprotein, mannosylated-bovine albumin, inhibited the attachment of *L. (L.) donovani* promastigotes to human macrophages by approximately 40%, suggesting that macrophage mannose–fucose receptors were involved in promastigote attachment (Blackwell *et al.*, 1985; Wilson and Pearson, 1986, 1988). The monosaccharide mannose, which is a poor inhibitor of the mannose–fucose receptor, had little effect. As expected, ligands of the mannose–fucose receptor do not inhibit the binding of *L. (L.) major* (which, in contrast to *L. (L.) donovani,* lacks a predominance of exposed mannosylated molecules on its surface) to macrophages.

Leishmania species promastigotes also bind to macrophage receptors for complement components. Many *Leishmania* species, including *L. (L.) major,* activate complement through the alternative pathway (Mosser and Edelson, 1984). In the case of *L. (L.) donovani,* complement is activated through the classical pathway by IgM antibodies, which are present even in the serum of persons who have had no known exposure to *Leishmania* (Pearson and Steigbigel, 1980; Navin *et al.*, 1989). When complement is activated by either the classical or alternative pathway, C3b and its cleavage product, iC3b, are deposited on the surface of promastigotes. C3b and iC3b can mediate attachment by binding to the macrophage type 1 complement receptor (CR1) or type 3 complement receptor (CR3), respectively. Promastigote binding to CR1 may be important because this receptor mediates phagocytosis without eliciting a potentially lethal phagocytic oxidative response (Da Silva *et al.*, 1989). Any promastigote surface molecule that is mannosylated and/or to which complement is bound can potentially be involved in parasite–macrophage binding.

Promastigotes have been observed to bind to macrophage CR3 even in the absence of serum. Blockade of human macrophage CR3 by monoclonal antibodies directed against either the iC3b binding site or an alternative, lectinlike binding site on the receptor inhibit attachment of *L. (L.) donovani* by approximately 60% (Wilson and Pearson, 1988). Concurrent use of macrophage anti-CR3 monoclonal antibodies and ligands of the mannose–fucose receptor decrease attachment of promastigotes to human macrophages by approximately 80% (Wilson and Pearson, 1988). When serum is not present, promastigotes may be opsonized by complement components produced and secreted by macrophages (Wozencraft *et al.*, 1986). Alternatively, some promastigote surface molecules may bind directly to CR3 in a complement-independent manner.

The *Leishmania* surface glycoprotein gp63 is involved in attachment of promastigotes to macrophages (Russell and Wilhelm, 1986; Wilson and Hardin, 1988; Chang, 1990). Preabsorption of purified gp63 to macrophages limits by approximately 50% the attachment of promastigotes; Fab fragments of polyclonal antibodies against gp63 also inhibit binding. By fixing complement, gp63 could potentially mediate attachment to macrophage complement receptors CR1 and CR3. When purified gp63 from *L. (L.) mexicana* was incorporated into liposomes and exposed to complement, C3b was fixed. The gene for gp63 has been cloned. The protein was initially thought to have an arginine–glycine–aspartate (RGD) sequence, which could have bound to an RGD binding site on macrophages (Russell and Wright, 1988), but the true amino acid sequence was subsequently found to lack an RGD sequence. In the case of *L. (L.) donovani, L. (L.) mexicana,* and other *Leishmania* species in which gp63 is highly mannosylated, gp63 can potentially mediate binding to macrophage mannose–fucose receptors.

Handman and Goding (1985) observed that LPG also serves as a ligand for the attachment of promastigotes to macrophages. Puentes *et al.* (1988) demonstrated that LPG from *L. (L.) major* is a major C3 receptor. Lipophosphoglycan can also bind directly to CR3 in the absence of complement (Talamdas-Rohana *et al.*, 1990) and to p150,95, another member of the CD18 complex of leukocyte integrins.

Other macrophage receptors are involved in the attachment of *Leishmania* promastigotes to macrophages. Promastigotes bind fibronectin and can attach by it to macrophage fibronectin receptors (Wyler *et al.*, 1985). Furthermore, attachment of *L. (L.) major* to macrophages is reduced when macrophage receptors for advanced glycosylation end products are blocked (Mosser *et al.*, 1987), suggesting that these receptors may play a role in parasite–phagocyte interactions. When antileishmanial antibodies are present, they too can be expected to mediate parasite attachment through macrophage Fc receptors as well as by activating complement.

In summary, the attachment of *Leishmania* promastigotes to macrophages is a complex event and appears to be mediated by multiple parasite surface antigens and macrophage receptor types. Much less is known about amastigote–macrophage interactions, but it is possible that the mechanisms of attachment are similar since the two stages of the parasite share many surface antigens.

10.2. Ingestion of Promastigotes

Ingestion of leishmania is dependent on the phagocytic apparatus of the macrophage. A macrophage pseudopod develops in response to attachment of a promastigote and progressively surrounds it (Pearson *et al.*, 1983a) (Figure 7). Electron microscopic studies suggest that contact occurs at discrete points along the parasite–phagocyte interface (Chang, 1979). The pseudopod extends around the parasite in a zipperlike manner and eventually fuses to form a vacuole.

10.3. Intracellular Survival

The attachment of promastigotes to mononuclear phagocytes elicits an oxidative microbicidal burst resulting in the reduction of oxygen to superoxide anion and the generation of toxic oxidants. Studies done in a phagocyte-free system indicate that promastigotes are potentially susceptible to the hydrogen peroxide–myeloperoxidase–halide microbicidal mechanism of phagocytes (Murray, 1981; Pearson and Steigbigel, 1981; Reiner and Kazura, 1982; Locksley and Klebanoff, 1983).

Promastigote survival depends on the magnitude of the oxidative responsive that they elicit and on the presence of myeloperoxidase (Locksley *et al.*, 1987). Circulating human monocytes are capable of a large oxidative burst, have lysosomal myeloperoxidase, and kill the majority of ingested promastigotes (Murray and Cartelli, 1983; Pearson *et al.*, 1981, 1982, 1983b,c). In contrast, promastigotes are not killed by monocytes from donors with chronic granulomatous disease of childhood, which are unable to mount an oxidative burst (Murray and Cartelli, 1983; Pearson *et al.*, 1983b). Human monocyte-derived macrophages cultured *in vitro* have reduced oxidative potency and lack myeloperoxidase (Locksley *et al.*, 1987). Promastigotes survive within them as well and convert to amastigotes (Pearson *et al.*, 1983b). Once inside macrophages, *Leishmania* do not stimulate a continuous oxidative response, but they do decrease the oxidative response to subsequent stimuli (Pearson *et al.*, 1983b). The surface-bound acid phosphatase of *Leishmania* can partially inhibit the magnitude of the phagocytic oxidative response (Remaley *et al.*, 1985) and thus potentially enhance the likelihood of parasite survival.

Leishmania amastigotes are resistant to killing by human monocytes, monocyte-derived macrophages, and rodent peritoneal macrophages unless those phagocytes are activated by cytokines such as interferon-gamma or direct contact with immune T cells, as discussed in Section 11. Amastigotes elicit a smaller oxidative response than promastigotes when ingested by nonactivated mononuclear cells, possibly because of their smaller surface area (Pearson *et al.*, 1983b). Both amastigotes and promastigotes possess superoxide dismutase (Meshnick and Eaton, 1981), but amastigotes have more catalase and glutathione peroxidase (Murray, 1982) and are consequently less susceptible to oxidants such as hydrogen peroxide (Haidaris and Bonventre, 1982; Murray, 1982; Pearson *et al.*, 1983b).

Amastigotes reside and multiply in phagosomes that fuse with lysosomes (Chang and Dwyer, 1976, 1978; Brazil, 1984). *Leishmania* thus resemble *Mycobacterium lepraemurium,* which permits phagosome–lysosome fusion in macrophages, but they differ from viable *Mycobacterium tuberculosis* and *Toxoplasma gondii,* which inhibit phagosome–lysosome fusion.

The pH in the parasitophorous vacuole in macrophages is approximately 5.0 (Rabinovitch *et al.*, 1982; Antoine *et al.*, 1990). *Leishmania* are well adapted for survival under these conditions. Amastigote metabolism of glucose and amino acids is optimal at approximately pH 5 (Mukkada *et al.*, 1985). Chang (1979) observed that debris trapped in phagolysosomes with amastigotes was degraded over time, suggesting that the parasites do not inhibit lysosomal enzymes. Furthermore, the swelling of *L. (L.) mexicana* containing phagolysosomes after treatment with a weak base such as leucine *O*-methyl ester suggests that substances are actively hydrolyzed in the parasitophorous vacuole (Rabinovitch *et al.*, 1986; Alfieri *et al.*, 1988, 1989).

11. IMMUNOLOGY OF LEISHMANIASIS

The outcome of a leishmanial infection is dependent on a complex interplay between the immune system of the host and the virulence characteristics of the parasite. The immunology of leishmaniasis has been discussed in detail elsewhere (Wilson and Pearson, 1990). In brief, inbred strains of mice have emerged as models for the study of the cellular immunology of leishmanial infections. The susceptibility or resistance of mice is genetically determined. In the case of *L. (L.) donovani,* susceptibility is determined by the *Lsh* gene on chromosome 1, the same gene that regulates susceptibility to two other intracellular pathogens, *Salmonella typhimurium* and *Mycobacterium bovis*. Genetic resistance to *L. donovani* is mediated at the level of the macrophage. Among susceptible strains

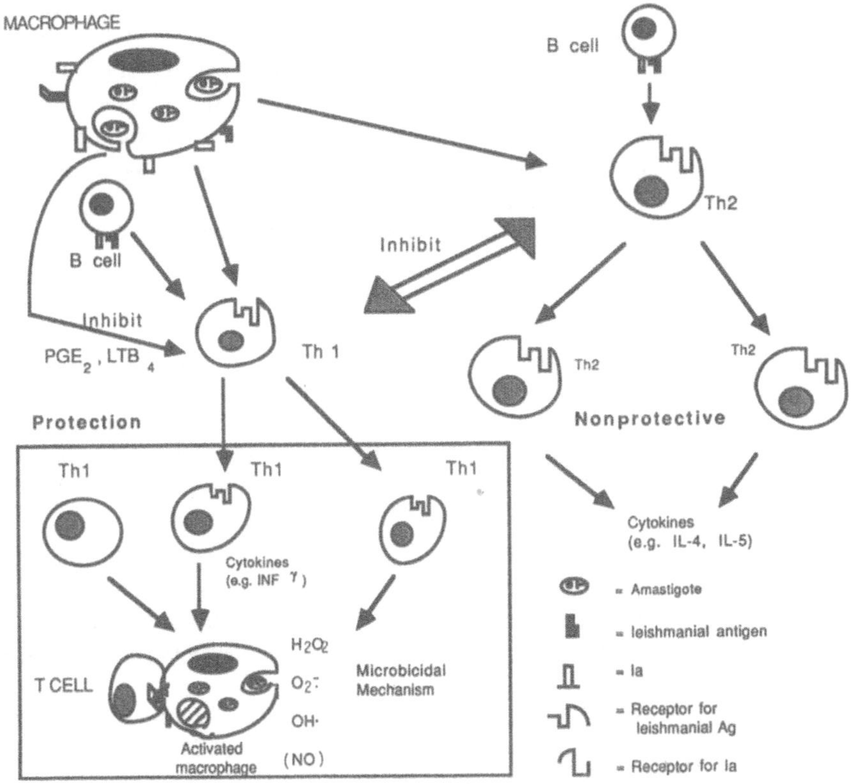

FIGURE 8. Proposed scheme of events leading to resolution or progression of leishmanial infection. Modified from Wilson and Pearson (1990).

of mice, some develop high parasite burdens that persist while in others the parasite burden decreases over time as a consequence of T-cell-mediated immune responses. Susceptibility to *L. (L.) major* is also genetically determined, but it is not mediated by the *Lsh* gene.

Protection against reinfection with either *L. (L.) major* or *L. (L.) donovani* in susceptible mice can be obtained by immunization. Protection is mediated in part by phenotypic T helper cells (CD4) of the Th1 type (Scott *et al.*, 1988; Heinzel *et al.*, 1989) (Figure 8). Th1 cells produce interleukin 2, a T cell growth factor, and interferon-gamma, a macrophage activating factor. Histocompatible murine *Leishmania*-specific CD4 cells can also activate macrophages to kill amastigotes by direct contact (Wyler *et al.*, 1987). Data from murine models indicate that cytotoxic/suppressor (CD8) cells also play a role in eradication of *L. (L.) donovani* amastigotes in established infections.

Interferon-gamma is the principal macrophage activating factor secreted by protective immune T cells (Murray *et al.*, 1989). Other cytokines such as granulocyte–macrophage colony stimulating factor (GM-CSF) can also contribute to macrophage activation (Weiser *et al.*, 1987). The mechanism(s) by which cytokine-activated macrophages kill amastigotes has been the subject of intense study. Superoxide anion-dependent microbicidal mechanisms, which are enhanced by interferon-gamma, may contribute to the killing of *Leishmania*, particularly at the time of ingestion, but they are not necessary for the killing of amastigotes by activated macrophages (Scott *et al.*, 1985).

Recent evidence suggests that the superoxide anion-independent killing of intracellular amastigotes that follows interferon-gamma activation is mediated by nitric oxide (Green *et al.*, 1990; Liew *et al.*, 1990). Nitric oxide is synthesized in macrophages from the terminal guanidine nitrogen of L-arginine. Nitric oxide may disrupt iron-dependent enzymatic pathways vital to parasite survival. Further study of this arginine metabolic pathway may lead to novel treatment strategies. The mechanism(s) by which amastigotes are killed when macrophages are activated by direct contact with *Leishmania*-specific CD4 cells has yet to be determined.

In genetically susceptible mice with progressive leishmanial infections, CD4 cells of the Th2 type come to dominate. These Th2 cells produce interleukin 4 but not interleukin 2 or interferon-gamma in response to parasite antigens. *Leishmania*-specific Th2 cells can inhibit the expansion of protective Th1 cells. *Leishmania*-infected macrophages can also suppress the development of potentially protective T cell subsets.

Why do protective Th1 cell populations proliferate and control infection in some *Leishmania*-susceptible mouse strains and not others? The full answer is not known, but genes at the major histocompatibility loci appear to be involved. Parasite antigens are also important. This was best illustrated by immunization studies using LPG from *L. (L.) major* (Handman *et al.*, 1986). When intact LPG with adjuvant was used as a vaccine in susceptible mice, protective T cells were elicited. When LPG was hydrolyzed by phospholipase C and then used as an immunogen, nonprotective CD4 cells were elicited. The data suggest that the presentation of parasite antigens by macrophages and/or other antigen-presenting cells such as B lymphocytes in the context of major histocompatibility antigens may be a critical factor in eliciting either protective Th1 cells or nonprotective Th2 cells. Further studies are needed to determine the genetic factors, the sequence of immune responses, and the parasite antigens that determine whether humans develop progressive leishmanial disease or asymptomatic, self-resolving infection.

12. REFERENCES

Alfieri, S. C., Ramazeilles, C., Zilberfarb, V., Galpin, I., Norman, S. E., and Rabinovitch, M., 1988, Proteinase inhibitors protect *Leishmania amazonensis* amastigotes from destruction by amino acid esters, *Mol. Biochem. Parasitol.* **29:**191–201.

Alfieri, S. C., Show, E., Zilberfarb, V., and Rabinovitch, M., 1989, *Leishmania amazonensis:* Involvement of cysteine proteinases in the killing of isolated amastigotes by L-leucine methyl ester, *Exp. Parasitol.* **68:**423–431.

Antoine, J. C., Prina, E., Jouanne, C., and Bongrand, P., 1990, Parasitophorous vacuoles of *Leishmania amazonensis*-infected macrophages maintain an acidic pH, *Infect. Immun.* **58:**779–787.

Arnot, D. E., and Barker, D. C., 1981, Biochemical identification of cutaneous leishmaniasis by analysis of kinetoplast DNA. II. Sequence homologies in *Leishmania* kDNA, *Mol. Biochem. Parasitol.* **3:**47–56.

Avila, J. L., Hernandez-Morales, D., Polegre, M. A., and Convit, J., 1989, On the acid phosphatase isoenzymes existing in American *Leishmania* promastigotes, *Comp. Biochem. Physiol.* **94B:**335–342.

Badaró, R., Jones, T. C., Lorenço, R., Cerf, B. J., Sampaio, D., Carvalho, E. M., Rocha, H., Teixeira, R., and Johnson, W. D., Jr., 1986a, A prospective study of visceral leishmaniasis in an endemic area of Brazil, *J. Infect. Dis.* **154:**639–649.

Badaró, R., Jones, T. C., Carvalho, E. M., Sampaio, D., Reed, S. G., Barral, A., Teixeira, R., and Johnson, W. D., Jr., 1986b, New perspectives on a subclinical form of visceral leishmaniasis, *J. Infect. Dis.* **154:**1003–1011.

Badaró, R., Carvalho, E. M., Rocha, H., Queiroz, A. C., and Jones, T. C., 1986c, *Leishmania donovani:* An opportunistic microbe associated with progressive disease in three immunocompromised patients, *Lancet* **1:**647–648.

Badaró, R., Falcoff, E., Badaró, F. S., Carvalho, E. M., Pedral-Sampaio, D., Barral, A., Carvalho, J. S., Barral-Netto, M., Brandely, M., Silva, L., Bina, J. C., Teixeira, R., Falcoff, R., Rocha, H., Ho, J. L., and Johnson, W. D., Jr., 1990, Treatment of visceral leishmaniasis with pentavalent antimony and interferon gamma, *N. Engl. J. Med.* **322:**16–21.

Bates, P. A., Kurtz, M. K., Gottlieb, M., and Dwyer, D. M., 1987, *Leishmania donovani:* Generation of monospecific antibody reagents to soluble acid phosphatase, *Exp. Parasitol.* **64:**157–164.

Benaim, G., and Romero, P. J., 1990, A calcium pump in plasma membrane vesicles from *Leishmania braziliensis, Biochim. Biophys. Acta* **1027:**79–84.

Berman, J. D., 1988a, Chemotherapy for leishmaniasis: Biochemical mechanisms, clinical efficacy, and future strategies, *Rev. Infect. Dis.* **10:**560–586.

Berman, J. D., 1988b, Inhibition of leishmanial protein kinase by antileishmanial drugs, *Am. J. Trop. Med. Hyg.* **38:**298–303.

Berman, J. D., Gallalee, J. V., and Best, J. M., 1987a, Sodium stibogluconate (pentostam) inhibition of glucose catabolism via the glycolytic pathway, and fatty acid β-oxidation in *Leishmania mexicana* amastigotes, *Biochem. Pharmacol.* **36:**197–201.

Berman, J. D., Gallalee, J. V., Best, J. M., and Hill, T., 1987b, Uptake, distribution, and oxidation of fatty acids by *Leishmania mexicana* amastigotes, *J. Parasitol.* **73:**555–560.

Blackwell, J. M., Ezekowitz, R. A. B., Roberts, M. B., Channon, J. Y., Sim, R. B., and Gordon, S., 1985, Macrophage complement and lectin-like receptors bind *Leishmania* in the absence of serum, *J. Exp. Med.* **162:**324–331.

Blum, J. J., 1987, Oxidation of fatty acids by *Leishmania braziliensis panamensis, J. Protozool.* **34:**169–174.

Brazil, R. P., 1978, Electrophoretic variation of the enzyme phosphoglucomutase in different strains of *Leishmania, Ann. Trop. Med. Parasitol.* **72:**289–291.

Brazil, R. P., 1984, *In vivo* fusion of lysosomes with parasitophorous vacuoles of *Leishmania*-infected macrophages, *Ann. Trop. Med. Parasitol.* **78:**87–91.

Bray, R. S., 1983, *Leishmania mexicana mexicana:* Attachment and uptake of promastigotes to and by macrophages *in vitro, J. Protozool.* **30:**314–322.

Button, L. L., and McMaster, W. R., 1988, Molecular cloning of the major surface antigen of *Leishmania, J. Exp. Med.* **167:**724–729.

Cairns, B. R., Collard, M. W., and Landfear, S. M., 1989, Developmentally regulated gene from *Leishmania* encodes a putative membrane transport protein, *Proc. Natl. Acad. Sci. USA* **86:**7682–7686.

Caltabiano, M. M., Koestler, T. P., Poste, G., and Greig, R. G., 1986, Induction of 32- and 34-kDa stress proteins by sodium arsenite, heavy metals, and thiol-reactive agents, *J. Biol. Chem.* **261:**13381–13386.

Carvalho, E. M., Badaró, R., Reed, S. G., Jones, T. C., and Johnson, W. D., Jr., 1985, Absence of gamma interferon and interleukin 2 production during active visceral leishmaniasis, *J. Clin. Invest.* **76:**2066–2069.

Carvalho, E. M., Bacellar, O., Barra, A., Badaró, R., and Johnson, W. D., Jr., 1989, Antigen specific immunosuppression in visceral leishmaniasis is cell mediated, *J. Clin. Invest.* **83:**860–864.

Chang, K.-P., 1979, *Leishmania donovani:* Promastigote–macrophage surface interactions *in vitro, Exp. Parasitol.* **48:**175–189.

Chang, K.-P., 1981, *Leishmania donovani*–macrophage binding mediated by surface glycoproteins/antigens: Characterization *in vitro* by a radioisotopic assay, *Mol. Biochem. Parasitol.* **4:**67–76.

Chang, K.-P., 1990, Cell biology of *Leishmania,* in *Modern Parasite Biology. Cellular, Immunological, and Molecular Aspects* (D. J. Wyler, ed.), pp. 79–90, W. H. Freeman and Company, New York.

Chang, K.-P., and Dwyer, D. M., 1976, Multiplication of a human parasite (*Leishmania donovani*) in phagolysosome of hamster macrophages *in vitro, Science* **193:**678–680.

Chang, K.-P., and Dwyer, D. M., 1978, *Leishmania donovani*–hamster macrophage interactions *in vitro:* Cell entry, intracellular survival and multiplication of amastigotes, *J. Exp. Med.* **147:**515–530.

Chang, K.-P., Fong, D., and Bray, R. S., 1985, Biology of *Leishmania* and leishmaniasis, in *Human Parasitic Diseases* (K.-P. Chang and R. S. Bray, eds.), Vol. 1, pp. 1–30, Elsevier Science Publishers, Amsterdam.

Coombs, G. H., Craft, J. A., and Hart, D. T., 1982, A comparative study of *Leishmania mexicana* amastigotes and promastigotes. Enzyme activities and subcellular locations, *Mol. Biochem. Parasitol.* **5:**199–211.

Darling, T. N., and Blum, J. J., 1987, *In vitro* reversible transformation of *Leishmania braziliensis panamensis* between promastigotes and ellipsoidal forms, *J. Protozool.* **34:**166–168.

Darling, T. N., and Blum, J. J., 1988, D-Lactate production by *Leishmania braziliensis* through the glyoxalase pathway, *Mol. Biochem. Parasitol.* **28:**121–128.

Darling, T. N., Davis, D. G., London, R. E., and Blum, J. J., 1987, Products of *Leishmania braziliensis* glucose catabolism: Release of D-lactate and, under anaerobic conditions, glycerol, *Proc. Natl. Acad. Sci. USA* **84:**7129–7133.

Darling, T. N., Davis, D. G., London, R. E., and Blum, J. J., 1989, Metabolic interactions between glucose, glycerol, alanine and acetate in *Leishmania braziliensis panamensis* promastigotes, *J. Protozool.* **36:**217–225.

Das, S., Saha, A. K., Remaley, A. T., Glew, R. H., Dowling, J. N., Kajiyoshi, M., and Gottlieb, M., 1986a, Hydrolysis of phosphoproteins and inositol phosphates by cell surface phosphatase of *Leishmania donovani, Mol. Biochem. Parasitol.* **20:**143–153.

Das, S., Saha, A. K., Mukhopadhyay, N. Y., and Glew, R. H., 1986b, A cyclic nucleotide-independent protein kinase in *Leishmania donovani, Biochem. J.* **240:**641–649.

Da Silva, R. P., Hall, B. F., Joiner, K. A., and Sacks, D. L., 1989, CR1, the C3b receptor, mediates binding of infective *Leishmania major* metacyclic promastigotes to human macrophages, *J. Immunol.* **143:**617–622.

Dwyer, D. M., 1974, Lectin binding saccharides on a parasitic protozoan, *Science* **184:**471–473.

Dwyer, D. M., 1977, *Leishmania donovani:* Surface membrane carbohydrates of promastigotes, *Exp. Parasitol.* **41:**341–358.

Dwyer, D. M., and Gottlieb, M., 1985, The biochemistry of *Leishmania* surface membranes, in *Human Parasitic Diseases* (K.-P. Chang and R. S. Bray, eds.), Vol. 1, pp. 31–47, Elsevier Science Publishers, Amsterdam.

Ebert, F., 1973, Charakterissierung von *Leishmania donovani*-Stammen mit der Disk-Electrophorese, *Z. Tropenmed Parasitol.* **24:**517–524.

Evans, T., Reis, M. F. F. E., de Alencar, J. E., Naidu, T. G., de Jesus, J. A., McAuliffe, J. F., and Pearson, R. D., 1985, American visceral leishmaniasis (kala-azar), *West. J. Med.* **142:**777–781.

Fernandez-Guerrero, M. L., Aguado, J. M., Buzon, L., Barros, C., Montalban, C., Martin, T., and Bouza, E., 1987, Visceral leishmaniasis in immunocompromised hosts, *Am. J. Med.* **83:**1098–1102.

Frommel, T. O., Button, L. L., Fujikura, Y., and McMaster, W. R., 1990, The major surface glycoprotein (Gp63) is present in both life stages of *Leishmania, Mol. Biochem. Parasitol.* **38:**25–32.

Gardener, P. J., Chance, M. L., and Peters, W., 1974, Biochemical taxonomy of *Leishmania.* II. Electrophoretic variation of malate dehydrogenase, *Ann. Trop. Med. Parasitol.* **68:**317–325.

Giannini, S. H., Schittini, M., Keithly, J. S., Warburton, P. W., Cantor, C. R., and Van der Ploeg, L. H. T., 1986, Karyotype analysis of *Leishmania* species and its use in classification and clinical diagnosis, *Science* **232:**762–765.

Glew, R. H., Saha, A. K., Das, S., and Remaley, A. T., 1988, Biochemistry of the *Leishmania* species, *Microbiol. Rev.* **52:**412–432.

Green, S. J., Meltzer, M. S., Hibbs, J. B., Jr., and Nacy, C. A., 1990, Activated macrophages destroy intracellular *Leishmania major* amastigotes by an L-arginine-dependent killing mechanism, *J. Immunol.* **144:**278–283.

Grimaldi, G., Jr., Tesh, R. B., and McMahon-Pratt, D., 1989, A review on the geographic distribution and epidemiology of leishmaniasis in the New World, *Am. J. Trop. Med. Hyg.* **41:**687–725.

Haidaris, C. G., and Bonventre, P. F., 1982, A role for oxygen-dependent mechanisms in killing of *Leishmania donovani* tissue forms by activated macrophages, *J. Immunol.* **129:**850–855.

Hammond, D. J., and Gutteridge, W. E., 1982, UMP synthesis in the kinetoplastida, *Biochim. Biophys. Acta* **718:**1–10.

Handman, E., and Goding, J. W., 1985, The *Leishmania* receptor for macrophages is a lipid-containing glycoconjugate, *EMBO J.* **4**:329–336.
Handman, E., Greenblat, C. L., and Goding, J. W., 1984, An amphipathic sulphated glycoconjugate of *Leishmania:* Characterization with monoclonal antibodies, *EMBO J.* **3**:2301–2306.
Handman, E., Schnur, L. F., Spithill, T. W., and Mitchell, G. F., 1986, Passive transfer of *Leishmania* lipopolysaccharide confers parasite survival in macrophages, *J. Immunol.* **137**:3608–3613.
Hart, D. T., and Coombs, G. H., 1982, *Leishmania mexicana:* Energy metabolism of amastigotes and promastigotes, *Exp. Parasitol.* **54**:397–409.
Hart, D. T., and Opperdoes, F. R., 1984, The occurrence of glycosomes (microbodies) in the promastigotes stage of four major *Leishmania* species, *Mol. Biochem. Parasitol.* **13**:159–172.
Hassan, H. F., and Coombs, G. H., 1985, *Leishmania mexicana:* Purine-metabolizing enzymes of amastigotes and promastigotes, *Exp. Parasitol.* **59**:139–150.
Hassan, H. F., and Coombs, G. H., 1987, Phosphomonoesterases of *Leishmania mexicana mexicana* and other flagellates, *Mol. Biochem. Parasitol.* **23**:285–296.
Heinzel, F. P., Sadick, M. D., Holaday, B. J., Coffman, R. L., and Locksley, R. M., 1989, Reciprocal expression of interferon gamma or interleukin 4 during the resolution or progression of murine leishmaniasis. Evidence for expansion of distinct helper T cell subsets, *J. Exp. Med.* **169**:59–72.
Hockmeyer, W. T., Kager, P. A., Rees, P. H., and Hendricks, L. D., 1981, The culture of *Leishmania donovani* in Schneider's insect medium: Its value in the diagnosis and management of patients with visceral leishmaniasis, *Trans. R. Soc. Trop. Med. Hyg.* **75**:861–863.
Holz, G. G., Jr., 1985, Lipids of leishmanias, in *Human Parasitic Diseases* (K.-P. Chang and R. S. Bray, eds.), Vol. 1, pp. 79–92, Elsevier Science Publishers, Amsterdam.
Hunter, K. W., Cook, C. L., and Hayunga, E. G., 1984, Leishmanial differentiation *in vitro:* Induction of heat shock proteins, *Biochem. Biophys. Res. Commun.* **125**:755–760.
Hurst, N. P., 1990, Stress (heat shock) proteins and rheumatic disease: New advance or just another band wagon? *Rheumatol. Int.* **9**:271–276.
Ivanetich, K. M., and Santi, D. V., 1990, Bifunctional thymidylate synthase–dihydrofolate reductase in protozoa, *Faseb J.* **4**:1591–1597.
Jackson, P. R., Wohlhieter, J. A., Jackson, J. E., Sayles, P., Diggs, C. L., and Hockmeyer, W. T., 1984, Restriction endonuclease analysis of *Leishmania* kinetoplast DNA characterizes parasites responsible for visceral and cutaneous disease, *Am. J. Trop. Med. Hyg.* **33**:808–819.
Jacobson, R. L., Slutzky, G. M., Greenblatt, C. L., and Schnur, L. F., 1982, Surface reaction of *Leishmania.* I. Lectin-mediated agglutination, *Ann. Trop. Med. Parasitol.* **76**:45–52.
Janovy, J., Jr., 1987, Physiology and biochemistry, in *The Leishmaniases in Biology and Medicine* (W. Peters and R. Killick-Kendrick, eds.), pp. 177–210, Academic Press, London.
Katakura, K., and Kobayashi, A. (1988), Acid phosphatase activity of virulent and avirulent clones of *Leishmania donovani* promastigotes, *Infect. Immun.* **56**:2856–2860.
Keegan, F. P., Sansone, L., and Blum, J. J., 1987, Oxidation of glucose, ribose, alanine and glutamate by *Leishmania braziliensis panamensis, J. Protozool.* **34**:174–179.
Keyse, S. M., and Tyrrell, R. M., 1987, Both near ultraviolet radiation and the oxidizing agent hydrogen peroxide induce a 32-kDa stress protein in normal human skin fibroblasts, *J. Biol. Chem.* **262**:14821–14825.
Klempner, M. S., Cendron, M., and Wyler, D. J., 1983, Attachment of plasma membrane vesicles of human macrophages to *Leishmania tropica* promastigotes, *J. Infect. Dis.* **148**:377–384.
Kornberg, H. L., 1966, The rate and control of the glyoxylate cycle in *Escherichia coli, Biochem. J.* **99**:1–11.
Kweider, M., Lemesre, J. L., Darcy, F., Kusnierz, J. P., Capron, A., and Santoro, F., 1987, Infectivity of *Leishmania braziliensis* promastigotes is dependent on the increasing expression of a 65,000-Dalton surface antigen, *J. Immunol.* **138**:299–305.

LaFon, S. W., Nelson, D. J., Berens, R. L., and Marr, J. J., 1982, Purine and pyrimidine salvage pathways in *Leishmania donovani*, *Biochem. Pharmacol.* **31**:231–238.

Lainson, R., and Shaw, J. J., 1987, Evolution, classification and geographic distribution, in *The Leishmaniases in Biology and Medicine* (W. Peters and R. Killick-Kendrick, eds.), pp. 1–120, Academic Press, London.

Lawrence, F., and Robert-Gero, M., 1985, Induction of heat shock proteins in promastigotes of three *Leishmania* species, *Proc. Natl. Acad. Sci. USA* **82**:4414–4417.

Lawyer, P. G., Ngumbi, P. M., Anjili, C. O., Odongo, S. O., Mebrahtu, Y. B., Githure, J., Koech, D. K., and Roberts, C. R., 1990, Development of *Leishmania major* in *Phlebotomus duboscqi* and *Sergentomyia schwetzi* (Diptera:Psychodidae), *Am. J. Trop. Med. Hyg.* **43**:31–43.

Lee, M. G.-S., Atkinson, B. L., Giannini, S. H., and Van der Ploeg, L. H. T., 1988, Structure and expression of hsp 70 gene family of *Leishmania major*, *Nucleic Acids Res.* **16**:9567–9585.

Liew, F. Y., Millott, S., Parkinson, C., Palmer, R. M., and Moncada, S., 1990, Macrophage killing of *Leishmania* parasite *in vivo* is mediated by nitric oxide from L-arginine, *J. Immunol.* **144**:4794–4797.

Locksley, R. M., and Klebanoff, S. J., 1983, Oxygen-dependent microbicidal systems of phagocytes and host defense against intracellular protozoa, *J. Cell. Biochem.* **22**:173–185.

Locksley, R. M., Nelson, C. S., Frankhauser, J. E., and Klebanoff, S. J., 1987, Loss of granule myeloperoxidase during *in vitro* culture of human monocytes correlates with decay in anti-protozoa activity, *Am. J. Trop. Med. Hyg.* **36**:541–548.

Looker, D. L., Berens, R. L., and Marr, J. J., 1983, Purine metabolism in *Leishmania donovani* amastigotes and promastigotes, *Mol. Biochem. Parasitol.* **9**:15–28.

Lovelace, J. K., and Gottlieb, M., 1986, Comparison of extracellular acid phosphatases from various isolates of *Leishmania*, *Am. J. Trop. Med. Hyg.* **35**:1121–1128.

Lovelace, J. K., Dwyer, D. M., and Gottlieb, M., 1986, Purification and characterization of the extracellular acid phosphatase of *Leishmania donovani*, *Mol. Biochem. Parasitol.* **20**:243–251.

Marr, J. J., and Berens, R. L., 1985, Purine and pyrimidine metabolism in *Leishmania*, in *Human Parasitic Diseases* (K.-P. Chang and R. S. Bray, eds.), Vol. 1, pp. 65–78, Elsevier Science Publishers, Amsterdam.

Marr, J. J., Berens, R. L., and Nelson, D. J., 1978, Purine metabolism in *Leishmania donovani* and *Leishmania braziliensis*, *Biochim. Biophys. Acta* **544**:360–371.

Martin, E., Simon, M. W., Schaefer, F. W., III, and Mukkada, A. J., 1976, Enzymes of carbohydrate metabolism in four human species of *Leishmania:* A comparative survey, *J. Protozool.* **23**:600–607.

Mauël, J., 1990, Macrophage–parasite interactions in *Leishmania* infections, *J. Leukocyte Biol.* **47**:187–193.

McMahon-Pratt, D., and David, J. R., 1981, Monoclonal antibodies that distinguish between New World species of *Leishmania*, *Nature* **291**:581–583.

Meade, J. C., Glaser, T. A., Bonventre, P. F., and Mukkada, A. J., 1984, Enzymes of carbohydrate metabolism in *Leishmania donovani* amastigotes, *J. Protozool.* **31**:156–161.

Meek, T. D., Garvey, E. P., and Santi, D. V., 1985, Purification and characterization of the bifunctional thymidylate synthetase-dihydrofolate reductase from methotrexate-resistant *Leishmania tropica*, *Biochemistry* **24**:678–686.

Meshnick, S. R., and Eaton, J. W., 1981, Leishmanial superoxide dismutase: A possible target for chemotherapy, *Biochem. Biophys. Res. Commun.* **102**:970–976.

Molyneux, D. H., and Killick-Kendrick, R., 1987, Morphology, ultrastructure and life cycles, in *The Leishmaniases in Biology and Medicine* (W. Peters and R. Killick-Kendrick, eds.), Vol. 1, pp. 121–176, Academic Press, London.

Molyneux, D. H., Killick-Kendrick, R., and Ashford, R. W., 1975, *Leishmania* in phlebotomid sandflies. III. The ultrastructure of *Leishmania mexicana amazonensis* in the midgut and pharynx of *Lutzomyia longipalpis*, *Proc. R. Soc. Lond. B* **190**:341–357.

Montalban, C., Martinez-Fernandez, R., Calleja, J. L., Garcia-Diaz, J. D., Rubio, R., Dronda, F., Moreno, S., Yebra, M., Barros, C., Cobo, J., Martinez, M. C., Ruiz, F., and Costa, J. R., 1989, Visceral leishmaniases (kala-azar) as an opportunistic infection in patients infected with the human immunodeficiency virus in Spain, *Rev. Infect. Dis.* **11:**655–660.

Mosser, D. M., and Edelson, P. J., 1984, Activation of the alternative complement pathway by *Leishmania* promastigotes: Parasite lysis and attachment to macrophages, *J. Immunol.* **132:**1501–1505.

Mosser, D. M., and Edelson, P. J., 1987, The third complement (C3) is responsible for the intracellular survival of *Leishmania major, Nature* **327:**329–331.

Mosser, D. M., Vlassara, H., Edelson, P. J., and Cerami, A., 1987, *Leishmania* promastigotes are recognized by the macrophage receptor for advanced glycosylation endproducts, *J. Exp. Med.* **165:**140–145.

Mukkada, A. J., 1985, Energy metabolism in *Leishmania,* in Human Parasitic Diseases (K. P. Chang and R. S. Bray, eds.), Vol. 1, pp. 49–64, Elsevier Science Publishers, Amsterdam.

Mukkada, A. J., Schaefer, F. W., III, Simon, M. W., and Neu, C., 1974, Delayed *in vitro* utilization of glucose by *Leishmania tropica* promastigotes, *J. Protozool.* **21:**393–397.

Mukkada, A. J., Meade, J. C., Glaser, T. A., and Bonventre, P. F., 1985, Enhanced metabolism of *Leishmania donovani* amastigotes at acid pH: An adaptation for intracellular growth, *Science* **229:**1099–1101.

Murray, H. W., 1981, Susceptibility of *Leishmania* to oxygen intermediates and killing by normal macrophages, *J. Exp. Med.* **153:**1302–1315.

Murray, H. W., 1982, Cell-mediated immune response in experimental leishmaniasis. II. Oxygen-dependent killing of intracellular *Leishmania donovani* amastigotes, *J. Immunol.* **129:**351–357.

Murray, H. W., and Cartelli, D. M., 1983, Killing of intracellular *Leishmania donovani* by human mononuclear phagocytes. Evidence of oxygen-dependent and -independent leishmanicidal activity, *J. Clin. Invest.* **72:**32–44.

Murray, H. W., Oca, M. J., Granger, A. M., and Schreiber, R. D., 1989, Requirement for T cells and effect of lymphokines in successful chemotherapy for an intracellular infection. Experimental visceral leishmaniasis, *J. Clin. Invest.* **83:**1253–1257.

Navin, T. R., Krug, E. C., and Pearson, R. D., 1989, Effect of immunoglobulin M from normal human serum on *Leishmania donovani* promastigote agglutination, complement-mediated killing, and phagocytosis by human monocytes, *Infect. Immun.* **57:**1343–1346.

Orlandi, P. A., Jr., and Turco, S. J., 1987, Structure of the lipid moiety of the *Leishmania donovani* lipophosphoglycan, *J. Biol. Chem.* **262:**10384–10391.

Padan, E., and Schuldiner, S., 1987, Intracellular pH and membrane potential as regulators in the prokaryotic cell, *J. Membrane Biol.* **95:**189–198.

Pan, A. A., and McMahon-Pratt, D., 1988, Monoclonal antibodies specific for the amastigote stage of *Leishmania pifanoi.* I. Characterization of antigens associated with stage- and species-specific determinants, *J. Immunol.* **140:**2406–2414.

Pearson, R. D., and de Sousa, A. Q., 1990, *Leishmania* species: Visceral (kala-azar), cutaneous, and mucosal leishmaniasis, in *Principles and Practice of Infectious Diseases* (G. L. Mandell, R. G. Douglas, Jr., and J. E. Bennett, eds.), pp. 2066–2077, Churchill Livingstone, New York.

Pearson, R. D., and Steigbigel, R. T., 1980, Mechanism of lethal effect of human serum upon *Leishmania donovani, J. Immunol.* **125:**2195–2201.

Pearson, R. D., and Steigbigel, R. T., 1981, Phagocytosis and killing of the protozoan *Leishmania donovani* by human polymorphonuclear leukocytes, *J. Immunol.* **127:**1438–1443.

Pearson, R. D., and Wilson, M. E., 1989, Host defenses against prototypical intracellular protozoans, the *Leishmania,* in *Parasitic Infections in the Compromised Host* (P. D. Walzer and R. M. Genta, eds.), pp. 31–81, Marcel Dekker, New York.

Pearson, R. D., Romito, R., Symes, P. H., and Harcus, J. L., 1981, Interaction of *Leishmania donovani* promastigotes with human monocyte-derived macrophages: Parasite entry, intracellular survival and multiplication, *Infect. Immun.* **32:**1249–1253.

Pearson, R. D., Harcus, J. L., Symes, P. H., Romito, R., and Donowitz, G. R., 1982, Failure of the phagocytic oxidative response to protect human monocyte-derived macrophages from infection by *Leishmania donovani, J. Immunol.* **129:**1282–1286.

Pearson, R. D., Sullivan, J. A., Roberts, D., Romito, R., and Mandell, G. L., 1983a, Interaction of *Leishmania donovani* promastigotes with human phagocytes, *Infect. Immun.* **40:**411–416.

Pearson, R. D., Harcus, J. L., Roberts, D., and Donowitz, G. R., 1983b, Differential survival of *Leishmania donovani* amastigotes in human monocytes, *J. Immunol.* **131:**1994–1999.

Pearson, R. D., Wheeler, D. A., Harrison, L. H., and Kay, H. D., 1983c, The immunobiology of leishmaniasis, *Rev. Infect. Dis.* **5:**907–927.

Pelham, H. R. B., 1986, Speculations on the functions of the major heat shock and glucose-regulated proteins, *Cell* **46:**959–961.

Pelham, H. R. B., 1989, The selectivity of secretion: Protein sorting in the endoplasmic reticulum, *Biochem. Soc. Trans.* **17:**795–802.

Puentes, S. M., Sacks, D. L., da Silva, R. P., and Joiner, K. A., 1988, Complement binding by two developmental stages of *Leishmania major* promastigotes varying in expression of a surface lipophosphoglycan, *J. Exp. Med.* **167:**887–902.

Rabinovitch, M., Dedet, J.-P., Ryter, A., Robineaux, R., Topper, G., and Brunet, E., 1982, Destruction of *Leishmania mexicana amazonensis* amastigotes within macrophages in culture by phenazine methosulfate and other electron carriers, *J. Exp. Med.* **155:**415–431.

Rabinovitch, M., Zilberfarb, V., and Ramazeilles, C., 1986, Destruction of *Leishmania mexicana amazonensis* amastigotes within macrophages by lysosomotropic amino acid esters, *J. Exp. Med.* **163:**520–535.

Rassam, M. B., Al-Mudhaffar, S. A., and Chance, M. L., 1979, Isoenzyme characterization of *Leishmania* species from Iraq, *Ann. Trop. Med. Parasitol.* **73:**527–534.

Reiner, N. E., and Kazura, J. W., 1982, Oxidant-mediated damage of *Leishmania donovani* promastigotes, *Infect. Immun.* **36:**1023–1027.

Remaley, A. T., Kuhns, D. B., Basford, R. E., Glew, R. H., and Kaplan, S. S., 1984, Leishmanial phosphatase blocks neutrophil O_2^- production, *J. Biol. Chem.* **259:**11173–11175.

Remaley, A. T., Das, S., Campbell, P. I., LaRocca, G. M., Pope, M. T., and Glew, R. H., 1985, Characterization of *Leishmania donovani* acid phosphatases, *J. Biol. Chem.* **260:**880–886.

Russell, D. G., and Wilhelm, H., 1986, The involvement of the major surface glycoprotein (gp63) of *Leishmania* promastigotes in attachment to macrophages, *J. Immunol.* **136:**2613–2620.

Russell, D. G., and Wright, S. D., 1988, Complement receptor type 3 (CR3) binds to an Arg-Gly-Asp-containing region of the major surface glycoprotein, gp63, of *Leishmania* promastigotes, *J. Exp. Med.* **168:**279–292.

Sacks, D. L., and da Silva, R. P., 1987, The generation of infective stage *Leishmania major* promastigotes is associated with the cell-surface expression and release of a developmentally regulated glycolipid, *J. Immunol.* **139:**3099–3106.

Sacks, D. L., and Perkins, P. V., 1984, Identification of an infective stage of *Leishmania* promastigotes, *Science* **223:**1417–1419.

Sacks, D. L., Hieny, S., and Sher, A., 1985, Identification of cell surface carbohydrate and antigenic changes between noninfective and infective developmental stages of *Leishmania major* promastigotes, *J. Immunol.* **135:**564–569.

Sadick, M. D., Locksley, R. M., and Raff, H. V., 1984, Development of cellular immunity in cutaneous leishmaniasis due to *Leishmania tropica, J. Infect. Dis.* **150:**135–138.

Schnur, L. F., Chance, M. L., Ebert, F., Thomas, S. C., and Peters, W., 1981, The biochemical and serological taxonomy of visceralizing *Leishmania, Ann. Trop. Med Parasitol.* **75:**131–144.

Schottelius, J., 1982, Lectin typing of *Leishmania*-strains from the New and Old World, in *Lectins,*

Biology, Biochemistry, Clinical Biochemistry, (T. C. Bog-Hansen, ed.), Vol. II, pp. 531–541, Walter de Gruyter, Berlin.

Scott, P., James, S., and Sher, A., 1985, The respiratory burst is not required for killing of intracellular and extracellular parasites by a lymphokine-activated macrophage cell line, *Eur. J. Immunol.* **15**:553–558.

Scott, P., Natovitz, P., Coffman, R. L., Pearce, E., and Sher, A., 1988, Immunoregulation of cutaneous leishmaniasis. T cell lines that transfer protective immunity or exacerbation belong to different T helper subsets and respond to distinct parasite antigens, *J. Exp. Med.* **168**:1675–1684.

Searle, S., Campos, A. J. R., Coulson, R. M., Spithill, T. W., and Smith, D. F., 1989, A family of heat shock protein 70-related genes are expressed in the promastigotes of *L. major, Nucleic Acids Res.* **17**:5081–5095.

Simon, M. W., Martin, E., and Mukkada, A. J., 1978, Evidence for a functional glyoxylate cycle in the *Leishmaniae, J. Bacteriol.* **135**:895–899.

Simon, M. W., and Mukkada, A. J., 1983, Intracellular protein degradation in *Leishmania tropica* promastigotes, *Mol. Biochem. Parasitol.* **7**:19–26.

Simon, M. W., Jayasimhulu, K., and Mukkada, A. J., 1983, The free amino acid pool in *Leishmania tropica* promastigotes, *Mol. Biochem. Parasitol.* **9**:47–57.

Sirawaraporn, W., Sertsrivanich, R., Booth, R. G., Hansch, C., Neal, R. A., and Santi, D. V., 1988, Selective inhibition of *Leishmania* dihydrofolate reductase and *Leishmania* growth by 5-benzyl-2,4-diaminopyrimidines, *Mol. Biochem. Parasitol.* **31**:79–86.

Spector, T., Jones, T. E., and Ellion, G. B., 1979, Specificity of adenylosuccinate synthethase and adenylosuccinate lyase from *Leishmania donovani, J. Biol. Chem.* **254**:8422–8426.

Steiger, R. F., and Meshnich, S. R., 1977, Amino-acid and glucose utilization of *Leishmania donovani* and *L. braziliensis, Trans. R. Soc. Trop. Med. Hyg.* **71**:441–443.

Stern, J. J., Oca, M. J., Rubin, B. Y., Anderson, S. L., and Murray, H. W., 1988, Role of L3T4$^+$ and Lyt-2$^+$ cells in experimental visceral leishmaniasis, *J. Immunol.* **140**:3971–3977.

Stinson, S., Sommer, J. R., and Blum, J. J., 1989, Morphology of *Leishmania braziliensis:* Changes during reversible heat-induced transformation from promastigote to an ellipsoidal form, *J. Parasitol.* **75**:431–440.

Talamas-Rohana, P., Wright, S. D., Lennartz, M. R., and Russell, D. G., 1990, Lipophosphoglycan from *Leishmania mexicana* promastigotes binds to members of the CR3, p150,95 and LFA-1 family of leukocyte integrins, *J. Immunol.* **144**:4817–4824.

Toye, P., and Remold, H., 1989, The influence of temperature and serum deprivation on the synthesis of heat-shock proteins and alpha and beta tubulin in promastigotes of *Leishmania major, Mol. Biochem. Parasitol.* **35**:1–10.

Turco, S. J., Wilkerson, M. A., and Clawson, D. R., 1984, Expression of an unusual acidic glycoconjugate in *Leishmania donovani, J. Biol. Chem.* **259**:3883–3889.

Turco, S. J., Hull, S. R., Orlandi, P. A., Jr., Shepherd, S. D., Homans, S. W., Dwek, R. A., and Rademacher, T. W., 1987, Structure of the major fragment of the *Leishmania donovani* lipophosphoglycan, *Biochemistry* **26**:6233–6238.

Tuttle, J. V., and Krenitsky, T. A., 1980, Purine phosphoribosyltransferases from *Leishmania donovani, J. Biol. Chem.* **255**:909–916.

Van der Ploeg, L. H. T., Giannini, S. H., and Cantor, C. R., 1985, Heat shock genes: Regulatory role for differentiation in parasitic protozoa, *Science* **228**:1443–1446.

Walters, L. L., Chaplin, G. L., Modi, G. B., and Tesh, R. B., 1989, Ultrastructural biology of *Leishmania (Viannia) panamensis* (= *Leishmania braziliensis panamensis*) in *Lutzomyia gomezi* (Diptera:Psychodidae): A natural host–parasite association, *Am. J. Trop. Med. Hyg.* **40**:19–39.

Weiser, W. Y., Van Niel, A., Clark, S. C., David, J. R., and Remold, H. G., 1987, Recombinant human granulocyte/macrophage colony-stimulating factor ativates intracellular killing of *Leishmania donovani* by human monocyte-derived macrophages, *J. Exp. Med.* **166:**1436–1446.

Wilson, M. E., and Hardin, K. K., 1988, The major concanavalin A-binding surface glycoprotein of *Leishmania donovani chagasi* promastigotes is involved in attachment to human macrophages, *J. Immunol.* **141:**265–272.

Wilson, M. E., and Hardin, K. K., 1990, The major *Leishmania donovani chagasi* surface glycoprotein in tunicamycin-resistant promastigotes, *J. Immunol.* **144:**4825–4834.

Wilson, M. E., and Pearson, R. D., 1984, Stage-specific variations in lectin binding to *Leishmania donovani. Infect. Immun.* **46:**128–134.

Wilson, M. E., and Pearson, R. D., 1986, Evidence that *Leishmania donovani* utilizes a mannose receptor on human mononuclear phagocytes to establish intracellular parasitism, *J. Immunol.* **136:**4681–4688.

Wilson, M. E., and Pearson, R. D., 1988, Roles of CR3 and mannose receptors in the attachment and ingestion of *Leishmania donovani* by human mononuclear phagocytes, *Infect. Immun.* **56:**363–369.

Wilson, M. E., and Pearson, R. D., 1990, Immunology of leishmaniasis, in *Modern Parasite Biology: Cellular, Immunological, and Molecular Aspects* (D. J. Wyler, ed.), pp. 200–221, W. H. Freeman and Company, New York.

Wilson, M. E., Innes, D. J., Sousa, A. Q., and Pearson, R. D., 1987, Early histopathology of experimental infection with *Leishmania donovani* in hamsters, *J. Parasitol.* **73:**55–63.

Winfield, J. B., 1989, Stress proteins, arthritis and autoimmunity, *Arthritis Rheum.* **32:**1497–1504.

Wirth, D. F., and McMahon-Pratt, D., 1982, Rapid identification of *Leishmania* species by specific hybridization of kinetoplast DNA in cutaneous lesions, *Proc. Natl. Acad. Sci. USA* **79:**6999–7003.

Wozencraft, A. O., Sayers, G., and Blackwell, J. M., 1986, Macrophage type 3 complement receptors mediate serum-independent binding of *Leishmania donovani.* Detection of macrophage-derived complement on the parasite surface by immunoelectron microscopy, *J. Exp. Med.* **164:**1332–1337.

Wyler, D. J., Sypek, J. P., and McDonald, J. A., 1985, *In vitro* parasite–monocyte interactions in human leishmaniasis: Possible role of fibronectin in parasite attachment, *Infect. Immun.* **49:**305–311.

Wyler, D. J., Beller, D. I., and Sypek, J. P., 1987, Macrophage activation for antileishmanial defense by an apparently novel mechanism, *J. Immunol.* **138:**1246–1249.

Zajtchuk, J. T., Casler, J. D., Netto, E. M., Grogl, M., Neafie, R. C., Hessel, C. R., De Magalhaes, A. V., and Marsden, P. D., 1989, Mucosal leishmaniasis in Brazil, *Laryngoscope* **99:**925–939.

Zenian, A., 1981, *Leishmania tropica:* Biochemical aspects of promastigotes attachment to macrophages *in vitro, Exp. Parasitol.* **51:**175–187.

Zilberstein, D., and Dwyer, D. M., 1984, Antidepressants cause lethal disruption of membrane function in human protozoan parasite *Leishmania, Science* **226:**977–979.

Zilberstein, D., and Dwyer, D. M., 1985, Protonmotive force-driven active transport of D-glucose and L-proline in the protozoan parasite *Leishmania donovani, Proc. Natl. Acad. Sci USA* **82:**1716–1720.

Zilberstein, D., and Dwyer, D. M., 1988, Identification of a surface membrane proton-translocating ATPase in promastigotes of the parasitic protozoan *Leishmania donovani, Biochem. J.* **256:**13–21.

Zilberstein, D., Philosoph, H., and Gepstein, A., 1989, Maintenance of cytoplasmic pH and proton motive force in promastigotes of *Leishmania donovani, Mol. Biochem. Parasitol.* **36:**109–118.

Chapter 2

Membrane Proteins and Enzymes of *Leishmania*

Pascal Schneider, Clément Bordier, and Robert Etges

1. INTRODUCTION

1.1. Biology of *Leishmania*

Leishmania are kinetoplastid protozoan parasites responsible for a spectrum of zoonotic, and less commonly anthroponotic, diseases of man and other mammals. The parasite is transmitted by a blood-feeding dipteran vector of the subfamily Phlebotominae, and undergoes radical morphological and metabolic changes as it passes from the insect vector to the mammalian host. In the latter, it is found as an obligately intracellular, small oval cell, 2–3 μm in diameter, with a truncated flagellum that barely extends beyond the plasma membrane. The intracellular, or amastigote, stage of the parasite specifically infects cells of the mononuclear phagocyte system, in which it lives and divides within the acidified phagolysosomes at pH 4.7–5.3 (Antoine *et al.*, 1990). The parasites disseminate within the mammal as amastigotes, released as a result of the destruction of the host cell, and are bound and engulfed by neighboring uninfected macrophages. Transmission from one mammal to another occurs through the sandfly vector:

Pascal Schneider and Robert Etges Institute of Biochemistry, University of Lausanne, CH-1066 Epalinges, Switzerland. **Clément Bordier** Affinity Products, CH-1023 Crissier-sur-Lausanne, Switzerland.

Subcellular Biochemistry, Volume 18: Intracellular Parasites, edited by J. L. Avila and J. R. Harris. Plenum Press, New York, 1992.

amastigotes within mononuclear cells are ingested during the blood meal of the female sandfly and differentiate within the presumably neutral-to-alkaline environment of the sandfly midgut to form elongated, motile, flagellate cells called promastigotes, which are 12–16 μm long and 1.5–3.5 μm wide. The promastigotes remain free in the midgut lumen or attach by means of their flagella to the microvillous epithelial cells of the insect midgut. The parasites divide actively, then colonize the chitinous foregut of the insect, where the morphologically distinct paramastigotes adhere by means of hemidesmosomes at the tips of their modified flagella (Killick-Kendrick *et al.*, 1974; reviewed in Molyneux and Killick-Kendrick, 1987). The promastigotes migrate anteriorly within the digestive tract of the sandfly, ultimately to be found in the proboscis and pharynx as infective metacyclic promastigotes, which are physiologically similar to the complement-resistant promastigotes arising in the stationary phase of culture *in vitro* (Killick-Kendrick, 1986). The small, nondividing, infective metacyclics, which swim actively in the lumen of the mouthparts, are transmitted to a new mammalian host during a subsequent blood meal of the sandfly (Killick-Kendrick, 1986). Components of *Phlebotomus* saliva, pharmacologically similar to mammalian calcitonin gene-related peptide (CGRP), cause prolonged vasodilation and vascular permeabilization at the site of the fly bite, not only providing the vector with a more abundant blood meal, but also favoring the influx of circulating monocytic cells (Titus and Ribeiro, 1988; Theodos, *et al.*, 1991). The *Leishmania* enhancing factor of sandfly salivary gland lysates, and to a lesser degree human CGRP, blocks the ability of macrophages to present antigen *in vitro* and prevents them from producing hydrogen peroxide in response to interferon-γ. Together, these activities would clearly favor the survival of the parasite at the early stages of infection (Titus and Ribeiro, 1990). Metacyclic promastigotes, in contrast to midgut stages that are thought to be equivalent to log-phase promastigotes grown *in vitro*, are relatively resistant to the lytic effects of complement (Sacks and Da Silva, 1987; Grimm *et al.*, 1991). Those parasites that escape such host defenses are rapidly engulfed by phagocytic cells in which they differentiate and continue to divide as amastigotes. Within the acidified phagolysosomes of the new host cell, *Leishmania* must resist the potentially microbicidal effects of lysosomal hydrolases (Chang and Fong, 1983), toxic oxygen metabolites resulting from the macrophage oxidative burst (Mauël, 1984a,b; Mauël *et al.*, 1984), arginine-derived nitric oxide and its metabolites (Green *et al.*, 1990; Mauël *et al.*, 1991), and cationic, channel-forming peptides (Selsted and Harwig, 1987).

Most biochemical investigations have been made on cultured promastigotes for reasons of convenience. They can be grown easily, but tend to lose infectivity after long-term culture *in vitro* unless they are regularly passaged in suitable mammalian hosts. Amastigotes, with rare exceptions, must be freshly and rapidly isolated from the tissues of their mammalian hosts and purified from contaminating host cells and debris under conditions that do not favor their transformation to the promastigote stage. In the case of several South American species of *Leishmania,* notably *L. pifanoi, L. panamensis,* and *L. braziliensis,* it is possible to cultivate amastigotelike organisms in long-term axenic culture (Pan, 1984; Eperon and McMahon-Pratt, 1989). Amastigotes of *L. mexicana* transformed *in vitro* from promastigotes or amastigotes isolated from lesions of experimentally infected BALB/c mice have been maintained successfully at high density for several weeks by one of the authors (R. E.). Although these "axenic amastigotes" are morphologically and antigenically more similar to authentic tissue-derived amastigotes than they are to promastigotes, it is premature to consider the axenically cultured cells to be identical to the obligately intracellular parasites isolated from infected macrophages. As a result of these complications, the bulk of available biochemical information on *Leishmania* membrane macromolecules refer only to the promastigote stage.

1.2. Organelles of *Leishmania*

The classical membrane-limited organelles of eukaryotic cells, such as the nucleus, mitochondria, endoplasmic reticulum, Golgi apparatus, and secretory vesicles, have all been identified in *Leishmania.* Membranes segregate intra- and extracellular material and allow the compartmentalization of different catabolic and biosynthetic pathways within the cell. Different concentrations of ions separated by a membrane generate a membrane potential that can be exploited as a source of energy for the selective transport of specific molecules across the membrane. In any parasitic organism, the plasma membrane represents the primary interface between various host defense mechanisms and the invader, and may therefore be presumed to be adapted to this highly specialized situation. Membrane specializations in a parasite such as *Leishmania,* which divides its existence between the neutral-to-alkaline hydrolytic environment of the insect midgut and the acidic hydrolytic environment of the mammalian phagolysosome, can be expected to be even more radical. The plasma membrane of *Leishmania* envelops the cell, including the flagellum, and is supported by an extensive microtubular cytoskeleton attached to the inner surface of the plasma membrane by a regular array of microtubule-associated proteins, which can be readily observed by electron microscopy (Bordier *et al.*, 1982; Seebeck *et al.*, 1990).

The flagellum of *Leishmania* promastigotes, like that of *Trypanosoma,* contains the 9 + 2 arrangement of microtubule doublets typical of the eukaryotic flagellar axoneme, as well as a peculiar structure of unknown function termed the paraxial rod. Like the subpellicular microtubules of the plasma membrane skeleton, the flagellar axoneme does not dissociate at low temperatures, and can thus be isolated as a morphologically intact structure. The membrane of this organelle includes, in addition to the major surface components of the plasma membrane discussed below, two low-molecular-weight surface proteins that appear to be involved in the establishment of flagellar interaction and attachment to the insect midgut epithelium (Ismach *et al.*, 1989; Warburg *et al.*, 1989).

The cold stability of the microtubular membrane skeleton and flagellar axoneme, which remain tightly associated to their respective membranes, allows the isolation of highly enriched plasma membranes and flagella following appropriate cell disruption and isopycnic gradient centrifugation (Dwyer, 1980). Purified pellicular membrane fractions of *L. donovani* promastigotes were used as starting material for the isolation of the membrane acid phosphatases and nucleotidases described in Sections 1.3 and 2.4. Similar membrane preparations of surface-iodinated *L. major* promastigotes prompted our work on the major 63-kDa surface antigen, which we subsequently identified as the promastigote surface metalloprotease (Etges *et al.*, 1985, 1986b; Bouvier *et al.*, 1989).

The megasome is thought to represent the lysosomal compartment in *Leishmania*, and can represent up to 14% of the cell volume in some species (Pupkis *et al.*, 1986). The flagellar pocket represents a specialized membrane domain to which endo- and exocytosis is restricted; it is the only portion of the membrane that is not intimately subtended by the microtubular membrane skeleton. *Leishmania*, like all members of the order Kinetoplastidae, possess a single, ramified tubular mitochondrion with a unique network of condensed, concatenated circular mDNA called the kinetoplast, from which the taxon is named. The peroxisomelike, nucleic acid-free organelle called the glycosome, in which glycolytic and several other enzymes are sequestered (Opperdoes, 1989), are also unique to the Kinetoplastidae.

1.3. Membrane Proteins of *Leishmania*

Virtually all *Leishmania* membrane proteins that have been isolated originate from promastigotes, which are readily cultivated in axenic culture and thus available in significant quantities. Little is known about the membrane proteins of the amastigote, except for those which are also expressed by the promastigote or which are known only by their antigenic, enzymatic, or transport activities.

Externally orientated plasma membrane proteins can be identified by labeling intact cells with nonpermeant reagents such as *N*-hydroxysuccinimidobiotin or radioactive iodine. If antibodies of known specificity are available, immunofluorescence observations of intact cells can provide additional information about the distribution of the molecule at the surface of the cell and possibly identify subpopulations of cells differing in their expression of given surface antigens. An assay to assess the internal versus external location of enzymatic activity was developed for *L. donovani* membrane acid phosphatase (Gottlieb and Dwyer, 1981a), 3′- and 5′-nucleotidases (Dwyer and Gottlieb, 1984), and an ATPase (Zilberstein and Dwyer, 1988), in which isolated pellicular membranes from disrupted cells were incubated with a suitable soluble substrate for the enzyme. Inorganic phosphate released by the activity of the phosphatase was precipitated as $Pb(PO_4)_2$ in the vicinity of the enzyme and detected by electron microscopy. As the internal face of the membrane remains tightly associated with cold-stable microtubules, it is easy to deduce the location of the enzyme relative to the membrane. Enzyme or protein location in organelles can be shown by copurification with marker enzymes in subcellular fractionation of the parasite or by electron microscopy using immunogold or lectin–gold labeling techniques.

The preparation of membranes, followed by their extraction with nonionic detergent and subsequent enrichment of extracted proteins by classical ion-exchange chromatographic methods, provides a valuable, general method for the isolation of membrane proteins. Antibodies or lectins known to react with surface macromolecules can be used to purify membrane components by affinity chromatography. In a more elegant method that capitalizes on the amphiphilic nature of membrane proteins, cells can be lysed in the nonionic detergent Triton X-114 and the integral membrane proteins are recovered in a detergent-enriched phase after a temperature-induced phase separation. This method has been applied analytically to identify membrane components of *Leishmania* and as a first step in large-scale purifications of the *Leishmania* surface metalloprotease, promastigote surface protease (Bordier, 1981, 1988; Etges *et al.*, 1985; Jähnig and Etges, 1988; Murray *et al.*, 1989b; Schneider *et al.*, 1990). Non-membrane-permeant reagents, such as $[^{125}I]$iodine, various derivatives of biotin, and galactose oxidase treatment followed by reduction with $NaB[^3H]_4$, can be used to mark surface molecules on living cells prior to their isolation. Membrane-associated enzyme activities can be used not only as intrinsic markers, but in rare cases they can be labeled with radioactive substrate analogues by photo cross-linking, as for the *Leishmania* glucose and folate transporters (Zilberstein *et al.*, 1986; Beck and Ullman, 1989). Finally, the membrane location of a protein can, with some degree of success, be deduced from the primary structure of the protein predicted from the gene sequence. The predicted amino acid sequence may indicate a hydrophobic polypeptide domain, multiple membrane spanning domains, or a signal for glycosyl-phosphatidylinositol addition at the COOH-terminus, or other similarities to membrane proteins characterized from other organisms (Bard, 1989; Cairns *et al.*, 1989; Murray *et al.*, 1989a; Meade *et al.*, 1987, 1989; Ouellette *et al.*, 1990; Stein *et al.*, 1990).

1.4. Metabolite Uptake and Energy Generation

Like any other eukaryotic cell, *Leishmania* needs to fulfill basic cellular functions, including the acquisition and concentration of nutrients, the excretion of metabolic waste products, energy generation, and the maintenance of a suitable membrane potential and intracellular pH. Metabolite transport involves

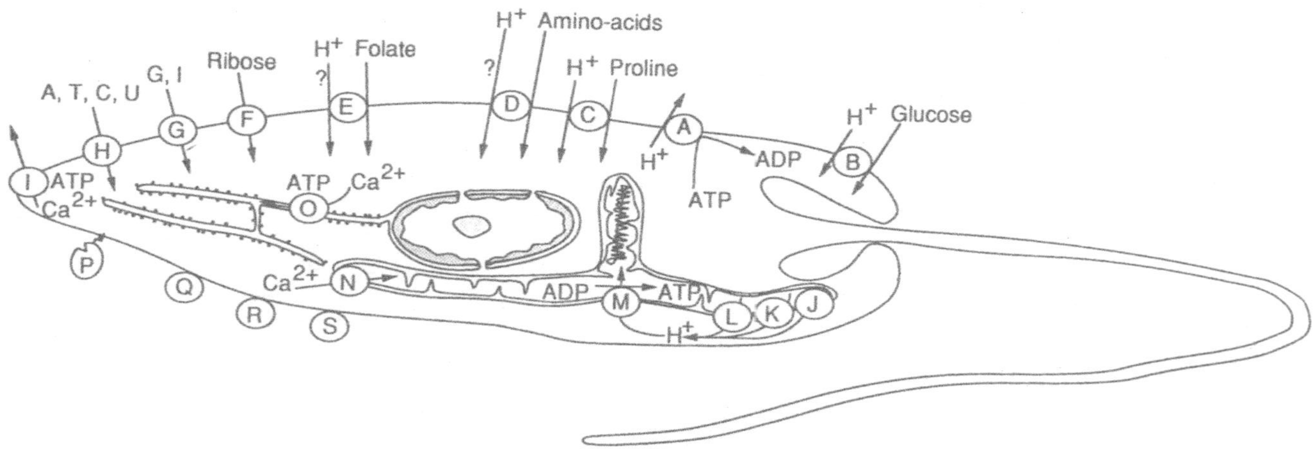

FIGURE 1. Membrane enzymes and transporters of *Leishmania*. The membrane enzymes and transporters identified in the promastigote stage of *Leishmania* are represented as circled letters: A, surface proton ATPase; B, D-glucose–H^+ symport; C, L-proline–H^+ symport; D, charged amino acid symport; E, folate–methotrexate transporter; F, ribose uniport; G, guanosine–inosine uniport; H, adenosine–pyrimidine uniport; I, surface Ca^{2+}-ATPase; J, NADH-Q-reductase; K, cytochrome C-reductase; L, cytochrome C-oxidase; M, mitochondrial F_0F_1 ATP synthase; N, mitochondrial Ca^{2+}-transporter; O, endoplasmic reticulum Ca^{2+}-ATPase; P, promastigote surface metalloprotease; Q, 3′-nucleotidase–nuclease; R, 5′-nucleotidase; S, membrane acid phosphatase.

specialized carrier proteins located in the membrane. Uniports allow the selective facilitated diffusion or group translocation of a given molecule only when the internal concentration of the solute is lower than the extracellular concentration. A favorable concentration gradient can be maintained by rapid transformation of that particular metabolite within the cell. Symports and antiports transport molecules against their concentration gradients, with simultaneous transport of ions down their respective electrochemical gradients. The electrochemical gradient depends on both the concentration gradient (i.e., Δ pH for a transporter coupled to proton translocation) and on the membrane potential, $\Delta\Psi$ (i.e., the overall difference in charges on both sides of the membrane). Transport of Na^+, K^+, Ca^{2+}, and H^+ up their respective electrochemical gradients is frequently achieved by the simultaneous hydrolysis of ATP or, in the case of mitochondrial H^+ transport, by a simultaneous electron transfer in the respiratory chain. The activities of the *Leishmania* transporters described here are shown in Figure 1.

2. PROTEINS AND ENZYMES OF THE PLASMA MEMBRANE

2.1. ATPases

Metabolic activities produce acidic components such as carbon dioxide and a variety of organic acids; excess protons must be exported in order to maintain a physiological intracellular pH. It was shown that the intracellular pH of *L. donovani* promastigotes varies only 0.3 pH units (6.4–6.7) when the extracellular pH changes nearly 2 pH units from 5.5 to 7.4 (Glaser *et al.*, 1988; Zilberstein *et al.*, 1989). Protons are extruded from the cytosol to the outside of the cell by a proton-translocating ATPase that requires Mg^{2+}-ATP (Zilberstein and Dwyer, 1988). This H^+-ATPase activity is located on the cytoplasmic side of the plasma membrane and has a pH optimum of 6.5, which corresponds to the cytosolic pH. It is not inhibited by the cardiotonic steroid inhibitor of the mammalian Na^+/K^+-ATPase, ouabain, by the mitochondrial F_0F_1 H^+-ATPase inhibitor, oligomycin, or by azide that blocks respiration. It is, however, inhibited by the phosphate transition-state analogue *ortho*-vanadate, which clearly distinguishes the *Leishmania* plasma membrane enzyme from the mitochondrial F_0F_1 H^+-ATPase. The proton-pumping ability of the *Leishmania* surface membrane ATPase was directly demonstrated by the acidification of inside-out surface membrane vesicles upon the addition of Mg^{2+}-ATP.

A high-affinity calmodulin-activated Ca^{2+}-ATPase was identified and purifed from the plasma membranes of *L. donovani* and *L. braziliensis* promastigotes. The *L. donovani* enzyme can be radioiodinated at the surface of intact cells, and consists of two subunits of 51 and 57 kDa. The vanadate-sensitive Ca^{2+}-ATPase has a K_m for Ca^{2+} of 35 nM, which is reduced to 12 nM in the presence of heterologous calmodulin (Ghosh *et al.*, 1990; Benaim and Romero, 1990).

Two tandemly linked *L. donovani* genes designated 1a and 1b were suggested to encode *ortho*-vanadate-sensitive cation-transporting ATPases. These genes, which were detected in four *Leishmania* species, are disposed head-to-tail on a small 750-kb chromosome. The genes were cloned, sequenced, and the deduced amino acid sequences were shown to be similar to both yeast and mammalian ATPases, with ten predicted transmembrane segments (Meade *et al.*, 1987, 1989, 1991). The two *Leishmania* gene products differ in 20 amino acids out of 974, with 15 changes occurring near the carboxyl terminus, adding three positive charges and a second potential glycosylation site in ATPase 1b. Most interestingly, the messenger for ATPase 1a was found at similar levels in both stages of the parasite's life cycle, whereas ATPase 1b mRNA was found to be expressed at a much higher level in amastigotes compared to log-phase promastigotes, and at an intermediate level in stationary-phase promastigotes. The specificities of the predicted ATPase activities are not known, but it is likely that they represent the surface H^+-ATPase. It would be tempting to speculate that the stage-specific expression of H^+-ATPase 1b represents one of the key adaptations of *Leishmania* to life in the acidified phagolysosome of the mammalian phagocyte. The ATPase genes could, however, encode the vanadate-sensitive organellar Ca^{2+}-ATPase described by Philosoph and Zilberstein (1989) or the plasma membrane Ca^{2+}-ATPase isolated and characterized by Ghosh *et al.* (1990). However, the abundance of the mRNA detected very likely rules out the plasma membrane Ca^{2+}-ATPase whose mRNA is normally very scarce (E. E. Strehler, personal communication).

The sequences of three related genes amplified in methotrexate-resistant *L. tarentolae* share significant similarities with those encoding mammalian *P*-glycoproteins (Ouellette *et al.*, 1990). These proteins are responsible for the ATP-dependent extrusion of many unrelated lipophilic drugs from mammalian cells, and perhaps the protozoan parasite *Plasmodium falciparum* (Wilson *et al.*, 1989). The putative *P*-glycoprotein of *L. tarentolae* would be composed of 12 membrane-spanning α-helices and a cytoplasmic nucleotide (presumably ATP) binding site. However, unlike multidrug-resistant mammalian cells and *Plasmodium*, methotrexate-resistant *L. tarentolae* are not resistant to other lipophilic drugs and appear to concentrate methotrexate to an extent similar to that observed in nonresistant cells. Further, the resistance is not reversed by the Ca^{2+} channel blocker verapamil (Ouellette *et al.*, 1990). However, a multidrug-resistant phenotype associated with H-region amplification was reported to occur in *L. major*, in which primaquine or terabinafine-resistant cells were significantly resistant to the unrelated methotrexate (Ellenberger and Beverly, 1989).

2.2. Symports and Antiports

Leishmania actively cotransports D-glucose into the cell by a proton motive force-driven system, that is, protons are imported in the cell down their electrochemical gradient in parallel to glucose. The nonmetabolizable analogue 2-deoxy-D-glucose, which uses the same transporter as glucose, is actively concentrated in the cell with a K_m of 24 μM to as much as 80 times the external concentration. Two thirds of the transported hexose was shown to be phosphorylated (Schaefer *et al.*, 1974). Transport of glucose is sensitive to sulfhydryl reagents, suggesting the presence of essential cysteine residues (Zilberstein and Dwyer, 1984b). Fructose apparently uses the same transporter (cited in Pastakia and Dwyer, 1987). D-Glucose added to metabolically inhibited *L. donovani* promastigotes resulted in a rapid carbonyl cyanide *p*-(trifluoromethoxy) phenylhydrazone (FCCP)-inhibitable H^+ influx. The absence of effect upon addition of L-glucose confirms the stereoselectivity of the D-glucose–H^+ symport. The $\Delta \Psi$-mediated component of glucose uptake was demonstrated using nonmetabolizing, potassium-loaded promastigotes: the artificially imposed membrane potential caused by the addition of the potassium ionophore valinomycin caused a rapid uptake of D-glucose (Zilberstein and Dwyer, 1985). The glucose transporter polypeptide in *Leishmania* surface membrane vesicles was photochemically cross-linked to labeled cytochalasin B, an inhibitor of the mammalian glucose transporter. The cross-linked protein is glycosylated, as assessed by lectin-binding assay, but its small 19-kDa size suggests that it represents a subunit or a degradation product of the intact transporter (Zilberstein *et al.*, 1986).

Another *Leishmania* symport has been identified from the sequence of a gene of *L. enriettii,* whose deduced sequence of 567 amino acids predicts 12 transmembrane segments and hydrophilic domains, but no *N*-glycosylation sites. It shows some similarity to both bacterial H^+–pentose transporters and the human erythrocyte glucose uniport, but not to the mammalian sodium–glucose symport. These findings are in accordance with the theory that glucose transport is driven by proton motive force in lower organisms, but at the level of helminths and higher multicellular organisms, glucose is transported by a novel, unrelated sodium-dependent transporter. The former, more primitive transporter is proposed to have lost its proton dependence and to have evolved as a uniport. The relation between this gene and the previously characterized glucose transporter remains to be established. Nevertheless, the mRNA transcribed from this gene is present at much higher levels in the promastigote form, suggesting that the leishmanial D-glucose–H^+ symport is developmentally regulated (Cairns *et al.*, 1989). Other species of *Leishmania* contain sequences that hybridize the cloned *L. enriettii* gene at high stringency (Stein *et al.*, 1990).

Different amino acids are transported by specific carriers. Proline, like D-glucose, is actively concentrated by a proton motive force-driven symport (Zilberstein and Dwyer, 1985). Transport was shown to occur even with an unfavorable proton gradient when the inside of the cell was made more acidic than the outside, demonstrating that the existing membrane potential $\Delta\Psi$ is sufficient by itself to allow proton-coupled import at relatively high external pH (Zilberstein *et al.*, 1989). The H^+–L-proline symport was shown by immunogold labeling to be present at the surface of the promastigote and in small amounts in the Golgi, as expected for a glycoprotein (D. Zilberstein, personal communication). The *Leishmania* H^+–L-proline symport displays an unusually high K_m of 645 μM (D. Zilberstein, personal communication). *Leishmania* transporters for proline, alanine, leucine, phenylalanine, methionine, and other amino acids or analogues were described (Simon and Mukkada, 1977; Bonay and Cohén, 1983; Lepley and Mukkada, 1983). From these results it appears that at least two transporters exist, each carrying either charged or neutral amino acids.

Exogenous folate, required for *Leishmania* growth, is actively concentrated by a high-affinity transporter. Inhibition of respiration results in drastic reduction of folate transport in *L. major* promastigotes, suggesting that the process might be proton motive force-driven (Ellenberger and Beverley, 1987). This transporter also carries methotrexate, a folate analogue that inhibits folate reductase. Folate transport activity is reduced tenfold when the promastigotes enter stationary phase, whereas only a twofold decrease is noted for glucose transport, reflecting the reduced requirement in thymidine synthesis in nondividing cells. A similar reduction in folate transport was shown for log-phase promastigotes in a folate-enriched medium. A mutant strain of *L. donovani* selected by its resistance to methotrexate was shown to be deficient in the folate–methotrexate transporter. This stable, nonreverting mutant transports folate at less than 1% of the wild-type rate (Kaur *et al.*, 1988). Photo cross-linking with a labeled ligand analogue showed that the folate–methotrexate transporter contains a 46 kDa polypeptide. Similar labeling of the methotrexate-resistant mutant occurred only at much higher concentrations of labeled ligand (Beck and Ullman, 1989).

2.3. Uniports

Uniports allow the facilitated diffusion of a metabolite down its concentration gradient. A favorable gradient is maintained by rapid metabolism of the solute within the cell.

Trials to show hexose or pentose accumulation in *L. donovani* show that only ribose, in addition to glucose, has a specific transporter. The ribose transporter has a K_m of 2 mM and is sensitive to sulfydryl reagents. Arabinose and xylose are poor substrates. Ribose uptake occurs by facilitated diffusion and group translocation, with two thirds of the internalized radiolabeled ribose recovered in a phosphorylated state.

Ribose-containing nucleosides do not appear to interact with the pentose transporter (Pastakia and Dwyer, 1987). Rather, nucleosides, produced by the combined actions of the surface 3′-nucleotidase–nuclease and 5′-nucleotidase, are taken up by two distinct high-affinity transporters, one for inosine and guanosine and the other for adenosine, cytosine, thymidine, and uracil (Aronow *et al.*, 1987). Mutants deficient in either of these nucleoside transporters were selected on the basis of resistance to toxic derivatives of corresponding nucleosides (Iovannisci *et al.*, 1984). Purines are taken up with a very high affinity, whereas pyrimidines have a lower affinity (K_m = 0.7 μM and ~ 5 μM, respectively). This transport is not likely to be proton motive force-driven, since 95% of incorporated [^{3}H]adenosine (somewhat less for uridine or inosine) is converted into nucleotide within 10 sec of transport. It was proposed that the driving force for nucleoside transport in *Leishmania* is the rapid group translocation of substrate that could even be coupled to transport.

2.4. Plasma Membrane Enzyme Activities

Metabolite transport across the plasma membrane is clearly necessary for the survival of *Leishmania* at each stage of its life cycle. The occurrence of free amino acids, sugars, and nucleosides is possible in the hydrolase-rich environments provided by both the neutral insect midgut and the acidified phagolysosome of the mammalian host; however, the precise nutrient composition of these compartments remains to be analyzed. Potential nutrients and growth factors like proteins, nucleic acid polymers, and complex carbohydrates would require extracellular processing and transformation to become suitable substrates for uptake by the amino acid, hexose, and nucleoside transporters previously described and would presumably complement intracellular degradation of endocytosed material in the parasite's own lysosomal compartment. In several instances, *Leishmania* surface enzymatic activities have been implicated in the establishment or maintenance of the host–parasite relationship independently of their potential roles in nutrient acquisition. Remarkably, for an organism that lives in intimate contact with host hydrolytic enzymes in the digestive tract of the sandfly and the phagolysosome of the macrophage, many of the enzymatic activities identified at the *Leishmania* surface are hydrolases (see Figure 1).

2.4.1. Surface Nucleotidases

A 3′-nucleotidase–nuclease activity is evenly distributed on the external surface of *L. donovani* promastigotes, including the flagella and the lining of the flagellar pocket, and was also detected on the amastigote (Dwyer and Gottlieb,

1984). 3′-Nucleotidase–nuclease is not found as a cell surface enzyme on mammalian cells. It was identified as a 43-kDa glycoprotein (Dwyer, 1987), whose activity is not affected by fluoride but is inhibited by EDTA. Optimum pH is 8.5 for nucleotidase activity and 6 for nuclease activity, with a preference for RNA over DNA (Gottlieb, 1989). 5′-Nucleotides and inorganic phosphate are thus generated from nucleic acids by the activity of this enzyme. The expression of the 3′-nucleotidase–nuclease of *Crithidia luciliae* is significantly increased when the organism is deprived of purines or inorganic phosphate (Gottlieb *et al.*, 1988).

A 5′-nucleotidase is active at the surface of intact *L. donovani* promastigotes and amastigotes, but is also known to occur as a surface membrane marker on many other cells (Gottlieb and Dwyer, 1983; Dwyer and Gottlieb, 1984; Hassan and Coombs, 1987). In contrast to the 3′-nucleotidase–nuclease, the 5′-nucleotidase is insensitive to EDTA, inhibited by fluoride, and hydrolyses both 5′-ribonucleotides and 5′-deoxyribonucleotides with an optimal pH of 6.8–8.0. The enzyme activity is present at the parasite surface including the flagella and the flagellar pocket. It is a surface-exposed glycoprotein of 72 kDa that shows immunological cross-reactivity with rat liver 5′-nucleotidase, suggesting that it might be evolutionarily conserved (Dwyer, 1987; Gottlieb and Dwyer, 1988). The combined activities of both 3′-nucleotidase–nuclease and 5′-nucleotidase would be sufficient to process nucleic acids and nucleotides to inorganic phosphate and nucleosides, which would be suitable substrates for their respective transporters.

2.4.2. Surface Acid Phosphatases

A tartrate-resistant membrane-bound acid phosphatase activity, accounting for over 50% of the total cellular acid phosphatase activity, was also shown by cytochemical methods to be present on the outer leaflet of *L. donovani* promastigote plasma membranes, as well as on the flagellum, the flagellar pocket membrane, and on the inner face of cytoplasmic vesicles in the vicinity of the flagellar pocket (Gottlieb and Dwyer, 1981a,b). Three membrane acid phosphatases that differ in size, isoelectric point, inhibitor sensitivity, and substrate specificity were characterized (Remaley *et al.*, 1985a). They display an optimal pH of 5–6, are inhibited by fluoride, and are unaffected by metal chelators. However, only the most abundant one, which accounts for 70% of the total membrane-bound acid phosphatase activity, is resistant to tartrate. The tartrate-resistant membrane acid phosphatase is a dimer of about 130 kDa composed of two subunits of 65–68 kDa. The enzyme displays a relatively narrow substrate specificity, hydrolysing phosphotyrosine, fructose-1,6-bisphosphate, ADP, AMP, pyridoxal-phosphate, and phosphorylated proteins like pyruvate kinase. The other two membrane acid phosphatases are sensitive to inhibition by tartrate and hydrolyse a broader range of substrates (Remaley *et al.*, 1985a). There is some evidence that the tartrate-resistant membrane acid phosphatase can inhibit the potentially leishmanicidal oxidative burst by dephosphorylating key neutrophil surface phosphoproteins, and thus contribute to the establishment of infection (Remaley *et al.*, 1985b). The amphiphilic membrane acid phosphatase purifed from *L. major* promastigotes was shown to be insensitive to the effects of the phosphatidylinositol-specific phospholipases C of *Trypanosoma brucei* and *Bacillus cereus*, suggesting that it is not anchored to the membrane by a glycosyl phosphatidylinositol anchor similar to that of the major membrane metalloprotease, promastigote surface protease (Etges, unpublished data). The membrane acid phosphatases are responsible, however, for only about 10% of the total acid phosphatase activity associated with cultured promastigotes; the remaining 90% is due to a heterogenously glycosylated, high-molecular-weight, tartrate-sensitive soluble enzyme (Gottlieb and Dwyer, 1982; Hassan and Coombs, 1987). Interestingly, the glycans of the secreted phosphatase include epitopes that are chemically and antigenically similar, if not identical, to the phosphodiester-linked di-, tri-, and tetrasaccharide repeats of *Leishmania* lipophosphoglycan, or LPG (Bates *et al.*, 1990; Turco, 1990).

2.4.3. Surface Metalloprotease

A membrane-bound metalloprotease is expressed at the promastigote surface, and has been called the promastigote surface protease, or PSP. Prior to its identification as an active proteolytic enzyme, the major surface antigen was called gp63. Although PSP has been shown to occur at the surface of all species of *Leishmania* that have been examined (Bouvier *et al.*, 1987; Etges, 1992), the results of efforts to demonstrate the active metalloprotease at the surface of amastigotes have been negative or inconclusive.

Surface labeling of promastigotes with [^{125}I]iodine revealed an abundant protein of about 63 kDa (Ramasamy *et al.*, 1983; Bouvier *et al.*, 1985; Colomer-Gould *et al.*, 1985; Etges *et al.*, 1985). Antibodies raised against promastigotes or against their membranes frequently recognize the protease (Fong and Chang, 1982; Soteriadou *et al.*, 1988). Peptide-mapping analyses showed that structurally related proteins are conserved among the recognized species of *Leishmania* (Colomer-Gould *et al.*, 1985; Etges *et al.*, 1985). Reactivity with polyclonal antibodies raised against chemically deglycosylated PSP, as well as surface location and surface-oriented proteolytic activity, was demonstrated for PSP in seven species of *Leishmania* (Bouvier *et al.*, 1987). Moreover, promastigotes of 15 species of *Leishmania* so far examined possess surface metalloprotease activity (Etges, 1992). The amount of PSP depends on the strain and species considered, and may not necessarily represent the major surface protein accessible to surface iodination. In *L. major*, PSP accounts for 70% of surface iodinatable proteins, 15% of [^{35}S]methionine-labeled membrane proteins, and up to 1% of the total cellular protein of the promastigote.

The metalloprotease is active at the surface of live and fixed promastigotes (Etges *et al.*, 1986b) and is able to hydrolyse a wide range of denatured polypeptides. The optimum pH for PSP activity was found to be neutral to basic for *L. major* PSP (Etges *et al.*, 1986b, 1987, 1989; Bouvier *et al.*, 1990) and *L. mexicana* PSP (Ip *et al.*, 1990), but was reported to be acidic for *L. amazonensis* PSP (Chaudhuri and Chang, 1988; Chaudhuri *et al.*, 1989). This striking difference may reflect the use of different assays that are not applicable over the extended pH range in which they were used. Specifically, the "acid" protease of *L. amazonensis,* which cleaves iodinated bovine serum albumin (BSA) at pH 4, does so with an insignificant specific activity. The same investigators showed that the enzyme was able to digest fibrinogen after sodium dodecyl sulfate–polyacrylamide gel electrophoresis (SDS-PAGE) separation only at neutral pH (Chaudhuri *et al.*, 1989), in accord with their earlier observation of a neutral-to-alkaline (pH 7.5–10), 43- to 68-kDa proteolytic activity detected by both fibrinogen and gelatin–SDS-PAGE analysis of *L. amazonensis* extracts (Fong and Chang, 1981). In an assay with defined synthetic peptide substrates, where conformational constraints of the substrates are unlikely to interfere with access to susceptible peptide bonds, both the *L. mexicana* and *L. major* enzymes show a neutral optimum pH (Ip *et al.*, 1990; Bouvier *et al.*, 1990). The reported acidic proteolytic activity of *L. amazonensis* PSP might be due to partial denaturation of otherwise resistant substrates that become sensitive to the residual proteolytic activity of PSP at acidic pH, which could also explain the low specific activity reported for the *L. amazonensis* enzyme. The alkaline elution of *L. amazonensis* PSP from the monoclonal antibody affinity column (20 mM lysine at pH 11; Chang and Chang, 1986) may remove the zinc from the enzyme, as was shown in the case of the membrane metalloprotease enkephalinase (Jongeneel *et al.*, 1989b). Recently, we have shown that the activity of PSP is not irreversibly affected by a 30-min preincubation between pH 4 and pH 11; at pH 11.5 and 12, enzymatic activity was not recovered upon neutralization (Bouvier *et al.*, 1990). Further, several unrelated proteins, including serum albumin, are cleaved inefficiently, if at all, at neutral pH in 1% SDS prior to SDS-PAGE analysis unless they are heat denatured or reduced and alkylated before addition of PSP (R. Etges, unpublished data).

Little is known about the relevant substrates of PSP *in vivo.* However, conservation of this proteolytic activity among *Leishmania* species, even in noninfective laboratory strains, suggests that PSP plays an essential role in the metabolism of the promastigote. Indeed, the LPG-deficient, noninfective strain of *L. major,* LRC-L119, produces significantly more PSP compared to several virulent stains of the same species (Murray *et al.*, 1989b; and R. Etges, unpublished data).

Promastigote surface protease is a zinc-containing metalloprotease that occurs as a dimer of 63-kDa monomers at the surface of the promastigote and in detergent solution. Each monomer of PSP contains one atom of zinc, as shown by atomic emission and atomic absorption spectroscopy, as well as by biosynthetic labeling with $[^{65}Zn]$-Cl_2 (Bouvier *et al.*, 1989). The proteolytic activity is abolished by dialysis against 1,10-phenanthroline, and can be restored upon addition of micromolar quantities of zinc (Chaudhuri *et al.*, 1989). In addition, a highly conserved amino acid sequence found in a variety of otherwise unrelated zinc metalloproteases was identified in PSP (Bouvier *et al.*, 1989; Chaudhuri *et al.*, 1989; Jongeneel *et al.*, 1989a). This "zinc signature" contains a glutamic acid and two histidine residues, which in the small, soluble bacterial metalloprotease thermolysin were shown to be involved in catalysis and zinc coordination, respectively (Monzingo and Matthews, 1984). However, the secondary structure of PSP contains approximately 50% antiparallel β-sheet, which is in striking contrast to the predominantly α-helical thermolysin (Jähnig and Etges, 1988; Bouvier *et al.*, 1989).

Like most metalloproteases, PSP is inhibited by metal chelating agents like 1,10-phenanthroline (but not its nonchelating analogue 1,7-phenanthroline) and peptide hydroxamate derivates, as well as divalent metal ions (Etges *et al.*, 1986b, 1989; Chaudhuri and Chang, 1988; Chaudhuri *et al.*, 1989; Bouvier *et al.*, 1990). Purified PSP was also shown to be inhibited by human α_2-macroglobulin. This inhibition does not occur at the surface of promastigotes, however, indicating that the active site of the membrane-bound PSP is not accessible to the relatively large α_2-macroglobulin *in vitro* (Heumann *et al.*, 1989). Thus far, the inhibitors of PSP that have been identified are too toxic for use with living cells (Etges *et al.*, 1989).

The glycosylation of PSP has been studied in some detail. Chemical deglycosylation with trifluoromethanesulfonic acid results in a molecular weight decrease from 63 kDa to 57 kDa with a concomitant loss of concanavalin A-binding capacity (Bouvier *et al.*, 1985). Immunoaffinity-purified PSP from *L. amazonensis* treated with endoglycosidase H or synthesized by promastigotes treated with the glycosylation inhibitor tunicamycin shows a decreased molecular weight of 54 kDa (Chang *et al.*, 1986). These findings are in agreement with the molecular weight of 53 kDa predicted from the genes for the mature polypeptides of *L. major* and *L. chagasi* PSP (Button and McMaster, 1988, 1990; Miller *et al.*, 1990). The predicted amino acid sequence contains two or three potential *N*-glycosylation sites. Structural analysis of the *N*-linked oligosaccharides of *L. mexicana* PSP revealed glycans consisting of four related biantennary oligomannoses with a unique terminal glucopyranosyl residue on the α_{1-3} arm in one case (Olafson *et al.*, 1990), representing the first instance that glucose was identified in the *N*-linked glycans of a mature surface glycoprotein. In contrast, *N*-linked glycans of *Trypanosoma brucei* variant surface glycoprotein (VSG) are very similar to mammalian high-mannose structures (Holder, 1985).

Promastigote surface protease is anchored to the promastigote membrane by a glycosyl phosphatidylinositol (GPI) anchor, and can be solubilized in detergent solution by the action of phosphatidylinositol-specific phospholipase C (PI-PLC) from bacterial or trypanosomal origins (Bordier *et al.*, 1986; Etges *et al.*, 1986a). *Leishmania* promastigotes themselves are devoid of PI-PLC activity able to remove GPI-anchored membrane component. After lipid removal, PSP exposes a new epitope, the cross-reacting determinant (CRD), which is common to many PI-PLC-solubilized GPI-anchored proteins (Bordier *et al.*, 1986; for a review, see Ferguson and Williams, 1988). An important feature recognized by anti-CRD antibodies consists of the inositol 1,2-cyclic phosphate that remains after PI-PLC action (Ferguson *et al.*, 1988; Zamze *et al.*, 1988).

All GPI anchors so far characterized have an identical carbohydrate core structure that is completely conserved in *Leishmania* PSP. In *Leishmania,* the anchoring lipid consist of a 1-0-alkyl-2-0-acyl glycerol with a marked preference for a fully saturated 24-carbon alkyl chain. The mature COOH-terminus of PSP to which the GPI anchor is attached is located 25 residues before the predicted COOH-terminus of the protein (Schneider *et al.*, 1990). Interestingly, the removal of the GPI anchor of PSP by treatment with phosphatidylinositol-specific phospholipase C reduces the antigenicity of the protein by nearly 100-fold (Rivier *et al.*, 1989).

Genes encoding PSP are linked at a single chromosome locus in *L. major* and consist in five repeats of 3.1 kb (with 1.8 kb of open reading frame) plus an additional gene 8 kb away, which all display conserved restriction maps (Button *et al.*, 1989). Similar tandemly linked genes occur in *L. chagasi* (Miller *et al.*, 1990). One of the PSP genes was sequenced in both organisms. The predicted amino acid sequences of both *L. major* and *L. chagasi* PSP are those of preproproteins with conventional signal sequences followed by prosequences of 100 amino acids. Messenger RNA for PSP is present at comparable levels in both promastigotes and amastigotes (Button *et al.*, 1989). The fact that some monoclonal antibodies fail to recognize amastigote PSP (Fong and Chang, 1982) suggests that PSP is either differentially processed or not expressed in amastigotes. Polyclonal antibodies against recombinant (thus not glycosylated) PSP recognize two polypeptides of slightly higher molecular weight in *L. major* amastigote extracts (Frommel *et al.*, 1990). The amastigote protein was shown to be structurally related to PSP by peptide map analysis, and the increase in molecular weight was suggested to be the result of quantitative and qualitative differences in glycosylation and perhaps by the mode of anchoring (Medina-Acosta *et al.*, 1989). Nothing is known about the metalloprotease activity in amastigotes, but its role is likely to be limited by the fact that purified *L. major* PSP exhibits less than 10% of its maximal activity at pH 5 as compared to pH 8.5 (Etges *et al.*, 1986b). Although PSP has been reported to be expressed (albeit in reduced quantities) by the amastigotes of *L. amazonensis* (Chaudhuri *et al.*, 1989), *L. mexicana* (Medina-Acosta *et al.*, 1989), and *L. major* (Frommel *et al.*, 1990), its presence at the surface of the cell remains uncertain. These findings are in striking contrast to those of Chang and Fong (1982), who clearly showed the abrupt cessation of PSP biosynthesis as promastigotes of *L. amazonensis* differentiated to amastigotes in macrophages *in vitro*. In any case, the role of a surface metalloprotease with a demonstrable neutral-to-alkaline pH optimum in the acidified phagolysosome is difficult to imagine.

2.4.4. Other Surface Antigens and Membrane Enzyme Activities

Several membrane proteins having no known enzymatic activities were characterized by use of monoclonal antibodies against *Leishmania* (Jaffe and Zalis, 1988; Jaffe *et al.*, 1990; White and McMahon-Pratt, 1988). In *L. major*, a family of proteins named the promastigote surface antigen-2 complex, or PSA-2, was shown to be anchored in the membrane by GPI anchors similar to that of PSP. In addition, a COOH-terminal GPI addition signal, structurally similar to that found in PSP, was predicted from the partially sequenced gene of the PSA-2 proteins (Murray *et al.*, 1989a). Another surface membrane antigen of 46 kDa is encoded by a gene family of about ten copies, and appears to be restricted to the *Leshmania amazonensis* group (i.e., *L. amazonensis, L. pifanoi,* and *L. garnhami,* but not *L. mexicana*) (Kahl and McMahon-Pratt, 1987; Champsi and McMahon-Pratt, 1988; Lohman *et al.*, 1990). The sequence deduced from the gene shows four repeats of a 24 amino acid segment, a COOH-terminal glycosylphosphatidylinositol anchor addition signal, and a cysteine-rich domain, reminiscent of either metal-chelating areas or disulfide-rich regions such as those found in snake venom toxins. The protein is *N*-glycosylated, as suggested by pulse-chase and endoglycosidase experiments. However, no consensus sequence for *N*-glycosylation is present in the gene, except an Asn-Ala-Cys sequence that was described once as an *N*-glycosylation site in a bovine protein (Lohman *et al.*, 1990). It is, however, possible that small differences exist between the cloned gene and the one that is really expressed.

A gene encoding a promastigote-specific protein was cloned and partially sequenced. It has a predicted myristylation signal at the *N*-terminus (Bard, 1989) very similar to that found in acylated *ras* proteins. This gene could correspond to the putative G-protein described in *L. donovani,* suggesting that a classical transmembrane signaling machinery exists in *Leishmania* (Cassel *et al.*, 1991).

Membrane-associated phospholipases A1, A2, and C were described in *L. donovani,* all of which show a preference for phosphatidylethanolamine. These enzymes are thought to be involved in membrane lipid restructuring and turnover (Dwyer, 1987). Again, no GPI-specific phospholipase activities have been identified in *Leishmania*.

3. MEMBRANE PROTEINS OF INTRACELLULAR ORGANELLES

The presence of a single large tubular mitochondrion containing the kinetoplast is a characteristic feature of the Kinetoplastidae. It contains two kinds of concatenated circular DNA, a large number of minicircles whose function is unknown and a few maxicircles that encode subunits of various mitochondrial enzymes.

3.1. Mitochondrial Proton Pumps

In the mitochondria of higher organisms, energy coupled to electron transfer from NADH to $FADH_2$ to dioxygen is used by three proton pumps (NADH Q-reductase, cytochrome C-reductase, and cytochrome C-oxidase) to extrude protons from the mitochondrial matrix to the cytosol, generating an electrochemical potential (Δ pH and $\Delta\Psi$) across the mitochondrial inner membrane. Several classical components of the respiratory chain were identified by spectrophotometric methods in *Leishmania,* namely cytochromes b, a^+, a_3, c_{555}, and flavoproteins (Martin and Mukkada, 1979). The mitochondrial genome of *L. tarentolae* encodes subunits 1, 4, and 5 of NADH Q-reductase, cytochrome *b* (which is part of the cytochrome reductase complex), subunits I, II, and III of cytochrome oxidase, and four open reading frames whose gene products are unknown. Some of the newly transcribed mitochondrial mRNAs undergo a peculiar form of co- or posttranscriptional maturation, called RNA editing, which consists of the addition or removal of several to many uridine residues within the mRNA, which are not coded in the mitochondrial genome. Cytochrome *b*, cytochrome C-oxidase subunits II and III, as well as two unidentified open reading frame mRNAs are edited in *L. tarentolae,* which is thus far the unique *Leishmania* species showing this phenomenon. Messenger editing is common, however, in *Trypanosoma* species (Simpson and Shaw, 1989). Terminal uridyl transferase and RNA ligase activities present in *L. tarentolae* mitochondria are likely to be involved in RNA editing (Bakalara *et al.*, 1989; Blum *et al.*, 1990). The fact that the phenomenon of RNA editing has not been described in other species of *Leishmania* may perhaps be explained by the recent reclassification of *L. tarentolae* as *Trypanosoma platydactyli* (Wallbanks *et al.*, 1985).

3.2. Mitochondrial F_0F_1 ATP Synthase

This enzyme takes advantage of the electrochemical potential created across the mitochondrial inner membrane by the proton pumps to couple ATP synthesis with proton influx. The highly conserved β-subunit of *L. donovani* F_1-ATPase cross-reacts immunologically with that of *Escherichia coli* (Zilberstein and Dwyer, 1988). The oligomycin-sensitive F-type enzyme has a pH optimum of 7.6, hydrolyses Mg^{2+}-ATP and Ca^{2+}-ATP, and displays kinetic characteristics similar to other mitochondrial ATPases (Rassam and Robert, 1988).

3.3. Mitochondrial Ca^{2+}-Transporter and ER Ca^{2+}-ATPase

The concentration of free Ca^{2+} is maintained at low levels in the cytoplasm of *L. donovani* (70–90 nM) in an energy-dependent manner. Two organellar transport systems were characterized, one of which is a mitochondrial low affinity, large capacity channel that takes advantage of the favorable electrochemical gradient generated by the respiratory chain for Ca^{2+} uptake (Benaim *et al.*, 1990). The other was shown to be a nonmitochondrial, ATP-dependent Ca^{2+}-ATPase that is likely to be located in the endoplasmic reticulum membrane, analogous to the mammalian Ca^{2+}-ATPase. The latter enzyme has a K_m of about 90 nM, a neutral pH optimum, and is sensitive to *ortho*-vanadate (Philosoph and Zilberstein, 1989). The properties and location of the different Ca^{2+} transporters described suggest that the internal Ca^{2+} regulation system of *Leishmania* is very similar to that of other eukaryotic cells (Carafoli *et al.*, 1990).

3.4. Glycosomal Integral Membrane Proteins

Two integral membrane proteins have been identified in the isolated glycosomes of *Trypanosoma brucei*. The 26 and 24 kDa amphiphilic proteins were identified by their partitioning behavior in Triton X-114 solution and by their sensitivity to the addition of proteases. The externally disposed, nonglycosylated proteins were suggested to be involved in the recognition and import of cytoplasmically synthesized protein precursors into the glycosome (Aman and Wang, 1987). Proteins with a similar function can be expected to be found in the glycosomes of *Leishmania*.

4. MEMBRANE PROTEIN ADAPTATIONS

Surface membrane molecules of *Leishmania* are under strong selective pressure in the diverse environments encountered by the parasite in its host and vector. In addition to nutrient uptake, parasite surface molecules are involved in the recognition of appropriate host tissue and cells, penetration into the macrophage (Chang and Dwyer, 1978), and for resisting the numerous microbicidal defenses of the host. Although membrane proteins play important roles in survival of the parasite (Chang, 1983; Alexander and Russell, 1985), they are not alone at the promastigote surface. Other, more abundant membrane molecules, such as the lipophosphoglycans (discussed by S. J. Turco in Chapter 3 of this volume), play crucial roles in the surface properties of *Leishmania*.

4.1. The Promastigote

The promastigote, or insect stage, is found in the sandfly midgut, which is characterized by a lower temperature compared to the mammalian host, neutral pH, and the presence of nutrients derived from the occasional blood meal or, more commonly, plant nectar taken by the vector. Promastigote enzymes and transporters on the parasite surface display high affinities for their substrates in order to enable the parasite to compete more efficiently with the host for essential nutrients. The proline transport system, however, has such a high K_m (645 μM) (D. Zilberstein, personal communication) that significant import would occur only in a proline-rich environment. If the proline transporter is expressed by both promastigote and amastigote, then entry in the insect gut could be accompanied by significant proline uptake. The rapid increase in the concentration of internal proline could act as a signal, together with temperature and pH, for differentiation from amastigote to promastigote.

The gene of a sugar transporter is transcribed at a higher level in the promastigote (Cairns *et al.*, 1989), suggesting that carbohydrate might be a more accessible substrate in the insect, which feeds on plant nectar between blood meals (Lewis and Domoney, 1966; Young *et al.*, 1980), than in the phagolysosome of the macrophage. This proposal is supported by the observation that promastigotes preferentially metabolize amino acids and carbohydrates, whereas amastigotes are more dependent on the β-oxidation of lipids for energy generation (Coombs *et al.*, 1982; Mukkada *et al.*, 1985).

The promastigote surface metalloprotease was shown by immunofluorescence to be present on *L. infantum* and *L. major* in the insect midgut (Grimm *et al.*, 1987; Davies *et al.*, 1990), and was suggested to be involved in protein degradation for nutritional purposes (Etges *et al.*, 1986b, 1987). This hypothesis is supported by the fact that the related trypanosomatids *Crithidia* and *Herpetomonas,* both of which have monogenetic life cycles in insect hosts only, also express a surface metalloprotease (Etges, 1992). The proteolytic activity could also be involved in the inactivation of insect microbicidal defenses, or be considered as a preadaptation to the parasitization of mammalian hosts.

In the sandfly midgut, promastigotes insert their flagella between microvilli of epithelial cells. A 13.2-kDa protein, localized uniquely on the flagellar membrane (Ismach *et al.*, 1989), might be of importance in this interaction, as monoclonal antibodies against the protein are able to inhibit the attachment of isolated flagella to insect gut epithelial cells *in vitro* (Warburg *et al.*, 1989). Promastigotes then colonize the insect foregut, which is covered by a chitin cuticle, and finally reach the anterior parts of the digestive tract, where infective, complement-resistant nondividing promastigotes are found free in the lumen of the pharynx and proboscis. The acquisition of infectivity, or metacyclogenesis (Sacks, 1989), represents a preadaptation to life in the mammalian host. Metacyclogenesis also occurs to a limited, strain-dependent degree *in vitro.* Although several species of *Leishmania* have been shown to develop resistance to serum complement-mediated lysis in stationary phase of culture *in vitro,* the pathway by which complement is activated, the *Leishmania* surface components to which activated C3 is bound, and the form of C3 bound to the surface (i.e., lytically active C3b or inactive iC3b) depend on the species of *Leishmania* under investigation (Fuhrman and Joiner, 1989). Only in the case of *L. major* has it been possible to separate metacyclic promastigotes from noninfective cells *in vitro* by agglutination of the noninfective cells with the galactose-specific lectin, peanut agglutinin (Sacks *et al.*, 1985). *Leishmania major* metacyclics are more resistant to complement-mediated lysis than log-phase cells. They are covered with a stage-specific lipophosphoglycan, or LPG, (Pimenta *et al.*, 1989) that activates the immunoglobulin-independent classical complement pathway (Sacks and Da Silva, 1987). This is in contrast to the alternative pathway activation of C3 of log-phase promastigotes of the same species. Perhaps the negative charge of the highly phosphorylated polysaccharide is involved in the deposition of hemolytically active, ester-linked C3b (Puentes *et al.*, 1988) far away from the membrane, not only preventing complement-mediated lysis, but also providing a suitable ligand for binding by the macrophage CR1 receptor for C3b. Nonimmune serum is able to induce promastigote lysis, but a short exposure to complement factors has been shown to be necessary for successful infection of macrophages *in vitro* (Mosser and Edelson, 1987).

With the exception of the yet-to-be elucidated, stage-specific modification of LPG of *Leishmania major* metacyclics, no surface markers have been identified for infective promastigotes. Although metacyclics exist in other species, they cannot be readily separated from noninfective cells. In experimental murine infections with unfractionated stationary-phase *L. major,* less than 1% of the injected promastigote inoculum survive (Müller *et al.*, 1989), suggesting that only a minor, preadapted subpopulation is able to survive the early stages of the infection. Stationary phase promastigotes of *L. donovani* and *L. mexicana* activate the alternative pathway of complement and fix iC3b and C3b, respectively (Russell, 1987; Puentes *et al.*, 1989). Promastigote surface protease, independent of its enzymatic activity, was shown to be the major acceptor of amide-linked C3b in log-phase *L. mexicana* promastigotes (Russell, 1987) and to be involved in the attachment of promastigotes to macrophages (Russell and Wilhelm, 1986; Wilson and Hardin, 1988). However, LPG is probably the physiologically significant complement acceptor in infective, stationary-phase cells, at least in the case of *L. major*. Opsonization and subsequent entry into the macrophage via complement receptors CR1 or CR3 would favor the survival of the parasite, as these receptors do not induce the microbicidal oxidative burst.

Loss of virulence following long-term culture *in vitro* has been shown to correlate with a decrease in the expression of the surface metalloprotease, which, in addition, is underglycosylated compared to PSP on infectious promastigotes (Kink and Chang, 1988; Wilson *et al.*, 1989). In contrast, tunicamycin-resistant parasites, which overexpress the first enzyme of the dolichol pathway of *N*-glycosylation, do not lose their virulence or their ability to glycosylate proteins in long-term culture, suggesting that glycoproteins like the surface protease somehow play a role in the establishment of the infection (Kink and Chang, 1988). In *L. braziliensis,* a greater average number of surface protease per cell was detected by immunofluorescence on stationary-phase promastigotes (Kweider *et al.*, 1987).

4.2. The Amastigote

The environment of the amastigote in the phagolysosome is characterized by an acidic pH of approximately 4.7–5.3 (Chang, 1980; Antoine *et al.*, 1990). The cytoplasm of the amastigote is nevertheless maintained at a nearly neutral pH by the active extrusion of protons (Glaser *et al.*, 1988). The plasma membrane proton ATPase, demonstrated in the promastigote by Zilberstein and Dwyer (1988), is likely to be present in the amastigote in order to maintain this large pH gradient. Its activity could be reinforced by the expression of an apparently amastigote-specific gene encoding a similar *ortho*-vanadate-sensitive ATPase 1b described by Meade *et al.* (1987), which could be a proton pump.

Amastigote nutrient uptake is enhanced at acidic pH (Mukkada *et al.*, 1985) and, although little is known about membrane transporters in the amastigote, it is likely that they will display an acidic optimum pH. Some surface enzymes such as 5′- and 3′-nucleotidase were detected on the amastigote (Dwyer and Gottlieb, 1983), and a modified form of the surface metalloprotease was reported to be expressed by the amastigote, although its quantity, surface location, mode of membrane attachment, and proteolytic activity have yet to be convincingly demonstrated (Chaudhuri *et al.*, 1989; Medina-Acosta *et al.*, 1989; Frommel *et al.*, 1990).

5. MEMBRANE PROTEINS AS THERAPEUTIC TARGETS IN LEISHMANIASIS

5.1. Chemotherapy

An effective leishmanicidal drug should minimally satisfy three criteria. It should have a deleterious effect on an essential function of the parasite; it should be as selective as possible in order to eliminate the parasite with minimal side effects on the host; and it must be able to reach the intracellular parasite, whose host macrophage may not be located in a well-vascularized tissue.

An ideal chemotherapeutic target would be a transport or enzymatic activity that is specific to the parasite. Concerning membrane proteins, the surface 3′-nucleotidase and the surface metalloprotease could represent interesting targets for drugs as they are specific to the parasite; thus far, however, it has not been demonstrated that these enzymes are essential for the parasite's survival in the mammal.

The nucleoside transporters are essential for the parasite as no *de novo* synthesis of purines occurs in trypanosomatids. The affinity of the *Leishmania* transporters for their ligand are two orders of magnitude higher than those of the host and display different substrate specificities for nucleoside analogues. Finally, the parasite transporters are not sensitive to inhibitors of mammalian nucleoside transport. In combination, toxic analogues of nucleosides would be preferentially taken up and metabolized by the parasite, while the simultaneous inhibition of the mammalian transporters would protect the host from the toxic analogue. Such an approach has been used successfully to cure mice infected with *Trypanosoma gambiense* (Ogbunude and Ikediobe, 1982).

Leishmania are also incapable of *de novo* synthesis of folate, and must therefore acquire exogenous folate for growth. Folate plays a major role in the synthesis of thymidine. Differences that exist between leishmanial and mammalian folate transport systems, such as distinct substrate specificity and differential sensitivity to inhibitors, suggest that development of specific inhibitors of the folate transport or of the bifunctional dihydrofolate reductase-thymidylate synthase may be possible (Ellenberger and Beverley, 1987).

Prolonged exposure of promastigotes to several leishmanicidal compounds results in the development of drug-resistant phenotypes *in vitro*. In several cases, membrane transporters were found to be altered in their affinities for their ligands (Iovannesci *et al.*, 1984; Kaur *et al.*, 1988). The amplified H-region-associated P-glycoprotein analogue may also play a role in the development of drug resistance in *Leishmania* (Ouellette *et al.*, 1990). Whatever the underlying mechanisms of resistance, the relative ease with which drug resistance is induced in these organisms places a serious limit on the future of chemotherapy in the treatment of leishmaniasis.

Membrane potential, essential for the proton motive force-driven import of metabolites, is the target of antidepressant drugs such as clomipramine and nitroimipramine (Zilberstein and Dwyer, 1984a). These compounds affect both proton gradient and membrane potential of promastigotes and have been shown to inhibit proline uptake at a drug concentration of only 100 μM. Intracellular amastigotes are even more sensitive than promastigotes to these compounds, while the macrophage remains unaffected. An improved toxicity was obtained with derivatives of imipramine that inhibits proline transport and pH gradient maintenance with a median inhibitory concentration (IC_{50}) of 5 μM (Zilberstein *et al.*, 1990).

5.2. Vaccine Development

The development of a vaccine would be invaluable to protect humans and domestic animals from leishmaniasis. Long-lasting protection of mice to challenge infection has been obtained using liposome-reconstituted *L. mexicana* surface metalloprotease plus LPG administered by intraperitoneal or subcutaneous injection (the protease alone was less efficient). CBA mice were shown to be immune to reinfection for at least seven months (Russell and Alexander, 1988). Partial protection of susceptible BALB/c mice was also achieved, but in this instance, the route of immunization was critical. Following this result, other investigations were conducted using purified PSP or recombinant PSP (Button *et al.*, 1991) for immunoprophylaxis against *L. major* infections. The fact that susceptible BALB/c mice immunized with recombinant PSP (Handman *et al.*, 1990) are not protected from challenge infection with virulent *L. major* promastigotes suggests that PSP is a poor vaccine candidate, although partial protection was achieved in CBA mice by subcutaneous inoculation of the amphiphilic form of PSP alone or with *C. parvum* or BCG adjuvants (D. Rivier, personal communication). Mice immunized orally with a vaccine strain of *Salmonella typhimurium* in which the gene of PSP was introduced developed smaller lesions than controls immunized with only the vector (Yang *et al.*, 1990). In the perspective of human vaccination, it is important to note that T lymphocytes from individuals infected with *L. major* fail to respond to the hydrophilic form of PSP *in vitro* (Jaffe *et al.*, 1990).

Some encouraging results were obtained with specific peptides of PSP believed to represent T-cell epitopes. Two peptides of 25 amino acids were able to activate T cells that, on the other hand, recognized epitopes expressed by antigen-presenting cells infected with promastigotes. CBA mice immunized intravenously with a mixture of these peptides with *Corynebacterium parvum* or liposomes developed smaller lesions (Yang *et al.*, 1991). A smaller peptide of 15 amino acids (whose sequence is included in one of the peptides described above) appears to stimulate the particular subset of T cells (Th1) that is believed to be important in cellular immunity against leishmaniasis in mice. Susceptible BALB/c mice immunized subcutaneously with this peptide in Poloxamer 407 developed lesions attenuated by a factor of four compared to controls after challenge with *L. major*. Half of the animals were subsequently cured after 12–16 weeks. The same immunization protocol using the *L. mexicana*–CBA mouse model seems to prevent disease for as long as three months after challenge (Jardim *et al.*, 1990). These results show that a molecularly defined vaccine for leishmaniasis is a likely possibility, even if no satisfactory prolonged protection has been reproducibly achieved yet. Other candidate surface antigens were investigated for their ability to vaccinate experimental animals. A mixture of integral membrane proteins and glycolipids of a strain of *L. major* deficient in LPG was able to protect mice, whereas the same extract treated with pronase (which does not affect glycolipids) could not. At least six glycoproteins were present in this mixture, including the surface metalloprotease (Murray *et al.*, 1989b). The purified 46-kDa membrane glycoprotein of *L. amazonensis*, when administered with the adjuvant *C. parvum*, was shown to protect susceptible strains of mice from subsequent infection, but only when low numbers of log-phase promastigotes were used in the challenge inoculum (Champsi and McMahon-Pratt, 1988).

5.3. Diagnostic Tools

The promastigote surface protease is recognized by sera from infected humans from various geographical areas (Colomer-Gould *et al.*, 1985) and by sera from naturally infected dogs in France (Etges *et al.*, 1985). However, attempts to develop an enzyme-linked immunosorbent assay (ELISA) based on the recombinant protease (synthesized as a denatured polypeptide in bacteria) were met with failure. The sensitivity of the test was low and was complicated by the fact that an important fraction of kala-azar patients produce little, if any, antibody against the protease (C. Bordier and L. L. Button, unpublished data). The choice of a more suitable antigen may eventually permit development of diagnostic assays that distinguish between uncomplicated cutaneous leishmaniasis and the more serious mucocutaneous or visceralizing forms. A candidate could be the 70- to 72-kDa proteins identified by Jaffee and Zalis (1988) that are specifically recognized by the sera of visceral leishmaniasis patients.

6. CONCLUSIONS

The membrane of *Leishmania* is of critical importance in the interaction of the parasites with their hosts. The study of membrane proteins leads to a better knowledge of the membrane functions and should allow the characterization of specific properties of the parasite that can be exploited for the development or improvement of vaccines, diagnostic tools, or effective, nontoxic leishmanicidal drugs. Despite considerable efforts, the molecular basis for the complex events leading to the establishment of infection is not completely understood. Apparently contradictory results have been obtained that can be explained by the multiplicity of uncontrolled variables that are encountered, including genetically defined species differences of both host and parasite *in vivo* and differences in differentiation state due to culture conditions of both types of cells *in vitro*.

A better understanding of host–parasite interactions that permit the establishment and propagation of infection within the mammal requires the study of the membrane proteins of the infective metacyclic promastigote and the amastigote. Current techniques used to isolate these stages of the parasite are inefficient and time consuming. Although the morphology of the amastigote remains unchanged for approximately 24 hr, the pattern of protein synthesis changes radically in less than 4 hr. Clearly, a more profound understanding of the stages of the parasite that are actually responsible for causing leishmaniasis is critical. It is hoped that this understanding will be attained when rapid, efficient methods for the isolation of these stages are developed.

ACKNOWLEDGMENTS. We thank Prof. Jacques Mauël and Dr. Ingrid Müller for their careful reading of this review and their many helpful suggestions, and Dr. Dan Zilberstein for sharing both his experience and unpublished results with us.

7. REFERENCES

Alexander, J., and Russell, D. G., 1985, Parasite antigens, their role in protection, diagnosis, and escape: The leishmaniases, *Curr. Top. Microbiol. Immunol.* **120**:43–67.

Aman, R. A., and Wang, C. C., 1987, Identification of two integral glycosomal membrane proteins in *Trypanosoma brucei, Mol. Biochem. Parasitol.* **25**:83–92.

Antoine, J. C., Prina, E., Jouanne, C., and Bongrand, P., 1990, Parasitophorous vacuoles of *Leishmania amazonensis*-infected macrophages maintain an acidic pH, *Infect. Immun.* **58**:779–787.

Aronow, B., Kaur, K., McCartan, K., and Ullman, B., 1987, Two high affinity nucleoside transporters in *Leishmania donovani, Mol. Biochem. Parasitol.* **22**:29–37.

Bakalara, N., Simpson, A. M., and Simpson, K., 1989, The *Leishmania* kinetoplast-mitochondrion contains terminal uridyltransferase and RNA ligase activities, *J. Biol. Chem.* **264**:18679–18686.

Bard, E., 1989, Molecular biology of *Leishmania, Biochem. Cell Biol.* **67**:516–524.

Bates, P. A., Hermes, I., and Dwyer, D. M., 1990, Golgi-mediated post-translational processing of secretory acid phosphatase by *Leishmania donovani* promastigotes, *Mol. Biochem. Parasitol.* **39**:247–256.

Beck, J. T., and Ullman, B., 1989, Affinity labeling of the folate–methotrexate transporter from *Leishmania donovani, Biochemistry* **28**:6931–6937.

Benaim, G., and Romero, P. J., 1990, A calcium pump in plasma membrane vesicles from *Leishmania braziliensis, Biochim. Biophys. Acta* **1027**:79–84.

Benaim, G., Bermudez, R., and Urbina, J. A., 1990, Ca^{2+} transport in isolated mitochondrial vesicles from *Leishmania braziliensis* promastigotes, *Mol. Biochem. Parasitol.* **39**:61–68.

Blum, B., Bakalara, N., and Simpson, L., 1990, A model for RNA editing in kinetoplastid mitochondria: "guide" RNA molecules transcribed from maxicircle DNA provide the edited information, *Cell* **60**:189–198.

Bonay, P., and Cohén, B. E., 1983, Neutral amino acid transport in *Leishmania* promastigotes, *Biochem. Biophys. Acta* **731**:222–228.

Bordier, C., 1981, Phase separation of integral membrane proteins in Triton X-114 solution, *J. Biol. Chem.* **256**:1604–1607.

Bordier, C., 1988, Analytical and preparative phase separation of glycolipid-anchored membrane proteins in Triton X-114 solution, in *Post-Translational Modification of Proteins by Lipids: A Laboratory Manual* (U. Brodbeck and C. Bordier, eds.), pp. 29–33, Springer Verlag, Berlin.

Bordier, C., Garavito, R. M., and Armbruster, B., 1982, Biochemical and structural analyses of microtubules in the pellicular membrane of *Leishmania tropica, J. Protozool.* **29**:560–565.

Bordier, C., Etges, R. J., Ward, J., Turner, M. J., and Cardoso de Almeida, M. L., 1986, *Leishmania* and *Trypanosoma* surface glycoproteins have a common glycophospholipid membrane anchor, *Proc. Natl. Acad. Sci. USA* **83**:5988–5991.

Bouvier, J., Etges, R. J., and Bordier, C., 1985, Identification and purification of membrane and soluble forms of the major surface protein of *Leishmania* promastigotes, *J. Biol. Chem.* **260**:15504–15509.

Bouvier, J., Etges, R., and Bordier, C., 1987, Identification of the promastigote surface protease in seven species of *Leishmania, Mol. Biochem. Parasitol.* **24**:73–79.

Bouvier, J., Bordier, C., Vogel, H., Reichelt, R., and Etges, R., 1989, Characterization of the promastigote surface protease of *Leishmania* as a membrane-bound zinc endopeptidase, *Mol. Biochem. Parasitol.* **37**:235–246.

Bouvier, J., Schneider, P., Etges, R., and Bordier, C., 1990, Peptide bond specificity of the membrane-bound metalloprotease of *Leishmania, Biochemistry* **29**:10113–10119.

Button, L. L., and McMaster, W. R., 1988, Molecular cloning of the major surface antigen of *Leishmania, J. Exp. Med.* **167**:724–729.

Button, L. L., and McMaster, W. R., 1990, Correction, *J. Exp. Med.* **171**:589.

Button, L. L., Russell, D. G., Klein, H. L., Medina-Acosta, E., Karess, R. E., and McMaster, W. R., 1989, Genes encoding the major surface glycoprotein in *Leishmania* are tandemly linked at a single chromosomal locus and are constitutively transcribed, *Mol. Biochem. Parasitol.* **32**:271–284.

Button, L. L., Reiner, N. E., and McMaster, W. R., 1991, Modification of GP63 genes from diverse species of Leishmania for expression of recombinant protein at high levels in *Escherichia coli, Mol. Biochem. Parasitol.* **44**:213–224.

Cairns, B. R., Collard, M. W., and Landfear, S. M., 1989, Developmentally regulated gene from *Leishmania* encodes a putative membrane transport protein, *Proc. Natl. Acad. Sci. USA* **86**:7682–7686.

Carafoli, E., Chiesi, M., and Gazzoti, P., 1990, The membrane carriers related to intracellular calcium regulation, in *Hypertension: pathophysiology, diagnosis, and management* (J. H. Laragh and B. M. Brenner, eds.), pp. 965–976, Raven Press, New York.

Cassel, D., Shoubi, S., Glusman, G., Cukierman, E., Rotman, M., and Zilberstein, D., 1991, *Leishmania donovani:* characterization of a 38-kDa membrane protein that cross-reacts with the mammalian G-protein transducin, *Exp. Parasitol.* **72**:411–417.

Champsi, J., and McMahon-Pratt, D., 1988, Membrane glycoprotein M-2 protects against *Leishmania amazonenesis* infection, *Infect. Immun.* **52**:3272–3279.

Chang, C. S., and Chang, K.-P., 1986, Monoclonal antibody affinity purification of a *Leishmania* membrane glycoprotein and its inhibition of *Leishmania*–macrophage binding, *Proc. Nat. Acad. Sci. USA* **83**:100–104.

Chang, C. S., Inserra, J. T., Kink, J. A., Fong, D., and Chang, K.-P., 1986, Expression and size heterogeneity of a 63 kDa membrane glycoprotein during growth and transformation of *Leishmania mexicana amazonensis, Mol. Biochem. Parasitol.* **18**:197–210.

Chang, K.-P., 1980, Endocytosis of *Leishmania*-infected macrophages. Fluorometry of pinocytic rate, lysosome–phagosome fusion, and intralysosomal pH, in *The Host–Invader Interplay* (H. Van den Bossche, ed.) pp. 231–234, Elsevier-North Holland, Amsterdam.

Chang, K.-P., 1983, Cellular and molecular mechanisms of intracellular symbiosis in leishmaniasis, *Int. Rev. Cytol.* **14**:267–305.

Chang, K.-P., and Dwyer, D. M., 1978, *Leishmania donovani*–hamster macrophage interaction *in vitro:* Cell entry, intracellular survival, and multiplication of amastigotes, *J. Exp. Med.* **147**:515–530.

Chang, K.-P., and Fong, D., 1982, Antigenic changes during intracellular differentiation of *Leishmania mexicana* in cultured macrophages, *Infect. Immun.* **36**:430–431.

Chang, K.-P., and Fong, D., 1983, Cell biology of host–parasite membrane interactions in leishmaniasis, in *Cytopathology of Parasitic Disease* CIBA Foundation Symposium 99, pp. 113–137, Pitman, London.

Chaudhuri, G., and Chang, K.-P., 1988, Acid protease activity of a major surface membrane glycoprotein (gp63) from *Leishmania mexicana* promastigotes, *Mol. Biochem. Parasitol.* **27**:43–52.

Chaudhuri, G., Chaudhuri, M., Pan, A., and Chang, K.-P., 1989, Surface acid proteinase (gp63) of *Leishmania mexicana:* A metallo-enzyme capable of protecting liposome-encapsulated proteins from phagolysosomal degradation by macrophages, *J. Biol. Chem.* **264**:7483–7489.

Colomer-Gould, V., Quintao, L. G., Keithly, J., and Nogueira, N., 1985, A common major surface antigen on amastigotes and promastigotes of *Leishmania* species, *J. Exp. Med.* **162**:902–916.

Coombs, G. H., Craft, J. A., and Hart, D. T., 1982, A comparative study of *Leishmania mexicana* amastigotes and promastigotes. Enzyme activities and subcellular locations, *Mol. Biochem. Parasitol.* **5**:199–211.

Davies, C. R., Cooper, A. M., Peacock, C., Lane, R. P., and Blackwell, J. M., 1990, Expression of LPG and GP63 by different developmental stages of *Leishmania major* in the sandfly Phlebotomus papatasi, *Parasitol.* **101**:337–343.

Dwyer, D. M., 1980, Isolation and partial characterization of surface membranes from *Leishmania donovani* promastigotes, *J. Protozool.* **27**:176–182.

Dwyer, D. M., 1987, The role of surface membrane enzymes and transporters in the survival of *Leishmania,* in *Host–Parasite Cellular and Molecular Interactions* (K.-P. Chang and D. Snary, eds.) NATO ASI Series, Vol. H11, pp. 175–182, Springer Verlag, Berlin, Heidelberg.

Dwyer, D. M., and Gottlieb, M., 1983, The surface membrane chemistry of *Leishmania:* Its possible role in parasite sequestration and survival, *J. Cell. Biochem.* **23**:35–45.

Dwyer, D. M., and Gottlieb, M., 1984, Surface membrane localization of 3'- and 5'-nucleotidase activities in *Leishmania donovani* promastigotes, *Mol. Biochem. Parasitol.* **10**:139–150.

Ellenberger, T. E., and Beverley, S. M., 1987, Biochemistry and regulation of folate and methotrexate transport in *Leishmania major, J. Biol. Chem.* **262**:10053–10058.

Ellenberger, T. E., and Beverley, S. M., 1989, Multiple drug resistance and conservative amplification of the H region in *Leishmania major, J. Biol. Chem.* **264**:15094–15103.

Eperon, S., and McMahon-Pratt, D., 1989, I. Extracellular cultivation and morphological characterization of amastigote-like forms of *Leishmania panamensis* and *L. braziliensis, J. Protozool.* **36**:502–510.

Etges, R., 1992, Identification of a surface metalloproteinase on 13 species of *Leishmania* isolated from humans, *Crithidia fasciculata,* and *Herpetomonas samuelpessoai, Acta Tropica* **50:** 205–217.

Etges, R. J., Bouvier, J., Hoffman, R., and Bordier, C., 1985, Evidence that the major surface proteins of three *Leishmania* species are structurally related, *Mol. Biochem. Parasitol.* **14**:141–149.

Etges, R., Bouvier, J., and Bordier, C., 1986a, The major surface protein of *Leishmania* promastigotes is anchored in the membrane by a myristic acid-labeled phospholipid, *EMBO J.* **3**:597–601.

Etges, R., Bouvier, J., and Bordier, C., 1986b, The major surface protein of *Leishmania* promastigotes is a protease, *J. Biol. Chem.* **261**:9098–9101.

Etges, R., Bouvier, J., and Bordier, C., 1987, The promastigote surface protease of *Leishmania,* in *Host–Parasite Cellular and Molecular Interactions* (K.-P. Chang and D. Snary, eds.), NATO ASI Series, Vol. H11, pp. 165–168, Springer Verlag, Berlin, Heidelberg.

Etges, R., Bouvier, J., and Bordier, C., 1989, The promastigote surface protease of *Leishmania:* pH optimum and effects of protease inhibitors, in *Leishmaniasis: The Current Status and New Strategies for Control* (D. T. Hart, ed.), NATO ASI Series A, vol. 163, pp. 627–633, Plenum Press, New York.

Ferguson, M. A. J., and Williams, A. J., 1988, Cell-surface anchoring of proteins via glycosylphosphatidylinositol structures, *Annu. Rev. Biochem.* **57**:285–320.

Ferguson, M. A. J., Homans, S. W., Dwek, R. A., and Rademacher, T. W., 1988, Glycosyl-phosphatidylinositol moiety that anchors *Trypanosoma brucei* variant surface glycoprotein to the membrane, *Science* **239**:753–759.

Fong, D., and Chang, K.-P., 1981, Protease activity of a parasitic protozoan, *Leishmania mexicana, J. Cell Biol.* **91**:43A.

Fong, D., and Chang, K.-P., 1982, Surface antigenic change during differentiation of a parasitic protozoan, *Leishmania mexicana:* Identification by monoclonal antibodies, *Proc. Natl. Acad. Sci. USA* **79**:7366–7370.

Frommel, T. O., Button, L. L., Fujikura, Y., and McMaster, W. R., 1990, The major surface glycoprotein (gp63) is present in both life stages of *Leishmania, Mol. Biochem. Parasitol.* **38**:25–32.

Fuhrman, S. A., and Joiner, K. A., 1989, Complement evasion by protozoa, *Exp. Parasitol.* **68**:474–481.

Ghosh, J., Ray, M., Sarkar, S., and Bhaduri, A., 1990, A high affinity Ca^{2+}-ATPase on the surface membrane of *Leishmania donovani* promastigote, *J. Biol. Chem.* **265**:11345–11351.

Glaser, T. A., Baatz, J. E., Kreishman, G. P., and Mukkada, A. J., 1988, pH homeostasis in *Leishmania donovani* amastigotes and promastigotes, *Proc. Natl. Acad. Sci. USA* **85**:7602–7606.

Gottlieb, M., 1989, The surface membrane 3′-nucleotidase/nuclease of trypanosomatid protozoan, *Parasitol. Today* **5**:257–260.

Gottlieb, M., and Dwyer, D. M., 1981a, Protozoan parasite of humans: Surface membrane with externally disposed acid phosphatase, *Science* **212**:939–941.

Gottlieb, M., and Dwyer, D. M., 1981b, *Leishmania donovani:* Surface membrane acid phosphatase activity of promastigotes, *Exp. Parasitol.* **52**:117–128.

Gottlieb, M., and Dwyer, D. M., 1982, Identification and partial characterization of an extracellular acid phosphatase activity of *Leishmania donovani* promastigotes, *Mol. Cell. Biol.* **2**:76–81.

Gottlieb, M., and Dwyer, D. M., 1983, Evidence for distinct 5′- and 3′-nucleotidase activities in the surface membrane fraction of *Leishmania donovani* promastigotes, *Mol. Biochem. Parasitol.* **7**:303–317.

Gottlieb, M., and Dwyer, D. M., 1988, Plasma membrane functions: A biochemical approach to understanding the biology of *Leishmania,* in *The Biology of Parasitism* (P. T. Englund and A. Sher, eds.), pp. 449–465, Alan R. Liss, New York.

Gottlieb, M., Mackow, M. C., and Neubert, T. A., 1988, *Crithidia luciliae:* Factors affecting the expression of 3′-nucleotidase/nuclease activity, *Exp. Parasitol.* **66**:108–117.

Green, S. J., Meltzer, M. S., Hibbs, J. B., and Nacy, C. A., 1990, Activated macrophages destroy intracellular *Leishmania major* amastigotes by an L-arginine-dependent killing mechanism, *J. Immunol.* **144**:278–283.

Grimm, F., Jenni, L., Bouvier, J., Etges, R. J., and Bordier, C., 1987, The promastigote surface protease of *Leishmania donovani infantum* in the midgut of *Phlebotomus perniciosus, Acta Trop.* **44**:375–377.

Grimm, F., Brun, R., and Jenni, L., 1991, Promastigote infectivity in *Leishmania infantum, Parasitol. Res.* **77**:185–191.

Handman, E., Button, L. L., and McMaster, R. W., 1990, *Leishmania major:* Production of recombinant gp63, its antigenicity and immunogenicity in mice, *Exp. Parasitol.* **70**:427–435.

Hassan, H. F., and Coombs, G. H., 1987, Phosphomonoesterases of *Leishmania mexicana* and other flagellates, *Mol. Biochem. Parasitol.* **23**:285–296.

Heumann, D., Burger, D., Vischer, T., de Colmenares, M., Bouvier, J., and Bordier, C., 1989, Molecular interactions of *Leishmania* promastigote surface protease with human α2-macroglobulin, *Mol. Biochem. Parasitol.* **33**:67–72.

Holder, A. A., 1985, Glycosylation of the variant surface antigens of *Trypanosoma brucei, Curr. Top. Microbiol. Immunol.* **117**:57–74.

Iovannisci, D. M., Kaur, K., Young, L., and Ullman, B., 1984, Genetic analysis of a nucleoside transport in *Leishmania donovani, Mol. Cell Biol.* **4:**1013–1019.

Ip, H. S., Orn, A., Russell, D. G., and Cross, G. A. M., 1990, *Leishmania mexicana mexicana* gp63 is a site-specific neutral endopeptidase, *Mol. Biochem. Parasitol.* **40:**163–172.

Ismach, R., Cianci, C. M. L., Caulfield, J. P., Langer, P. J., Hein, A., and McMahon-Pratt, D., 1989, Flagellar membrane and paraxial rod proteins of *Leishmania:* Characterization employing monoclonal antibodies, *J. Protozool.* **36:**617–624.

Jaffe, C. L., and Zalis, M., 1988, Purification of two *Leishmania donovani* membrane proteins recognized by sera from patients with visceral leishmaniasis, *Mol. Biochem. Parasitol.* **27:**53–62.

Jaffe, C. L., Shor, R., Trau, H., and Passwell, J. H., 1990, Parasite antigens recognized by patients with cutaneous leishmaniasis, *Clin. Exp. Immunol.* **80:**77–82.

Jähnig, F., and Etges, R., 1988, Secondary structure of the promastigote surface protease of *Leishmania, FEBS Lett.* **241:**79–82.

Jardim, A., Alexander, J., Teh, H. S., Ou, D., and Olafson, R. W., 1990, Immunoprotective *Leishmania major* synthetic T cell epitopes, *J. Exp. Med.* **172:**645–648.

Jongeneel, C. V., Bouvier, J., and Bairoch, A., 1989a, A unique signature identifies a family of zinc-dependent metallopeptidases, *FEBS Lett.* **242:**211–214.

Jongeneel, C. V., Quackenbush, E. J., Ronco, P., Verroust, P., Carrel, S., and Letarte, M., 1989b, Common acute lymphoblastic leukemia antigen expressed on leukemia and melanoma cells has neutral endopeptidase activity, *J. Clin. Invest.* **83:**713–717.

Kahl, L. P., and McMahon-Pratt, D., 1987, Structural and antigenic characterization of a species- and promastigote-specific *Leishmania mexicana amazonenesis* membrane protein, *J. Immunol.* **138:**1587–1595.

Kaur, K., Coons, T., Emmett, K., and Ullman, B., 1988, Methotrexate-resistant *Leishmania donovani* genetically deficient in the folate–methotrexate transporter, *J. Biol. Chem.* **263:**7020–7028.

Killick-Kendrick, R., 1986, The transmission of leishmaniasis by the bite of the sandfly, *J. R. Army Med. Corps* **132:**134–140.

Killick-Kendrick, R., Molyneux, D. H., and Ashford, R. W., 1974, *Leishmania* in phlebotomid sandflies. I. Modifications of the flagellum in association with attachment to the midgut and oesophageal valve of the sandfly, *Proc. R. Soc. Lond.* **187:**409–419.

Kink, J. A., and Chang, K.-P., 1988, *N*-Glycosylation as a biochemical basis for virulence in *Leishmania mexicana amazonensis, Mol. Biochem. Parasitol.* **27:**181–190.

Kweider, M., Lemesre, J.-L., Darcy, F., Kusnierz, J. P., Capron, A., and Santoro, F., 1987, Infectivity of *Leishmania braziliensis* promastigotes is dependent on the increasing expression of a 65000-dalton surface antigen, *J. Immunol.* **138:**299–305.

Lepley, P. R., and Mukkada, A. J., 1983, Characteristics of an uptake system for α-aminoisobutyric acid in *Leishmania tropica* promastigotes, *J. Protozool.* **30:**41–46.

Lewis, D. J., and Domoney, C. R., 1966, Sugar meals in Phlebotominae and Simulidae, *Proc. R. Entomol. Soc. Lond. (A)* **41:**175–179.

Lohman, K. L., Langer, P. J., and McMahon-Pratt, D., 1990, Molecular cloning and characterization of the immunologically protective surface glycoprotein GP46/M-2 of *Leishmania amazonensis, Proc. Natl. Acad. Sci. USA* **87:**8393–8397.

Martin, E., and Mukkada, A. J., 1979, Respiratory chain components of *Leishmania tropica* promastigotes, *J. Protozool.* **26:**138–142.

Mauël J., 1984a, Mechanisms of survival or protozoan parasites in mononuclear phagocytes, *Parasitology* **88:**579–592.

Mauël, J., 1984b, Intracellular parasite killing induced by electron carriers. I. Effects of electron carriers on intracellular *Leishmania* ssp. in macrophages from different genetic backgrounds, *Mol. Biochem. Parasitol.* **13:**83–96.

Mauël, J., Schnyder, J., and Baggiolini, M., 1984, Intracellular parasite killing induced by electron carriers. II. Correlation between parasite killing and the induction of the oxidative events in macrophages, *Mol. Biochem. Parasitol.* **13:**97–110.

Mauël, J., Ransijn, A., and Buchmüller-Rouiller, Y., 1990, Killing of *Leishmania* parasites in activated murine macrophages is based on an L-arginine-dependent process that produces nitrogen derivatives, *J. Leukocyte Biol.* **49:**73–82.

Meade, J. C., Shaw, J., Lemaster, S., Gallagher, G., and Stringer, J. R., 1987, Structure and expression of a tandem gene pair in *Leishmania donovani* that encodes a protein structurally homologous to eukaryotic cation-transporting ATPases, *Mol. Cell. Biol.* **7:**3937–3946.

Meade, J. C., Hudson, K. M., Stringer, S. L., and Stringer, J. R., 1989, A tandem pair of *Leishmania donovani* cation transporting ATPase genes encode isoforms that are differentially expressed, *Mol. Biochem. Parasitol.* **33:**81–92.

Meade, J. C., Coombs, G., Steele, P. E., and Stringer, J. R., 1991, Conservation of cation-transporting ATPase genes in *Leishmania, Mol. Biochem. Parasitol.* **45:**29–38.

Medina-Acosta, E., Karess, R. E., Schwartz, H., and Russell, D. G., 1989, The promastigote surface protease (gp63) of *Leishmania* is expressed but differentially processed and localized in the amastigote stage, *Mol. Biochem. Parasitol.* **37:**263–274.

Miller, R. A., Reed, S. G., and Parsons, M., 1990, *Leishmania* gp63 molecule implicated in cellular adhesion lacks an Arg-Gly-Asp sequence, *Mol. Biochem. Parasitol.* **39:**267–274.

Molyneux, D. H., and Killick-Kendrick, R., 1987, Morphology and ultrastructure, in *The Leishmaniases in Biology and Medicine* (W. Peters and R. Killick-Kendrick, eds.), Vol. 1, pp. 121–176, Academic Press, London.

Monzingo, A. F., and Matthews, B. W., 1984, Binding of *N*-carboxymethyl dipeptide inhibitors to thermolysin determined by X-ray crystallography: A novel class of transition-state analogues for zinc peptidases, *Biochemistry* **23:**5724–5729.

Mosser, D. M., and Edelson, P. J., 1987, The third component of complement (C3) is responsible for the intracellular survival of *Leishmania major, Nature* **327:**329–331.

Mukkada, A. J., Meade, J. C., Glaser, T. A., and Bonventre, P. F., 1985, Enhanced metabolism of *Leishmania donovani* amastigotes at acid pH: An adaptation for intracellular growth, *Science* **229:**1099–1101.

Müller, I., Titus, R., Caldumbide, I., and Louis, J. A., 1989, T-cell responses in resistance and susceptibility to experimental infection with *Leishmania major,* in *Frontiers in Infectious Diseases; New Strategies in Parasitology* (K. P. W. J. McAdam, ed.), pp. 158–175, Churchill-Livingston, London.

Murray, P. J., Spithill, T. W., and Handman, E., 1989a, The PSA-2 glycoprotein complex of *Leishmania major* is a glycosylphosphatidylinositol-linked promastigote surface antigen, *J. Immunol.* **143:**4221–4226.

Murray, P. J., Spithill, T. W., and Handman, E., 1989b, Characterization of integral membrane proteins of *Leishmania major* by Triton X-114 fractionation and analyses of vaccination effects on mice, *Infect. Immun.* **57:**2203–2209.

Ogbunude, P. O. J., and Ikediobe, C. O., 1982, Effect of nitrobenzylthioinosinate on the toxicity of tubercidin and ethidium against *Trypanosoma gambiense, Acta Trop.* **39:**219–224.

Olafson, R. W., Thomas, J. R., Ferguson, M. A. J., Dwek, R. A., Chaudhuri, M., Chang, K.-P., and Rademacher, T. W., 1990, Structures of the *N*-linked oligosaccharides of gp63, the major surface glycoprotein from *Leishmania mexicana amazonensis, J. Biol. Chem.* **265:**12240–12247.

Opperdoes, F. R., 1989, The glycosome of *Leishmania* as a possible target for chemotherapeutic attack, in *Leishmaniasis: The Current Status and New Strategies for Control* (D. T. Hart, ed.), NATO ASI Series A, Vol. 163, pp. 859–863, Plenum Press, New York.

Ouellette, M., Fase-Fowler, F., and Borst, P., 1990, The amplified H-circle of methotrexate-resistant *Leishmania tarentolae* contains a novel P-glycoprotein gene, *EMBO J.* **9:**1027–1033.

Pan, A. A., 1984, *Leishmania mexicana:* Serial cultivation of intracellular stages in a cell-free medium, *Exp. Parasitol.* **58:**72–80.

Pastakia, K. B., and Dwyer, D. M., 1987, Identification and characterization of a ribose transport system in *Leishmania donovani* promastigotes, *Mol. Biochem. Parasitol.* **26:**175–182.

Philosoph, H., and Zilberstein, D., 1989, Regulation of intracellular calcium in promastigotes of the human protozoan parasite *Leishmania donovani, J. Biol. Chem.* **264:**10420–10424.

Pimenta, P. F. P., Da Silva, R. P., Sacks, D. L., and Da Silva, P. P., 1989, Cell surface nanoanatomy of *Leishmania major* as revealed by fracture-flip. A surface meshwork of 44 nm fusiform filaments identifies infective developmental stage promastigotes, *Eur. J. Cell. Biol.* **48:**180–190.

Puentes, S. M., Sacks, D. L., Da Silva, R. P., and Joiner, K. A., 1988, Complement binding by two developmental stages of *Leishmania major* promastigotes varying in expression of a surface lipophosphoglycan, *J. Exp. Med.* **167:**887–902.

Puentes, S. M., Dwyer, D. M., Bates, P. A., and Joiner, K. A., 1989, Binding and release of C3 from *Leishmania donovani* promastigotes during incubation in normal human serum, *J. Immunol.* **143:**3743–3749.

Pupkis, M. F., Tetley, L., and Coombs, G. H., 1986, *Leishmania mexicana:* Amastigote hydrolases in unusual lysosomes, *Exp. Parasitol.* **62:**29–39.

Ramasamy, R., Kar, S. K., and Jamnadas, H., 1983, Cross-reacting surface antigens on *Leishmania* promastigotes, *Int. J. Parasitol.* **13:**337–341.

Rassam, M. B. A.-K., and Robert, Z. J., 1988, Solubilization and kinetic characterization of mitochondrial adenosine triphosphatase from *Leishmania donovani* promastigotes, *Mol. Biochem. Parasitol.* **29:**153–158.

Remaley, A. T., Das, S., Campbell, P. I., La Rocca, G. M., Pope, M. T., and Glew, R. H., 1985a, Characterization of *Leishmania donovani* acid phosphatases, *J. Biol. Chem.* **260:**880–886.

Remaley, A. T., Glew, R. H., Douglas, B. K., Basford, R. E., Waggoner, A. S., Ernst, L. A., and Pope, M., 1985b, *Leishmania donovani:* Surface membrane acid phosphatase blocks neutrophil oxidative metabolite production, *Exp. Parasitol.* **60:**331–341.

Rivier, D., Didisheim, S., Etges, R., Bordier, C., and Mauël, J., 1989, Studies on antigenicity of amphiphilic and hydrophilic forms of *Leishmania major* promastigote surface protease, in *Leishmaniasis: The Current Status and New Strategies for Control* (D. T. Hart, ed.) NATO ASI Series A, Vol. 163, pp. 423–430, Plenum Press, New York.

Russell, D. G., 1987, The macrophage-attachment glycoprotein gp63 is the predominant C3-acceptor site on *Leishmania mexicana* promastigotes, *Eur. J. Biochem.* **164:**213–221.

Russell, D. G., and Alexander, J., 1988, Effective immunization against cutaneous leishmaniasis with defined membrane antigens reconstituted into liposomes, *J. Immunol.* **140:**1274–1279.

Russell, D. G., and Wilhelm, H., 1986, The involvement of the major surface glycoprotein (gp63) of *Leishmania* promastigotes in attachment to macrophages, *J. Immunol.* **136:**2613–1620.

Sacks, D. L., 1989, Metacyclogenesis in *Leishmania* promastigotes, *Exp. Parasitol.* **69:**100–103.

Sacks, D., and Da Silva, R. P., 1987, The generation of infective stage *Leishmania major* promastigotes is associated with the cell-surface expression and release of a developmentally regulated glycolipid, *J. Immunol.* **139:**3099–3106.

Sacks, D. L., Hieny, S., and Sher, A., 1985, Identification of cell surface carbohydrate and antigenic changes between noninfective and infective developmental stages of *Leishmania major* promastigotes, *J. Immunol.* **135:**564–569.

Schaefer, F. W. III, Martin, E., and Mukkada, A. J., 1974, The glucose transport system in *Leishmania tropica* promastigotes, *J. Protozool.* **21:**592–596.

Schneider, P., Ferguson, M. A. J., McConville, M. J., Mehlert, A., Homans, S. W., and Bordier, C., 1990, Structure of the glycosyl-phosphatidylinositol membrane anchor of the *Leishmania major* promastigote surface protease, *J. Biol. Chem.* **265:**16955–16964.

Seebeck, T., Hemphill, A., and Lawson, D., 1990, The cytoskeleton of trypanosomes, *Parasitol. Today* **6**:49–52.

Selsted, M. E., and Harwig, S. S. L., 1987, Purification, primary structure, and antimicrobial activities of a guinea pig neutrophil defensin, *Infect. Immun.* **55**:2281–2286.

Simon, M. W., and Mukkada, A. J., 1977, *Leishmania tropica:* Regulation and specificity of the methionine transport system in promastigotes, *Exp. Parasitol.* **42**:97–105.

Simpson, L., and Shaw, J., 1989, RNA editing and the mitochondrial cryptogenes of kinetoplastid protozoa, *Cell* **57**:355–366.

Soteriadou, K. P., Tzinia, A. K., Hadziautoniou, M. G., and Tzartos, S. J., 1988, Identification of monomeric and oligomeric forms of a major *Leishmania infantum* antigen by using monoclonal antibodies, *Infect. Immun.* **56**:1180–1186.

Stein, D. A., Cairns, B. R., and Landfear, S. M., 1990, Developmentally regulated transporter in *Leishmania* is encoded by a family of clustered genes, *Nucl. Acids Res.* **18**:1549–1557.

Theodos, C. M., Ribeiro, J. M., and Titus, R. G., 1991, Analysis of enhancing effect of sandfly saliva on *Leishmania* infection in mice, *Infect. Immun.* **59**:1592–1598.

Titus, R. G., and Ribeiro, J. M. C., 1988, Salivary gland lysates from the sandfly *Lutzomyia longipalpis* enhance *Leishmania* infectivity, *Science* **239**:1306–1308.

Titus, R. G., and Ribeiro, J. M. C., 1990, The role of vector saliva in transmission of arthropod-borne diseases, *Parasitol. Today* **6**:157–160.

Turco, S. J., 1990, The leishmanial lipophosphoglycan: A multifunctional molecule, *Exp. Parasitol.* **70**:241–245.

Wallbanks, K. R., Maazoun, R., Canning, E. U., and Rioux, J.-A., 1985, The identity of *Leishmania tarentolae* Wenyon (1921), *Parasitology* **90**:67–78.

Warburg, A., Tesh, R. B., and McMahon-Pratt, D., 1989, Studies on the attachment of *Leishmania* flagella to sandfly midgut epithelium, *J. Protozool.* **36**:613–617.

White, A. C., and McMahon-Pratt, D., 1988, Purification and characterization of an 80-kDa membrane protein from *Leishmania donovani, Infect. Immun.* **56**:2385–2391.

Wilson, G. M., Serrano, A. E., Wasley, A., Bogenschutz, M. P., Shankar, A. H., and Wirth, D. F., 1989, Amplification of a gene related to mammalian *mdr* genes in drug-resistant *Plasmodium falciparum, Science* **244**:1184–1186.

Wilson, M. E., and Hardin, K. K., 1988, The major concanavalin A-binding surface glycoprotein of *Leishmania donovani chagasi* promastigotes is involved in attachment to human macrophages, *J. Immunol.* **141**:265–272.

Wilson, M. E., Hardin, K. K., and Donelson, J. E., 1989, Expression of the major surface glycoprotein of *Leishmania donovani chagasi* in virulent and attenuated promastigotes, *J. Immunol.* **143**:678–684.

Young, C. J., Turner, D. P., Killick-Kendrick, R., Rioux, J.-A., and Leaney, A. J., 1980, Fructose in wild caught *Phlebotomus ariasi* and the possible relevance of sugars taken by sandflies to the transmision of leishmaniasis, *Trans. R. Soc. Trop. Med. Hyg.* **74**:363–366.

Yang, D. M., Fairweather, N., Button, L. L., McMaster, W. R., Kahl, L. P., and Liew, F. Y., 1990, Oral *Salmonella typhimurium* (AroA$^-$) vaccine expressing a major leishmanial surface protein (gp63) preferentially induces T helper 1 cells and protective immunity against leishmaniasis, *J. Immunol.* **145**:2281–2285.

Yang, D. M., Rogers, M. V., and Liew, F. Y., 1991, Identification and characterization of host-protective T-cell epitopes of a major surface glycoprotein (gp63) from *Leishmania major, Immunol.* **72**:3–9.

Zamze, S. E., Ferguson, M. A. J., Collins, R., Dwek, R. A., and Rademacher, T. W., 1988, Characterization of the cross-reacting determinant (CRD) of the glycosyl-phosphatidylinositol membrane anchor of *Trypanosoma brucei* variant surface glycoprotein, *Eur. J. Biochem.* **176**:527–534.

Zilberstein, D., and Dwyer, D. M., 1984a, Antidepressants cause lethal disruption of membrane function in the human protozoan parasite *Leishmania, Science* **226:**977–979.

Zilberstein, D., and Dwyer, D. M., 1984b, Glucose transport in *Leishmania donovani* promastigotes, *Mol. Biochem. Parasitol.* **12:**327–336.

Zilberstein, D., and Dwyer, D. M., 1985, Proton motive force-driven active transport of D-glucose and L-proline in the protozoan parasite *Leishmania donovani, Proc. Natl. Acad. Sci. USA* **82:**1716–1720.

Zilberstein, D., and Dwyer, D. M., 1988, Identification of a surface membrane proton-translocating ATPase in promastigotes of the parasitic protozoan *Leishmania donovani, Biochem. J.* **256:**13–21.

Zilberstein, D., Dwyer, D. M., Matthaei, S., and Horuk, R., 1986, Identification and biochemical characterization of the plasma membrane glucose transporter of *Leishmania donovani, J. Biol. Chem.* **261:**15053-15057.

Zilberstein, D., Philosoph, H., and Gepstein, A., 1989, Maintenance of cytoplasmic pH and proton motive force in promastigote of *Leishmania donovani, Mol. Biochem. Parasitol.* **36:**109–118.

Zilberstein, D., Liveanu, V., and Gepstein, A., 1990, Tricyclic drugs reduce proton motive force in *Leishmania donovani* promastigotes, *Biochem. Pharmacol.* **39:**935–940.

Chapter 3

The Lipophosphoglycan of *Leishmania*

Salvatore J. Turco

1. INTRODUCTION

Parasites of the genus *Leishmania* have the striking ability to survive in hydrolytic environments encountered throughout their digenetic life cycle. In one part of the life cycle, the parasite assumes an extracellular, flagellate promastigote form in the alimentary tract of its insect vector, the phlebotomine sandfly. Upon inoculation into a human or other suitable host, the parasite proliferates as an intracellular, nonflagellate amastigote in the phagolysosomes of phagocytic cells. Thus, to survive, the *Leishmania* must avoid destruction in (1) the sandfly gut where the parasite could be vulnerable to a variety of digestive enzymes, (2) the bloodstream of the host where the organism transiently exists and would be exposed to the lytic complement pathway, and, most spectacularly, (3) in the phagolysosome of host macrophages where the parasite would be subject to a number of hydrolytic enzymes and the microbicidal oxidative burst. Little is known about the molecular details of how this pathogen survives in obviously hostile environments. Cell surface glycoconjugates undoubtedly play a key role in the survival of the *Leishmania* parasite throughout its existence. The major cell surface glycoconjugate of the promastigote form is a polydisperse lipid-contain-

Salvatore J. Turco Department of Biochemistry, University of Kentucky Medical Center, Lexington, Kentucky 40536.

Subcellular Biochemistry, Volume 18: Intracellular Parasites, edited by J. L. Avila and J. R. Harris. Plenum Press, New York, 1992.

ing polysaccharide called lipophosphoglycan (LPG). Current information is reviewed that suggests that this interesting molecule is multifunctional.

2. STRUCTURE OF LPG

Lipophosphoglycan can be extracted from leishmanial parasites by sequential organic solvent extraction and further purified by gel filtration and methanol precipitation (Orlandi and Turco, 1987). Alternatively, organic solvent-extracted LPG can be purified by hydrophobic chromatography on phenyl–Sepharose (Turco *et al.*, 1989) or on octyl–Sepharose (McConville *et al.*, 1987). Minimally, purification yields 1 mg of LPG per 10^{10} parasites. The LPG preparations can be examined for purity of the glycoconjugate by sodium dodecyl sulfate-polyacylamide gel electrophoresis followed by periodic acid-Schiff base staining. Lipophosphoglycan prepared with these procedures is estimated to be greater than 99% pure. Although contaminating proteins cannot be detected by standard colorimetric or electrophoretic methods, gas-phase microsequencing routinely reveal several contaminating peptides. Exhaustive chromatographic and enzymatic attempts to remove the peptides have been unsuccessful (R. W. Olafson and S. J. Turco, unpublished observations). The significance, if any, of the peptides is unknown. Lipophosphoglycan is insoluble in acidic solutions, partially soluble in water, and soluble in water-containing, polar organic solvents or in aquesous solutions having an elevated pH, such as 10% NH_4OH (Turco *et al.*, 1984; McConville *et al.*, 1987).

Structurally, LPG is a heterogeneous glycoconjugate averaging 9–20 kDa and is a tripartite molecule, consisting of (1) repeating phosphorylated saccharides containing 2–6 sugars, (2) a phosphosaccharide core, and (3) a phosphatidylinositol lipid anchor. A generalized structure of the leishmanial LPG is schematically represented in Figure 1. Analysis of LPG from several species of *Leishmania* promastigotes have indicated variabilities of sugar composition and sequence in the repeating phosphorylated saccharide units whereas the core region and lipid anchor appear to be structurally conserved.

The structure of the *L. donovani* LPG is shown in Figure 2. The repeating phosphorylated saccharide units are $PO_4 \rightarrow 6Gal\,(\beta_1 \rightarrow 4)Man\alpha_1$ disaccharides (Turco *et al.*, 1987). There is an average of 16 of these repeat units linked together in a linear array by α-glycosidic linkages between the mannose residue of one unit and the phosphate group of another. One of the structural hallmarks of LPG is its extreme lability to dilute acid; conditions of 0.02 N HCl, for 15 min, at 60 °C hydrolyzes LPG into low-molecular-weight fragments. It is the mannosyl(α_1)phosphate bonds in the repeating units that hydrolyze with the mild acid (Turco *et al.*, 1984, 1989). Another interesting feature of the repeating units of the *L. donovani* LPG is the presence of the 4-substituted mannose residue. Since

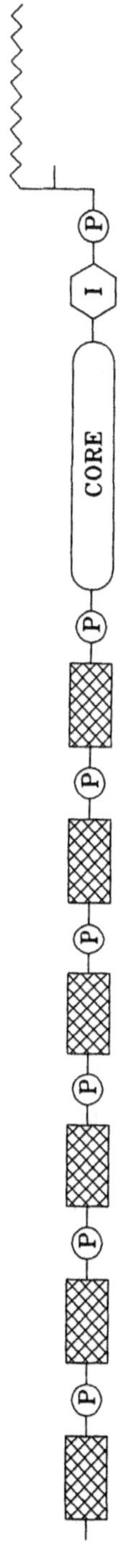

FIGURE 1. Generalized structure of the leishmania LPG.

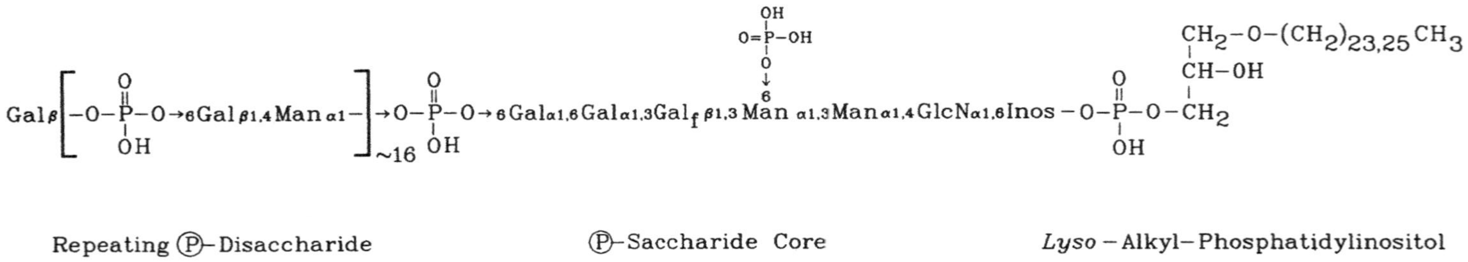

FIGURE 2. Structure of the LPG of *Leishmania donovani*.

this type of sugar substitution is not found in any other known eukaryotic glycoconjugate, its presence in LPG may have important implications for the parasite, as discussed below. In future investigations, it will be interesting to determine whether these repeating phosphorylated saccharide units constitute any secondary or tertiary structure in the glycoconjugate.

The repeating phosphorylated saccharide units of the *L. major* LPG are more complex compared to the *L. donovani* LPG. The units are a series of small phosphorylated oligosaccharides (2–6 sugars) composed of the hexoses galactose, mannose, glucose, and the pentose arabinose (McConville *et al.*, 1987); the complete structures and arrangement of these units are not yet known. In *L. mexicana* and *L. tropica,* the LPG molecule is more similar to the LPG from *L. donovani* in that the repeating units are mostly phosphorylated disaccharide units of $PO_4 \rightarrow 6Gal(\beta_1 \rightarrow 4)Man\alpha_1$ with lesser amounts of uncharacterized phosphorylated trisaccharide units (S. J. Turco and D. L. Sacks, unpublished observations; M. A. J. Ferguson, personal communication).

The repeating phosphorylated saccharide units are attached to the phosphosaccharide core region of LPG. In *L. donovani* (Turco *et al.*, 1989) and *L. major* (M. McConville, personal communication), the core region contains phosphate and consists of one glucosamine, two mannose, and three galactose residues. An unusual feature of the core is the presence of an internal galactofuranose, which is not common as a sugar component in eukaryotic glycoconjugates. Mild acidic conditions that are used to cleave the extremely labile mannosyl(α_1)phosphate bonds in the repeating units of LPG do not appreciably hydrolyze the labile galactofuranosylmannose bond in the core region. Interestingly, the phosphosaccharide core portion of LPG contains the sequence of Gal($\alpha_1 \rightarrow$ 3)Gal, which is believed to be the epitope for circulating antibodies in patients with leishmaniasis (Towbin *et al.*, 1987; Avila *et al.*, 1988a,b). Another noteworthy feature of the core region is that the hexosamine is unacetylated; this provides a convenient hydrolysis point in the molecule enabling LPG to be susceptible to nitrous acid deamination (Orlandi and Turco, 1987; McConville *et al.*, 1987).

The polysaccharide portion of LPG is anchored by an unusual phospholipid derivative 1-0-alkyl-2-*lyso*-phosphatidylinositol. In *L. donovani* (Orlandi and Turco, 1987), *L. major* (McConville *et al.*, 1987), and probably all species of *Leishmania,* the aliphatic chain consists of either a C_{24} or C_{26} saturated, unbranched hydrocarbon. Similar to many glycosyl-phosphatidylinositol (GPI)-anchored proteins (reviewed by Ferguson and Williams, 1988; Low, 1989), LPG can be hydrolyzed by bacterial phosphatidylinositol-specific phospholipase C yielding the 1-0-alkylglycerol and the entire polysaccharide chain (termed phosphoglycan) as products. As with all other reported GPI-anchored proteins with defined structures, LPG contains the Man($\alpha_1 \rightarrow$ 4)GlcN($\alpha_1 \rightarrow$ 6)*myo*-inositol sequence. Since this sequence is present in such widely diverse macromolecules as LPG, the variant surface glycoprotein of *Trypanosoma brucei* (Ferguson *et al.*, 1988), the promastigote surface protease gp63 of *L. major* (M. Ferguson, personal communication) and the thy-1 glycoprotein of rat brain (Homans *et al.*, 1988), this would indicate that it might be conserved in all GPI anchors of macromolecules.

All of the structural information known about LPG has been obtained from LPG isolated from the promastigote form of the parasite. Very little has been reported regarding the LPG from amastigotes, or even whether the amastigotes indeed synthesize any LPG. Although there have been several reports that suggest that this intracellular form of the parasite expresses the LPG, as detected by immunological means (Schnur *et al.*, 1972; Handman *et al.*, 1987), the identity of the detected antigen was not further characterized. The latter is important in view of recent evidence that LPG shares common epitopes with other glycoconjugates in promastigotes (see Section 3).

3. CELLULAR ASPECTS

Lipophosphoglycan is expressed on the cell surface of axenically grown promastigotes mainly in late log-phase and in the stationary phase of growth (King *et al.*, 1987). The copy number is relatively high (approximately 1.25×10^6 molecules of LPG/cell) accounting for about 25–30% of the cell surface (Orlandi and Turco, 1987). In addition, LPG can be found in the culture medium of parasites in two structurally distinct forms (El-On *et al.*, 1979; Slutzky *et al.*, 1979; King *et al.*, 1987). One form of LPG establishes a very tight complex with albumin in the medium, and analysis of this form indicates that the LPG retains its lipid moiety. One probable interpretation is that the lipid portion of LPG interacts with a hydrophobic binding pocket of albumin facilitating its release from the surface of the parasite. The other form of LPG found in conditioned medium is a hydrophilic form (originally referred to as the "excreted factor") in which the lipid is absent from the glycoconjugate. The structure of the hydrophilic form and the mechanism of its release are not known. Although a phospholipase cleavage to initiate its release has been suggested (Handman and Goding, 1985; Elhay *et al.*, 1988), there has been no definitive evidence as yet. It is just as possible that the hydrophilic form may be secreted as a polysaccharide analogous to bacterial secretion of polysaccharides (Braatz and Heath, 1974).

During entry of a *Leishmania* parasite into a host macrophage, LPG can be visualized on the surface of the macrophage by immunofluorescence using anti-LPG monoclonal antibodies (Tolson *et al.*, 1990). Lipophosphoglycan is present as early as 5–10 min postinfection and is localized to the immediate area of internalization of the promastigote. Subsequently, the immunodetected LPG is evenly distributed over the entire macrophage surface and continues to be present as long as two days postinfection. Only parasitized macrophages exhibit the LPG epitope whereas nonparasitized neighboring macrophages do not. The ability of LPG to transfer easily from the parasite to the macrophage surface may be due to the interaction of the glycoconjugate with a macrophage receptor or may be a reflection of the *lyso* nature of the phospholipid anchor. The latter may also contribute to the rapid distribution of LPG over the surface of the macrophage. Although lateral diffusion for the membrane form of the variant surface glycoprotein (VSG) (containing dimyristoylphosphatidylinositol in the GPI anchor) has been shown to be comparable to other membrane-bound glycoproteins (Bulow *et al.*, 1988), the relative behavior of a *lyso* derivative as in LPG is not known.

4. DEVELOPMENTAL MODIFICATION OF LPG DURING METACYCLOGENESIS

Following ingestion of a *Leishmania*-infected blood meal by a sandfly, the parasite attaches to the epithelial cells that line the midgut of the insect. While attached, the promastigotes divide and are avirulent. Eventually, the parasites cease multiplying, detach from the epithelial cells, and migrate to the mouthparts of the sandfly. During this detachment and migration event, the promastigote becomes infectious in a process called metacyclogenesis (reviewed by Sacks, 1989) and the infectious forms of the parasite are therefore named metacyclic promastigotes. This sequential development of promastigotes from a dividing noninfective stage to a resting infective stage has been observed for promastigotes growing both within the sandfly alimentary tract and within axenic culture (Giannini, 1974; Keithly, 1976; Sacks and Perkins, 1984; Sacks *et al.*, 1985; Howard *et al.*, 1987; Kweider *et al.*, 1987).

Sacks and Perkins (1984) and Sacks *et al.* (1985) have observed that the differentiation of *L. major* promastigotes into a metacyclic stage during growth is accompanied by changes in cell surface carbohydrates, which can be detected by the lectin peanut agglutinin or by stage-specific monoclonal antibodies. These antibodies were found to recognize a major surface glycolipid that is expressed in two developmental forms, one found exclusively on log-phase promastigotes and the other on metacyclic promastigotes purified from stationary cultures. Further studies have revealed the glycolipid to be LPG (Sacks *et al.*, 1990). Comparison of the *L. major* LPG derived from these stages revealed that the two characteristic structural features of this molecule, known to be conserved between species, are also developmentally conserved. Both contain the *lyso*-1-0-alkylphosphatidylinositol lipid anchor, and the polysaccharide fragments of each contain a polymer of repeating acid labile, phosphorylated saccharide units.

Within these shared basic structures, two major developmental differences were identified. First, the composition of at least some of the saccharide units were altered. Although the structures are not yet known, the difference may account for the expression of a novel epitope on metacyclic LPG (Sacks and Da Silva, 1987). The second difference, and the more striking, was an approximate doubling in size displayed by the metacyclic version of LPG, due to an elongation of the polysaccharide portion of the molecule (Sacks *et al.*, 1990). The relative increase in size of metacyclic LPG is consistent with recent freeze-fracture electron microscopic studies of the metacyclic promastigote cell surface that have revealed the presence of densely packed filamentous structures not present on noninfectious promastigotes (Pimenta *et al.*, 1989). These studies also showed a greater than twofold thickening of a surface coat which could be specifically labeled with the monoclonal antibody against the metacyclic form of the LPG. A finding similar to that demonstrated with *L. major* has also been observed with the metacyclic promatigotes of *L. donovani* in axenic culture (S. J. Turco and D. L. Sacks, unpublished observations). Interestingly, the LPG from axenically grown *L. mexicana* was not found to differ in size in various phases of growth. However, a structural change in the repeating phosphorylated saccharide units was detected (S. J. Turco and D. L. Sacks, unpublished observations).

In the sandfly, these developmentally regulated changes in LPG structure might mediate attachment and release of the parasite from the midgut cells and also its subsequent migration. Following inoculation of the metacyclic promastigote into a host, the developmentally altered LPG might also preadapt the promastigote for survival. As reviewed by Sacks (1989), metacyclic promastigotes are resistant to complement-mediated killing, yet they activate complement as efficiently as serum-sensitive noninfectious promastigotes. In each case, C3 deposition on the parasite surface can be found covalently associated with the respective forms of LPG (Puentes *et al.*, 1988). It seems likely that the elongated LPG molecule on metacyclic promastigotes (at least with *L. donovani* and *L. major*) protects the cell from complement-mediated damage by sterically hindering access of macromolecules to the cell membrane, thereby preventing channel formation and lysis by the C5–9 membrane attack complex (Figure 3). Formation and release of C5–9 by *L. major* has recently been demonstrated (Puentes *et al.*, 1990). Thus, metacyclic *Leishmania* spp. that express elongated LPG appear to have evolved a mechanism of complement resistance similar to that reported in gram-negative bacteria, for which the presence of smooth lipopolysaccharide containing long polysaccharide chains correlates with serum resistance and virulence (Joiner *et al.*, 1986). How *Leishmania* spp. that do not express elongated LPG (e.g., *L. mexicana*) may avoid complement-mediated destruction is open to question. Furthermore, whether these developmental modifications of LPG during metacyclogenesis also account for the diverse tissue

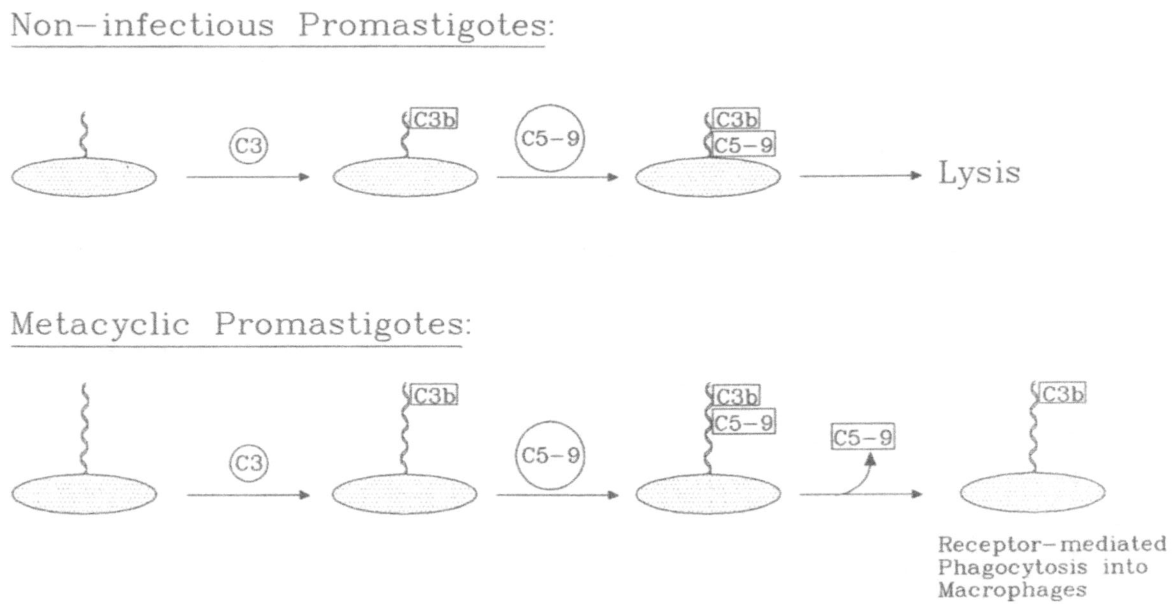

FIGURE 3. Proposed involvement of LPG in *Leishmania major*'s encounter with complement components. One molecule of LPG is schematically represented on the surface of a parasite and the length of the glycoconjugate in metacyclic *L. major* is proposed to be approximately twice that in the noninfectious form (Sacks *et al.*, 1990).

tropisms of the various *Leishmania* species and consequent pathogenesis is unknown.

5. PARTICIPATION OF LPG IN PARASITE–MACROPHAGE SURFACE INTERACTIONS

The attachment and entry of *Leishmania* promastigotes into host macrophages is receptor mediated (reviewed by Russell and Talamas-Rohana, 1989; Sacks, 1989). Lipophosphoglycan, as a ligand that interacts with a macrophage receptor, was implicated by the work of Handman and Goding (1985). They showed that addition of anti-LPG antibodies to *L. major* promastigotes blocked their attachment to macrophages. The other major *Leishmania* macromolecule reported to be involved in host–parasite interactions is gp63 (Russell and Talamas-Rohana, 1989). Elucidation of the receptor on the macrophage for these two glycoconjugates has been somewhat controversial and is due, in part, to whether experiments are conducted in the presence or absence of serum and to the particular species of *Leishmania* used.

In *L. major,* the metacyclic promastigote possessing the elongated version of LPG is the form that produces successful parasitism of the macrophage. As mentioned earlier, the metacyclic LPG provides an effective barrier against direct complement-mediated killing. It activates complement efficiently, which results in extensive deposition of C3b on the LPG. Da Silva *et al.* (1989) have shown that the C3b-opsonized parasites are proficient in the attachment to and uptake by macrophages via the CR1 receptor. The primary use of CR1, which does not trigger the oxidative burst (Wright and Silverstein, 1983), may be critical for their subsequent intracellular survival. The situation for *L. donovani* may be similar since its LPG also undergoes elongation during metacyclogenesis. Although Puentes *et al.* (1988) have shown that the majority of C3 deposited on the surface of this particular species was the hemolytically inactive iC3b, sufficient C3b was present following serum incubation to promote receptor-mediated phagocytosis.

Although preliminary indications are that *L. mexicana* does not express the elongated version of LPG, nevertheless, this glycoconjugate has been implicated in macrophage interactions. Using beads coated with *L. mexicana* LPG, LPG was shown to bind directly to the lectinlike lipopolysaccharide (LPS) binding site present on the CD18 family of macrophage surface receptors, CR3, LFA-1, and p150,95 (Talamas-Rohana *et al.*, 1990). The involvement of the CR3 receptor in *L. donovani* promastigote–macrophage interactions was originally reported by Blackwell (1985). Since these experiments were conducted in the absence of serum, it was proposed that macrophages were secreting complement components, which resulted in activation of the alternative pathway and deposition of iC3b on the parasite surface. Conceivably, the LPG in *L. mexicana* might serve as an acceptor for the iC3b. This complex, in conjugation with gp63, could facilitate entry of the parasite into the macrophage.

6. INTRACELLULAR FUNCTIONS OF LPG

6.1. Requirement of LPG for Intracellular Survival

Upon entry of the promastigote into the phagolysosome of the macrophage, the importance of LPG for intracellular survival has been demonstrated using variant species of *Leishmania* defective in LPG expression. Handman *et al.* (1986) showed that an avirulent strain of *L. major* deficient in LPG was rapidly phagocytized by macrophages *in vitro*. However, in contrast to virulent strains of *L. major,* they were then killed over a period of 18 hr. Passive transfer of purified LPG from a virulent clone of *L. major* into the avirulent strain conferred to them the ability to survive in macrophages.

A similar observation was made with variants of *L. donovani* defective in LPG expression (McNeely and Turco, 1990). The variants were selected for resistance to the cytotoxic lectin ricin agglutinin (King and Turco, 1988) and were shown to be partially deficient (clone RT5) or totally lacking (clone C3PO) in the synthesis and expression of LPG. RT5 and C3PO variant cells were examined for their ability to enter and maintain an infection in the monocytes. The results indicated that both variant cells could easily enter the phagocytic cells, most likely through an unfamiliar alternative receptor such as the mannose–fucose receptor. Upon entry, the variant cells were unable to survive and were completely destroyed within several days postinfection. Significantly, adding purified LPG to these variant cells allowed them to survive within monocytes for an extended period of time. From the results with the LPG-deficient strains of *L. major* and *L. donovani,* expression of LPG appears to be crucial in the overall infection process.

6.2. Effects of LPG on Action of Hydrolases

The molecular basis for the role of LPG in intracellular survival of the parasite is unknown and is only now beginning to be investigated. A great deal of interpretations will hinge on the form of LPG, if any, that amastigotes express. Since the promastigote form of LPG is observable for several days postinfection, it may have an important adverse influence on macrophage functions during this time frame.

For successful parasitism and survival with impunity, the parasite must be able to avoid destruction by hydrolytic enzymes and the microbicidal oxidative burst (also called the respiratory burst). Lipophosphoglycan might be important in enabling the parasite to avoid destruction by both. Regarding the former, it is well-known that phagocytosis of the parasite results in fusion of the parasitophorous vacuole with lysosomal granules, resulting in confrontation of the parasite with host hydrolases. This phagocytic event displayed by *Leishmania* parasites is unlike that of *Trypanosoma cruzi*, which, following entry into a macrophage, lyses the phagolysosomal membrane before fusion with the lysosomal granules. It is also unlike that of *Toxoplasma* spp. in which the endocytic vesicle containing the parasite does not fuse with granules. In *Leishmania* spp, which encounter host degradative enzymes, it is possible that the unique structure of leishmanial LPG may offer protection as follows:

1. The highly anionic nature of LPG may afford protection for the parasite against enzymatic attack. Precedence for such a possibility comes from the work of Rogers (1946), who showed that both heparin and chondroitin sulfate, due to their anionic character, inhibit the action of streptococcal hyaluronidase.

2. Lipophosphoglycan may act as a cell-surface barrier preventing lysosomal hydrolases from attacking the parasite. Lysosomes may not possess the necessary glycosidases and lipases that are capable of degrading the LPG on the surface of the invading promastigote, and consequently this might preclude access of the parasite from the action of other degradative enzymes. Specifically, a lysosomal enzyme may not be available that is capable of cleaving the unique Gal($\beta_1 \rightarrow$ 4)Man linkage in the repeating phosphorylated saccharide units of LPG. There is precedence for the lack of lysosomal hydrolases capable of cleaving specific linkages. As an example, lysosomes in mammalian cells do not possess a glycosidase capable of cleaving the simple disaccharide sucrose. Also, there may not be a hydrolase competent to degrade the ether-linked hydrocarbon of the LPG lipid anchor. In a related observation, Dwyer and co-workers (Wassef *et al.*, 1985) have shown that *L. donovani* contains relatively large amounts of ether-linked phospholipids in its surface membrane. This may provide additional protection against degradative substances in the lysosome as well as in the alimentary tract of the insect vector.

3. Lipophosphoglycan might act as a direct inhibitor of lysosomal enzymes. The effect of partially purified excreted factor (LPG) from *L. donovani* promastigotes on the activity of four enzymes from lysed peritoneal exudate cells of mice has been examined (El-On *et al.*, 1980). There was no demonstrable effect on acid phosphatase, β-glucuronidase, or *N*-acetylglucosaminidase. However, β-galactosidase was inhibited up to 72%. The significance of this observation is not known, but additional exhaustive studies on this possibility are clearly warranted.

6.3. Effects of LPG on the Oxidative Burst

Although avoiding degradation by lysosomal hydrolases is obviously important for intracellular survival of the parasite, circumvention of the microbicidal oxidative burst is presumably a more ominous challenge. Ironically, *Leishmania* parasites are extremely sensitive to hydrogen peroxide and are rapidly destroyed by it *in vitro* (Murray, 1981; Pearson and Steigbigel, 1981). Murray (1981) has shown that *Leishmania* parasites do possess a superoxide dismutase that is capable of converting the superoxide anion to hydrogen peroxide. In contrast, they do not possess adequate amounts of scavenging enzymes, such as catalase or glutathione peroxidase, which can inactivate the toxic oxygen metabolites of the burst. Therefore, the promastigotes must avoid damage from the oxidative burst in another fashion. Entry of the promastigote into a macrophage through a nontriggering macrophage receptor such CR1 or CR3 (Wright and Silverstein, 1983) is clearly advantageous. During the phagocytic process, LPG may also be actively involved in helping to thwart activation of the potentially lethal oxidative burst.

Normally the oxidative burst is dormant and requires activation during entry of a microbe into a macrophage. The oxidative burst is initiated by a one-electron transfer from NADPH to oxygen, forming the superoxide anion that is released at the outer surface of the plasma membrane into the extracellular space or into phagocytic vacuoles. This step is catalyzed by the NADPH oxidase (reviewed by Baggiolini and Wyman, 1990), a large complex of molecules containing membrane-bound components and one or two cytoplasmic subunits. The oxidase is inactive in resting macrophages. Upon chemotactic stimulation or phagocytosis, the cytosolic components are translocated to the membrane, forming an active NADPH oxidase system. Current information indicates that one step in the induction of the NADPH oxidase occurs by phosphorylation of one or more of its subunits by the diacylglycerol, phospholipid, and Ca^{2+}-dependent protein kinase (protein kinase C). Exactly how the protein kinase C plays a role in activating the system is obscure. One or more of the subunits of the NADPH oxidase has been reported to be phosphorylated by the kinase, including one of the cytosolic subunits (reviewed by Baggiolini and Wyman, 1990). Besides the protein kinase C-activation of the burst, another Ca^{2+}-independent, non-protein kinase C signal transduction sequence is also believed to be operational. It has been shown that both sequences must be functional for the activation of the oxidative burst (Dewald *et al.*, 1988).

Since stimulation of the oxidative burst is mediated by protein kinase C, this enzyme might be an important molecule for the parasite to target for inhibition, thereby ensuring its survival. Lipophosphoglycan was found to be a potent *in vitro* inhibitor of protein kinase C purified from rat brain (McNeely and Turco, 1987; McNeely *et al.*, 1989). The glycoconjugate inhibited the enzyme and was found to be a competitive inhibitor with respect to diacylglycerol (K_I of approximately 1 μM) and a noncompetitive inhibitor with respect to phosphatidylserine. Significantly, LPG had no effect on protein kinases A and M. Since the latter kinase represents the catalytic domain of protein kinase C and cannot be stimulated by diacylglycerol, phosphatidylserine, or calcium, this would suggest that LPG interacts with the regulatory domain of protein kinase C. Additional information indicated that the particular region of LPG that was responsible for the *in vitro* inhibition of the enzyme was the 1-0-alkylglycerol portion. In an *in vivo* experiment, phagocytosis of LPG-coated beads by resting monocytes resulted in an impaired ability of the monocytes to consume oxygen when subsequently stimulated with phorbol myristate acetate, an inducer of protein kinase C (McNeely and Turco, 1990). Evidently, LPG presented to the monocytes in this manner can effectively interact with the monocyte's protein kinase C or another factor involved with the oxidative burst mechanism. The details remain to be elucidated and will depend on whether LPG or a cleaved fragment of LPG is the active inhibitor. If inhibition is due to the lipid fragment of LPG only, then it is possible to rationalize how the lipid fragment would be able to traverse the macrophage cell membrane and interact with membrane-bound protein kinase C or another component. If intact LPG is the inhibiting molecule, the problem of topography must be explained. Nevertheless, these preliminary indications are consistent with the hypothesis that protein kinase C is a key target for the parasites to overcome within host macrophages.

Lipophosphoglycan is reported to have a second property that can adversely affect the oxidative burst. It appears to be a scavenger of the oxygen metabolites generated during the burst. The property was demonstrated by directly testing LPG's capacity to scavenge superoxide anions and hydroxyl radicals, by an *in vitro* biochemical assay using electron spin resonance spectroscopy and spin trapping to monitor scavenging (Chan *et al.*, 1989). In this study, the phosphorylated disaccharide portion of LPG was found to be responsible for the scavenging phenomenon. That LPG can scavenge is not surprising since carbohydrates, such as mannitol, are common scavengers; the hydroxyl groups of sugars can be easily oxidized by hydrogen peroxide or hydroxyl radicals to ketones (reviewed by Green, 1980). Although not yet tested, the metacyclic form of LPG might, in fact, scavenge more effectively because of the increased numbers of sugars per molecule, which could act as a more efficient sink for reactive oxygen intermediates. Using living monocytes stimulated for the production of the oxidative burst with either phorbol myristate acetate or opsinized zymosan, LPG was shown to exhibit a powerful quenching effect on the burst, as measured by luminol chemiluminescence (McNeely and Turco, 1990). Addition of LPG to the monocytes following their stimulation resulted in an immediate and significant attenuation of the burst, indicating an effect on the products of the burst that had already been excreted into the medium. This quenching is most likely a direct effect of scavenging of toxic oxygen radicals produced during the oxidative response. Other human pathogens that produce polysaccharide-containing molecules to scavenge free radicals have been reported. A polysaccharide product of *Pseudomonas aeruginosa* characterized as alginate (composed of mannuronic and glucuronic acid residues) was shown to scavenge oxygen radicals produced during activation of the oxidative burst in macrophages (Simpson *et al.*, 1989). *Mycobacterium leprae*, the etiologic agent of leprosy, synthesizes an unusual glycolipid (phenolic glycolipid I), which also scavenges superoxide ions and hydroxyl radicals (Chan *et al.*, 1989).

Thus, LPG is proposed to play an important intracellular, protective role for the parasite by effectively disarming the host cell's oxidative burst. It is probable that LPG has at least two functions in this role: (1) as an inhibitor of the macrophage's protein kinase C, thereby keeping the burst dormant or at a low level, and (2) as a scavenger of the microbicidal elements of the oxidative burst.

These effects of isolated LPG on monocytes are consistent with several observations reported with *Leishmania*-infected macrophages. The presence of intracellular leishmanial parasites inhibited the oxidative burst in lymphokine-activated and lipopolysaccharide-stimulated mouse peritoneal macrophages; inert intracellular particle, in contrast, resulted in a strong oxidative response (Buchmuller-Rouiller and Mauel, 1987). It was concluded that the impairment of macrophage oxidative metabolism by the intracellular parasite emanated from a specific leishmanial effect. In another study, Eilam *et al.* (1985) found in mouse macrophages that the rate of cytolysis of erythrocytes coated with excreted factor (LPG) was significantly lower than that of uncoated controls.

The observation that LPG inhibits protein kinase C *in vitro* may extend to other protein kinase C-mediated pathways, in addition to the respiratory burst. In a recent example, *L. donovani*-infected mouse bone marrow-derived macrophages had a reduced ability to express the *c-fos* gene in response to lipopolysaccharide or diacylglycerol (Descoteaux and Matlashewski, 1989). Since induction of this particular gene is mediated by protein kinase C, the authors suggested that LPG might be involved in abrogating the effect of the kinase.

7. IMMUNOLOGICAL IMPLICATIONS OF LPG

While all *Leishmania* promastigotes examined synthesize LPG, it is antigenically polymorphic. After the excreted factor (LPG) was shown to be species specific, it became the basis for serotyping *Leishmania* strains and is a valuable tool in diagnostic, demographic, and ecological studies of leishmaniasis (Schnur *et al.*, 1972). Monoclonal antibodies have been generated using LPG as an immunogen and some of these antibodies have been useful for serotyping

(Greenblatt *et al.*, 1983). Some monoclonal antibodies have been shown to have cross-reactivity with LPG from different species of *Leishmania* (de Ibarra *et al.*, 1982; Greenblatt *et al.*, 1983; Handman *et al.*, 1984). This may not be surprising if the minimal portion of repeating units, the phosphorylated disaccharide of PO_4 → Gal → Man, is found to be shared among some or all LPGs. This particular phosphorylated disaccharide unit was demonstrated to be the epitope of two monoclonal antibodies (CA7AE and BF9CC) raised against the *L. donovani* LPG (Tolson *et al.*, 1989). Two other monoclonal antibodies (L157 and L98) have been obtained that recognize the phosphosaccharide core region of LPG of *L. donovani* (Tolson *et al.*, 1989). Immunofluorescence experiments using living and formaldehyde-fixed *L. donovani* promastigotes showed that the CA7AE monoclonal antibodies bound to the entire surface and flagellum, whereas the LPG core-specific monoclonal antibodies were not bound. This result indicated that in the orientation of LPG on the cell surface, there is differential accessibility of the repeating phosphorylated disaccharide and phosphosaccharide-core epitopes. Several other useful monoclonal antibodies are available, although the epitopes for them are not yet mapped. One of these is monoclonal antibody WIC-79.3, which was prepared by de Ibarra *et al.* (1982) and has been commonly used to detect and isolate the excreted factor (Handman *et al.*, 1984; Handman and Mitchell, 1985). Another is monoclonal antibody 3F12 that specifically recognizes the LPG from metacyclic *L. major* (Sacks and Da Silva, 1987). In addition, Jaffe *et al.* (1989, 1990) used several *L. tropica*-specific monoclonal antibodies (T11, T13–T15) to demonstrate that LPG and secreted acid phosphatase share common epitopes.

Handman and Mitchell (1985) were the first to show that immunization of an animal with LPG protects against leishmaniasis. In this study, LPG was purified from detergent-solubilized *L. major* promastigotes and purified by monoclonal antibody affinity chromatography. Genetically resistant mice could be totally protected from cutaneous disease by intraperitoneal immunization with LPG and the adjuvent *Corynebacterium parvum*. A high but not absolute level of resistance was also induced in susceptible mice. Other studies have shown that vaccination of genetically susceptible mice with purified LPG could protect against challenge when LPG was reconstituted into liposomes (McConville *et al.*, 1987; Russell and Alexander, 1988). Interestingly, immunization of mice with the hydrophilic excreted factor released by promastigotes into culture medium is nonprotective, and when used with Freund's complete adjuvant can exacerbate lesion development (Handman and Mitchell, 1985).

To analyze whether T cells are able to recognize LPG, Moll *et al.* (1989) studied both humoral and cellular responses to purified *L. major* LPG. The data suggested that immunization of mice with highly purified LPG induced an increase in frequency of *L. major*-reactive T cells that are able to secrete lymphokines and an increase in the production of immunoglobulin G antibodies to LPG. Furthermore, genetically resistant mice infected with *L. major* were able to develop a specific delayed-type hypersensitivity response in the ear to LPG. Furthermore, injecting the phosphoglycan portion of LPG failed to elicit the response, suggesting that proper presentation of the carbohydrate epitope is important for protection and that the lipid portion of LPG is required for this presentation. The authors concluded that T cells can recognize and respond to glycolipid antigens, such as LPG, even though such antigens appear not to be potent T-cell stimulators in mice. However, the authors were unable to detect any T-cell responsiveness to LPG *in vitro* and did not investigate the possible participation of contaminating peptides in their highly purified (99.5%) LPG preparation.

Another group of investigators have also begun to examine whether LPG epitopes can be recognized by T cells (Jardim *et al.*, 1991). Lymphocytes from mice immunized with *L. donovani* LPG were specifically stimulated to proliferate *in vitro* by highly purified LPG and by its delipidated form PG. The response was dose-dependent and was eliminated by removal of the $CD4^+$ $CD8^-$ subset by treatment with specific antibodies and complement. Pronase-treated LPG or PG elicited a stimulatory response as well as the untreated controls. In contrast, selective chemical modification of LPG or PG carbohydrates abrogated the proliferative response. In the most critical aspect, cleavage of glycosidic and phosphodiester linkages in LPG by anhydrous trifluoromethanesulfonic acid treatment almost completely eliminated the ability of the glycolipid to stimulate cell proliferation. This chemical treatment is commonly used in solid-phase peptide synthesis and does not affect peptide bonds (Tam *et al.*, 1986). From these results, the authors argue that *Leishmania*-protective $CD4^+$ $CD8^-$ effector T cells can recognize complex carbohydrate epitopes, although the involvement of LPG-associated peptides in T-cell recognition is a possibility.

If T lymphocytes can recognize such structures as LPG, the role of complex carbohydrates as immunogens in cell-mediated immunity may be more important than previously thought, and may dramatically influence host responses to microbes, tumor cells, and self-components in autoimmunity. Clearly much more effort in this area is warranted since this raises very interesting questions regarding antigen processing and Major Histocompatibility Complex (MHC) associations.

8. OTHER UNUSUAL PARASITIC GLYCOCONJUGATES

8.1. LPG-like Glycoconjugates in *Leishmania*

In *Leishmania* the various molecular portions of LPG have been found to exist as separate entities or as constituents of other macromolecules. Regarding the former, the phosphosaccharide core-alkylglycerol portion of LPG is found abundantly in *L. major* as a family of molecules termed glycosyl phosphatidylinositol antigens (GPIs) (Rosen *et al.*, 1988) or glycoinositol-phospholipids (GIPLs) (Elhay *et al.*, 1988; McConville and Bacic, 1989); these molecules are not linked to either protein or polysaccharide. The structures of the individual members of the family have been recently elucidated and all have a mannose- and galactose-containing glycan moiety that is attached by an unacetylated glucosamine to either alkylacyl-PI or *lyso*-1-0-alkyl-PI (phosphatidylinositol) (McConville and Bacic, 1989, 1990; Rosen *et al.*, 1989; McConville *et al.*, 1990). The carbohydrate portions of these molecules are exactly as the phosphosaccharide core of LPG, including the salient Man($\alpha_1 \rightarrow$ 4)GlcN($\alpha_1 \rightarrow$ 6)*myo*-inositol and, in the larger GIPLs, the unusual galactofuranose. These findings would suggest that the GIPLs are precursors in the biosynthetic pathway leading to LPG. Although the pathways of biosynthesis of LPG and the GIPLs are not yet known, the qualitative and quantitative differences in the respective lipid portions would indicate that extensive lipid remodeling must take place if there is precursor–product relationship. In particular, there is an increase in the proportion of 24:0 and 26:0 alkyl chains (present in the LPG lipid anchor) and a concomitant decrease in the proportions of the 18:0 and 16:0 hydrocarbons (prevalent in the GIPL lipid anchors) (McConville and Bacic, 1989; Rosen *et al.*, 1989; McConville *et al.*, 1990). Extensive lipid remodeling has already been described in the biosynthetic pathway of the variant surface glycoprotein lipid anchor of *T. brucei,* which contains dimyristoyl-PI (reviewed by Doering *et al.*, 1990). Furthermore, acylalkyl substitutions in the glycerol backbone have been demonstrated in the bovine erythrocyte acetylcholinesterase membrane anchor and other proteins (reviewed by Ferguson and Williams, 1988; Low, 1989). Clearly, if some of the GIPLs are indeed precursors to LPG, deacylation of the fatty acid on C2 of the glycerol by an endogenous phospholipase A_2 and alkyl chain modification would have to occur before LPG is expressed.

Besides the possiblity of being precursors to LPG, some of the GIPLs may have other biological roles. Several of the GIPLs are relatively abundant on the surface of the parasite. It has been suggested (McConville and Bacic, 1989, 1990; Rosen *et al.*, 1989; McConville *et al.*, 1990), therefore, that they might be involved in parasite–host cell interactions or in transmembrane signaling analogous to that reported for the conpositionally similar insulin-sensitive glycolipid (Low and Saltiel, 1988). Furthermore, the GIPLs, as proposed for LPG (Turco, 1990), may play a protective role for the parasite through modulation of protein kinase C activity (McNeely *et al.*, 1989). The location of some of the GIPLs on the surface of the parasite and their antigenicity in mice and rabbits indicate that these glycolipids may be important antigens in the course of human infection (Slutzky *et al.*, 1985; Frankenburg *et al.*, 1988; Rosen *et al.*, 1988; McConville *et al.*, 1990). It has also been proposed (McConville *et al.*, 1990) that the elevated anti-Gal antibodies commonly seen in leishmaniasis patients (Towbin *et al.*, 1987; Avila *et al.*, 1988a,b) may be directed against the terminal Gal($\alpha_1 \rightarrow$

3)Gal residues in some of the surface GIPLs. Although the significance is unknown, the GIPLs have been reported to be released from the parasite surface in a water-soluble form by a presumed endogenous phospholipase (Elhay *et al.*, 1988).

Interestingly, the prominent feature of LPG, the repeating phosphorylated saccharides, has been shown to constitute a carbohydrate component of an acid phosphatase secreted by *L. donovani* (Bates *et al.*, 1989; D. M. Dwyer, unpublished observations) and *L. tropica* (Jaffe *et al.*, 1990a,b). Since LPG, PG, and acid phosphatase (and possibly other glycoconjugates) can be released into extracellular medium and share the repeating phosphorylated saccharide epitope, this knowledge complicates the interpretation of immunological results concerning the originally described excreted factor.

8.2. Other Unusual Lipid-Containing Glycoconjugates in Parasites

The existence of unusual lipid-containing complex carbohydrate seems to be common in parasites. As an example, a glycolipid of the epimastigote form of *T. cruzi* called lipopeptidophosphoglycan (Lederkremer *et al.*, 1977) contains a small glycan attached by a nonacetylated glucosamine residue to an inositol phosphoceramide. This material is not found in trypomastigotes or amastigotes (Zingales *et al.*, 1982). Two unusual features of the carbohydrate portion of the lipopeptidophosphoglycan are that it contains aminophosphonate (Ferguson *et al.*, 1982; Mendonca-Previato *et al.*, 1983) and terminal galactofuranose. It also contains the typical hexopyranoses galactose, mannose, and glucose (Lederkremer *et al.*, 1985; Previato *et al.*, 1990). The presence of galactose and mannose has also been reported in a polysaccharide antigen in the plasma of mice infected with *T. cruzi* (Gottlieb, 1977). Its relationship to the lipopeptidophosphoglycan is not known. As with LPG, GIPLs, and glycosyl-phosphatidylinositol-anchored proteins, the lipopeptidophosphoglycan contains the distinguishing Man($\alpha_1 \rightarrow$ 4)GlcN($\alpha_1 \rightarrow$ 6)*myo*-inositol sequence (Previato *et al.*, 1990). The lipopeptidophosphoglycan is antigenic; its terminal sequence of Gal_f($\beta_1 \rightarrow$ 3)Man structure is involved in its antigenticity (Mendonca-Previato *et al.*, 1983). The biological significance of the glycophosphosphingolipids in *T. cruzi* is not yet known. Previato *et al.* (1990) suggest that these kinds of compounds that have been described mainly in plants, yeast, and fungi (Laine and Hsieh, 1987) may be functionally analogous to the gangliosides in animal cells.

Other interesting glycoconjugates have been described but are far less characterized. In *L. donovani*, several complex carbohydrates other than LPG have been reported (Palatnik *et al.*, 1989): a lipopeptidophosphoglycan (LPPD), a fucose–mannose glycoproteic ligand (FML), and a phosphate mannogalactan (PMGL). Since analysis of the LPPD indicated the presence of fatty acids similar to results reported by Kaneshiro *et al.* (1986) and contains mostly 3-substituted mannose residues, it is believed to be distinct from LPG. The FML and PMGL have been shown to behave as separate specific ligands for parasite internalization by mouse peritoneal macrophages (Palatnik *et al.*, 1989). In *L. alderi* a lipopeptidephosphomannan was found to consist mainly of short oligosaccharides with 6-substituted mannopyranose and, in minor amounts, galactopyranose (Palatnik *et al.*, 1985). An apparent analogous compound called lipopeptidephosphoxylan (LPPX) has been detected in *Leptomonas samueli* and *Herpetomonas samuelpessoai* (Palatnik *et al.*, 1987). A lipophosphoglycan of undefined structure has been reported in the bloodstream forms of *Trypanosoma brucei* (Hublart *et al.*, 1988), but it may be a biosynthetic precursor to the glycolipid anchor of the variant surface glycoprotein (reviewed by Doering *et al.*, 1990).

9. FUTURE CONSIDERATIONS

Among the many unanswered questions concerning the leishmanial LPG, several important ones are: (1) What are the distinguishing structural features of LPG among various *Leishmania* spp. and do the structural differences in LPG during metacyclogenesis account for the diverse tissue tropisms of the assorted *Leishmania* spp.? (2) What is the "excreted factor" released from the surface of promastigotes and what is its biological role, if any? (3) Can a firm answer be obtained as to whether or not LPG can be recognized by T cells and can a segment of LPG be used as a molecularly defined vaccine? (4) What is the identity of the amastigote version of LPG and what is its function? (5) How is LPG synthesized by the parasite and can any chemotherapeutic drugs be targeted against its assembly? From information gathered from such studies, it will be possible to determine whether the structural polymorphisms of LPG contribute, in any manner, to the pathological differences associated with the various leishmaniases.

10. REFERENCES

Avila, J. L., Rojas, M., and Towbin, H., 1988a, Serological activity against galactosyl-α(1 $\rightarrow$ 3)galactose in sera from patients with several kinetoplastida infections, *J. Clin. Microbiol.* **26**:126–132.

Avila, J. L., Rojas, M., and Garcia, L., 1988b, Persistence of elevated levels of galactosyl-α(1 $\rightarrow$ 3)galactose antibodies in sera from patients cured of visceral leishmaniasis, *J. Clin. Microbiol.* **26**:1842–1847.

Baggiolini, M., and Wyman, M. P., 1990, Turning on the respiratory burst, *Trends in Biochem. Sci.* **15**:69–72.

Bates, P. A., Hermes, I., and Dwyer, D. M., 1989, Golgi-mediated posttranslational processing of secretory acid phosphatase by *Leishmania donovani* promastigotes, *Mol. Biochem. Parasitol.* **68:**335–346.

Blackwell, J. M., 1985, Receptors and recognition mechanisms of *Leishmania* species, *Trans. R. Soc. Trop. Med. Hyg.* **79:**606–612.

Braatz, J. A., and Heath, E. C., 1974, The role of polysaccharide in the secretion of protein in *Micrococcus sodonensis, J. Biol. Chem.* **249:**2536–2547.

Buchmuller-Rouiller, Y., and Mauel, J., 1987, Impairment of the oxidative metabolism of mouse peritoneal macrophages by intracellular *Leishmania* spp., *Infect. Immun.* **55:**587–593.

Bulow, R., Overath, P., and Davous, J., 1988, Rapid lateral diffusion of the variant surface glycoprotein in the coat of *Trypanosoma brucei, Biochemistry* **27:**2384–2388.

Chan, J., Fujiwara, T., Brennan, P., McNeil, M., Turco, S. J., Sibille, J., Snapper, M., Aisen, P., and Bloom, B. R., 1989, Microbial glycolipids: Possible virulence factors that scavenge oxygen radicals, *Proc. Nat. Acad. Sci. USA* **86:**2453–2457.

Da Silva, R. P., Hall, B. F., Joiner, K. A., and Sacks, D. L., 1989, CR1, the C3b receptor, mediates binding of infective *Leishmania major* metacyclic promastigotes to human macrophages, *J. Immunol.* **143:**617–622.

de Ibarra, A., Howard, J. G., and Snary, D., 1982, Monoclonal antibodies to *Leishmania tropica major:* Specificities and antigen location, *Parasitology* **85:**523–531.

Descoteaux, A., and Matlashewski, G., 1989, *c-fos* and tumor necrosis factor gene expression in *Leishmania donovani*-infected macrophages, *Mol. Cell. Biol.* **9:**5223–5227.

Dewald, B., Thelen, M., and Baggiolini, M., 1988, Two transduction sequences are necessary for neutrophil activation by receptor agonists, *J. Biol. Chem.* **263:**16179–16184.

Doering, T. L., Masterson, W. J., Hart, G. W., and Englund, P. T., 1990, Biosynthesis of glycosyl phosphatidylinositol membrane anchors, *J. Biol. Chem.* **265:**611–614.

Eilam, Y., El-On, J., and Spira, D. T., 1985, *Leishmania major:* Excreted factor, calcium ions, and the survival of amastigotes, *Exp. Parasitol.* **59:**161–168.

Elhay, M. J., McConville, M. J., and Handman, E., 1988, Immunochemical characterization of a glycoinositol-phospholipid membrane antigen of *Leishmania major, J. Immunol.* **141:**1326–1331.

El-On, J., Schnur, L. F., and Greenblatt, C. L., 1979, *Leishmania donovani:* Physiochemical, immunological and biological characterization of excreted factor from promastigotes, *Exp. Parasitol.* **47:**254–269.

El-On, J., Bradley, D. J., and Freeman, J. C., 1980, *Leishmania donovani:* Action of excreted factor on hydrolytic enzyme activity of macrophages from mice with genetically different resistance to infection, *Exp. Parasitol.* **49:**167–174.

Ferguson, M. A. J., and Williams, A. F., 1988, Cell-surface anchoring of proteins via glycosylphosphatidylinositol structures, *Annu. Rev. Biochem.* **57:**285–320.

Ferguson, M. A. J., Allen, A. K., and Snary, D., 1982, The detection of phosphonolipids in the protozoan *Trypanosoma cruzi, Biochem. J.* **207:**171–174.

Ferguson, M. A. J., Homans, S. W., Dwek, R. A., and Rademacher, T. W., 1988, Glycosyl-phosphatidylinositol moiety that anchors *Trypanosoma brucei* variant surface glycoprotein to the membrane, *Science* **239:**753–759.

Frankenburg, S., Rosen, G., and Londner, M. V., 1988, Cell-mediated responses and protection elicited by a carbohydrate-lipid-containing fraction extracted from *Leishmania major* promastigotes, *Cell. Immunol.* **111:**287–295.

Giannini, M. S., 1974, Effects of promastigote growth phase, frequency of subculture, and host age on promastigote-initiated infections in *Leishmania donovani* in the golden hamster, *J. Protozool.* **21:**521–527.

Gottlieb, M., 1977, A carbohydrate-containing antigen from *Trypanosoma cruzi* and its detection in the circulation of infected mice, *J. Immunol.* **119:**465–470.

Green, J. W., 1980, Oxidative reactions and degradations, in *The Carbohydrates* (W. Pigman, ed.), Vol. 1B, pp. 1101–1166, Academic Press, New York.

Greenblatt, C. L., Slutzky, G. M., de Ibarra, A. A., and Snary, D., 1983, Monoclonal antibodies for serotyping of *Leishmania* strains, *J. Clin. Microbiol.* **18**:191–193.

Handman, E., and Goding, J. W., 1985, The *Leishmania* receptor for macrophages is a lipid-containing glycoconjugte, *EMBO J.* **4**:329–336.

Handman, E., and Mitchell, G. F., 1985, Immunization with *Leishmania* receptors for macrophages protects mice against cutaneous leishmaniasis, *Proc. Natl. Acad. Sci. USA* **82**:5910–5914.

Handman, E., Greenblatt, C. L., and Goding, J. W., 1984, An amphipathic sulphate glycoconjugate of *Leishmania:* Characterization with monoclonal antibodies, *EMBO J.* **3**:2301–2306.

Handman, E., Schnur, L. F., Spithill, T. W., and Mitchell, G. F., 1986, Passive transfer of *Leishmania* lipopolysaccharide confers parasite survival in macrophages, *J. Immunol.* **137**:3608–3613.

Handman, E., Mitchell, G. F., and Goding, J. W., 1987, *Leishmania major:* A very sensitive dot-blot ELISA for detection of parasites in cutaneous lesions, *Mol. Biol. Med.* **5**:377–383.

Howard, M. K., Sayers, G., and Miles, M. M., 1987, *Leishmania donovani* metacyclic promastigotes: Transformation *in vitro,* lectin agglutination, complement resistance and infectivity, *Exp. Parasitol.* **64**:147–156.

Hublart, M., Tetaert, D., Mendonca-Previato, L., and Degand, P., 1988, Presence of a lipophosphoglycan in two variants of *Trypanosoma brucei brucei, Biochem. Biophys. Res. Commun.* **153**:1257–1266.

Jaffe, C. J., Martinez, L. P., and Sarfstein, R., 1989, *Leishmania tropica:* Characterization of a lipophosphoglycan-like antigen recognized by species-specific monoclonal antibodies, *Exp. Parasitol.* **70**:12–24.

Jaffe, C. L., Perez, L., and Schnur, L. F., 1990, Lipophosphoglycan and secreted acid phosphatase of *Leishmania tropica* share species-specific epitopes, *Mol. Biochem. Parasitol.* **41**:233–240.

Jardim, A., Tolson, D. L., Turco, S. J., Pearson, T. W., and Olafson, R. W., 1991, The Leishmania donovani lipophosphoglycan T-lymphocyte reactive component is a tightly associated protein complex, *J. Immunol.* **147**:3538–3544.

Joiner, K. A., Grossman, N., Schmetz, M., and Leive, L., 1986, C3 binds preferentially to long-chain lipopolysaccharide during alternative pathway activation by *Salmonella montevideo, J. Immunol.* **136**:710–715.

Kaneshiro, E. S., Keka, J. A., and Lester, R. L., 1986, Characterization of inositol lipids from *Leishmania donovani* promastigotes. Identification of inositol sphingo-phospholipids, *J. Lipid Res.* **27**:1294–1303.

Keithly, J. S., 1976, Infectivity of *Leishmania donovani* amastigotes and promastigotes for golden hamsters, *J. Protozool.* **23**:244–248.

King, D. L., and Turco, S. J., 1988, A ricin agglutinin-resistant clone of *Leishmania donovani, Mol. Biochem. Parasitol.* **28**:285–294.

King, D. L., Chang, Y. D., and Turco, S. J., 1987, Cell surface lipophosphoglycan of *Leishmania donovani, Mol. Biochem. Parasitol.* **24**:47–53.

Kweider, M., Lemesre, J., Darcy, F., Kusnierz, J. P., Capron, A., and Santoro, F., 1987, Infectivity of *Leishmania braziliensis* promastigotes is dependent on the increasing expression of a 65,000 dalton surface antigen, *J. Immunol.* **138**:299–305.

Laine, R. A., and Hsieh, T., 1987, Inositol-containing sphingolipids, *Meth. Enzymol.* **138**:186–195.

Lederkremer, R. M., Tanaka, C. T., Alves, M. J. M., and Colli, W., 1977, Lipopeptidophosphoglycan from *Trypanosoma cruzi, Eur. J. Biochem.* **74**:265–267.

Lederkremer, R. M., Casal, O. L., Alves, M. J. M., and Colli, W., 1985, *Trypanosoma cruzi:* Amino acid and phosphorus linkages in the lipopeptidophosphoglycan, *Biochem. Int.* **10**:89–96.

Low, M. G., 1989, The glycosyl-phosphatidylinositol anchor of membrane proteins, *Biochim. Biophys. Acta* **988**:427–454.

Low, M. G., and Saltiel, A. R., 1988, Structural and functional roles of glycosyl-phosphatidylinositol in membranes, *Science* **239**:268–275.

McConville, M. J., and Bacic, A., 1989, A family of glycoinositol phospholipids from *Leishmania major, J. Biol. Chem.* **264**:757–766.

McConville, M. J., and Bacic, A., 1990, The glycoinositolphospholipids profiles of two *Leishmania major* strains that differ in lipophosphoglycan expression, *Mol. Biochem. Parasitol.* **38**:57–68.

McConville, M. J., Bacic, A., Mitchell, G. F., and Handman, E., 1987, Lipophosphoglycan of *Leishmania major* that vaccinates against cutaneous leishmaniasis contains an alkylglycerophosphatidylinositol lipid anchor, *Proc. Natl. Acad. Sci. USA* **84**:8941–8945.

McConville, M. J., Homans, S. W., Thomas-Oates, J. E., Dell, A., and Bacic, A., 1990, Structures of the glycoinositolphospholipids from *Leishmania major:* A family of novel galactosylfuranose-containing glycolipids, *J. Biol. Chem.* **265**:7385–7394.

McNeely, T. B., and Turco, S. J., 1987, Inhibition of protein kinase C by the *Leishmania donovani* lipophosphoglycan, *Biochem. Biophys. Res. Commun.* **148**:653–657.

McNeely, T. B., and Turco, S. J., 1990, Requirement of lipophosphoglycan for intracellular survival of *Leishmania donovani* within human monocytes, *J. Immunol.* **144**:2745–2750.

McNeely, T. B., Rosen, G., Londner, M. V., and Turco, S. J., 1989, Inhibitory effects on protein kinase C by lipophosphoglycan and glycosyl-phosphatidylinositol antigens of *Leishmania, Biochem. J.* **259**:601–604.

Mendonca-Previato, L., Gorin, P. A. J., Braga, A. F., Scharfstein, J., and Previato, J. O., 1983, Chemical structure and antigenic aspects of complexes obtained from epimastigotes of *Trypanosoma cruzi, Biochemistry* **22**:4980–4987.

Moll, H., Mitchell, G. F., McConville, M. J., and Handman, E., 1989, Evidence for T-cell recognition in mice of a purified lipophosphoglycan from *Leishmania major, Infect. Immunol.* **57**:3349–3356.

Murray, H. W., 1981, Susceptibility of *Leishmania* to oxygen intermediates and killing by normal macrophages, *J. Exp. Med.* **153**:1302–1315.

Orlandi, P. A., and Turco, S. J., 1987, Structure of the lipid moiety of the *Leishmania donovani* lipophosphoglycan, *J. Biol. Chem.* **262**:10384–10391.

Palatnik, C. B., Previato, J. O., Gorin, P. A. J., and Previato, L. M., 1985, Partial chemical characterization of the carbohydrate moieties in *Leishmania adleri* glycoconjugates, *Mol. Biochem. Parasitol.* **14**:41–45.

Palatnik, C. B., Previato, L. O., Gorin, P. A. J., and Mendonca-Previato, L., 1987, *Leptomonas samueli* glycoconjugates. Comparison with *Herpetomonas samuelipessoai, Comp. Biochem. Physiol.* **86B**:593–599.

Palatnik, C. B., Borojevic, R., Previato, J. O., and Mendonca-Previato, L., 1989, Inhibition of *Leishmania donovani* promastigote internalization into murine macrophages by chemically defined parasite glycoconjugate ligands, *Infect. Immun.* **57**:754–763.

Pearson, R. D., and Steigbigel, R. T., 1981, Phagocytosis and killing of the protozoan *Leishmania donovani* by human polymorphonuclear leukocytes, *J. Immunol.* **127**:1438–1442.

Pimenta, P. F., Da Silva, R. P., Sacks, D. L., and Da Silva, P., 1989, Cell surface nanoanatomy of *Leishmania major* as revealed by fracture flip. A surface meshwork of 44 nm fusiform filaments identifies infective developmental stage promastigotes, *Eur. J. Cell Biol.* **48**:180–190.

Previato, J. O., Gorin, P. A. J., Mazurek, M., Xavier, M. T., Fournet, B., Wieruszesk, J. M., and Mendonca-Previato, L., 1990, Primary structure of the oligosaccharide chain of lipopeptidophosphoglycan of epimastigote forms of *Trypanosoma cruzi, J. Biol. Chem.* **265**:2518–2526.

Puentes, S. M., Sacks, D. L., Da Silva, R. P., and Joiner, K. A., 1988, Complement binding by two developmental stages of *Leishmania major* promastigotes varying in expression of a surface lipophosphoglycan, *J. Exp. Med.* **167**:887–902.

Puentes, S. M., Da Silva, R. P., Sacks, D. L., Hammer, C. H., and Joiner, K. A., 1990, Serum resistance of metacyclic stage *Leishmania major* promastigotes is due to release of C5B-9, *J. Immunol.* **145**:4311–4316.

Rogers, H. J., 1946, The influence of hydrolysates of hyaluronate upon hyaluronidase production by micro-organisms, *Biochem. J.* **40**:583–588.

Rosen, G., Londner, M. V., Sevlever, D., and Greenblatt, C. L., 1988, *Leishmania major:* Glycolipid antigens recognized by immune human sera, *Mol. Biochem. Parasitol.* **27**:93–100.

Rosen, G., Pahlsson, P., Londner, M. V., Westerman, M. E., and Nilsson, B., 1989, Structural analysis of glycosyl-phosphatidylinositol antigens of *Leishmania major, J. Biol. Chem.* **264**:10457–10463.

Russell, D. G., and Alexander, J., 1988, Effective immunization against cutaneous leishmaniasis with defined membrane antigens reconstituted into liposomes, *J. Immunol.* **140**:1274–1279.

Russell, D. G., and Talamas-Rohana, P., 1989, *Leishmania* and the macrophage: A marriage of inconvenience, *Immunol. Today* **10**:328–334.

Sacks, D. L., 1989, Metacyclogenesis in *Leishmania* promastigotes, *Exp. Parasitol.* **69**:100–103.

Sacks, D. L., and Da Silva, R. P., 1987, The generation of infective stage *Leishmania major* promastigotes is associated with the cell-surface expression and release of a developmentally regulated glycolipid, *J. Immunol.* **139**:3099–3106.

Sacks, D. L., and Perkins, P. V., 1984, Identification of an infective stage of *Leishmania* promastigotes, *Science* **223**:1417–1419.

Sacks, D. L., Hieny, S., and Sher, A., 1985, Identification of cell surface carbohydrate and antigenic changes between noninfective and infective developmental stages of *Leishmania major* promastigotes, *J. Immunol.* **135**:564–569.

Sacks, D. L., Brodin, T., and Turco, S. J., 1990, Developmental modification of the lipophosphoglycan from *Leishmania major* promastigotes during metacyclogenesis, *Mol. Biochem. Parasitol.* **42**:225–234.

Schnur, L. F., Zuckerman, A., and Greenblatt, C. L., 1972, Leishmanial serotypes as distinguished by the gel diffusion of factors excreted *in vitro* and *in vivo, Isr. J. Med. Sci.* **8**:932–942.

Simpson, J. A., Smith, E., and Dean, R. T., 1989, Scavenging by alginate of free radicals released by macrophages, *Free Radical Biol. Med.* **6**:347–353.

Slutzky, G. M., El-On, J., and Greenblatt, C. L., 1979, Leishmanial excreted factor: Protein-bound and free forms from promastigote cultures of *Leishmania tropica* and *Leishmania donovani, Infect. Immun.* **26**:916–924.

Slutzky, G. M., Londner, M. V., and Greenblatt, C. L., 1985, Lipid and lipopolysaccharide-like antigens of *Leishmania* promastigotes, *J. Protozool.* **32**:347–352.

Talamas-Rohana, P., Wright, S. D., Lennaretz, M. R., and Russell, D. G., 1990, Lipophosphoglycan from *Leishmania mexicana* promastigotes binds to members of the CR3, p150, 95, and LFA-1 family of leukocyte integrins, *J. Immunol.* **144**:4817–4824.

Tam, J. P., Heath, W. F., and Merrifield, R. B., 1986, Mechanisms for the removal of benzyl protecting groups in synthetic peptides by trifluormethane sulfonic acid-trifluoracetic acid-dimethyl sulfide, *J. Am. Chem. Soc.* **108**:5242–5251.

Tolson, D. L., Turco, S. J., Beecroft, R. P., and Pearson, T. W., 1989, Immunochemical and cell surface arrangement of the *Leishmania donovani* lipophosphoglycan determined using monoclonal antibodies, *Mol. Biochem. Parasitol.* **35**:109–118.

Towbin, H., Rosenfelder, G., Wieslander, J., Avila, J. L., Rojas, M., Szarfman, A., Esser, K., Nowack, H., and Timpl, R., 1987, Circulating antibodies to mouse laminin in Chagas' disease, American cutaneous leishmaniasis, and normal individuals recognize terminal galactosyl($\alpha_1 \rightarrow$ 3)galactose epitopes, *J. Exp. Med.* **166**:419–432.

Turco, S. J., 1990, The leishmanial lipophosphoglycan: A multifunctional molecule, *Exp. Parasitol.* **70:**241–245.

Turco, S. J., Wilkerson, M. A., and Clawson, D. R., 1984, Expression of an unusual acidic glycoconjugate in *Leishmania donovani, J. Biol. Chem.* **259:**3883–3889.

Turco, S. J., Hull, S. R., Orlandi, P. A., Jr., Shepherd, S. D., Homans, S. W., Dwek, R. A., and Rademacher, T. W., 1987, Structure of the major carbohydrate fragment of the *Leishmania donovani* lipophosphoglycan, *Biochemistry* **26:**6233–6238.

Turco, S. J., Orlandi, P. A., Jr., Homans, S. W., Ferguson, M. A. J., Dwek, R. A., and Rademacher, T. W., 1989, Structure of the phosphosaccharide-inositol core of the *Leishmania donovani* lipophosphoglycan, *J. Biol. Chem.* **264:**6711–6715.

Wassef, M. K., Fioretti, T. B., and Dwyer, D. M., 1985, Lipid analysis of isolated surface membranes of *Leishmania donovani* promastigotes, *Lipids* **20:**108–115.

Wright, S. D., and Silverstein, S. C., 1983, Receptors for C3b and C3bi promote phagocytosis but not the release of toxic oxygen from human phagocytes, *J. Exp. Med.* **158:**2016–2019.

Zingales, B., Martin, N. F., Lederkremer, R. M., and Colli, W., 1982, Endogenous and surface labeling of glycoconjugates from the three differentiation stages of *Trypanosoma cruzi, FEBS Lett.* **142:**238–242.

Chapter 4

Role of T Cells in Immunity to the Intracellular Pathogen, *Leishmania major*

Richard G. Titus, Cynthia M. Theodos, Paul B. Kimsey, Anuraj Shankar, Laurie Hall, Mary McGurn, and Laura Povinelli

1. GENERAL INTRODUCTION

Members of the genus *Leishmania* are protozoan parasites that exist as flagellated promastigotes in the vector for the parasite, the phlebotomine sandfly. Man acquires the disease when bitten by an infected sandfly. The life cycle of *Leishmania* is schematically depicted in Figure 1. Within man, the parasite is an obligate intracellular parasite of macrophages in which it transforms into its nonflagellated amastigote form. The amastigotes are found in phagolysosomes in the macrophage where they not only survive but replicate until specific immunity intervenes and activates the macrophage to destroy the parasite. For a more detailed description of the biology of *Leishmania* and the epidemiology and

Richard G. Titus, Cynthia M. Theodos, Anuraj Shankar, Laurie Hall, Mary McGurn, and Laura Povinelli Department of Tropical Public Health, Harvard School of Public Health, Boston, Massachusetts 02115. **Paul B. Kimsey** Division of Comparative Medicine, Massachusetts Institute of Technology, Cambridge, Massachusetts 02139.

Subcellular Biochemistry, Volume 18: Intracellular Parasites, edited by J. L. Avila and J. R. Harris. Plenum Press, New York, 1992.

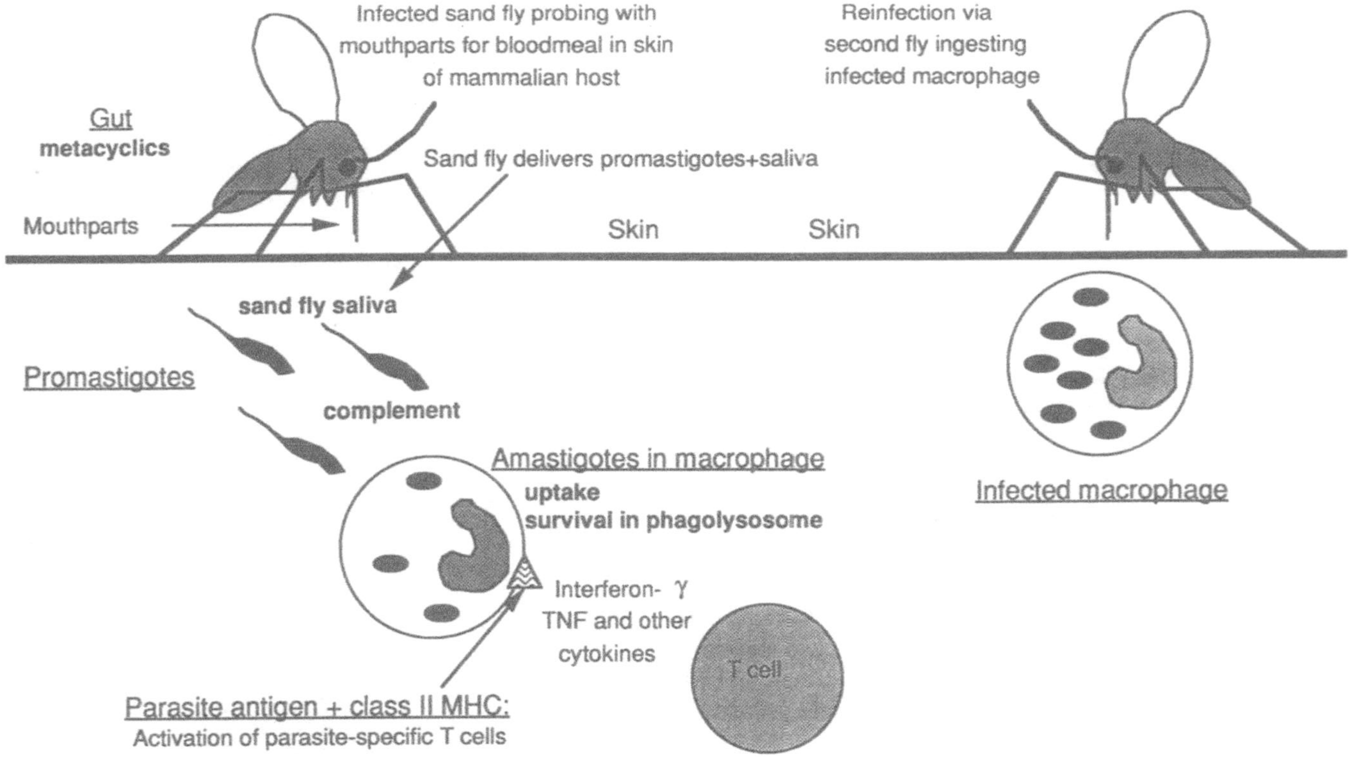

FIGURE 1. Life cycle of *Leishmania*. The vector for *Leishmania*, the phlebotomine sand fly, injects the promastigote form of the parasite into the mammalian host's skin while probing for a blood meal. The promastigote is engulfed by macrophages and transforms into the amastigote form of the parasite. In the macrophage, *Leishmania* resides within the phagolysosome. Sand flies become reinfected when they ingest an infected macrophage while taking a blood meal. The items in bold in the figure are subjects discussed further in the text.

clinical aspects of leishmaniasis, the reader is referred to any of several surveys on the subjects, among which are Howard (1986), Molyneux and Killick-Kendrick (1987), and Neva and Sacks (1990).

It is estimated that the incidence of leishmaniasis in the developing world is 1,000,000 cases/year (Walsh, 1990). Leishmaniasis is a complex of diseases caused by a number of different species of *Leishmania*. The clinical manifestations of the disease are determined by the species of *Leishmania* that infects the host and by the response of the host to the parasite (Neva and Sacks, 1990). *Leishmania* are generally classified based on the nature of the disease they produce and on the geographic region in which the disease occurs. Thus, cutaneous leishmaniasis (CL) can be caused by *L. major* in the Old World and *L. mexicana* or *L. braziliensis* in the New World; mucocutaneous leishmaniasis (MCL) can be caused by *L. braziliensis* in the New World; and visceral leishmaniasis (VL) can be caused by *L. donovani* in the Old World (see Table I) (Neva and Sacks, 1990).

Leishmaniasis, like leprosy, has often been thought of as a spectrum of diseases, with VL and MCL at the poles of the spectrum and CL in the middle (Neva and Sacks, 1990). The clinical features of CL, MCL, and VL are characteristic for each disease (Table I). In CL, cutaneous lesions are often solitary; however, more than one lesion can occur. Lesions can persist for over a year but eventually spontaneously resolve with the patient becoming solidly immune to reinfection. In MCL, the parasite induces hideous disfiguring lesions in nasooropharyngeal tissues. In VL the parasite replicates in the liver, spleen, bone marrow, and lymph nodes, and if the patient is left untreated, the disease is often fatal. Treatment still relies heavily on the use of pentavalent antimonials (Neva and Sacks, 1990).

Like the clinical manifestations, each form of leishmaniasis elicits a characteristic immune response. In VL, the patient produces large amounts of parasite-

Table I
Clinical Forms of Leishmaniasis and the Organisms Responsible

Human Disease	Organism
Visceral leishmaniasis	*L. donovani*
Cutaneous leishmaniasis	*L. major*
(Old World)	*L. aethiopica*
Cutaneous leishmaniasis	*L. mexicana mexicana*
(New World)	*L. mexicana amazonensis*
	L. braziliensis panamensis
	L. braziliensis guyanensis
Mucocutaneous leishmaniasis (MCL)	*L. braziliensis braziliensis*

specific antibody but lacks parasite-specific cell-mediated responses [i.e., such patients do not mount a delayed-type hypersensitivity (DTH) response to the intradermal injection of leishmanin, a suspension of phenolized promastigotes]. As mentioned above, patients with VL often die if they are not drug treated. These facts have been interpreted to mean that an effective cell-mediated response (DTH) is necessary for the resolution of leishmaniasis. In the middle of the spectrum of leishmanial diseases lies CL. In patients with CL, parasite-specific DTH reactivity is often present and these patients will spontaneously resolve their lesions and develop solid immunity to reinfection. At the other end of the spectrum of leishmanial diseases lies MCL. In the lesions of patients with MCL, very few parasites can be found. Despite this, lesions persist and host tissue damage occurs. It is often observed that MCL patients have heightened parasite-specific DTH reactivity. It is therefore possible that, although potentially curative, the heightened DTH reactivity of MCL patients leads to the host tissue damage observed. Thus, the degree of parasite-specific DTH reactivity of patients with leishmaniasis is not a perfect indicator of whether the individual's infection will resolve or whether the individual will be immune to reinfection. In fact, although it was stated above that parasite-specific DTH reactivity often correlates with resistance in CL patients, this is not always true (Mauel and Behin, 1987). In addition, in certain experimental animal models for leishmaniasis, DTH reactivity is not always an accurate measure of immunity to the parasite (Liew, 1989). Nevertheless, the concept that immunity in leishmaniasis is a cell-mediated phenomenon has led several investigators to examine cell-mediated responses in experimental animal models for the disease. These studies are uncovering a fascinating mosaic of immunoregulation in which $CD4^+$ parasite-specific T cells appear to play a pivotal role.

The subject of this review will be to describe the role of parasite-specific T cells in a well-studied experimental system for leishmaniasis, namely, experimental murine cutaneous leishmaniasis induced with *L. major*. There are, however, several other aspects of the biology of *L. major* that are being investigated. Some of these are indicated by bold type in Figure 1. After the sandfly ingests a *Leishmania*-infected macrophage, the parasites are released from the macrophage in the fly's gut. The parasites then undergo developmental changes, resulting in the generation of highly infective "metacyclic" forms of the parasite (Sacks and Perkins, 1984). These metacyclic parasites express a different form of their surface lipophosphoglycan (LPG) molecule that makes them more resistant to destruction by complement (Franke *et al.*, 1985; Puentes *et al.*, 1988) and by macrophages (Sacks and Perkins, 1984). The LPG molecule may also influence the manner in which the parasite is taken up by macrophages (Da Silva *et al.*, 1989) and may be capable of inhibiting macrophage activation through its ability to inhibit protein kinase C activity (McNeely and Turco, 1987). Interestingly, sensitization of mice with the sugar portion of the LPG molecule has been shown to have an exacerbative effect on the course of leishmaniasis, whereas sensitization with the entire LPG molecule is protective (Handman *et al.*, 1987). In addition to the surface LPG molecule, a major surface glycoprotein, gp63, has also been described (Bordier, 1987). Gp63 may also influence the manner in which the parasite is taken up by macrophages (Russell and Talamas-Rohana, 1989), and, due to the fact that it is a potent protease, it may be important in defending the parasite against destruction by the host.

Finally, it is being found that the sandfly, the vector for *Leishmania*, is not simply a "flying syringe." We have shown that a substance in the saliva of the sandfly markedly enhances infection with *Leishmania* in mice (Titus and Ribeiro, 1988). This substance is related to the mammalian neuropeptide, calcitonin gene-related peptide (CGRP), and both sandfly CGRP and mammalian CGRP are potent immunosuppressives (Nong *et al.*, 1989; Ribeiro *et al.*, 1989). When low numbers of parasites (10–100) are injected into mice, the parasite either survives poorly or does not survive at all. However, when these numbers of parasites are injected with sandfly saliva, survival of the parasites is greatly enhanced (Titus and Ribeiro, 1988). Since 10–100 parasites is approximately the number of parasites that the sandfly injects into the mammalian host (Warburg and Schlein, 1986), it is possible that sandfly saliva is essential for the parasite to establish an infection in human beings. These observations offer the exciting prospect that immunization of man against vector saliva may represent a means for controlling leishmaniasis. Specifically, in individuals vaccinated against sandfly saliva, neutralization of the immunosuppressive effect of the saliva might occur and infection by *Leishmania* might be prevented (Titus and Ribeiro, 1990).

2. T CELLS AND LEISHMANIASIS

2.1. Early Work

2.1.1. Immunogenetics in Murine Leishmaniasis

Resolution of an infection with *Leishmania* and resistance to reinfection with the parasite are dependent on the activities of parasite-specific T cells (Howard, 1986). However, the innate susceptibility or resistance of mice to infection with *Leishmania* is more complex. When inbred mice of various H-2 haplotypes were analyzed for their innate susceptibility/resistance to infection with *L. major,* the locus (loci) responsible did not map to the H-2 region (Behin *et al.*, 1979; Handman *et al.*, 1979; Howard *et al.*, 1980a). For example, BALB/c (H-2^d) mice are genetically susceptible to infection with *L. major;* however, B10.D2 (also H-2^d) mice are resistant to infection. A possible explanation for these observations comes from work with *L. donovani.* Work with this parasite has revealed that a locus on chromosome 1 in the mouse (termed Lsh) is involved in the innate susceptibility/resistance of mice to infection with *L. donovani* (Bradley, 1987; Blackwell *et al.*, 1980). Interestingly, loci controlling the host's innate susceptibility/resistance to two other pathogens, namely, *Salmonella typhimurium* and *Mycobacterium bovis*, also map to chromosome 1 in the mouse (Howard, 1986). Since *L. donovani, S. typhimurium,* and *M. bovis* are all obligate intracellular pathogens of macrophages, the loci controlling relative resistance to these microbes may encode for products that affect macrophage functions. Thus, the innate ability of a host to resist infection with *Leishmania* may be influenced by the manner in which the macrophage responds to infection with the parasite. It should be noted that loci (e.g., the Scl locus) (Blackwell *et al.*, 1985) that control the host's relative resistance to infection with *L. major* have also been described. The Scl locus does not map to mouse chromosome 1 and is thus distinct from the Lsh locus. Nevertheless, it is possible that the Scl locus, or other loci involved in resistance to *L. major,* may also influence macrophage functions.

There are several reports in the literature of the differential responsiveness of macrophages to infection with *Leishmania*. It has been observed that when glass-adherent cells were isolated from murine skin, one subset of these cells was permissive for replication of intracellular *Leishmania* (Gorczynski and MacRae, 1982). It is known that when macrophages obtained from genetically susceptible mice are infected with *L. major,* the cells express less major histocompatibility complex (MHC) class II (I-A) than macrophages from resistant animals (Heinzel *et al.*, 1988). Lower levels of I-A may compromise the ability of the cells to induce a protective immune response. In addition, susceptible macrophages secrete high levels of interleukin-1 (IL-1) (Cillari *et al.*, 1989) and secretion of prostaglandin E_2 by the cells might also influence the development of immunity (Cillari *et al.*, 1986; Farrell and Kirkpatrick, 1987). It should be noted, however, that in contrast to *L. major,* infecting macrophages with *L. donovani* inhibits their ability to produce IL-1 (Reiner, 1987). Thus, different species of *Leishmania* can affect macrophage function in different ways. Finally, once infected, macrophages may vary in their ability to be activated by products of an immune response, such as interferon-γ (IFN-γ) (Buchmuller-Rouiller and Mauel, 1986; Scott and Sher, 1986).

2.1.2. T Cells Are Protective

An important seminal experiment by Howard *et al.* (1980b) examined resistance to infection with *L. major* using mouse radiation-induced bone marrow chimeras. In these experiments, it was shown that the resistance or susceptibility of a mouse was determined exclusively by cells derived from the donor bone marrow. The fact that T cells are involved in resistance to *Leishmania* was most clearly demonstrated by Mitchell and colleagues (Mitchell *et al.*, 1982). In their studies it was shown that athymic nu/nu mice (animals that are exquisitely susceptible to infection with *L. major*) could be protected from infection by adoptive transfer of Lyt-1^+ splenic T cells from euthymic syngeneic mice. Interestingly, when BALB/c nu/nu mice were injected with low numbers (10^6–10^7) of lymphoid cells, the animals became resistant to a subsequent challenge with *L. major*. In contrast, transfer of high numbers ($> 10^8$) of these lymphoid cells yielded mice that were susceptible to infection. The simplest interpretation of these data is that two kinds of parasite-specific T cells exist in the BALB/c mouse. One of these T cells is curative and the other is exacerbative. In the uninfected BALB/c mouse, the curative cell is at a higher frequency. When low numbers of splenic T cells are adoptively transferred from normal BALB/c mice to BALB/c nu/nu mice, too few of the potentially exacerbative T cells are transferred and the cell is effectively inoperative. Thus, when the animal is challenged with *L. major*, the curative cell prevails and protects the mouse. In contrast, when high numbers of T cells are adoptively transferred, the exacerbative cell is somehow able to override the effects of the protective cell, and the animal is susceptible to challenge with the parasite. This hypothesis was in fact advanced by Mitchell and Handman (1985). At the time, however, the differences between the two T-cell types and the mechanism behind this phenomenon were not understood.

Further work by Howard, Liew, and colleagues confirmed that Lyt-1^+2^- T cells mediated host resistance to *L. major* in mice (Liew *et al.*, 1982); yet, T cells with this same surface phenotype were also reported to inhibit parasite-specific cell-mediated immune responses (Liew *et al.*, 1982, 1985a,b; Liew, 1983). These latter cells, termed "suppressor cells" by Howard and Liew, did not have the classical surface phenotype of suppressor cells (i.e., Lyt-1^-2^+, I-J^+), but instead bore the surface phenotype of the helper/inducer T cell lineage (i.e., Lyt-1^+2^-, I-J^-).

Studies by Louis, Titus, and colleagues showed that L3T4^+ (i.e., $CD4^+$) parasite-specific T-cell lines could be generated from mice, which, on adoptive transfer to normal syngeneic recipients, would exacerbate the course of cutaneous leishmaniasis (Titus *et al.*, 1984b). Typical results of such an experiment are shown in Figure 2. Mice that received $CD4^+$ *L. major*-specific T cells followed by challenge with *L. major* developed markedly enhanced cutaneous lesions when compared to control animals. While the results shown in Figure 2 were obtained using genetically susceptible BALB/c mice, the same exacerbative phenomenon could also be observed using genetically resistant CBA/T6 mice (Titus *et al.*, 1985b). In addition, not only was lesion size enhanced, but the number of parasites present in the lesion was found to be greatly elevated (Titus *et al.*, 1985b). Also, $CD4^+$ ovalbumin (OVA)-specific T cells (which are non-cross-reactive with *L. major*) did not exacerbate cutaneous leishmaniasis. Thus,

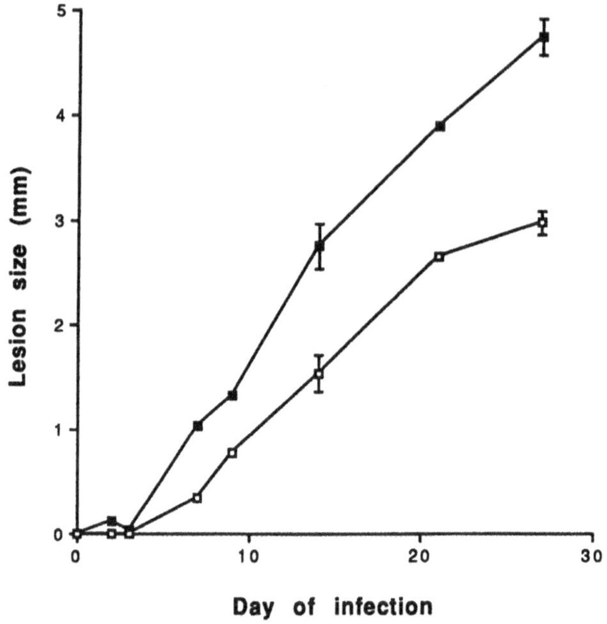

FIGURE 2. Exacerbation of infection with *Leishmania major* by $CD4^+$ parasite-specific T cells. BALB/c mice were sensitized to *L. major* by the injection of *L. major* antigens in complete Freund's adjuvant. A $CD4^+$ parasite-specific T-cell line was derived from the mice by techniques described in detail in Titus *et al.* (1984b). To determine the effect of the T cells on the course of cutaneous leishmaniasis, 10^7 of the cells were injected intravenously into normal BALB/c mice and the animals were challenged 24 hr later with 10^7 *L. major* promastigotes subcutaneously in the hind footpad. Lesions that developed were monitored by determining with a vernier caliper the increase in the thickness of the infected footpad compared to the thickness of the contralateral uninfected footpad. The figure depicts mean lesion size ± SE; ■, T cells; □, control.

exacerbation of cutaneous leishmaniasis was specific for *L. major* T cells. However, the OVA T cells were capable of exacerbating the disease when the animals that received these T cells were challenged with a mixture of *L. major* and OVA (Titus *et al.*, 1984b). This result suggested that (1) at least one step in the exacerbative process was nonspecific and (2) *any* $CD4^+$ T cell might be capable of exacerbating disease provided it was activated in the developing lesion and mimicked the activities of an *L. major*-specific exacerbative T cell. In fact, in light of our current knowledge of cutaneous leishmaniasis (see Section 3), it is likely that the activity that the OVA T cell mimicked was to secrete the same panel of lymphokines that the parasite-specific T cell secreted.

In contrast to the protective role played by T cells in experimental murine leishmaniasis, parasite-specific antibody, which is produced in large quantities by infected mice, appeared to have no effect on the outcome of the disease

(Howard, 1986). The clearest experiments examining the role of parasite-specific antibody in cutaneous leishmaniasis were those of Sacks *et al.* (1984) and Scott *et al.* (1986). In these experiments genetically resistant mice could heal cutaneous lesions of *L. major* despite the fact that the ability of the mice to produce parasite-specific antibody was completely suppressed. Moreover, in genetically susceptible mice, elimination of B cells was beneficial for the mice, implying that B cells might play a role in inducing parasite-specific T cells that are detrimental to the host.

Taken together, these early studies led to the following conclusions:

1. Innate susceptibility to *Leishmania* is governed by genes lying outside the H-2 region. These genes may encode for various macrophage functions. Later in the course of disease, macrophages are involved in triggering specific T cells in an MHC-restricted fashion.
2. Bone-marrow-derived cells determine the outcome of infection in experimental murine cutaneous leishmaniasis. At least T cells, B cells, and macrophages are involved in this process; however, the exact role of other bone-marrow-derived cells has not been thoroughly investigated. For example, it is interesting that *Leishmania* parasites are rapidly destroyed by degranulating polymorphonuclear leukocytes (Chang, 1981; Pearson and Steigbigel, 1981), but the role of these cells in the pathogenesis of leishmaniasis has not been thoroughly analyzed.
3. Parasite-specific T cells can be induced that are either detrimental or protective for the host. These T cells have similar surface phenotypes.
4. Antibody, although produced in large quantities by infected mice, does not appear to play a protective role. In genetically susceptible mice, parasite-specific B cells may exacerbate disease. The role of B cells as antigen-presenting cells in the disease has not been investigated.

2.2. $CD4^+$, $CD8^+$, and $\gamma\delta$-TCR^+ T Cells in Leishmaniasis

2.2.1. $CD4^+$ and $CD8^+$ T Cells

In order to investigate the potential role of $CD4^+$ and $CD8^+$ T cells in experimental murine cutaneous leishmaniasis, *L. major*-infected mice were treated with monoclonal antibody (mAb) specific for either the $CD4^+$ or $CD8^+$ T-cell subsets. Typical results of such an experiment are depicted in Figure 3. When genetically susceptible BALB/c mice were treated with an anti-CD4 mAb and infected with *L. major*, the course of infection in the mice was far less severe (Figure 3A) (Titus *et al.*, 1985a). In fact, the animals ultimately healed their cutaneous lesions and became resistant to reinfection (Titus *et al.*, 1988). This therapeutic effect on cutaneous leishmaniasis was obtained when 60 to 80% of all

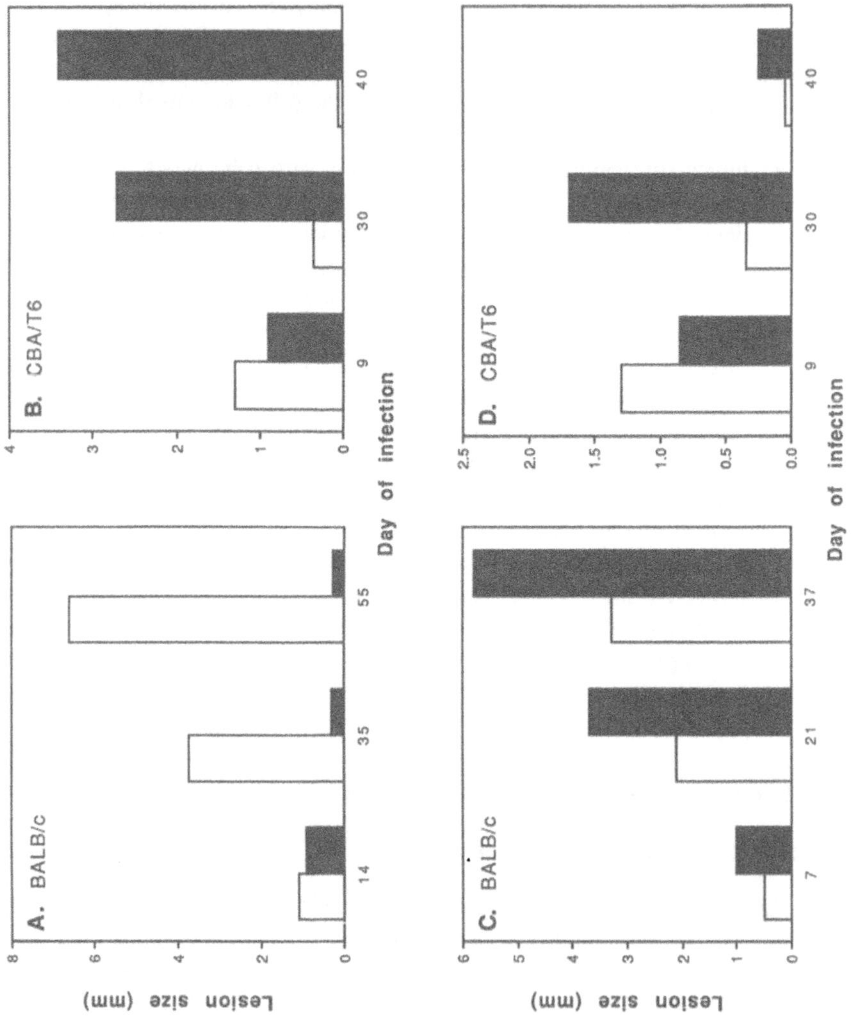
A. BALB/c
B. CBA/T6
C. BALB/c
D. CBA/T6
Lesion size (mm)
Lesion size (mm)
Day of infection
Day of infection
14
35
55
9
30
40
7
21
37
9
30
40
0
2
4
6
8
0
1
2
3
4
0
1
2
3
4
5
6
0.0
0.5
1.0
1.5
2.0
2.5

$CD4^+$ cells were depleted from the mice. If > 95% of $CD4^+$ T cells were depleted from the animals, the treated mice were once again exquisitely susceptible to infection with *L. major* (Titus *et al.*, 1988). These results suggest that a parasite-specific $CD4^+$ T cell is induced in BALB/c mice that can cause exacerbation of the infection. However, there is also a parasite-specific $CD4^+$ T cell present in the mice that is involved in protection since depletion of nearly 100% of all $CD4^+$ T cells results in a mouse fully susceptible to *L. major*. As a final note regarding the experiments discussed in this paragraph and in the rest of this section, in all of the experiments depicted in Figure 3 both lesion size and the number of parasites present in the lesion were determined (Titus *et al.*, 1985b). In all cases, parasite numbers directly correlated with lesion size (Titus *et al.*, 1985a, 1988).

Experiments were also conducted in which genetically resistant mice were treated with anti-CD4 mAb. As can be seen in Figure 3B, treating CBA/T6 mice with an anti-CD4 mAb made the animals fully susceptible to infection (Titus *et al.*, 1987). The results depicted in Figure 3B were obtained with mice in which > 95% of all $CD4^+$ T cells were eliminated. In some experiments the animals were only partially depleted of $CD4^+$ T cells. In these mice, lesions healed at a slower rate but the animals did resolve their infection (Louis *et al.*, 1986). This result is different than the result obtained in susceptible BALB/c mice in which partial depletion of $CD4^+$ T cells was beneficial for the mice, presumably due to the elimination of an exacerbative $CD4^+$ T cell (see above). The simplest interpretation of the results discussed thus far is that in genetically susceptible BALB/c mice infected with *L. major*, the predominant $CD4^+$ T cell that is elicited is one that is exacerbative. In contrast, in the genetically resistant CBA/T6 mouse infected with *L. major*, the predominant $CD4^+$ T cell is a protective cell. This is not meant to imply, however, that exacerbative $CD4^+$ T cells do not exist in the CBA/T6 mouse. Such cells have been isolated from CBA/T6 mice and these T cells are identical phenotypically and functionally with exacerbative $CD4^+$ T cells isolated from BALB/c mice (see Section 2.1.2. and Titus *et al.*, 1985b).

BALB/c and CBA mice were also treated with an anti-CD8 mAb and infected with *L. major* to determine the role of $CD8^+$ T cells in experimental

---

FIGURE 3. The effect of depleting $CD4^+$ or $CD8^+$ T cells from mice infected with *Leishmania major*. Genetically susceptible BALB/c or genetically resistant CBA/T6 mice were infected with *L. major* and then treated with an anti-CD4 or CD8 antibody to determine the effect of depleting these T-cell subsets on the course of cutaneous leishmaniasis. For details of the exact protocols used, see Titus *et al.* (1985a, 1987). (A) BALB/c mice depleted of 70% of their $CD4^+$ T cells; (B) > 95% depletion of $CD4^+$ T cells in CBA/T6 mice; (C) 65% depletion of $CD8^+$ T cells; (D) 85% depletion of $CD8^+$ T cells. Lesions were monitored as described in the legend of Figure 2. Symbols: □, control; ■ (A,B), anti-CD4, (C,D), anti-CD8. The data presented here are modified from Titus *et al.* (1985a, 1987).

murine cutaneous leishmaniasis (Titus *et al.*, 1987). As can be seen in Figure 3C,D, depletion of $CD8^+$ cells in either mouse was detrimental for the course of infection. However, the effect of the anti-CD8 treatment was not nearly as dramatic as the effect of anti-CD4 treatment. Treatment with an anti-CD4 antibody could convert genetically susceptible mice into resistant mice and genetically resistant mice into susceptible mice (Figure 3A,B). When genetically susceptible mice were treated with anti-CD8, the animal developed lesions that were somewhat larger (approximately 1.5 times) than control lesions (Figure 3C). The genetically resistant mouse treated with anti-CD8 mAb developed more severe lesions, especially around 30 days of infection (Figure 3D). However, despite severe depletion (85%) of $CD8^+$ cells, the resistant mouse still healed its lesions (Figure 3D) (Titus *et al.*, 1987). Therefore, the overall role of parasite-specific $CD8^+$ T cells in experimental murine cutaneous leishmaniasis is not fully understood. Hill *et al.* (1989) have reported that $CD8^+$ T cells can play an important protective role in susceptible BALB/c mice. In addition, $CD8^+$ cells have been shown to play a role in vaccine-induced immunity against *L. major* (Farrell *et al.*, 1989).

In conclusion, it is clear that $CD4^+$ parasite-specific T cells can both mediate protection in mice infected with *L. major* and can cause progression of disease. In Section 3, the possible mechanisms for selective activation of these cells and their mode of action will be discussed.

2.2.2. γδ-T-Cell Receptor (TCR) T Cells

It has recently been reported that 20% of $CD3^+$ cells in the lesions of patients with localized American cutaneous leishmaniasis are γδ-TCR-bearing T cells (Modlin *et al.*, 1989). This observation suggests that γδ-TCR T cells may play a role in the pathogenesis of cutaneous leishmaniasis. In preliminary experiments to investigate this possibility, we have injected both genetically susceptible (BALB/c) and genetically resistant (C57BI/6) mice with a pan-anti-γδ-TCR mAb (UC7-13D5, a gift from Dr. Jeffrey Bluestone, University of Chicago). Mice were injected with the pan-anti-γδ-TCR mAb (5 μg mAb subcutaneously in the hind footpad) 24 hr before challenging the same site with *L. major*. Additional injections of the antibody were given at 4-day intervals thereafter. Results in Figure 4 show that little effect on lesion development was seen as a result of this treatment. These results were obtained using susceptible BALB/c mice; similar results were seen using resistant C57BI/6 mice. We are currently investigating whether injection of larger amounts of the anti-γδ-TCR mAb might have an effect on lesion development and whether treatment of mice in this fashion effectively depletes γδ-TCR-bearing cells from the lesion site.

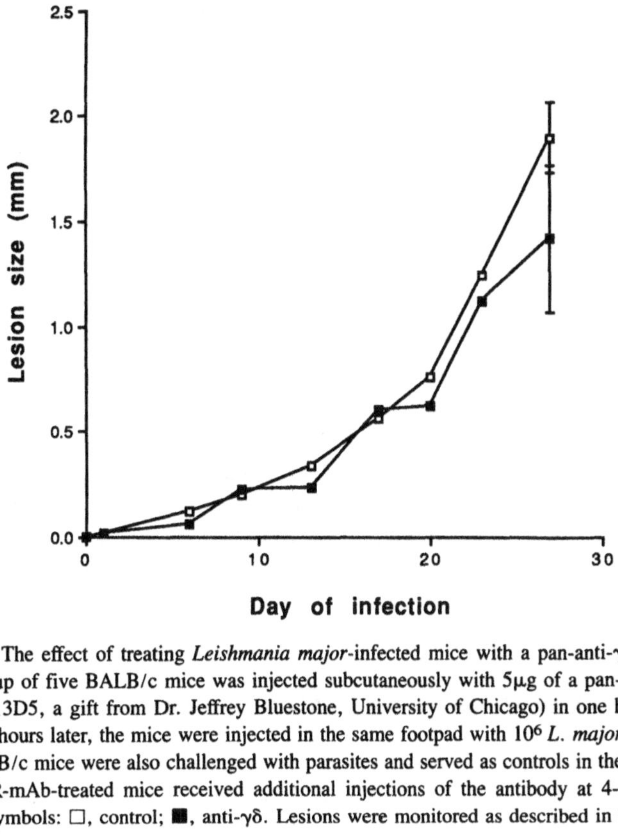

FIGURE 4. The effect of treating *Leishmania major*-infected mice with a pan-anti-γδ TCR antibody. A group of five BALB/c mice was injected subcutaneously with 5μg of a pan-anti-γδ-TCR mAb (UC7-13D5, a gift from Dr. Jeffrey Bluestone, University of Chicago) in one hind footpad. Twenty-four hours later, the mice were injected in the same footpad with 10^6 *L. major*. A group of normal BALB/c mice were also challenged with parasites and served as controls in the experiment. Anti-γδ-TCR-mAb-treated mice received additional injections of the antibody at 4-day intervals thereafter. Symbols: □, control; ■, anti-γδ. Lesions were monitored as described in the legend of Figure 2.

3. CYTOKINES AND LEISHMANIASIS

3.1. Introduction

Mosmann, Coffman, and colleagues have recently shown that $CD4^+$ T cells may, in fact, consist of subpopulations that secrete different cytokines. The clearest definition of these subsets has emerged from analyses of mouse T-cell clones, many of which can be divided into two nonoverlapping groups. One, called Th1, produces IL-2, IFN-γ, and lymphotoxin and is most important for cell-mediated immunity, and the other subset, called Th2, produces IL-4, IL-5, and IL-6 and plays a major role in the induction of humoral immunity (Bottomly, 1988; Swain *et al.*, 1988; Mosmann and Coffman, 1989). Such distinctions may not apply equally to all species, since the majority of human T-cell clones produce combinations of lymphokines that do not permit classification into Th1 and Th2 subsets (Maggi *et al.*, 1988; Paliard *et al.*, 1988). Moreover, even in mice, precursors of T-cell clones or recently activated T cells may produce various combinations of cytokines (Mosmann and Coffman, 1989). Nevertheless, in a given disease process, preferential induction of $CD4^+$ T cells that secrete distinct sets of cytokines may contribute to the outcome of that infection.

As detailed above, $CD4^+$ parasite-specific T cells appear to be capable of either ameliorating or exacerbating experimental murine cutaneous leishmaniasis. The distinction between these cells may be the antigen each cell recognizes and/or the cytokine the cell is triggered to secrete upon antigen activation. Thus, cutaneous leishmaniasis offers an ideal model system to test whether Th1- or Th2-type T cells plays a role in determining the outcome of the disease. In addition, the availability of recombinant cytokines and neutralizing anticytokine antibodies make it feasible to administer these biologic response modifiers to alter the outcome of diseases, including infections such as leishmaniasis. Recently, a number of laboratories, including our own, have begun to examine the role of $CD4^+$ T-cell subsets and various cytokines in the course of cutaneous leishmaniasis.

3.2. Role of Interferon-γ, Interleukin-2, and Interleukin-4

A number of groups have addressed the question of the role of IFN-γ, IL-4, and IL-2 in cutaneous leishmaniasis. Solbach *et al.* (1987) reported that a progressive impairment of the response to and the ability to generate IL-2 occurred in genetically susceptible mice infected with *L. major*. In contrast, this did not occur in genetically resistant mice. Essentially simultaneously, Scott *et al.* (1988) reported on the effect of adoptively transferring $CD4^+$ parasite-specific T-cell lines that secreted different panels of lymphokines and Heinzel *et al.* (1989) reported on the levels of cytokine-specific mRNA produced by mice infected with *L. major*. In the work of Scott *et al.* (1988), *L. major*-specific T-cell lines were derived by sensitizing mice to chromatographically separated fractions of a soluble preparation of parasite antigens. A $CD4^+$ parasite-specific T-cell line derived from susceptible BALB/c mice sensitized to one antigen fraction (9) secreted IFN-γ and IL-2, but not IL-4. This cell protected when adoptively transferred to sublethally irradiated BALB/c mice—an animal that is otherwise susceptible to infection. In contrast, a cell line obtained from mice sensitized to antigen fraction 1 produced only IL-4 and exacerbated infection when adoptively transferred. In recent work, this same approach was performed using $CD4^+$ T-cell clones derived from the T-cell lines and the same results were obtained (Scott *et al.*, 1990).

In the work of Heinzel *et al.* (1989), mice that were genetically resistant to infection with *L. major* (C57Bl/6) and those genetically susceptible to infection (BALB/c) were infected with the parasite. At intervals thereafter, the lymph nodes draining the lesion site were examined for the presence of mRNA specific for IFN-γ, IL-2, or IL-4. It was found that infected C57Bl/6 mice produced high amounts of IFN-γ-specific mRNA, moderate amounts of IL-2-specific mRNA, and no IL-4-specific mRNA. In contrast, susceptible BALB/c mice produced low amounts of IFN-γ-specific mRNA, moderate amounts of IL-2-specific mRNA, and high amounts of IL-4-specific mRNA. Taken collectively, these results show that production of IFN-γ in response to infection with *L. major* correlates with the ability of the infected host to heal whereas production of IL-4 correlates with progression of the disease.

In addition to these studies, Muller and Louis (1989) reported that protective $CD4^+$ *L. major*-specific T-cell clones could be derived from the spleens of susceptible BALB/c mice that had been infected with the parasite intravenously. Interestingly, when an anti-IFN-γ-specific mAb was administered to mice that had received these T cells, the protective effect of the T cell was rapidly abrogated. These observations corroborate the results discussed above and argue that IFN-γ plays a protective role in experimental murine cutaneous leishmaniasis.

In studies conducted in this laboratory, we have examined the cytokines released by $CD4^+$ T cells isolated from mice infected with *L. major* (Boom *et al.*, 1990). Resistant and susceptible mice were infected with *L. major* and at intervals thereafter the lymph nodes draining the lesion were removed and restimulated *in vitro* with *L. major* parasites. Four days later, the T-cell blasts were recovered from this primary culture and the cells were rested for 5 to 10 days on irradiated spleen cell feeder layers. T-cell blasts isolated from these rested cultures were parasite-specific and were composed of 75 to 80% $CD4^+$ and < 5% $CD8^+$. These T cells were then restimulated with parasites and the supernatants of the cultures were analyzed for the presence of IFN-γ, IL-2, and IL-4. The resistant mouse that was used in these studies was the B10.D2 mouse (H-2^d haplotype), which is congenic with BALB/c mice (H-2^d haplotype). The use of congenic mouse strains made it possible to restimulate T cells derived from resistant or susceptible mice with either syngeneic or congenic antigen-presenting cells (APC). For example, resistant B10.D2 $CD4^+$ parasite-specific T cells could be restimulated with either resistant B10.D2 APC or with susceptible BALB/c APC. The purpose of these experiments was to determine whether the APC used *in vitro* would alter the profile of cytokines secreted by the T cells. The results of these studies may be summarized as follows:

1. Parasite-specific $CD4^+$ T cells isolated from B10.D2 mice produced primarily IFN-γ and IL-2 in response to restimulation with *L. major in vitro*. These cells also produced low levels of IL-4.

2. In contrast, parasite-specific CD4$^+$ T cells derived from susceptible BALB/c mice produced more IL-4, less IL-2, and little IFN-γ in response to stimulation with the parasite *in vitro*.
3. These patterns of lymphokine secretion were the same whether the T cells were cultured *in vitro* on syngeneic or congenic APC. Thus, the profile of cytokines secreted by the T cells could not be altered by the APC used to restimulate the T cells *in vitro*. However, since the T cells were originally isolated from *L. major*-infected mice, it is possible that the T cells became committed to secreting a certain set of lymphokines in the infected animal and this could not be altered by restimulating the cells with congenic APC *in vitro*.

Interferon-γ has long been known to be able to activate macrophages to kill *Leishmania in vitro* (Murray *et al.*, 1983; Titus *et al.*, 1984a; Nacy *et al.*, 1985). However, the role of IFN-γ in cutaneous leishmaniasis may be much more complex than simply activation of macrophages. Belosevic *et al.* (1989) have reported that treating *L. major*-infected genetically resistant mice with a neutralizing anti-IFN-γ antibody abolishes their resistance to infection. However, this effect was seen only when the mice were treated with anti-IFN-γ beginning in the first week of infection—treatment beyond the first week had no effect. Interferon-γ is known to have an antiproliferative effect on many cell types, including Th2-type CD4$^+$ T cells (Gajewski *et al.*, 1989). Therefore, it is possible that production of IFN-γ by mice infected with *L. major* influences the character of the immune response that develops against the parasite; that is, IFN-γ may prevent the outgrowth of parasite-specific Th2-type T cells. If this were the case, this would also inhibit IL-4 production since Th2-type T cells are a major source of IL-4. Inhibition of IL-4 production could, in turn, be beneficial for the animals since the work of Scott *et al.* (1988) and Heinzel *et al.* (1989) discussed in Section 3.1.1 showed that IL-4 production by mice infected with *L. major* correlated with progression of cutaneous leishmaniasis. In fact, Scott *et al.* (1989) have recently shown that if susceptible BALB/c mice are injected with a mixture of a soluble preparation of *L. major* antigens plus IFN-γ and are subsequently challenged with the parasite, the course of lesion development in the animals is less severe than in mice preinjected with antigen alone or in mice that were left untreated. This result argues that in addition to activating macrophages to destroy *L. major* in infected mice, IFN-γ may influence the character of the immune response that develops toward the parasite such that a protective T-cell response is preferentially elicited.

It should be stressed, however, that there are likely to be other cytokines in addition to IFN-γ that play a role in resolution of experimental murine cutaneous leishmaniasis. Sadick *et al.* (1990) have shown that when genetically susceptible BALB/c mice were treated with IFN-γ and infected with *L. major*, the treatment had no effect on the disease outcome. This result was probably not due to treating with insufficient amounts of IFN-γ; the dose of IFN-γ used approached a level that was toxic for the mice and it was a dose that had been used to successfully alter the course of infection in mice infected with other pathogens. These and other observations led Sadick *et al.* (1990) to propose that there is a non-IFN-γ–mediated mechanism for healing in cutaneous leishmaniasis. Recent evidence suggests that tumor necrosis factor (TNF) and perhaps other cytokines may be involved in this non-IFN-γ mechanism for healing. These experiments will be discussed further in Sections 3.3 and 3.4. In addition, we have recently derived clonal $CD4^+$ parasite-specific T cells from a $CD4^+$ T-cell line that exacerbates experimental murine cutaneous leishmaniasis (Titus *et al.*, 1984b) (see Section 2.1.2). These clonal lines also exacerbate cutaneous leishmaniasis when adoptively transferred to normal syngeneic recipients. Typical results for one T-cell clone are shown in Figure 5 (Titus *et al.*, 1991). When stimulated with parasite

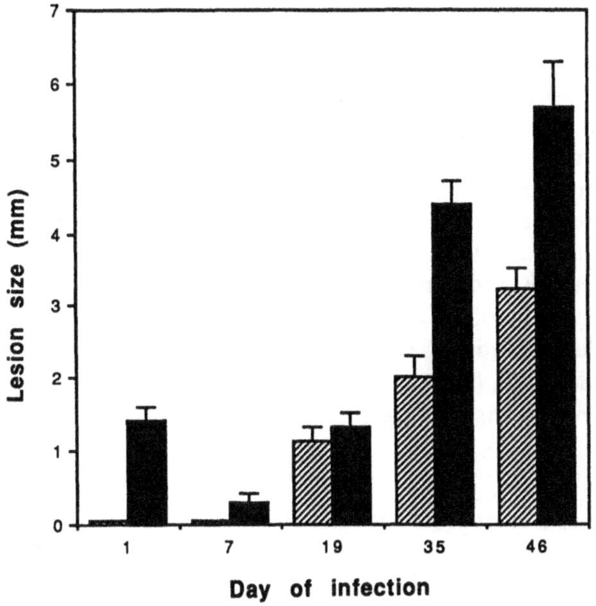

FIGURE 5. Exacerbation of infection with *Leishmania major* by a $CD4^+$ parasite-specific T-cell clone. Groups of five BALB/c mice each were injected subcutaneously in the hind footpad with 10^7 *L. major* promastigotes or with a mixture of 10^7 *L. major* and 2×10^6 *L. major*-specific $CD4^+$ T-cell clone Bd12. Symbols: ▨, control; ■, T cells. Methods used for deriving parasite-specific T-cell clones have been described in detail elsewhere (Louis *et al.*, 1982). Lesion development was followed as described in the legend in Figure 2.

antigens *in vitro,* these T-cell clones produced large amounts of IFN-γ, moderate amounts of IL-2, and no IL-4 and therefore belong to the Th1 subset of the mouse $CD4^+$ T-cell lineage. These results also suggest, as mentioned, that IFN-γ may not be the only cytokine responsible for resistance in experimental murine cutaneous leishmaniasis and that other cytokines (TNF, etc.) may also be involved.

In all of the studies previously described, IL-4 production was consistently shown to correlate with susceptibility to infection with *L. major*. The mechanism by which IL-4 causes progression of experimental murine cutaneous leishmaniasis is not fully understood. In fact, in some systems IL-4 can act as an activator of macrophages (Crawford *et al.*, 1987; Stuart *et al.*, 1988) and there has been one report of IL-4 having a therapeutic effect on the outcome of leishmaniasis (Carter *et al.*, 1989). Hopefully, as the effects of IL-4 on macrophage function are elucidated, it will become clearer how IL-4 production can be detrimental in leishmaniasis. For example, it has been recently reported that IL-4 can inhibit H_2O_2 (Lehn *et al.*, 1989), superoxide (Abramson and Gallin, 1990), and IL-1 and TNF (Essner *et al.*, 1989) production by human monocytes. In addition, IL-4 can inhibit IFN-γ production in human mixed lymphocyte cultures (Vercelli *et al.*, 1990). All of these effects may contribute to the additional observation that whereas treatment of human monocytes with IFN-γ activates the cells to destroy *L. donovani,* treating the cells with a combination of IFN-γ and IL-4 does not lead to activation (Lehn *et al.*, 1989). The same effect of IL-4 can be seen using murine macrophages infected with *L. major* if the macrophages are pretreated with IL-4 (Scott *et al.*, 1989). Interestingly, when *L. major*-infected macrophages were pretreated with a combination of IFN-γ and IL-4 (Belosevic *et al.*, 1988), or if IL-4 was added to the macrophages after IFN-γ (Scott *et al.*, 1989), IL-4 enhanced the ability of IFN-γ to activate the macrophages. Thus, depending on the source of the macrophage, the parasite used and the timing of the addition of IL-4, this lymphokine has different effects on macrophage function. This fact, as well as other unexplored aspects of the mode of action of IL-4, may explain how IL-4 can be protective in some systems, yet detrimental in leishmaniasis.

3.3. Role of Tumor Necrosis Factor

We have been investigating the role of TNF in experimental murine cutaneous leishmaniasis (Titus *et al.*, 1989). When genetically resistant (C3H) or susceptible (BALB/c) mice were treated with recombinant human TNF (rHuTNF) and challenged with *L. major* promastigotes, the cutaneous lesions that developed were significantly smaller and contained fewer parasites than lesions on untreated control mice. In addition, the opposite result was obtained when mice were treated with a neutralizing rabbit anti-TNF antibody: lesions

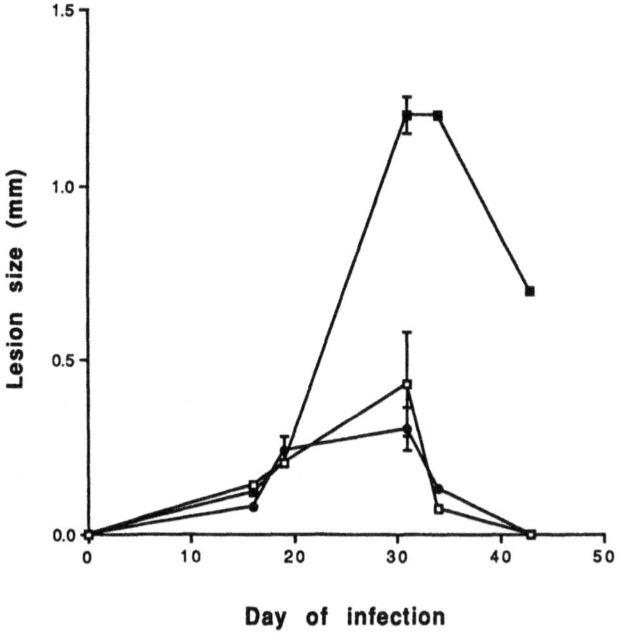

FIGURE 6. Effect of treating *Leishmania major*-infected mice with a neutralizing anti-tumor necrosis factor antibody. Groups of five C3H mice each were injected subcutaneously in one hind footpad with 2×10^5 *L. major*. Seven days later, the mice were injected intraperitoneally with 200 μg of a neutralizing rabbit anti-mouse TNF. Injections with anti-TNF (■) were repeated every 2 days thereafter. Control mice received an equivalent amount of normal rabbit immunoglobulin (●) or were left untreated (□). Lesions were monitored as described in the legend of Figure 2.

were larger and contained greatly elevated numbers of *L. major*. Typical results of such an experiment are depicted in Figure 6. It should be mentioned that results similar to these have also been recently reported by Liew *et al.* (1990).

We also investigated the ability of mice infected with *L. major* to produce TNF during the course of infection. We infected genetically resistant and susceptible mice with *L. major* and the lymph node cells draining the lesion were recovered at varying intervals of infection and restimulated with *L. major in vitro*. It was found that lymph node cells isolated from resistant mice produced large amounts of TNF, especially at the time cutaneous lesions were resolving on the mice. In contrast, susceptible mice produced low levels of TNF throughout the disease. Thus, the ability of the host to produce TNF in response to infection correlated with the genetic resistance/susceptibility of the host (Titus *et al.*, 1989).

We have also been examining the possible mechanism of action of the protective effect of TNF in cutaneous leishmaniasis. We found that although TNF had no direct effect on the parasite itself, TNF activated macrophages infected with *L. major* to destroy the parasite *in vitro* (Theodos *et al.*, 1991). Both purified native murine TNF and rHuTNF could activate macrophages; however, the dose of TNF used to activate the macrophages was critical. When present in the macrophage cultures at a final concentration of 100 to 200 units/ml, TNF effectively activated the macrophages to kill *L. major*. In contrast, when concentrations in excess of 1000 units/ml were used, parasite replication within the macrophages was initially inhibited. However, at later time points, the number of parasites within the macrophages was no different than in control cultures. While the exact mechanism underlying these observations is not known, it is possible that excessive amounts of TNF can inhibit a macrophage function(s) that is important in destruction of the parasite. Thus, in leishmaniasis, TNF could be a double-edged sword: TNF could be protective when produced in moderate amounts but detrimental when present in high amounts. In fact, TNF is known to have cytotoxic effects on various cell types, and when produced in large quantities by the host, TNF can induce a wasting syndrome (cachexia) which can lead to death of the host (Beutler and Cerami, 1988).

Finally, TNF appears to play a protective role in one other system in which the mechanisms of immunity to *L. major* have been analyzed. In the work of Heinzel *et al.* (1989) and Sadick *et al.* (1990), genetically susceptible BALB/c mice were treated with an anti-IL-4 mAb to determine whether this treatment would be therapeutic for the course of cutaneous leishmaniasis in the animals. As discussed in Section 3.2, IL-4 production by the host consistently associates with susceptibility to infection with *L. major*. Heinzel and Sadick found that treatment with anti-IL-4 antibody allowed BALB/c mice to heal cutaneous lesions of *L. major*. They next reasoned that the mechanism by which anti-IL-4 treatment protected BALB/c mice was through its ability to permit IFN-γ production in the treated animals. Thus, they treated BALB/c mice with a combination of anti-IL-4 and anti-IFN-γ antibody and predicted that the animals would now be susceptible to infection with *L. major*. Contrary to what they predicted, the animals were still resistant. In recent experiments (R. Locksley, personal communication), BALB/c mice were treated with a combination of anti-IL-4 and anti-TNF antibody and the resistance conferred by anti-IL-4 treatment was totally abolished by the anti-TNF antibody. This indicates that in this system for inducing resistance in BALB/c mice to infection with *L. major*, TNF plays a pivotal role.

3.4. Role of Other Cytokines

The role of other cytokines in experimental murine cutaneous leishmaniasis is also being investigated. A brief summary of this work follows.

3.4.1. Interleukin-1

Genetically susceptible mice produce large amounts of IL-1 in response to infection with *L. major* (Cillari *et al.*, 1989). Production of IL-1 by susceptible mice may be important in determining the nature of the parasite-specific T-cell response generated in the animals. As mentioned above, the response of genetically susceptible mice is predominantly of the Th2 type and IL-1 is an important co-factor for activation of Th2 cells (Kurt-Jones *et al.*, 1987; Lichtman *et al.*, 1988).

3.4.2. Interleukin-2

In addition to the experiments discussed above regarding IL-2 production by mice infected with *L. major* (see Section 3.1.1), it has been reported that IL-2 can enhance replication of *L. mexicana amazonensis* promastigotes *in vitro* and can exacerbate lesions induced with the parasite in mice (Mazingue *et al.*, 1989). It is not known whether this same result is obtained with *L. major*.

3.4.3. Interleukin-3

Interleukin-3 has been shown to be produced by mice susceptible to infection with *L. major* and to also exacerbate infection with *L. major* in mice (Feng *et al.*, 1988; Lelchuk *et al.*, 1988). The exact mechanism behind this effect is not known, but may be related to the ability of IL-3 to increase the number of immature monocytes present in the developing lesion. Such cells may be "safe targets" (i.e., permissive cells) in which the parasite can replicate, a hypothesis advanced by Mirkovich *et al.* (1986).

3.4.4. Granulocyte Monocyte-Colony Stimulating Factor

Granulocyte monocyte-colony stimulating factor (GM-CSF) has been shown in various settings to either have a beneficial (Handman and Burgess, 1979), detrimental (Greil *et al.*, 1988), or no effect (Corcoran *et al.*, 1988) on experimental murine cutaneous leishmaniasis. More work is required to elucidate its exact role in the disease.

3.4.5. Migration Inhibition Factor

Human migration inhibition factor (MIF) has been recently cloned (Weiser *et al.*, 1989). Functional studies have revealed that MIF is able to activate human monocytes to destroy *L. donovani in vitro* (Pozzi *et al.*, 1990). In addition, MIF is also a potent adjuvant in mice (Weiser *et al.*, 1990) and, thus, if the molecule is produced in leishmanial lesions, it may dramatically influence the character of the immune response engendered against the parasite. In fact, we have preliminary evidence that protective parasite-specific CD4$^+$ T cells secrete MIF activity when the cells are activated with the parasite *in vitro*. The cells that produce MIF are cloned *L. major*-specific T cells that were isolated from genetically resistant C3H mice.

3.4.6. Macrophage Inflammatory Proteins 1 and 2

Macrophage inflammatory proteins 1 and 2 (MIP 1 and 2), which are products of macrophages (Wolpe and Cerami, 1989), are proteins that are capable of eliciting many aspects of an inflammatory response when they are injected into the skin of mice or rabbits. Recently, in collaboration with B. Sherry and A. Cerami (Rockefeller University), we have been investigating the effect of MIP 1 and 2 on killing of *L. major* by murine macrophages. Both MIP 1 and 2 are potent activators of macrophages in this assay system. We are currently studying MIP 1 and 2 further to determine their role in infection with *L. major*.

4. VACCINATION

Several antigens of *L. major* are being studied as potential vaccine candidates. Among these are the LPG molecule of the parasite, gp63, and other potential protective antigens that have been characterized as antigens recognized by protective T cells. We have been analyzing an avirulent *L. major* parasite that is able to induce a potent state of immunity even in genetically susceptible BALB/c mice. In collaboration with Dr. Thierry Boon (International Institute of Cellular and Molecular Pathology, Brussels, Belgium), stable avirulent clones of *L. major* were generated by chemical mutagenesis with nitrosoguanidine. When 10^6 of these mutagenized parasites were injected subcutaneously into normally highly susceptible BALB/c mice, no cutaneous lesions developed. In fact, *in vitro* these parasites were spontaneously destroyed by the macrophages within 72 to 96 hr following phagocytosis in the absence of activating signals such as IFN-γ (Marchand *et al.*, 1987). Furthermore, vaccination of genetically susceptible BALB/c mice with avirulent parasites (vir$^-$79) resulted in the induction of immunity to challenge with virulent *L. major* as evidenced by a reduction in lesion size and parasite numbers within the lesion (McGurn *et al.*, 1990). Typical results of such an experiment are depicted in Table II. Finally, we also characterized the cell responsible for protection. Spleen cells from vaccinated mice were depleted of CD4$^+$ or CD8$^+$ T cells by treatment with CD4-specific or CD8-specific monoclonal antibody plus complement. The cells were then adoptively transferred to normal syngeneic recipients and these animals were chal-

Table II
Lesion Size and Number of Parasites Present in the Lesions of BALB/c Mice Vaccinated with Avirulent *Leishmania major* and Challenged with Virulent Parasites[a]

	Lesion size, mm ± SE		Number of parasites	
Day of infection	Control	Immune	Control	Immune
7	0 ± 0	0 ± 0	0.1×10^6	0.3×10^6
17	0.04 ± 0.05	0.14 ± 0.12	73.7×10^6	2.4×10^6
27	3.16 ± 0.14	1.06 ± 0.27	209×10^6	1.4×10^6
48	4.16 ± 0.14	1.00 ± 0.13	884×10^6	1.1×10^6

[a]Groups of 4 BALB/c mice each were injected with 10^7 vir$^-$79 intravenously. One week later, these mice and normal control mice were challenged with 10^6 virulent *L. major* subcutaneously in the hind footpad. Lesions that developed were monitored by determining with a vernier caliper the increase in the thickness of the infected footpad compared to the thickness of the contralateral uninfected footpad. Lesion size is given as the mean ± SE. In addition to lesion size, the table gives the number of parasites in the lesion as determined by a limiting dilution assay (Titus *et al.*, 1985b). Data modified from McGurn *et al.* (1990)

lenged with virulent *L. major*. This experiment revealed that protection could be adoptively transferred only with $CD4^+$ T cells from the vaccinated animals (McGurn *et al.*, 1990).

We have recently been characterizing the avirulent parasite (vir$^-$79) in order to attempt to determine the reason for its loss of virulence. In addition, we have derived T-cell lines from the vaccinated animals in order to determine the basis for the immunizing effect of the avirulent parasite. We have examined vir$^-$79 for its ability to produce two important surface molecules of the parasite: gp63 and the LPG molecule. Parasites were biosynthetically labeled with [^{3}H]myristate followed by sodium dodecyl sulfate-polyacrylamide gel electrophoresis autoradiography (Bordier, 1987). Using this method, gp63 was found to be present in vir$^-$79 at levels equivalent to that found in virulent clones of *L. major*. We next examined whether the gp63 present on vir$^-$79 was fully functional. Since gp63 exhibits protease activity, *L. major* promastigotes will degrade azocasein (Etges *et al.*, 1986). Using this assay, we found no difference between vir$^-$79 and virulent clones of *L. major*.

As previously mentioned, the infectivity of *Leishmania* promastigotes is associated with the ability of the parasites to transform into an infective metacyclic stage during the stationary phase of growth. It is known that there is a change in the LPG molecule that is expressed on infective metacyclic parasites, which results in the loss of agglutination of the parasites by the lectin peanut agglutinin (PNA) (Puentes *et al.*, 1988). Since an inability to undergo metacyclogenesis may explain the loss of virulence of vir$^-$79, we examined the ability of the parasite to transform into the metacyclic stage. Our results demonstrated that a very small percentage (1 to 2%) of the stationary phase vir$^-$79 parasites were PNA$^-$ compared to 30 to 50% of the stationary-phase virulent *L. major* parasites. Interestingly, when these PNA$^-$ vir$^-$79 parasites were injected subcutaneously into genetically susceptible BALB/c mice, these parasites were poorly infective and, in fact, elicited resistance in the animals to a subsequent challenge with virulent *L. major*. This latter result suggests that although PNA$^-$ vir$^-$79 parasites express an LPG molecule that is not agglutinable with PNA, this molecule is not identical to the LPG molecule present on virulent metacyclic *L. major* parasites. In conclusion, the only defect that we are aware of in vir$^-$79 is an inability to express the virulent form of the LPG molecule. This defect may be the explanation for the ability of the parasite to vaccinate mice against virulent *L. major*.

We have recently been able to derive CD4$^+$ parasite-specific T-cell lines from these immunized BALB/c mice that are capable of adoptively transferring a substantial degree of protection to normal syngeneic recipients as evidenced by both reduced lesion size (approximately a fourfold reduction in lesion size) and a greatly reduced parasite burden in the lesions (approximately a 20-fold reduction in parasite numbers) (C. M. Theodos, P. Kimsey, A. Shankar, L. Hall, and R. G. Titus, manuscript submitted for publication). The protective CD4$^+$ parasite-specific T cells secrete more IL-2 than IL-4, a finding similar to that of Scott *et al.* (1988). However, these cells are unique in that they secrete low levels of IFN-γ but high levels of TNF. This is interesting since, as discussed in Section 3.3, TNF is host protective in murine cutaneous leishmaniasis and likely mediates this protective effect through its ability to activate macrophages to destroy *L. major*.

Finally, the protective CD4$^+$ parasite-specific T cells that we have derived from BALB/c mice immunized with avirulent parasites proliferate *in vitro* in response to living promastigotes but not to frozen and thawed promastigote antigen preparations. These observations are in agreement with a recent report of Muller and Louis (1989) in which they described CD4$^+$ parasite-specific protective T cells that also did not proliferate in response to *Leishmania* antigen but would respond to the living parasite. In contrast, the CD4$^+$ parasite-specific T cells that we have described previously that have an exacerbative effect on the course of cutaneous leishmaniasis proliferate strongly in response to soluble antigen preparations of the parasite (Titus *et al.*, 1984a). There are two simple interpretations of these data. One is that our protective T cells are recognizing antigens expressed only by living promastigotes. Alternatively, since stimulation of these T cells *in vitro* is performed with APC, which includes macrophages, living promastigotes would transform rapidly into amastigotes inside macrophages and thus the protective T cells may be responding to amastigote-specific antigens. Possible candidates include heat shock proteins of the parasite since transformation of the promastigote to amastigote is accompanied by a drastic change in temperature, from 26 °C to approximately 34 to 37 °C. In fact, heat shock proteins have been described in a number of microbes, including *Leishmania* (Kaufmann, 1990).

5. CONCLUSIONS

Leishmaniasis is caused by a group of organisms that induce in the infected host a variety of diseases that are often chronic and are associated with a complex array of clinical manifestations. The evidence is now compelling that T cells, especially $CD4^+$ parasite-specific T cells, are important in the development of immunity to the parasite. These T cells either produce or induce the production of various cytokines that are likely to be the mediators that are responsible for either progression or resolution of infection. There is evidence for the involvement of many cytokines in the disease. For the most part, all of this evidence is based on correlations showing that the cytokine is produced by infected animals or that T cells from the infected animals can produce this cytokine *in vitro*. Further work should expand on examining the effect of treating *Leishmania*-infected animals with each of these cytokines or with neutralizing antibodies to each cytokine. The most important question, however, is to determine the mechanism by which T cells that secrete the protective cytokines are selectively activated in the infected host. These experiments are complex, and the answers to the experiments are likely to be as complex as the parasite itself. Some of the first experiments performed that addressed the mechanisms of immunity to *Leishmania* were those in which *Leishmania*-infected macrophages were activated *in vitro* to kill *Leishmania* with what were then called "supernatant factors." The principal factor in those supernatants that was responsible for macrophage activation was IFN-γ. Since then, many cytokines have been shown to be able to activate macrophages to kill *Leishmania*. Often, activation is not efficient unless the macrophages are treated with the cytokines before they are infected with *Leishmania*. Furthermore, a given cytokine (including IFN-γ) cannot by itself activate the macrophage to kill *Leishmania*, and synergistic and inhibitory effects among the cytokines are common (Belosevic *et al.*, 1988). These observations imply that multiple activating signals may be required to activate macrophages to kill *Leishmania*. If the complexity of this *in vitro* system is reflective of what occurs in the infected animal, much work remains to be done to piece together the puzzle that constitutes immunity to *Leishmania*.

6. REFERENCES

Abramson, S. L., and Gallin, J. I., 1990, IL-4 inhibits superoxide production by human mononuclear phagocytes, *J. Immunol.* **144**:625–630.

Behin, R., Mauel, J., and Sordat, B., 1979, *Leishmania tropica:* Pathogenicity and *in vitro* macrophage function in strains of inbred mice, *Exp. Parasitol.* **48:**81–91.

Belosevic, M., Davis, C. E., Meltzer, M. S., and Nacy, C. A., 1988, Regulation of activated macrophage antimicrobial activities. Identification of lymphokines that cooperate with IFN-γ for induction of resistance to infection, *J. Immunol.* **141:**890–896.

Belosevic, M., Finbloom, D. S., van der Meide, P. H., Slayter, M. V., and Nacy, C. A., 1989, Administration of monoclonal anti-IFN-gamma antibodies *in vivo* abrogates natural resistance of C3H/HeN mice to infection with *Leishmania major, J. Immunol.* **143:**266–274.

Beutler, B., and Cerami, A., 1988, Tumor necrosis, cachexia, shock and inflammation: A common mediator, *Annu. Rev Biochem.* **57:**505–518.

Boom, W. H., Liebster, L., Abbas, A. K., and Titus, R. G., 1990, Patterns of lymphokine secretion in murine leishmaniasis: Correlation with disease progression or resolution, *Infect. Immun.* **58:**3863–3870.

Blackwell, J. M., Freeman, J., and Bradley, D. J., 1980, Influence of H-2 complex on acquired resistance to *Leishmania donovani* infection in mice, *Nature* **283:**72–74.

Blackwell, J. M., Roberts, B., and Alexander, J., 1985, Response of BALB/c mice to leishmanial infection, *Curr. Top. Microbiol. Immunol.* **122:**97–106.

Bordier, C., 1987, The promastigote surface protease of *Leishmania, Parasitol. Today* **3:**151–153.

Bottomly, K., 1988, A functional dichotomy in CD4$^+$ T lymphocytes, *Immunol. Today* **9:**268–272.

Bradley, D. J., 1987, Genetics of susceptibility and resistance in the vertebrate host, in *The Leishmaniases in Biology and Medicine* (W. Peters and R. Killick-Kendrick, eds.), pp. 551–581, Academic Press, London.

Buchmuller-Rouiller, Y., and Mauel, J., 1986, Correlation between enhanced oxidative metabolism and leishmanicidal activity in activated macrophages from healer and nonhealer mouse strains, *J. Immunol.* **136:**3884–3890.

Carter, K. C., Gallagher, G., Baillie, A. J., and Alexander, J., 1989, The induction of protective immunity to *Leishmania major* in the BALB/c mouse by interleukin 4 treatment, *Eur. J. Immunol.* **19:**779–782.

Chang, K.-P., 1981, Leishmanicidal mechanisms of human polymorphonuclear phagocytes, *Am. J. Trop. Med. Hyg.* **30:**322–333.

Cillari, E., Liew, F. Y., and Lelchuk, R., 1986, Suppression of interleukin-2 production by macrophages in genetically susceptible mice infected with *Leishmania major, Infect. Immun.* **54:**386–394.

Cillari, E., Dieli, M., Maltese, E., Milano, S., Salerno, A., and Liew, F. Y., 1989, Enhancement of macrophage IL-1 production by *Leishmania major* infection *in vitro* and its inhibition by IFN-γ, *J. Immunol.* **143:**2001–2005.

Corcoran, L. M., Metcalf, D., Edwards, S. J., and Handman, E., 1988, GM-CSF produced by recombinant vaccinia virus or in GM-CSF transgenic mice has no effect *in vivo* on murine cutaneous leishmaniasis, *J. Parasitol.* **74:**763–767.

Crawford, R. M., Finbloom, D. S., Ohara, J., Paul, W. E., and Meltzer, M. S., 1987, B cell stimulatory factor-1 (interleukin 4) activates macrophages for increased tumoricidal activity and expression of Ia antigens, *J. Immunol.* **139:**135–141.

Da Silva, R. P., Hall, B. F., Joiner, K. A., and Sacks, D. L., 1989, CR1, the C3b receptor, mediates binding of infective *Leishmania major* metacyclic promastigotes to human macrophages, *J. Immunol.* **143:**617–622.

Essner, R., Rhoades, K., McBride, W. H., Morton, D. L., and Economou, J. S., 1989, IL-4 downregulates IL-1 and TNF gene expression in human monocytes, *J. Immunol.* **142:**3857–3861.

Etges, R., Bouvier, J., and Bordier, C., 1986, The major surface protein of *Leishmania* promastigotes is a protease, *J. Biol. Chem.* **261:**9098–9101.

Farrell, J. P., and Kirkpatrick, C. E., 1987, Experimental cutaneous leishmaniasis. II. A possible role for prostaglandins in exacerbation of disease in *Leishmania major*-infected BALB/c mice, *J. Immunol.* **138**:902–907.

Farrell, J. P., Muller, I., and Louis, J. A., 1989, A role for Lyt-2$^+$ T cells in resistance to cutaneous leishmaniasis in immunized mice, *J. Immunol.* **142**:2052–2055.

Feng, Z. Y., Louis, J., Kindler, V., Pedrazzini, T., Eliason, J. F., Behin, R., and Vassalli, P., 1988, Aggravation of experimental cutaneous leishmaniasis in mice by administration of interleukin 3, *Eur. J. Immunol.* **18**:1245–1251.

Franke, E. D., McGreevy, P. B., Katz, S. P., and Sacks, D. L., 1985, Growth cycle-dependent generation of complement-resistant *Leishmania* promastigotes, *J. Immunol.* **134**:2713–2718.

Gajewski, T. F., Schell, S. R., Nau, G. N., and Fitch, F. W., 1989, Regulation of T-cell activation: Differences among T-cell subsets, *Immunol. Rev.* **111**:79–109.

Gorczynski, R. M., and MacRae, S., 1982, Analysis of subpopulation of glass-adherent mouse skin cells controlling resistance/susceptibility to infection with *Leishmania tropica*, and correlation with the development of independent proliferative signals to Lyt-1$^+$/Lyt-2$^+$ T lymphocytes, *Cell. Immunol.* **67**:74–89.

Greil, J., Bodenhofer, B., Rollinghoff, M., and Solbach, W., 1988, Application of recombinant granulocyte-macrophage colony-stimulating factor has a detrimental effect in experimental murine leishmaniasis, *Eur. J. Immunol.* **18**:1527–1533.

Handman, E., and Burgess, A. W., 1979, Stimulating by granulocyte-macrophage colony-stimulating factor of *Leishmania tropica* killing by macrophages, *J. Immunol.* **122**:1134–1137.

Handman, E., Ceredig, R., and Mitchell, G. F., 1979, Murine cutaneous leishmaniasis: Disease patterns in intact and nude mice of various genotypes and examination of some differences between normal and infected macrophages, *Aust. J. Exp. Biol. Med. Sci.* **57**:9–29.

Handman, E., McConville, M. J., and Goding, J. W., 1987, Carbohydrate antigens as possible parasite vaccines, *Immunol. Today* **8**:181–185.

Heinzel, F. P., Sadick, M. D., and Locksley, R. M., 1988, *Leishmania major:* Analysis of lymphocyte and macrophage cellular phenotypes during infection of susceptible and resistant mice, *Exp. Parasitol.* **65**:258–268.

Heinzel, F. P., Sadick, M. D., Holaday, B. J., Coffman, R. L., and Locksley, R. M., 1989, Reciprocal expression of interferon γ or interleukin 4 during the resolution or progression of murine leishmaniasis. Evidence for expansion of distinct helper T cell subsets, *J. Exp. Med.* **169**:59–72.

Hill, J. O., Awwad, M., and North, R. J., 1989, Elimination of CD4$^+$ suppressor cells from susceptible BALB/c mice releases CD8$^+$ T lymphocytes to mediate protective immunity against *Leishmania*, *J. Exp. Med.* **169**:1819–1827.

Howard, J. G., 1986, Immunological regulation and control of experimental leishmaniasis, *Int. Rev. Exp. Parasitol.* **28**:79–116.

Howard, J. G., Hale, C., and Chan-Liew, W. L., 1980a, Immunological regulation of experimental cutaneous leishmaniasis. I. Immunogenetic aspects of susceptibility to *Leishmania tropica* in mice, *Parasite Immunol.* **2**:303–314.

Howard, J. G., Hale, C., and Liew, F. Y., 1980b, Genetically determined susceptibility to *Leishmania tropica* infection is expressed by haematopoietic donor cells in mouse radiation chimeras, *Nature* **288**:161–162.

Kaufmann, S. H. E., 1990, Heat shock proteins in the immune response, *Immunol. Today* **11**:129–136.

Kurt-Jones, E. A., Hamberg, S., Ohara, Y., Paul, W. E., and Abbas, A. K., 1987, Heterogeneity of helper/inducer T lymphocytes. I. Lymphokine production and lymphokine responsiveness, *J. Exp. Med.* **166**:1774–1787.

Lehn, M., Weiser, W., Englehorn, S., Gillis, S., and Remold, H., 1989, Interleukin 4 inhibits H_2O_2 production and anti-leishmania capacity of human cultured monocytes mediated by interferon-gamma, *J. Immunol.* **143**:3020–3024.

Lelchuk, R., Graveley, R., and Liew, F. Y., 1988, Susceptibility to murine cutaneous leishmaniasis correlates with the capacity to generate interleukin 3 in response to *Leishmania* antigen *in vitro*, *Cell. Immunol.* **111**:66–74.

Lichtman, A. H., Chin, J., Schmidt, J. A., and Abbas, A. K., 1988, Role of interleukin-1 in the activation of T lymphocytes, *Proc. Natl. Acad. Sci. USA* **85**:9699–9703.

Liew, F. Y., 1983, Specific suppression of responses to *Leishmania tropica* by a cloned T cell line, *Nature* **305**:630–632.

Liew, F. Y., 1989, Functional heterogeneity of CD4$^+$ T cells in leishmaniasis, *Immunol. Today* **10**:40–45.

Liew, F. Y., Hale, C., and Howard, J. G., 1982, Immunologic regulation of experimental cutaneous leishmaniasis. V. Characterization of effector and specific suppressor T cells, *J. Immunol.* **128**:1917–1922.

Liew, F. Y., Hale, C., and Howard, J. G., 1985a, Prophylactic immunization against experimental leishmaniasis. IV. Subcutaneous immunization prevents the induction of protective immunity against fatal *Leishmania major* infection, *J. Immunol.* **135**:2095–2101.

Liew, F. Y., Singleton, A., Cilarri, E., and Howard, J. G., 1985b, Prophylactic immunization against experimental leishmaniasis. V. Mechanisms of the antiprotective blocking effect induced by subcutaneous immunization against *Leishmania major* infection, *J. Immunol.* **135**:2102–2107.

Liew, F. Y., Parkinson, C., Millott, S., Severn, A., and Carrier, M., 1990, Tumour necrosis factor (TNFα) in leishmaniasis. I. TNFα mediates host protection against cutaneous leishmaniasis, *Immunology* **69**:570–573.

Louis, J. A., Zubler, R. H., Coutinho, S. G., Lima, G., Behin, R., Mauel, J., and Engers, H. D., 1982, The *in vivo* generation and functional analysis of murine T cell populations and clones specific for a protozoan parasite, *Leishmania tropica*, *Immunol. Rev.* **61**:215–243.

Louis, J. A., Mendonca, S., Titus, R. G., Cerottini, J. C., Cerny, A., Zinkernagel, R., Milon, G., and Marchal, G., 1986, The role of specific T cell subpopulations in murine cutaneous leishmaniasis, in *Progress in Immunology, VI* (B. Cinader and R. G. Miller, eds.), pp. 762–769, Academic Press, New York.

Maggi, E., DelPrete, G., Macchia, D., Parzanchi, P., Tiri, A., Chretien, L., Ricci, M., and Romagnani, S., 1988, Profiles of lymphokine activities and helper function for IgE in human T cell clones, *Eur. J. Immunol.* **18**:1045–1050.

Marchand, M., Daoud, S., Titus, R. G., Louis, J., and Boon, T., 1987, Variants with reduced virulence derived from *Leishmania major* after mutagen treatment, *Parasite Immunol.* **9**:81–92.

Mauel, J., and Behin, R., 1987, Immunity: Clinical and experimental, in *The Leishmaniases in Biology and Medicine* (W. Peters and R. Killick-Kendrick, eds.), pp. 731–792, Academic Press, London.

Mazingue, C., Cottrez-Detoeuf, F., Louis, J., Kweider, M., Auriault, C., and Capron, A., 1989, *In vitro* and *in vivo* effects of interleukin 2 on the protozoan parasite *Leishmania*, *Eur. J. Immunol.* **19**:487–491.

McGurn, M., Boon, T., Louis, J. A., and Titus, R. G., 1990, *Leishmania major:* Nature of immunity induced by immunization with a mutagenized avirulent clone of the parasite in mice, *Exp. Parasitol.* **71**:81–89.

McNeely, T. B., and Turco, S. J., 1987, Inhibition of protein kinase C activity by the *Leishmania donovani* lipophosphoglycan, *Biochem. Biophys. Res. Comm.* **148**:653–657.

Mirkovich, A. M., Galelli, A., Allison, A. C., and Modabber, F. Z., 1986, Increased myelopoesis during *L. major* infection in mice: Generation of "safe targets" a possible way to evade the effector immune mechanism, *Clin. Exp. Immunol.* **64:**1–7.

Mitchell, G. F., and Handman, E., 1985, T lymphocytes recognize *Leishmania* glycoconjugates, *Parasitol. Today* **1:**61–64.

Mitchell, G. F., Anders, R. F., Brown, G. V., Hardman, E., Roberts-Thomson, I. C., Chapman, C. B., Forsyth, K. P., Kahl, L. P., and Cruise, K. M., 1982, Analysis of infection characteristics and antiparasite immune responses in resistant compared with susceptible hosts, *Immunol. Rev.* **61:**137–188.

Modlin, R. L., Pirmez, C., Hofman, F. M., Torigian, V., Uyemura, K., Rea, T. H., Bloom, B. R., and Brenner, M. B., 1989, Lymphocytes bearing antigen-specific γδ T-cell receptors accumulate in human infectious disease lesions, *Nature* **339:**544–548.

Molyneux, D. H., and Killick-Kendrick, R., 1987, Morphology, ultrastructure and life cycles, in *The Leishmaniases in Biology and Medicine* (W. Peters and R. Killick-Kendrick, eds.), pp. 121–176, Academic Press, London.

Mosmann, T. R., and Coffman, R. L., 1989, Th1 and Th2 cells: Different patterns of lymphokine secretion lead to different functional properties, *Annu. Rev. Immunol.* **7:**145–173.

Muller, I., and Louis, J. A., 1989, Immunity to experimental infection with *Leishmania major:* Generation of protective L3T4$^+$ T cell clones recognizing antigen(s) associated with live parasites, *Eur. J. Immunol.* **19:**865–871.

Murray, H. W., Rubin, B. Y., and Rothermel, C. D., 1983, Killing of intracellular *Leishmania donovani* by lymphokine-stimulated human monocytes. Evidence that interferon γ is the activating lymphokine, *J. Clin. Invest.* **72:**1506–1510.

Nacy, C. A., Fortier, A. H., Meltzer, M. S., Buchmeier, N. A., and Gray, P. W., 1985, Macrophage activation to kill *Leishmania major:* Activation of macrophages for intracellular destruction of amastigotes can be induced by both interferon-γ and non-interferon lymphokines, *J. Immunol.* **135:**3505–3511.

Neva, F., and Sacks, D., 1990, Leishmaniasis, in *Tropical and Geographic Medicine* (K. S. Warren and A. A. S. Mahmoud, eds.), pp. 296–308, McGraw-Hill, New York.

Nong, Y.-H., Titus, R. G., Ribeiro, J. M. C., and Remold, H. G., 1989, Peptides encoded by the calcitonin gene inhibit macrophage function, *J. Immunol.* **143:**45–49.

Paliard, X., deWaal Malefyt, R., Yessel, H., Blanchard, D., Chretien, J., Abrams, J., deVries, J. E., and Spits, H., 1988, Simultaneous production of IL-2, IL-4 and IFN-gamma by activated human CD4$^+$ and CD8$^+$ T cell clones, *J. Immunol.* **141:**849–855.

Pearson, R. D., and Steigbigel, R. T., 1981, Phagocytosis and killing of the protozoan *Leishmania donovani* by human polymorphonuclear leukocytes, *J. Immunol.* **127:**1438–1443.

Pozzi, L. M., David, J. R., and Wieser, W. Y., 1990, Recombinant human macrophage migration inhibitory factor (MIF) activates intracellular killing of *L. donovani* by human monocyte-derived macrophages, *FASEB J.* **4:**A2031.

Puentes, S. M., Sacks, D. L., Da Silva, R. P., and Joiner, K. A., 1988, Complement binding to two developmental stages of *Leishmania major* promastigotes varying in expression of a surface lipophosphoglycan, *J. Exp. Med.* **167:**887–902.

Reiner, N. E., 1987, Parasite accessory cell interactions in murine leishmaniasis. I. Evasion and stimulus-dependent suppression of the macrophage interleukin 1 response by *Leishmania donovani, J. Immunol.* **138:**1919–1925.

Ribeiro, J. M. C., Vachereau, A., Modi, G. B., and Tesh, R. B., 1989, A novel vasodilatory peptide from the salivary glands of the sand fly *Lutzomyia longipalpis, Science* **243:**212–214.

Russell, D. G., and Talamas-Rohana, P., 1989, *Leishmania* and the macrophage: A marriage of inconvenience, *Immunol. Today* **10:**328–333.

Sacks, D. L., and Perkins, P. V., 1984, Identification of an infective stage of *Leishmania* promastigotes, *Science* **223:**1417–1419.

Sacks, D. L., Scott, P. A., Asofsky, R., and Sher, F. A., 1984, Cutaneous leishmaniasis in anti-IgM-treated mice: Enhanced resistance due to functional depletion of a B cell-dependent T cell involved in the suppressor pathway, *J. Immunol.* **132:**2072–2077.

Sadick, M. D., Heinzel, F. P., Holaday, B. J., Pu, R. T., Dawkins, R. S., and Locksley, R. M., 1990, Cure of murine leishmaniasis with anti-interleukin 4 monoclonal antibody. Evidence for a T cell-dependent, interferon γ-independent mechanism, *J. Exp. Med.* **171:**115–127.

Scott, P., and Sher, A., 1986, A spectrum in the susceptibility of leishmanial strains to intracellular killing by murine macrophages, *J. Immunol.* **136:**1461–1466.

Scott, P., Natovitz, P., and Sher, A., 1986, B lymphocytes are required for the generation of T cells that mediate healing of cutaneous leishmaniasis, *J. Immunol.* **137:**1017–1021.

Scott, P., Natovitz, P., Coffman, R. L., Pearce, E., and Sher, A., 1988, Immunoregulation of cutaneous leishmaniasis. T cell lines that transfer protective immunity or exacerbation belong to different T helper subsets and respond to distinct parasite antigens, *J. Exp. Med.* **168:**1675–1684.

Scott, P., Pearce, E., Cheever, A. W., Coffman, R. L., and Sher, A., 1989, Role of cytokines and $CD4^+$ T-cell subsets in the regulation of parasitic immunity and disease, *Immunol. Rev.* **112:**161–182.

Scott, P., Caspar, P., and Sher, A., 1990, Protection against *Leishmania major* in BALB/c mice by adoptive transfer of a T cell clone recognizing a low molecular weight antigen released by promastigotes, *J. Immunol.* **144:**1075–1079.

Solbach, W., Lohoff, M., Streck, H., Rohwer, P., and Rollinghoff, M., 1987, Kinetics of cell-mediated immunity developing during the course of *Leishmania major* infection in "healer" and "non-healer" mice: Progressive impairment of response to and generation of interleukin-2, *Immunology* **62:**485–492.

Stuart, P. M., Zlotnik, A., and Woodward, J. G., 1988, Induction of class I and class II MHC antigen expression on murine bone marrow-derived macrophages by IL-4, *J. Immunol.* **140:**1542–1547.

Swain, S. L., McKenzie, D. T., Dutton, R. W., Tonkonogy, S. L., and English, M., 1988, The role of IL4 and IL5: Characterization of a distinct helper T cell subset that makes IL4 and IL5 (Th2) and requires priming before induction of lymphokine secretion, *Immunol. Rev.* **102:**77–105.

Theodos, C. M., Povinelli, L., Molina, R., Sherry, B., and Titus, R. G., 1991, Role of tumor necrosis factor in macrophage leishmanicidal activity *in vitro* and resistance to cutaneous leishmaniasis *in vivo, Infect. Immunity* **59:**2839–2842.

Titus, R. G., and Ribeiro, J. M. C., 1988, Salivary gland lysates from the sand fly *Lutzomyia longipalpis* enhance *Leishmania* infectivity, *Science* **239:**1306–1308.

Titus, R. G., and Ribeiro, J. M. C., 1990, The role of vector saliva in transmission of arthropod-borne disease, *Parasitol. Today* **6:**157–160.

Titus, R. G., Kelso, A., and Louis, J. A., 1984a, Intracellular destruction of *Leishmania tropica* by macrophages activated with macrophage activating factor/interferon, *Clin. Exp. Immunol.* **55:**157–165.

Titus, R. G., Lima, G. C., Engers, H. D., and Louis, J. A., 1984b, Exacerbation of murine cutaneous leishmaniasis by adoptive transfer of parasite-specific helper T cell populations capable of mediating *Leishmania major*-specific delayed-type hypersensitivity, *J. Immunol.* **133:**1594–1600.

Titus, R. G., Ceredig, R., Cerottini, J.-C., and Louis, J. A., 1985a, Therapeutic effect of anti-L3T4 monoclonal antibody GK1.5 on cutaneous leishmaniasis in genetically susceptible BALB/c mice, *J. Immunol.* **135:**2108–2114.

Titus, R. G., Marchand, M., Boon, T., and Louis, J. A., 1985b, A limiting dilution assay for quantifying *Leishmania major* in tissues of infected mice, *Parasite Immunol.* **7:**545–555.

Titus, R. G., Milon, G., Marchal, G., Vassalli, P., Cerottini, J.-C., and Louis, J. A., 1987, Involvement of specific $Lyt\text{-}2^+$ T cells in the immunological control of experimentally induced murine cutaneous leishmaniasis, *Eur. J. Immunol.* **17:**1429–1433.

Titus, R., Marchand, M., Milon, G., Marchal, G., Boon, T., and Louis, J., 1988, Cutaneous leishmaniasis: Relative role of T cell subsets, in *Host Defenses and Immunomodulation to Intracellular Pathogens* (T. K. Eisenstein, W. E. Bullock, and N. Hanna, eds.), pp. 121–134, Plenum Press, New York.

Titus, R. G., Sherry, B., and Cerami, A., 1989, Tumor necrosis factor plays a protective role in experimental murine cutaneous leishmaniasis, *J. Exp. Med.* **170**:2097–2104.

Titus, R. G., Muller, I., Kimsey, P., Cerney, A., Behin, R., Zinkernagel, R., and Louis, J., 1991, Exacerbation of experimental murine cutaneous leishmaniasis with CD4$^+$ *Leishmania major*-specific T cell lines of clones which secrete interferon-gamma and mediate parasite-specific delayed-type hypersensitivity, *Eur. J. Immunol.* **21**:559–567.

Vercelli, D., Jabara, H. H., Lauener, R. P., and Geha, R. S., 1990, IL-4 inhibits the synthesis of IFN-γ and induces the synthesis of IgE in human mixed lymphocyte cultures, *J. Immunol.* **144**:570–573.

Walsh, J. A., 1990, Estimating the burden of illness in the tropics, in *Tropical and Geographic Medicine* (K. S. Warren and A. A. S. Mahmoud, eds.), pp. 296–308, McGraw-Hill, New York.

Waburg, A., and Schlein, Y., 1986, The effect of post-blood meal nutrition of *Phlebotomus papatasi* on the transmission of *Leishmania major, Am. J. Trop. Med. Hyg.* **36**:926–930.

Weiser, W. Y., Temple, P. A., Witek-Giannotti, J., Remold, H. G., Clark, S., and David, J. R., 1989, Molecular cloning of a CDNA encoding a human macrophage migration inhibitory factor, *Proc. Natl. Acad. Sci. USA* **86**:7522–7526.

Weiser, W. Y., Titus, R. G., Pozzi, L. M., and David, J. R., 1990, Recombinant human macrophage migration inhibitory factor (MIF) has adjuvant-like activity, *FASEB J.* **4**:A1725.

Wolpe, S. D., and Cerami, A., 1989, Macrophage inflammatory proteins 1 and 2: Members of a novel superfamily of cytokines, *FASEB J.* **3**:2565–2573.

Chapter 5

Molecular Karyotype Analysis in *Leishmania*

Patrick Bastien, Christine Blaineau, and Michel Pagès

1. INTRODUCTION

1.1. The Parasite

Leishmania is a flagellated protozoan parasite belonging to the order Kinetoplastida and to the family Trypanosomatidae. A total number of 25 species has been described,* grouped in at least 12 complexes (Lainson and Shaw, 1972, 1987; Rioux, 1986; Rioux *et al.*, 1990). They are responsible for a wide spectrum of human diseases, mostly in the tropics and subtropics, ranging from sometimes relatively benign cutaneous disease (*L. major, L. tropica, L. mex-*

*The term "species" is not strictly acceptable in its biological sense for organisms in which sexual processes have not been identified (see Lainson and Shaw, 1987). However, we shall use it in this review for clarity. Moreover, in accordance with the recommendations of the Second International Symposium on Leishmaniases (Anon., 1986), only binomials (e.g., *L. guyanensis*) will be used, the old "subspecies" (e.g., *L. braziliensis guyanensis*) being considered as species (see also Grimaldi *et al.*, 1989, p. 689).

Patrick Bastien and Christine Blaineau Laboratoire d'Ecologie Médicale et Pathologie Parasitaire, Annexe de la Faculté de Médecine, 34000 Montpellier, France. **Michel Pagès** Centre de Recherche de Biochimie Macromoleculaire du CNRS et Unité 249 de l'INSERM, 34033 Montpellier Cedex, France.

Subcellular Biochemistry, Volume 18: Intracellular Parasites, edited by J. L. Avila and J. R. Harris. Plenum Press, New York, 1992.

icana, L. panamensis) to mutilating mucocutaneous and diffuse cutaneous lesions (*L. braziliensis, L. amazonensis, L. aethiopica*), and potentially fatal visceral disease affecting the hematopoietic organs (*L. donovani, L. infantum, L. chagasi*). Leishmaniasis has been estimated to affect 20 million people, mainly in poor countries from both the Old and the New World (Barker, 1987).

All *Leishmania* are transmitted by the bite of sandfly vectors. The parasite is digenetic and undergoes a complex life cycle: nonflagellated immotile intracellular amastigotes are taken from the skin of the vertebrate host and ingested with a blood meal by the vector; within the alimentary tract of the latter, they transform into motile promastigotes, and go through a complex series of developmental stages, giving rise to the migration into the foregut of infective metacyclic forms; these are inoculated into a human or a mammal reservoir, where they immediately enter macrophages and transform back into amastigotes (see Molyneux and Killick-Kendrick, 1987 for a review). Intense mitotic multiplication occurs by binary fission at both stages: in the phagolysosome of macrophages as amastigotes and in the sandfly's midgut as promastigotes. The parasites can be isolated either from infected organs or lesions or from an infected sandfly, and can then be cultivated *in vitro* in a variety of culture media (see Chang and Hendricks, 1985).

The characterization of *Leishmania* species and strains* can be done using a variety of criteria, among which the clinical, epidemiological, and biochemical are the most used. The latter particularly includes isoenzyme profiles (Chance, 1986), kinetoplast DNA restriction patterns (Barker *et al.*, 1986; Pacheco *et al.*, 1986; Barker, 1987), and patterns of recognition by monoclonal antibodies (McMahon-Pratt *et al.*, 1986). The classification of the genus has evolved gradually, but has come to be based on isoenzymic patterns, particularly with the help of numerical analysis (Lanotte *et al.*, 1981; Le Blancq *et al.*, 1986; Rioux *et al.*, 1990).

Our current understanding of the genetic properties and genome organization of *Leishmania* is limited. No conventional genetic analysis is possible since a sexual cycle has not yet been identified. Genetic exchange has been shown to occur in the related protozoan *Trypanosoma brucei* in the vector (Jenni *et al.*, 1986) and it is suspected, but it has yet to be demonstrated in *Leishmania* (Maazoun *et al.*, 1981; Rioux *et al.*, 1985; Evans *et al.*, 1987; Lanotte and

*The terms "isolate," "stock," and "line" have been clearly defined for trypanosomes (Anon., 1978), and these definitions apply to *Leishmania* (Anon., 1986). Unlike *Plasmodium* or some *Trypanosoma*, isolates of *Leishmania* cannot be examined directly, when the primary isolate still has to be amplified in culture to provide sufficient material for biochemical studies. The term "stock" implies the absence of characterization, but the vast majority of stocks used in PFGE studies have been previously characterized by one or more of the methods mentioned here. Therefore, in this chapter we shall generally use the term "strain": this could be defined as a set of populations from a group of *Leishmania* of a given species present in a given culture (or host) and characterized by the possession of several designated characters.

Rioux, 1990). Lastly, a conventional karyotype could never be described due to the fact that the chromosomes, like in many protozoans, do not condense at any phase of the mitotic cycle (Raikov, 1982).

1.2. Pulsed Field Electrophoresis

The karyotype is classically defined in higher eukaryotes as the microscopic picture of the whole chromosomal complement in metaphase, the chromosomes being individually recognized and arranged in order of size. By analogy, the term "karyotype" has been used for yeast and protozoans for the picture of the chromosomal DNA molecules separated by pulsed field electrophoresis (PFE). This original technique, described for the first time in 1983 for the separation of the chromosomes of the yeast *Saccharomyces cerevisiae* (Schwartz and Cantor, 1984; Carle and Olson, 1984), allows the resolution of DNA molecules larger than 50 and as large as 10,000 kilobasepairs (kb). This has dramatically improved the analysis of large DNA fragments that are very difficult to separate in conventional electrophoresis in the range 20–750 kb (Fangman, 1978). In PFE, the DNA molecules are subjected to electric fields that alternate between two approximately perpendicular orientations. The principle is to force large molecules to turn corners periodically in the gel matrix, leading to a zigzag "reptation" (see Richards, 1989) involving cycles of elongation and contraction of the molecule (Schwartz and Koval, 1989). The net mobility of DNA becomes strongly size dependent because of the time the molecule takes to reorient in the new electric field. Thus the resolution of linear DNA is a complex function of the length of the molecule, the strength of the applied electric field, the pulse time (i.e., the time interval between each change of orientation of the electric field), and the agarose concentration and temperature (Vollrath and Davis, 1987; Mathew *et al.*, 1988a,b). The resolution of DNAs within a particular size range can be optimized by adjusting the pulse times, with larger molecules being resolved at increased pulse times (Schwartz and Cantor, 1984; Carle and Olson, 1984; Mathew *et al.*, 1988b). The migration is also dependent on the topology: thus, circular DNAs can enter pulsed field gels but their mobility is insensitive to pulse time (Garvey and Santi, 1986; Mathew *et al.*, 1988c; Hightower and Santi, 1989).

One of the other critical variables is the electric field geometry of the apparatus used (Cantor *et al.*, 1988). A number of different electrode geometries have been used, generating: either (1) single or double nonhomogeneous electric fields: pulsed-field gradient gel electrophoresis (PFGGE) (Schwartz and Cantor, 1984), orthogonal-field-alternation gel electrophoresis (OFAGE) (Carle and Olson, 1984), transverse-alternating-field electrophoresis (TAFE) (Gardiner *et al.*, 1986); or (2) homogeneous fields: field-inversion gel electrophoresis (FIGE) (Carle *et al.*, 1986), and electrophoresis with "contour-clamped homogeneous electric fields" (CHEF) (Chu *et al.*, 1986). In this review, we shall use the term

PFE (pulsed field electrophoresis) as a general term designating any of these techniques. It should be noted that the complex electrode arrangements used in PFE lead to the migration patterns being different for each of the variations of the technique described above (Carle and Olson, 1987), hence to variations from lab to lab in the general aspect of the patterns and in the response to identical electrophoretic conditions. Most of the figures presented here and the work carried out in our laboratory were on a homemade CHEF-like apparatus generating two homogeneous electric fields at 120° (Bellis *et al.*, 1987).

1.3. Molecular Karyotype and Chromosomes

Among its many utilizations, PFE has been used for the separation of chromosomal DNA from a wide range of lower eukaryotes, such as *Saccharomyces* (Schwartz and Cantor, 1984), *Trypanosoma* (Van der Ploeg *et al.*, 1984a), *Plasmodium* (Kemp *et al.*, 1985), *Tetrahymena* (Altschuler and Yao, 1985), *Leishmania* (Spithill and Samaras, 1985), *Candida* (Snell and Wil2kins, 1986), *Schizosaccharomyces* (Smith *et al.*, 1987), *Neurospora* (Orbach *et al.*, 1988), *Histoplasma* (Steele *et al.*, 1989), and *Eimeria* (Shirley *et al.*, 1990).

Thc banding pattern obtained by PFE of total intact genomic DNA from these organisms has been termed "molecular" karyotype or "electrophoretic" karyotype. This recent acceptance of the term "karyotype" can pose difficulties: the bands visualized in PFE result from a large number of cells (6×10^7 per well in our laboratory), and assumptions have to be made about the homogeneity of the strain population. Another question one can legitimately ask is that of the identity between the molecules separated by PFE and "true" chromosomes. However, sufficient evidence has been provided to justify such a correlation. In the yeast *Saccharomyces cerevisiae,* for which much genetic information is available, the chromosome sizes are consistent with the genetic linkage maps, thus providing unambiguous evidence that each gel band consists of one or more intact chromosomal DNAs (Schwartz and Cantor, 1984; Carle and Olson, 1985). The same correlation with estimated linkage maps was obtained for the seven multimegabase chromosomes of *Neurospora crassa* (Orbach *et al.*, 1988). Lastly, in *Schizosaccharomyces pombe,* the three expected chromosomes (3500–5700 kb) were also visualized in PFE (Smith *et al.*, 1987; Vollrath and Davis, 1987). In *Plasmodium falciparum,* physical macrorestriction maps have been constructed for five distinct chromosomes isolated from PFE gels (Wellems *et al.*, 1987; Corcoran *et al.*, 1988; Sinnis and Wellems, 1988; Foote and Kemp, 1989); also, a constant number of 14 chromosomes is found per haploid genome (Kemp *et al.*, 1987), in agreement with the observed number of kinetochores seen by electron microscopy (Prensier and Slomianny, 1986). Lastly, in trypanosomatids, this was supported by the hybridization of telomeric probes to all chromosomal bands and by the high reproducibility of the PFE banding patterns, both for *Trypanosoma* (Van der Ploeg *et al.*, 1984a,b; Johnson and Borst, 1986) and *Leishmania* (Spithill and Samaras, 1987a; Ellis and Crampton, 1988; Pagès *et al.*, 1989). Therefore, although it has not been formally demonstrated for this organism, it seems legitimate to identify the chromosomal bands of *Leishmania* observed in PFE as intact chromosomes.

It should also be noted that other uncertainties regarding the authenticity of the chromosomal pattern observed in PFE have been removed: (1) Van der Ploeg *et al.* (1984a) have excluded any major contribution of random degradation or aggregation of DNA to the banding patterns observed; (2) similarly, Carle and Olson (1987) ruled out any electrophoretic shearing under the right experimental conditions; (3) preferential trapping of very large chromosomes (> 2000 kb) has been suspected for *S. cerevisiae* (Carle and Olson, 1985) and *T. brucei* (Van der Ploeg *et al.*, 1984a, 1989; Gibson and Borst, 1986) and some molecules can indeed get trapped with each other (Schwartz and Koval, 1989); however, this is only partial, and the recent separation of chromosomes as large as 10 megabasepairs (Mb) with specific electrophoretic conditions (Orbach *et al.*, 1988; Van der Ploeg *et al.*, 1989) shows this not to be a real hindrance; (4) mitochondrial kinetoplast DNA does not interfere with the PFE banding pattern, since it was shown to remain mostly in the slot and be partly degraded as a fast moving smear (Schwartz and Cantor, 1984; Van der Ploeg *et al.*, 1984a; Samaras and Spithill, 1987).

Thus, the molecular karyotype does represent the picture of intact chromosomal molecules arranged in order of increasing size. It should be noted that this does not imply any assumption about the number as well as the nature of the chromosomes thus observed. It must also be kept in mind that a good visualization of the whole karyotype of *Leishmania* requires the run of several gels at various pulse times, the fine resolution of the very large-sized DNAs implying the compression or the loss of the smaller ones, and vice versa.

The term "chromosome" has often been used by PFE workers in its wider sense, corresponding to a band individualized in PFE experimental patterns with the best possible resolution. However, "chromosomal band" seems more appropriate since it does not make any assumption on sequence homologies. In this review, a distinction will be made between "chromosome" and "chromosomal band" each time a chromosome has been specifically identified, in particular when homologous chromosomes of a different size migrate as different bands.

1.4. Aims and Scope of This Chapter

Despite numerous studies on the molecular karyotype of *Leishmania* strains, it is not yet clear what exactly is the karyotype of *Leishmania,* what a chromosomal band actually does represent, what are the chromosomal homologies between strains and between species, or if the chromosomal organization is conserved at the species/genus level. The forefront feature noted in all studies was the high degree of polymorphism observed between the karyotypes of strains and species, and it seems impossible to define a karyotype common to the whole genus *Leishmania.* We shall therefore first present the karyotypic features of two cloned strains of different *Leishmania* species and the identification of chromosomes by specific DNA probes. In the following section, we shall review the dramatic polymorphisms observed between *Leishmania* species, strains, and clones as regards the chromosome sizes and perhaps their number. The interest of PFE for the taxonomy of *Leishmania* will be examined in the light of such polymorphisms. We shall then discuss our current understanding of the possible mechanisms leading to size variations of homologous chromosomes and analyze the implications of these polymorphisms on the fundamental biology of the parasite.

2. THE MOLECULAR KARYOTYPE OF *LEISHMANIA*

Pulsed field electrophoresis allows the resolution of about 25 chromosomal bands in *Leishmania* cloned strains, in the size range of 250–2600 kb. All molecular karyotypes share certain characteristics such as the grouping of chromosomes in four main size classes, nonstoichiometrically staining bands, reproducibility, and stability. The number of chromosomes and the genome size are difficult to infer from PFE patterns, in view of our ignorance of the parasite's ploidy and of the actual contents of individual bands. Nevertheless, our understanding of such questions should progress rapidly, with the prerequisite of the identification of individual chromosomes becoming increasingly accessible thanks to "chromosome-specific" DNA probes.

2.1. The Complete Karyotype of Two Cloned Strains of *Leishmania*

Few authors have reported the fine resolution of all or most of the chromosomal DNAs in *Leishmania,* the difficulty residing in the very large ones (> 1800 kb). Spithill and Samaras (1987b) were the first to establish an apparently complete karyotype of a clone of *L. major.* They could resolve 23 bands ranging in size from 250 to > 2000 kb. Six of these bands were present in greater abundance than the others, suggesting that the 23 bands were an underestimate of the chromosome number. Later, Pagès *et al.* (1989) also described an apparently complete karyotype of several *L. infantum* strains and clones.

Figure 1A,B details the karyotype of one clone each of *L. major* and *L. infantum,* respectively, clearly resolved on several gels by using different pulse times (= intervals) for each gel. The diagrams presented in Figure 2 summarize this information. One of the first striking features is the complexity of the

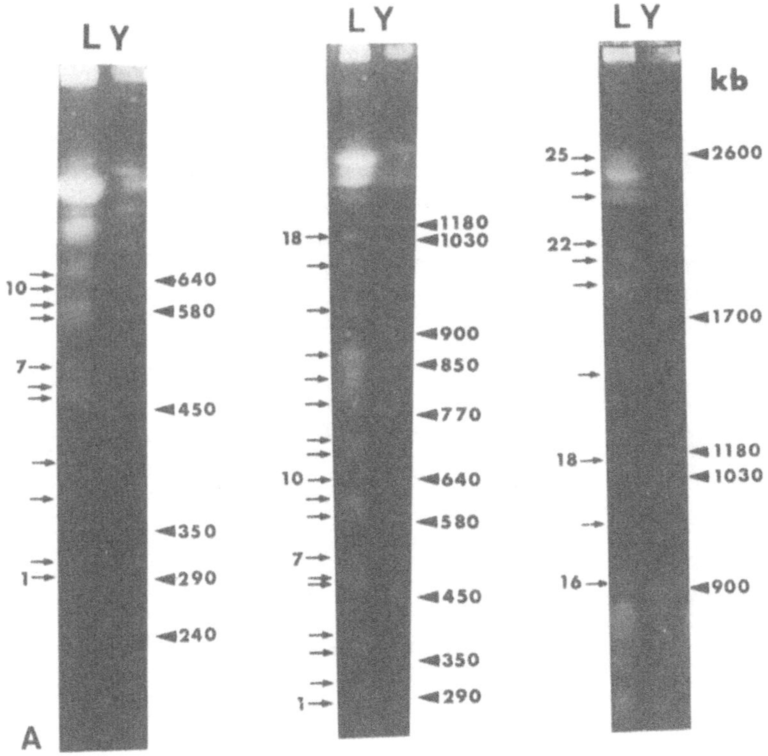

FIGURE 1. Resolution of the whole molecular karyotype of two cloned strains of *Leishmania*. The chromosomes were separated in order of increasing size on various PFE gels from left to right. L = *Leishmania;* Y = yeast (*Saccharomyces cerevisiae*) chromosomal DNA. The figures and arrows on the left of each gel indicate the numbers given to the chromosomal bands of *Leishmania*. The figures and arrowheads on the right show the sizes of the chromosomes of yeast in kilobasepair. (A) The karyotype of *L. major* (cloned MHOM/IL/81/Friedlin) was resolved at 150 V using three different combinations of pulse times, that is, from left to right: 90 sec over 22 hr + 60 sec/46 hr; 150 sec/22 hr + 120 sec/38 hr + 90 sec/22 hr; and 150 sec/90 hr; bands 2, 4, 10 and 22 are hardly visible on these photographs because of their weak intensity, but are constantly present (P. Bastien and D. F. Smith, unpublished data). (B) The karyotype of *L. infantum* (LEM 1317 = cloned MHOM/FR/82/LEM356) was resolved at 150 V using four different combinations of pulse times, that is, from left to right: 60 sec + 40 sec + 30 sec; 80 sec + 70 sec + 60 sec; 120 sec + 100 sec + 90 sec; and 150 sec + 130 sec + 120 sec, each pulse time being used during 24 hr each. The equipment used in A was the LKB Pulsaphor, and in B a homemade CHEF-like apparatus. Note the size variations of yeast chromosomes between the two strains used in A and B (334 and 288C, respectively); also, the largest yeast chromosome appears split into two bands in B (? on third gel from the left).

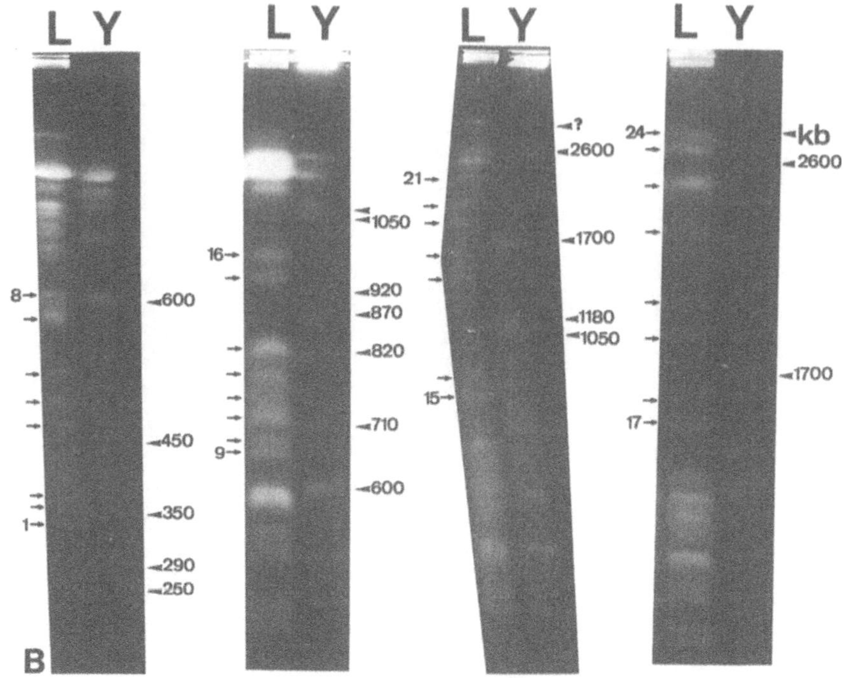

FIGURE 1 (Continued)

banding pattern compared, for instance, to yeast. The chromosome sizes can be estimated against those of *S. cerevisiae,* which are known as approximately 250–2600 kb (Frutos *et al.*, 1989).* It can be seen from Figure 1 that the chromosome size range in *Leishmania* is similar to that of yeast. Four main size classes can be arbitrarily distinguished†: for these clones of *L. major* and *L. infantum,* respectively, seven or six small bands of 250–500 kb, ten close medium-sized bands of 550–1000 kb, two intermediate bands of 1000–1600 kb, and six large bands of > 1800 kb (Figure 2). Actually, these large-sized chromosomes, which have

*It must be kept in mind that *S. cerevisiae* strains exhibit a sometimes dramatic polymorphism in the chromosome sizes (De Jonge *et al.*, 1986; Mathew *et al.*, 1988a). Moreover, the heaviest band is often split into two under the electrophoretic conditions for large chromosomes (see Figure 1B). It is therefore recommended to keep to one marker strain that has been well characterized: in our laboratory, we have characterized and used strain 288C (Frutos *et al.*, 1989).

†The DNA band sizes determined from replicate samples run at different pulse frequencies or on different gels show 5% coefficients of variation (Giannini *et al.*, 1990; P. Bastien, C. Blaineau, and M. Pagès, unpublished data). All the sizes given here are therefore approximate (± 5%).

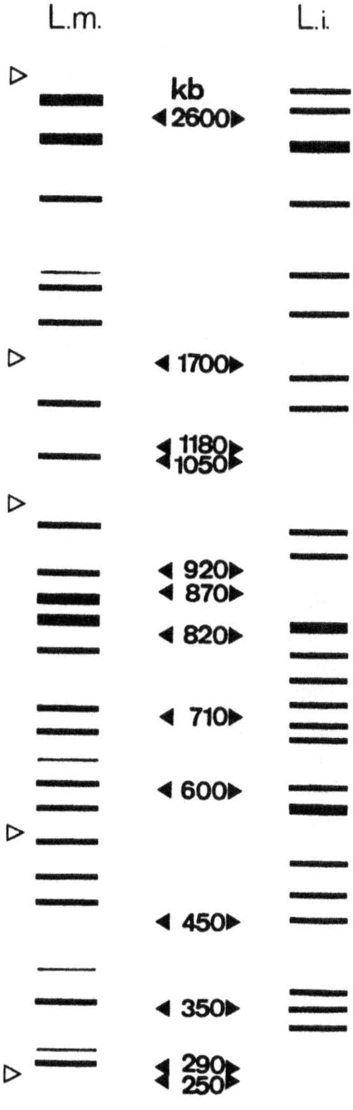

FIGURE 2. Diagrammatic representation of the karyotypes of the two cloned strains electrophoresed in Figure 1. L.m. = *L. major* Friedlin; L.i. = *L. infantum* LEM 1317. The molecular weight marker (in kb) is yeast chromosomal DNA (*S. cerevisiae* strain 288C). The scale is not directly proportional; rather, these diagrams try to represent what is actually seen on several gels run at different pulse times for the best resolution of different size classes. The sizes indicated here, therefore, may not tally with the picture of one particular gel run for another size class. Four size classes can be arbitrarily grouped (open arrowheads), which are best resolved by different running conditions. Lines are of varying thickness according to the staining intensity of the band; thick lines represent bands with a double intensity.

been counted as five (Samaras and Spithill, 1987; Pagès *et al.*, 1989) or six (Kidane *et al.*, 1989) or left unresolved as a compression zone (Comeau *et al.*, 1986; Bishop, 1990), might be better separated into nine bands, but it is not clear yet whether any DNA is larger than the largest yeast chromosome (P. Bastien and D. F. Smith, unpublished data). A total number of 25 and 24 chromosomal bands is thus observed; this is of the same order as given in previous reports: 20

(Comeau *et al.*, 1986; Giannini *et al.*, 1986), 22 (Bishop and Miles, 1987), and 24 (Scholler *et al.*, 1986; Kidane *et al.*, 1989) for different species of *Leishmania*. If the improved resolution of megabase chromosomes is confirmed, it would total 28 for *L. major*. No multiple minichromosomes (40–150 kb) similar to those found in *T. brucei* (Van der Ploeg *et al.*, 1984a) have been found in *Leishmania*.‡

A characteristic of the *Leishmania* karyotype is the nonstoichiometric staining intensities of the different chromosomal bands. At least three different intensity gradations can be seen even when every band has been resolved (Figures 1 and 2). Galindo and Ramirez (1989) apparently measured gradations from 1 to 13, but they had only separated 18 bands, thus perhaps leaving many bands unresolved. This nonstoichiometric staining remains unexplained. It cannot be simply due to a clonal heterogeneity of the strains, since identical intensity variations are observed when the strains are cloned (Pagès *et al.*, 1989; unpublished data). It could easily be accounted for by the comigration of identical or different molecules of the same size and by the presence of distinct-sized homologous chromosomes. This has been clearly demonstrated for the three smallest chromosomes (Iovannisci and Beverley, 1989; Bastien *et al.*, 1990; Blaineau *et al.*, 1991), but might not be sufficient to explain the extent of these variations. Several authors have also invoked aneuploidy (Scholler *et al.*, 1986; Bishop and Miles, 1987; Spithill and Samaras, 1987a; Galindo and Ramirez, 1989; Pagès *et al.*, 1989; Giannini *et al.*, 1990). Finally, two alternative explanations, neither of which excludes those above, have been proposed. One invokes a model of genetic exchange akin to conjugation (Pagès *et al.*, 1989) and the other implies a mosaic structure of the strain population (Bastien *et al.*, 1990; see also Section 5).

2.2. Stability of the Molecular Karyotype

A notable feature of the molecular karyotype of a *Leishmania* strain/clone is its high reproducibility in similar electrophoretic conditions, as has been shown for the number, sizes, and intensities of the chromosomal bands. Another distinctive feature is stability, which has been demonstrated: (1) over prolonged periods of *in vitro* culture, up to several months (Pagès *et al.*, 1989) or even years (Giannini *et al.*, 1986); (2) among several cryostabilates of the same strain (P.

‡Circular DNAs (not found in these clones) have also been described in *Leishmania* karyotypes: they are extrachromosomal supercoiled molecules, usually (Beverley *et al.*, 1988) but not always (Hightower *et al.*, 1988; Gajendran *et al.*, 1989) associated with drug resistance. They exhibit unusual mobility properties, such as sharp light banding pattern, laterally skewed migration (relative to linear chromosomal DNA), and pulse-time-independent mobility (Garvey and Santi, 1986; Mathew *et al.*, 1988c). Supercoiled molecules ranging in size from 4 to 91 kb are resolved over the same region as linear chromosomes from 260 to 2200 kb (Hightower and Santi, 1989).

Bastien, C. Blaineau, and M. Pagès, unpublished data); (3) among the different stages of the life cycle of the parasite: the karyotypes of axenically cultured amastigotes or amastigotes from skin lesions were found identical to those of culture promastigotes (Giannini *et al.*, 1986); (4) upon passage through rodents (Dujardin *et al.*, 1989; Giannini *et al.*, 1990); and (5) even through a complete life cycle involving mice and vector (Giannini *et al.*, 1990).

Thus, the molecular karyotype is usually a stable characteristic of a cloned *Leishmania* strain, and often it can actually be used for strain identification. It should nevertheless be noted that rare examples of chromosome size modifications, affecting one or two chromosomes only, have been seen in strains maintained in culture (Dujardin *et al.*, 1989; Iovannisci and Beverley, 1989; Beverley and Coburn, 1990); in our experience, this occurred in just one strain out of 22 of *L. infantum* and only in aging cultures with rarefaction of the medium. Whether this represents true chromosomal rearrangements in subpopulations followed by their selection by *in vitro* culture conditions or the selection of preexisting variants remains to be determined.

2.3. The Karyotype of a *Leishmania* Strain Is Representative of One Clone

Although chromosome size polymorphisms have been described between clones derived from the same strain (see Section 3.3), several authors have also noted identical patterns between a "parental" strain and the clones derived from it (Giannini *et al.*, 1986; Scholler *et al.*, 1986). In each strain of *L. infantum* where more than one clone was examined (11 strains), we have found at least one clone with a karyotype indistinguishable from that of the parental strain (Bastien *et al.*, 1990; unpublished data), including in strains with complex patterns believed to be due to numerous different-sized homologous chromosomes. It is noteworthy that this identity equally includes the nonstoichiometric staining intensities of the chromosomal bands. Therefore, the karyotype of a *Leishmania* strain appears to be a "clonal" karyotype.

2.4. Genome Size

The actual number of chromosomes in *Leishmania* is difficult to infer from PFE banding patterns, particularly in view of the nonstoichiometric staining intensities of the chromosomal bands. Densitometric analysis of such patterns has led to tentative estimates of 26 to 33 distinct chromosomes (depending on the strain) in *L. panamensis* and *L. guyanensis* (Scholler *et al.*, 1986) and 96 DNA molecules in *L. mexicana* (Galindo and Ramirez, 1989); these discrepancies seem to be due to interpretations of the technique. Pagès *et al.* (1989) estimated 24 to 31 chromosomes (depending on the strain) in *L. infantum* by assuming there were only two staining intensities (representing single and double chromosomes). Likewise, by counting each double band as two chromosomes, the number of chromosomes for the clones of *L. major* and *L. infantum* presented in Figure 2 would be 29 and 27, respectively.

Pulsed field electrophoresis was then used to estimate the amount of nuclear DNA per haploid genome. This has been calculated as 23–27 Mb in *L. panamensis* and *L. guyanensis* (Scholler *et al.*, 1989), 26 Mb in *L. infantum* (Pagès *et al.*, 1989), 31 Mb in *L. major* (P. Bastien and D. F. Smith, unpublished data), and 67 Mb in *L. mexicana* (Galindo and Ramirez, 1989). These results are consistent with each other, except for Galindo and Ramirez who made different assumptions on staining intensities and on ploidy. Other more sensitive methods, based on kinetic, chemical, or colorimetric DNA analyses, have yielded values of DNA content per nucleus of 91 Mb for *L. donovani* (Leon *et al.*, 1978), 155 Mb for *L. tropica* (Vedel and Robert-Géro, 1987), and 206 Mb for *L. garnhami* (Villalba and Ramirez, 1982). As stated in Section 1, we believe that all the chromosomal DNA species migrate out of the wells in the PFE conditions used for the separation of large molecules. The large differences in DNA content obtained by PFE and other methods is unlikely, therefore, to be due to technical artifacts. They could be accounted for by the presence of numerous different or homologous molecules comigrating in the gel, keeping in mind our ignorance of the ploidy of *Leishmania* and a possible aneuploidy or polyploidy (see Section 6.2). Variations in total DNA content of up to 40% have been shown by microfluorometry between clones and strains of *Trypanosoma cruzi* (Dvorak *et al.*, 1982; Kooy *et al.*, 1989).

2.5. Recognition of the Different Chromosomes and Chromosomal Homologies

The grading of chromosomal bands in order of increasing size is not sufficient to define the chromosomal organization of *Leishmania*. It is essential to find out the sequence homologies between the different chromosomal bands and chromosome classes. To attain these objectives, the first step should be the reliable identification of each chromosome, with the purpose of establishing an "electrophoretic linkage map" for *Leishmania*. Several types of DNA probes are available to specifically identify a chromosome: protein-coding sequences, or gene probes; random "anonymous" sequences; or whole or part of chromosomes directly excised from PFE gels.

Most authors have used gene probes (see Section 3.5 for references). Figure 3A shows 17 out of 24 chromosomes of a clone of *L. major* identified by marker gene probes (Samaras and Spithill, 1987; T. W. Spithill, P. J. Murray, and N. Samaras, personal communication). The advantage of this method is the clearly defined specificity of the mapping. One of its disadvantages is that multicopy

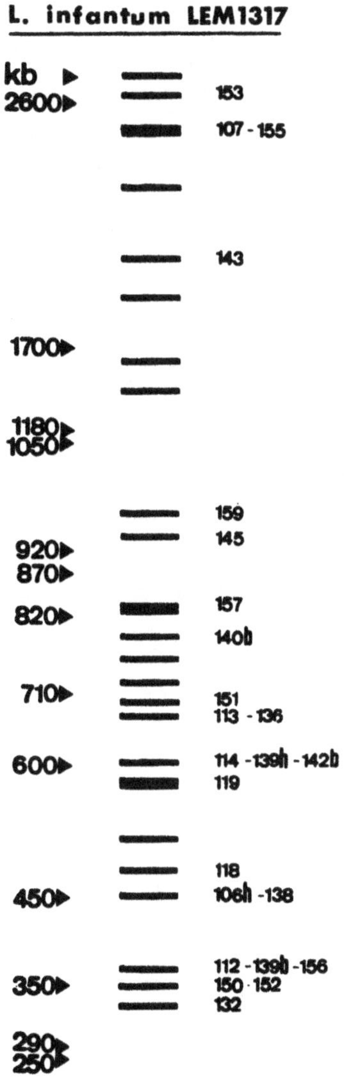

FIGURE 3 (*Continued*)

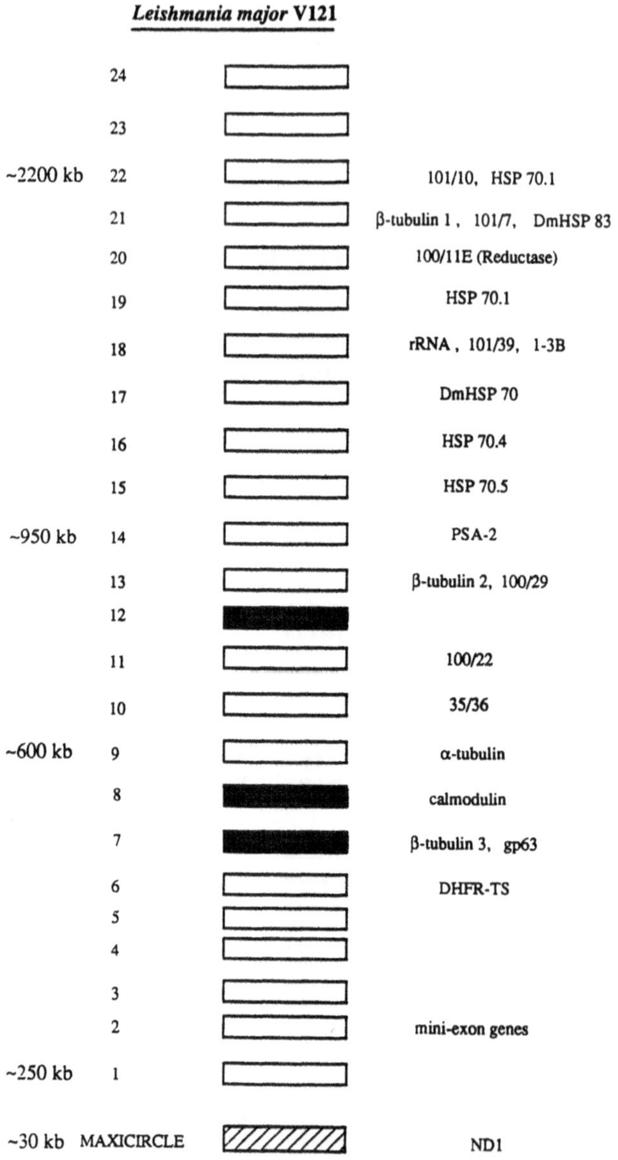

FIGURE 3. Identification of chromosomes and chromosomal homologies by Southern analysis, using as probes either (A) protein-encoding sequences or (B) single- or low-copy "anonymous" cloned DNA fragments. In A, 24 gene probes were mapped to 17 chromosomes of a cloned strain of *L. major*, LRC-L137/7/V121 (original strain = MHOM/IL/67/JerichoII). PSA-2 is a cDNA for the PSA2 surface protein of *L. major*, and probes 35/36, 100/22, 100/29, 101/39, and 1-3B are uncharacterized cDNAs (T. W. Spithill, P. J. Murray, and N. Samaras, personal communication). In B, 16 chromosomes of the *L. infantum* clone shown in Figure 1 were specifically identified with a total of 24 anonymous DNA probes (P. Bastien, C. Blaineau, and M. Pagès, unpublished data); the code name of these probes is made of the prefix ISA followed by the number indicated. Bands with a double ethidium bromide-staining intensity are shaded in A and drawn as thick lines in B.

genes are not necessarily restricted to one chromosome, as has been demonstrated for a homologous β-tubulin gene probe (Spithill and Samaras, 1987b). Homologies of gene probes with several chromosomal sites could also be due to shared noncoding repetitive sequences included in the probe, like the common (CA/GT)n sequences (Landfear *et al.*, 1986; Kapler *et al.*, 1987). Indeed, Galindo and Ramirez (1989) have shown the reduction of artifactual hybridizations to additional chromosomes after removal of (CA/GT)n sequences from two *Leishmania* gene probes. However, some of their results must be taken cautiously since the hybridization sites they identified with the original probes did not agree completely with those found by Scholler *et al.* (1986) and Samaras and Spithill (1987). These discrepancies (which were not mentioned by the authors) could simply be due to lower stringency posthybridizational washes and do not nullify their main finding. Repetitive sequences do constitute 25–38% of the *Leishmania* genome (Leon *et al.*, 1978) and therefore will frequently occur in cloned DNA fragments. Another disadvantage of mapping with gene probes is their limited availability: in nine publications on this subject, only 14 probes have been used, of which six hybridized exclusively to large chromosomes that were not resolved.

The other method of mapping makes use of anonymous cloned DNA restriction fragments, either from a total genomic DNA library (Evans *et al.*, 1987; Pagès *et al.*, 1989) or from a specific chromosome directly excised from the PFE gel (Bishop, 1990). Such sequences offer two advantages: their almost unlimited availability and the ease of selection of single-copy or low-copy sequences. Indeed, the latter should be the ones used for the identification of chromosomes. Figure 3B shows the karyotype of the clone of *L. infantum* presented in Figure 1, where 16 of 24 chromosomes could each be specifically identified by 1 to 3 such probes (P. Bastien, C. Blaineau, and M. Pagès, unpublished data).

Clearly, a combination of single- or low-copy number gene-coding and anonymous probes would give the best results for our purpose. But the confirmation of the "true" identity between two chromosomes (i.e., different-sized homologues within the same or between different strains) should involve an assessment of sequence homology usually entailing restriction mapping.

Finally, several authors have used whole chromosomes or chromosome fragments as probes, after excision or elution from the gel and radiolabeling (Bishop and Miles, 1987; Samaras and Spithill, 1987; Spithill and Samaras, 1987a; Iovannisci and Beverley, 1989; Bastien *et al.*, 1990; Giannini *et al.*, 1990). This is not the method of choice for the recognition of individual chromosomes, since hybridization usually occurs on several chromosomes, probably due to the presence of shared (repetitive) sequences.* Nevertheless, this method can

*The only exception we know of is a probe eluted from a 900-kb chromosomal band, which hybridized essentially to the same band, in a strain of *L. major* (Giannini *et al.*, 1990).

give valuable information on the distribution of repetitive sequence families among chromosomes. Thus, a whole "chromosome 1" probe from a strain of *L. donovani* hybridized only to four/six chromosomal bands in the same karyotype: bands number 1, 4, 13/15, and 21/22 (Bishop and Miles, 1987). Similar results were obtained with a strain of *L. panamensis* (Samaras and Spithill, 1987).* Finally, when an entire large chromosome (1750 kb) from a clone of *L. infantum* was used as a probe on the same karyotype, we surprisingly observed hybridization on only 11 chromosomal bands of 20 (P. Bastien, C. Blaineau, and M. Pagès, unpublished data); in particular, the probe did not see three of the six large megabase chromosomes. These data indicate that in *Leishmania* (1) repetitive sequences are generally not chromosome-specific, but are distributed among several different chromosomes; and (2) the repetitive DNA sequences present on a chromosome (even if a large one) are not necessarily shared by all chromosomes (Samaras and Spithill, 1987). This organization is shared by other protozoans (Van der Ploeg *et al.*, 1985; Gibson and Borst, 1986; Gibson and Miles, 1986), as well as by many other eukaryotes (Jelinek and Schmid, 1982).

3. MOLECULAR KARYOTYPE POLYMORPHISMS IN *LEISHMANIA*

The most striking feature of molecular karyotypes in *Leishmania* is the high degree of polymorphism observed between strains and species. When the PFE technique made it possible to karyotype protozoan parasites, one of the first questions addressed was the conservation of genome organization between and within genera, with particular reference to species of uncertain status. Initial PFE studies on *Leishmania* therefore concentrated on comparisons between different species and demonstrated extensive polymorphisms, some of which were suspected to be species-specific. We shall see, however, that the latter is in most cases comparable to the *intra*specific polymorphism; only extreme differences and extreme similarities could reflect respectively different species or a recent common origin. We shall also show that the clonal heterogeneity of the strains only play a very minor role in PFE patterns and their polymorphisms. The chromosome size polymorphisms, and the chromosomal homologies within the same or between different karyotypes, will then be more precisely examined: it will be shown that the extent of these size variations is responsible for much of the polymorphism and how it precludes simple banding patterns comparisons. In view of this data, we shall then try to show why, in our opinion, PFE is of limited usefulness for any classification of *Leishmania*.

*In some instances, chromosome I probes have given very different results on account of the distinct nature of chromosome I in the strains examined; this will be treated in Section 4.6. Whole chromosomes have proved very useful for the study of such minichromosomes and of circular DNAs.

3.1. Polymorphisms between Different Species*

Numerous studies have centered on this problem. Soon after the advent of PFE, most of the major recognized species of *Leishmania* were examined, but each research group compared only one or a few strains of each species. Moreover, the complete karyotype was seldom considered for these studies, and a "compression zone" was left unresolved (> 1000–1800 kb, depending on the technique). Some comparisons between species focused solely on the six smallest bands of the karyotype (Bishop and Miles, 1987).

In most studies, the PFE karyotypes were found to differ greatly between *Leishmania* species (Comeau *et al.*, 1986; Giannini *et al.*, 1986; Scholler *et al.*, 1986; Samaras and Spithill, 1987). This interspecific polymorphism affects the number, size, and staining intensity of the chromosomal bands (Figures 2 and 4). It can sometimes give the appearance of different clusters of bands in different molecular weight regions (Samaras and Spithill, 1987). Any chromosomal size polymorphism is nevertheless much more evident in the small chromosomes, which are easily resolved by PFE and where size variations as little as 10 kb can be reproducibly demonstrated.† In contrast, the differences in size between large chromosomes (> 1800 kb), which are more difficult to resolve, have to be considerable (≥ 100 kb?) if they are to be detected. The chromosomal distribution of several genes was also studied and confirmed chromosome size variations of 30–200 kb between homologues in strains of different species. Moreover, some genes such as the β-tubulin isogenes were localized to several chromosomal bands, which appeared to vary in size and number between species (Scholler *et al.*, 1986; Bishop and Miles, 1987; Dujardin *et al.*, 1989). Such polymorphism led several authors to postulate the existence of species-specific karyotypic features (Giannini *et al.*, 1986, 1990; Bishop and Miles, 1987; Dujardin *et al.*, 1989). This controversial point will be examined in further detail in Section 3.5.

3.2. Polymorphisms within the Same Species

Before being able to define species-specific karyotypic features, it seems preferable to assess the intraspecific polymorphism based on a sufficiently large number of strains,‡ keeping in mind that the actual assessment of karyotypic

*See note p. 131.

†Chromosome sizes can be estimated with 5% precision (see note p. 138). This observation has led Giannini *et al.* (1990) to consider "minor" size differences (< 25 kb) between the chromosomes of different strains as not significant. In our experience, repeated lane-to-lane comparisons of different strains does allow one to confirm the constancy of minor (≥ 10 kb) size differences between neighboring-sized chromosomal bands. Likewise, Iovannisci and Beverley (1989) have reported size variations of 10, 25, and 35 kb for their chromosome II in *L. major*.

‡The actual minimum number of strains needed to reveal a "representative" strain from PFE patterns remains to be defined.

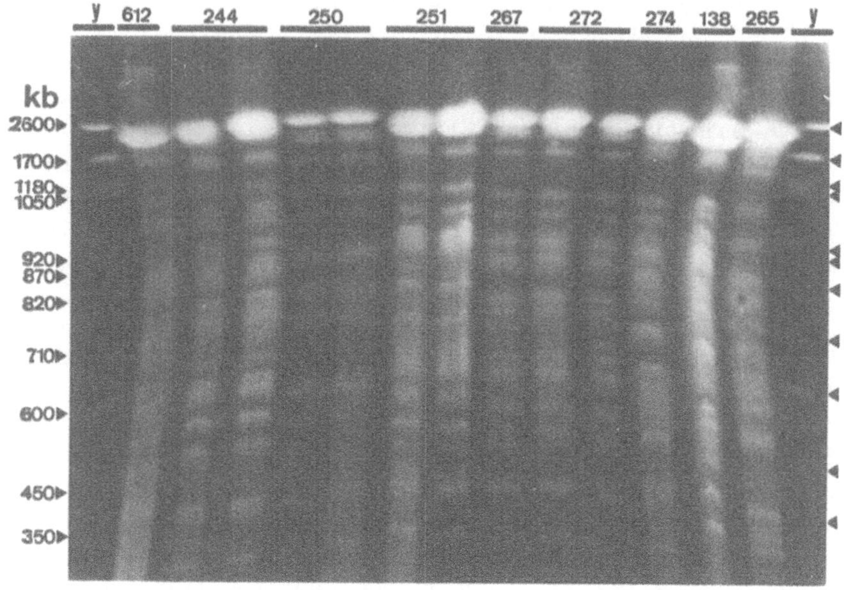

FIGURE 4. Inter- and intraspecific karyotype polymorphisms in *Leishmania*. Ethidium bromide-stained PFE gel of four different species of *Leishmania* and six sympatric strains of *L. infantum*. All strains are termed LEM followed by the number indicated: LEM 612 = *L. tropica*; LEM 138 = *L. donovani*; LEM 265 = *L. major*; LEM 244, 250, 251, 267, 272, and 274 = *L. infantum*; y = yeast chromosomal DNA. A high degree of polymorphism can be noted in the number, size, and intensity of the chromosomal bands, between different species as well as within the same species. Only two strains were found with indistinguishable karyotypes: LEM 267 and 272 (see text). The compression zone above 1700 kb can be better resolved using longer pulse times and also shows some polymorphisms (not shown). World Health Organization code numbers of the strains: 612 = MHOM/IQ/65/AV.SINAII; 138 = MHOM/IN/00/Devi; 265 = MHOM/MA/81/LEM265; all *L. infantum* = MHOM/FR/81/LEM (followed by) 244, 250, and so on.

diversity and the significance given to the differences between karyotypes may be controversial.

Giannini *et al.* (1986) first reported differences in the karyotypes of two *L. major* strains from Senegal and Egypt, but estimated they had seven tenths of the bands in common, as opposed to one tenth usually found between different species. They also observed "considerable intraspecies variation in the karyotypes of the New World *Leishmania,*" and found only one third of the chromosomes in common between a strain and a recent "isolate" of *L. panamensis*. Scholler *et al.* (1986) described large variations between three strains of *L. amazonensis* and less between two strains of *L. braziliensis*. Comeau *et al.* (1986) and Samaras and Spithill (1987) also detailed the numerous karyotypic differences seen between two other strains of *L. amazonensis* and five strains of *L. major*, respectively. Finally, Bishop and Miles (1987) mentioned intraspecific differences in *L. donovani, L. infantum,* and *L. chagasi*. All, except Comeau *et al.*, found the karyotypes more similar among strains of the same species than between different species. However, when Pagès *et al.* (1989) examined a large number of strains of *L. infantum,* they observed a dramatic polymorphism that appeared just as important as that between different species. Like the latter, it affects the number, size, and staining intensity of the chromosomal bands in apparently any size class of the karyotype.

Figures 4 and 5 show a gel and a diagram of various *L. infantum* strains examined in our laboratory, belonging to the same zymodeme and coming from the same (Figure 4) or different (Figure 5) geographic areas (Pagès *et al.*, 1989; unpublished data). The first conclusion to be drawn from these data is that virtually any strain can be identified by its distinctive PFE pattern. This was also noted by several groups (Comeau *et al.*, 1986; Giannini *et al.*, 1986; Scholler *et al.*, 1986; Bishop and Miles, 1987; Spithill and Samaras, 1987a; Pagès *et al.*, 1989). So far, only two strains of *L. infantum* have been found with indistinguishable karyotypes (Figure 4): they were isolated in the same village and on the same day (M. Pagès *et al.*, unpublished data). They could have originated from the same host source or alternatively from a human error in the manipulation of the strains.* The only other report of two indistinguishable karyotypes came from Bishop and Miles (1987) about two *L. donovani* strains that had been isolated in the same locality in India at a two-year interval. The second conclusion is the incapacity to delineate a species-specific karyotype, a conclusion also drawn by Comeau *et al.* (1986) and Scholler *et al.* (1986). The third is the apparent absence of any karyotypic features particular even to a zymodeme or to a visceral–cutaneous phenotype. When these 11 strains of *L. infantum* of the same zymodeme are compared, the polymorphisms observed appear just as great as when strains from different zymodemes are examined (Pagès *et al.*, 1989). A similar degree of polymorphism is found when strains isolated from visceral or cutaneous lesions or from human or dog are compared (for example, see strains 244 and 250 in Figure 4; 1163 and 332 in Figure 5).

Lastly, it has also been stated that within the same species *Leishmania* strains from the same geographic area are karyotypically more similar than those from different areas (Bishop and Miles, 1987; Samaras and Spithill, 1987; Bard, 1989). Thus, Samaras and Spithill (1987) found nine tenths of the bands in common between three *L. major* strains from Israel as opposed to six tenths

*It must be emphasized that each strain used should be properly checked by isoenzyme profiles or another characterization technique to ensure its identity prior to every molecular study. Unexplained and unnoticed mixing of strains is relatively frequent and occurs in the best laboratories; too many colleagues around the world have seen months or years of work ruined by such 'misidentifications' (see also Lainson and Shaw, 1990).

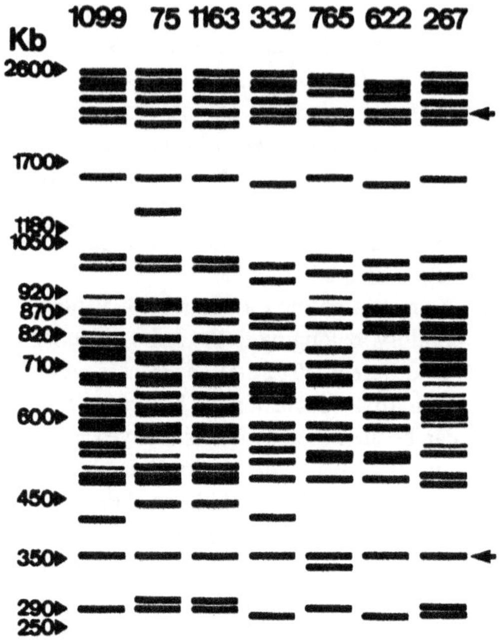

FIGURE 5. Karyotype polymorphisms within *L. infantum* zymodeme MON-1. Diagrammatic representation of the karyotypes (see Figure 2) of six strains of various geographic origins. Numerous chromosomal polymorphisms can be noted, except for LEM 75 and 1163 (respectively, parental strain and derived clone), which have almost identical karyotypes. Only two bands are strictly size-conserved among these six strains (arrows at right). From Pagès *et al.* (1989). Reproduced by permission of Elsevier Science Publishers.

between strains from Israel and Sudan–Soviet Union. However, when we examined a number of *L. infantum* strains originating from a restricted endemic area (Pyrénées Orientales, Southern France, and Catalonia, Northern Spain), eight of these strains exhibited extremely polymorphic karyotypes, with only two tenths to four tenths of the bands shared, irrespective of their host, place of isolation, or isoenzymic identity (Figures 4 and 7A): they belonged to three zymodemes, and several of them were isolated in the same 30-km-long valley (two in the same village). On the other hand, eight strains showed a surprisingly great karyotypic homogeneity, with seven tenths to nine tenths of the chromosomal bands in common (Figures 6 and 7B). Even more surprising was the fact that (1) these strains also belonged to three different zymodemes and (2) they were not all isolated in nearby villages, but, quite the opposite, in localities up to 120 km away. Thus, it is probably true that strains which have nine tenths of their bands in common can only be found in the same geographic area, and they are most

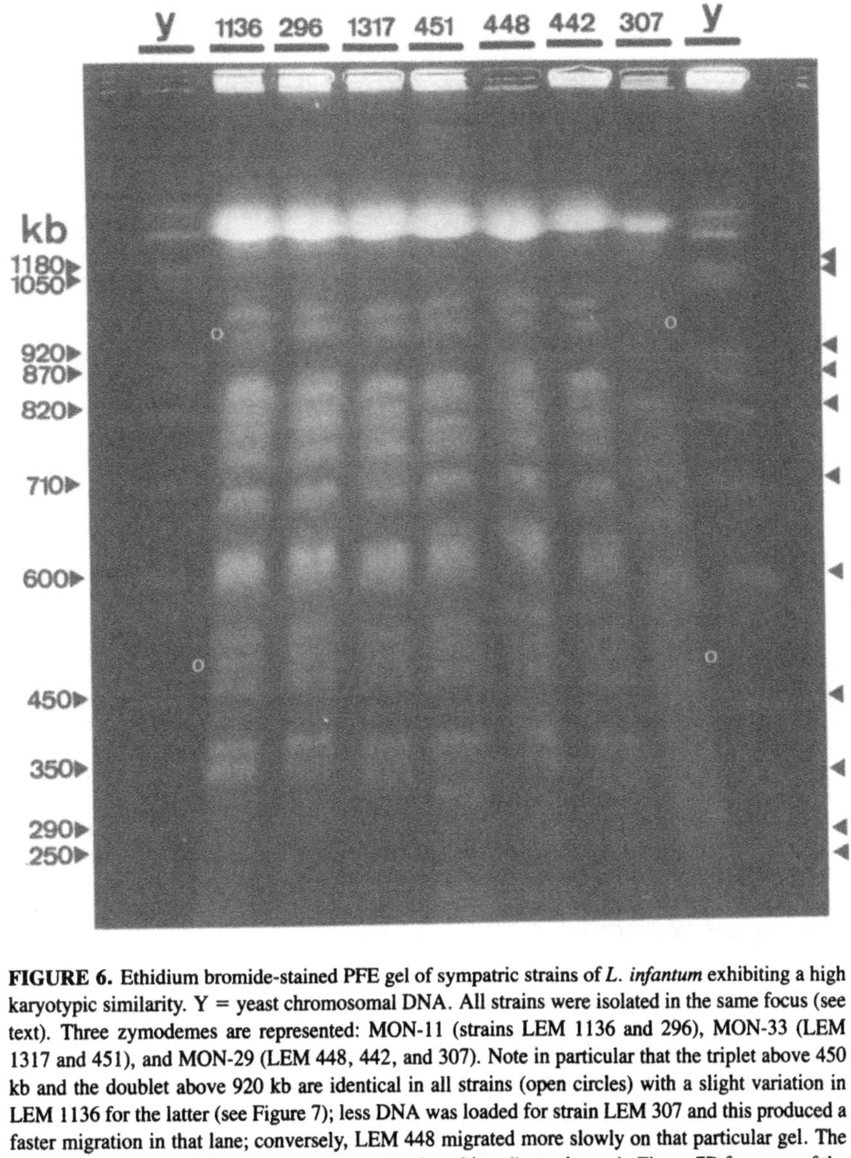

FIGURE 6. Ethidium bromide-stained PFE gel of sympatric strains of *L. infantum* exhibiting a high karyotypic similarity. Y = yeast chromosomal DNA. All strains were isolated in the same focus (see text). Three zymodemes are represented: MON-11 (strains LEM 1136 and 296), MON-33 (LEM 1317 and 451), and MON-29 (LEM 448, 442, and 307). Note in particular that the triplet above 450 kb and the doublet above 920 kb are identical in all strains (open circles) with a slight variation in LEM 1136 for the latter (see Figure 7); less DNA was loaded for strain LEM 307 and this produced a faster migration in that lane; conversely, LEM 448 migrated more slowly on that particular gel. The chromosomes above 1180 kb in size (not resolved on this gel) are shown in Figure 7B for some of the strains. The gel was run at 150 V with pulse times of 80 sec, 60 sec, and 40 sec for 24 hr each. World Health Organization code numbers of the strains: 1136 = clone of MHOM/FR/80/LEM189; 296 = MCAN/FR/81/LEM296; 1317 = clone of MHOM/FR/82/LEM356; 451 = MHOM/FR/83/LEM451; 448 = MHOM/FR/83/LEM448; 442 = MHOM/FR/83/LEM442; 307 = MHOM/ES/81/LEM307.

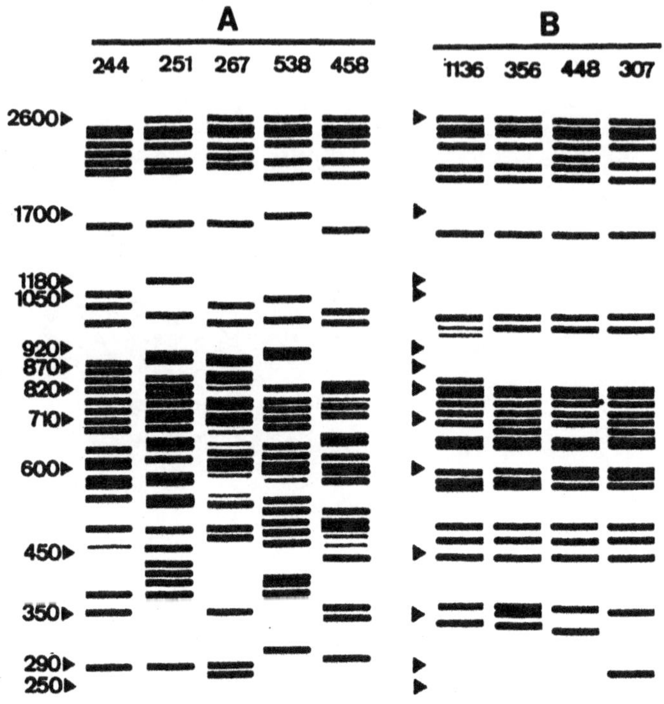

FIGURE 7. Polymorphism and homogeneity of karyotypes within the same geographic area. These are diagrammatic representations of the karyotypes of some of the strains of *L. infantum* electrophoresed in Figures 4 and 6 (356 = parental strain of clone 1317). All strains were isolated in the same focus (see text) in close, or sometimes identical, localities. (A) Strains with numerous polymorphisms in all chromosome size classes. (B) Strains exhibiting a remarkable karyotypic similarity. Three different zymodemes are represented in each panel: in A, zymodemes MON-1 (strains LEM 244, 251, 267), MON-34 (LEM 538), and MON-28 (LEM 458); in B, MON-11 (LEM 1136), MON-33 (LEM 356), and MON-29 (LEM 448 and 307). World Health Organization code numbers of the strains: see Figures 4 and 6, except strains 538 = MHOM/FR/84/LEM538 and 458 = MHOM/ES/83/BCN2.

probably of recent common origin. But it does not follow that all strains isolated in the same locality will necessarily be similar karyotypically.

Therefore, much polymorphism is observed among strains of the same *Leishmania* species, including those that are identical by all other criteria (geographic origin, host and life cycle, pathogenicity, isoenzymic profiles). In our opinion, such polymorphism is independent of the category (species, zymodeme) to which the strains belong, with the possible exception of two extremes: on the one hand, an excessive polymorphism ($< 1/10$ of bands in common) may reflect a very distant origin, hence distinct species; on the other hand, extensive similarities ($> 9/10$ of bands in common) could signify some genetic relatedness and a common geographic origin.

3.3. Interclonal (Intrastrain) Polymorphisms

Most of the *Leishmania* strains examined by PFE were not cloned, so that some doubt persisted as to the role of clonal heterogeneity in the observed polymorphisms (Scholler *et al.*, 1986; Giannini *et al.*, 1990) and in the non-stoichiometric staining intensities of the chromosomal bands (Bishop and Miles, 1987). However, we have seen in Section 2.4 that numerous clones have been reported with a karyotype indistinguishable from that of the parental strain in various species. On the other hand, a karyotype polymorphism between sibling clones (derived from the same strain) has also been described for five strains in three species: *L. major* (Spithill and Samaras, 1987a; Iovannisci and Beverley,

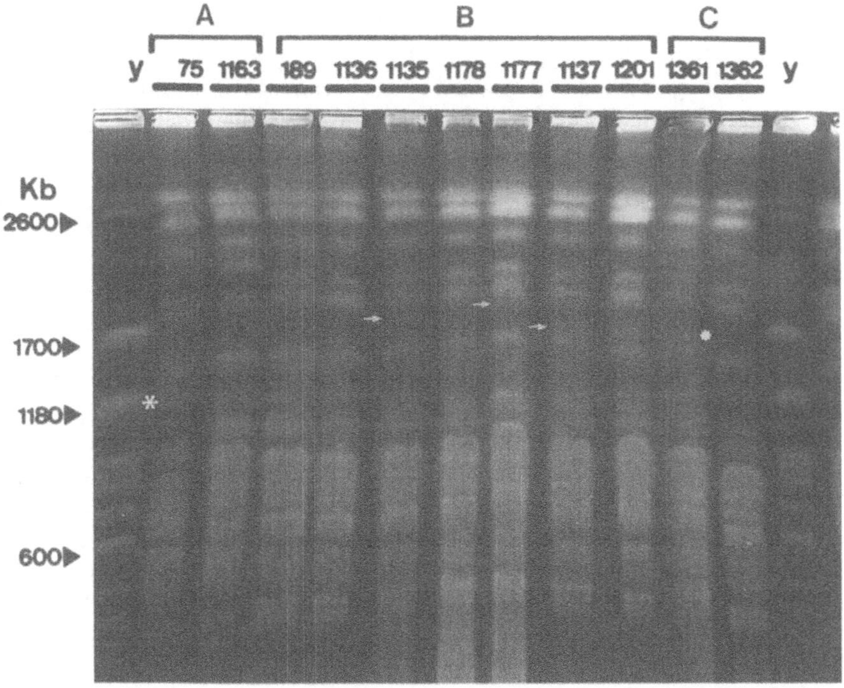

FIGURE 8. Intrastrain karyotype polymorphisms in *L. infantum* in the large chromosome size classes (ethidium bromide-stained PFE gel). (A) LEM 1163 is a clone of LEM 75 and has an 1100-kb chromosomal band missing compared with the parent (asterisk). (B) Strain LEM 189 and six derived clones: LEM 1136 is indistinguishable from the parental strain; LEM 1135 and 1178, 1177, and 1137 and 1201 all present an additional chromosomal band of 1750, 1850, and 1650 kb, respectively (arrows). (C) Clone LEM 1137 was eventually subcloned: the extra 1650-kb band was still found in eight subclones (e.g., LEM 1361) but was lost in subclone LEM 1362 (dot). From Bastien *et al.* (1990). Reproduced by permission of Elsevier Science Publishers.

1989), *L. amazonensis* (Comeau *et al.*, 1986; Scholler *et al.*, 1986), and *L. panamensis* (Giannini *et al.*, 1990). In *L. infantum*, the analysis of 36 clones derived from eight strains has shown a chromosomal polymorphism in a high proportion of clones (50%), although this was quite variable with the strain (Bastien *et al.*, 1990). However, if not rare, this interclonal polymorphism is always limited in comparison with the interstrain polymorphism. As shown in Figures 8 and 9, three types of variations are found in *L. infantum* (Bastien *et al.*, 1990): (1) disappearance of one band (Figure 8A); (2) presence of one or two additional bands, usually but not always with a lower staining intensity (Figures 8B and 9); and (3) size modification of one band (Figure 9). The fact that the clones differ from the parents by only one or two bands of 24 argues for a close genetic relatedness between both and a relative homogeneity of the strain. It must be emphasized that in contrast with *T. cruzi* (Engman *et al.*, 1987) there are no reports for *Leishmania* of marked differences between clones and either their parental strain or their sibling clones. This suggests, then, relative homogeneity of the original *Leishmania* strain rather than the selection of particular clones by *in vitro* culture. The high proportion of variant clones we found in PFE might look surprising: it could be explained by the occurrence of modifications during or shortly after cloning or by a particular structure of the strain populations in culture (see Section 5).

In any case, it is clear from the data on parental and variant clones that this heterogeneity is limited and does not play any significant role in the interstrain karytope polymorphisms in *Leishmania*.

3.4. Chromosome Size Polymorphisms and Size-Variable Homologues

Most of the authors mentioned in this chapter have evoked size differences of homologous chromosomes *between* the different strains/species to explain the karyotype polymorphisms observed. This was supported by the hybridization patterns obtained with gene probes that hybridized to chromosomal DNAs exhibiting size variations between strains (see Section 3.5).

The first mention of different-sized homologous chromosomes *within* the same strain came from Scholler *et al.* (1986), who observed hybridization of the β-tubulin probe to two closely migrating chromosomal bands (≈ 600 kb) in one strain each of *L. braziliensis* and *L. guyanensis*, instead of a single band of

FIGURE 9. Interclonal polymorphisms of chromosome I in *L. infantum*. (A) Ethidium bromide-stained PFE gel showing, from left to right: strain LEM 251 bracketing four derived clones, with LEM 1284 showing an additional chromosome I (open circle); strain LEM 189 and one of its variant clones with a different-sized version of chromosome I (arrowheads). (B) Southern blot of the same gel showing the identification of chromosome I with the chromosome-specific DNA probe ISA132. From Bastien *et al.* (1990). Reproduced by permission of Elsevier Science Publishers.

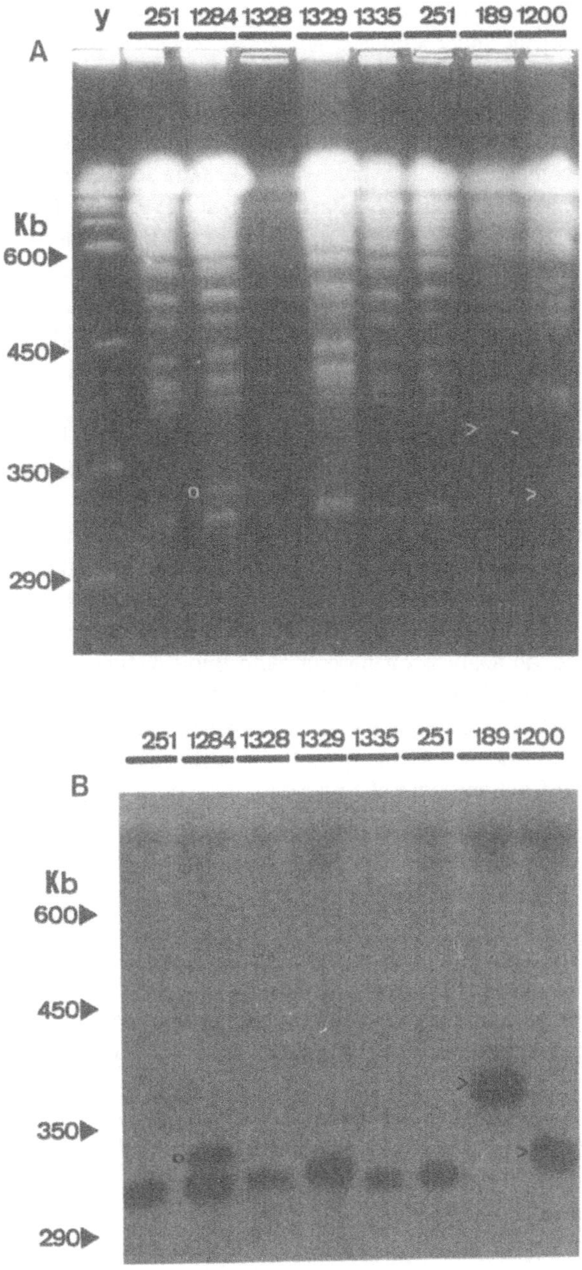
A
y
251
1284
1328
1329
1335
251
189
1200
Kb
600
450
350
290
B
251
1284
1328
1329
1335
251
189
1200
Kb
600
450
350
290

similar size in the other strains/species examined. Similarly, the 5′-spliced leader sequence gene was found to hybridize to a doublet (400–500 kb) in two strains of *L. tropica* and one strain each of *L. donovani* and *L. amazonensis*, instead of one band as in all the other strains examined (Comeau *et al.*, 1986; Samaras and Spithill, 1987). Finally, two independent single-copy DNA probes gave the same pattern of hybridization (a doublet of bands ≈ 500 kb, differing by 40 kb) in 3 of 21 strains of *L. infantum* (Pagès *et al.*, 1989). Interestingly, when one of these strains was cloned, three clones of five had lost the doublet and exhibited a single band of hybridization instead (Bastien *et al.*, 1990). Conversely, in another strain, a doublet of chromosome I homologues (≈ 300 kb) appeared in one clone (Figure 9). The most likely interpretation of these data is that homologous chromosomes in *Leishmania* are size-variable, not only between different karyotypes but within the same one. Iovannisci and Beverley (1989) clearly demonstrated such size variations of homologous chromosomes: they observed a 35-kb size increase in one of the homologues of chromosomes II in a variant line of *L. major*, due to an amplification of the miniexon gene array. Different versions of chromosome II were found in clones of this line, yielding in each case a doublet of homologues, ranging in size from 350 to 400 kb.

We have summarized in Figure 10 the analysis of such size-variable homologues for the three smallest chromosomes (termed I, II, and III) in seven cloned strains of *L. infantum*. The homologies were deduced after identification of each chromosome by one to three specific DNA probes (see Figure 3) and confirmed by the restriction analysis of these chromosomes in five clones (Bastien *et al.*, 1990; Blaineau *et al.*, 1991; P. Bastien, C. Blaineau, and M. Pagès, unpublished data). All three chromosomes vary in size and every strain has a distinctive pattern. The difficulty of interpretation of the PFE karyotypes in different strains/species without such data appears evident: a single chromosomal band may contain two distinct chromosomes, a pair of homologous chromosomes, or just one of two (or more?) homologues. In this way, we found that the first two bands in one clone, ranging in size from 350 to 390 kb, were equivalent to the first six bands, ranging in size from 280 to 440 kb, in another! Moreover, these differences are not necessarily or evidently correlated with changes in the staining intensities, which could help to distinguish a reduction in the chromosome copy number present in one band. For example, the staining intensity of a band representing the comigration of chromosomes II and III (as verified by complete restriction digestion) was not higher than that of the smallest band where only chromosome I was found (Blaineau *et al.*, 1991). Likewise, the appearance of an additional larger version of chromosome I in clone LEM 1284 (Figure 9) is not reflected in any reduction in intensity of the "original" smaller chromosome I band. It should be noted that this contrasts with the densitometric analysis of similar gels by Iovannisci and Beverley (1989), who observed the expected

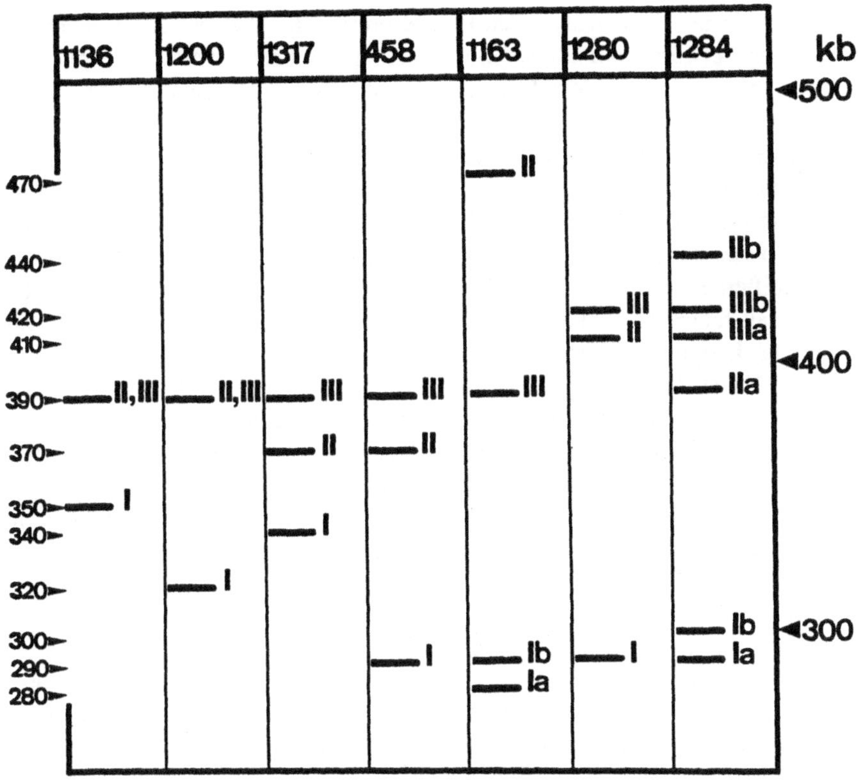

FIGURE 10. Homologies of the three smallest chromosomes among seven cloned sympatric strains of *L. infantum*. The chromosomes were termed I, II, and III, and different-sized homologues in the same karyotype were termed a and b. Sizes are indicated on both sides in kilobasepairs. Each strain shows a distinctive pattern. In LEM 1284, all three chromosomes are present as different-sized homologues. Size variations of homologues of chromosomes I and II among the strains can reach up to 25% of the chromosome length (e.g., chromosome II in LEM 458 and 1163).

reduction and increase in intensity of the second and third band, respectively, when one chromosome II homologue increased in size to that of chromosome III.

The extent of the size variations between homologues must be emphasized: thus, between different strains, chromosome I can vary by 70 kb (280–350 kb) and chromosome II by 100 kb, which is about 25% of their size. Even within the same karyotype (clone LEM 1284), the chromosome II homologues can differ by 60 kb. Likewise, Iovannisci and Beverley (1989) have shown a 50-kb variation for their chromosome II homologues in clones of *L. major*. We have also found variations of over 200 kb for homologues of a large chromosome (1650–1850 kb) between different clones of a strain of *L. infantum* (P. Bastien, C. Blaineau, and M. Pagès, unpublished data). It is noteworthy that the same degree of variation between homologous chromosomes has been reported in other protozoa: rearrangements involving hundreds of kilobasepairs have been described in *Trypanosoma,* and in particular a deletion of 100 kb on a 300-kb chromosome (Van der Ploeg *et al.*, 1984c); in *Plasmodium,* variations of up to 20% in chromosome size have been shown to result from homologous recombination for chromosomes I (600–730 kb) and II (850–1050 kb) (Corcoran *et al.*, 1988).

To summarize, a major difficulty arises when comparing similar-sized bands from different karyotypes: chromosomal DNA molecules of a markedly different size may represent homologues, and conversely, similar-sized molecules may have no homology (Comeau *et al.*, 1986; Scholler *et al.*, 1986; Bishop and Miles, 1987). The karyotype polymorphisms in *Leishmania,* then, appear to be largely accounted for by the presence of size-variable homologous chromosomes. Some of the *L. infantum* strains (see LEM 251, Figures 4 and 7), which have a particularly complex PFE pattern with numerous and often ill-defined bands, probably possess numerous "pairs" of different-sized homologues, as shown here for the three smallest chromosomes in clone LEM 1284. Conversely, "simple" patterns could be due to the comigration of different or homologous chromosomes in several size classes, like clone 1136 in Figure 10. Aneuploidy, that is, differing amounts of homologues (mono- or polysomy), could also account for such variations (Section 6.2). It is clear from these data that the proper understanding of the chromosomal organization and of the significance of karyotype polymorphisms in *Leishmania* will require each band to be precisely identified and related to a defined chromosome.

3.5. PFE and the Taxonomy of *Leishmania*

Several authors have stated the interest of PFE for the species identification of *Leishmania* (Giannini *et al.*, 1986, 1990; Bishop and Miles, 1987; Dujardin *et al.*, 1989), including for species belonging to the same complex (subspecies). This was based on two types of results: (1) the presence of shared size-conserved chromosomes within the same species, and (2) species-specific hybridization patterns with some gene probes such as β-tubulin.

3.5.1. PFE Patterns

One of the methods that can be used for comparison purposes is to estimate the proportion of size-conserved chromosomal bands in common between strains/species. For clarity, we shall express tthese in approximate percentages (to the nearest 10%), in order to better compare the results presented by different authors. Scholler *et al.* (1986) did not give any figures, but stated that although they could not determine any species-specific karyotype, the differences were greater between species than within the same species. As regards distinct species, Giannini *et al.* (1986) reported less than 10% of chromosomes in common between very distant species and 12–30% between species from the same complex (respectively, *L. mexicana*/*L. amazonensis* and *L. braziliensis*/*L. guyanensis*). *Leishmania panamensis* and *L. chagasi,* which are also very distant, shared 20% of their chromosomes. Samaras and Spithill (1987) found 40–60% of chromosomes in common among three strains from distinct Old World species and 20–30% between *L. major* and *L. mexicana.* Likewise, Bishop and Miles (1987) found that strains of the New world *L. chagasi* shared more similarities between each other than with the very close Old World species *L. infantum.* However, one strain of *L. chagasi* was actually most similar to one of *L. donovani,* and these comparisons were on the six smallest bands only. Furthermore, the karyotypic diversity shown for these bands (in particular for the three smallest chromosomes) between different species is remarkably comparable to that shown between different sympatric strains of *L. infantum* (Figure 10).

As regards the same species, Giannini *et al.* (1986) found, by contrast with distinct species, that two strains of *L. major* had 70% of chromosomes in common; but two strains of *L. panamensis* (a reference strain and a recent isolate) had only 30% in common. Samaras and Spithill (1987) then found 60–90% of the chromosomal bands shared among five strains of *L. major*. Later, Giannini *et al.* (1990) also found 70–80% of the band shared among five other strains of *L. major*. However, two of these *L. major* strains were reported as being derived from the same strain, and only 15–17 DNA bands were resolved for the whole karyotype; moreover, if one compares the karyotype of one of these *L. major* with a strain of *L. donovani* in the same article, 65% of the bands are found in common, which still seems a high proportion. Within *L. infantum,* we have found 30–60% of the bands in common between strains belonging to different zymodemes, and 30–65% within the same zymodeme (MON-1). However, if all the MON-1 strains examined are taken together, only 10% of the bands are size-conserved (Figure 5).

It seems difficult, therefore, to establish any clear-cut differences in the proportion of chromosomes in common between strains either of the same or of different species. In our opinion, this method appears of limited use for any identification or classification, the only exception perhaps regarding strains having 90% of bands in common, which seem to originate always from the same species and same area (Section 3.2).

Another method that has been used for species comparisons is the determination of a pattern of "size-conserved" chromosomal bands characteristic of a species (or subspecies). Thus, Dujardin *et al.* (1989) found six specific marker chromosomes for species of the complex *L. braziliensis,* including the first and second chromosomal bands. This is surprising in view of the great size polymorphism shown for the smallest chromosomes within several Old World species

(Section 3.4). It is possible that these chromosomes could be highly variable in size in Old World species and less so in the New World, but further clarification is needed of this point. In *L. infantum,* we have noted two size-conserved bands in 11 MON-1 strains, one of 350 kb and one of 2200 kb (Pagès *et al.*, 1989). However, bands of the same size have also been found in strains of *L. major* (Figures 2 and 4), and we are as yet incapable of attributing any marker chromosomes to *L. infantum.*

It must be remembered that some authors do not consider minor size variations (< 25 kb) as significant, while others do.* These divergent views on the quantification or interpretation of the size variations might explain some of the discrepancies between the results above, but not the general inconsistencies of all the data put together, which call for three major observations. The first is the small number of strains generally examined within one species and the extrapolation of the features of one or two strains to a whole species. The second is that the karyotypes compared were often only partially resolved. The third criticism concerns the principle of the method itself, that is, the comparison of bands of which very little is known. Since bands of equivalent size can actually contain different chromosomes (Section 3.4), such comparisons based on size should be taken very cautiously.

3.5.2. Southern Analysis: Chromosomal Location of Genes

All the authors mentioned in this chapter have compared the hybridization patterns of several gene probes onto the karyotypes of different species—the hypothesis being that extensive genome reorganizations might have occurred during species evolution. However, the same criticisms as above (as regards the number of strains examined and the resolution of karyotypes) apply to these studies. Moreover, some authors have only reported the number of the chromosomal band seen by a probe instead of its actual size, which, in view of the high degree of karyotype polymorphism, hinders comparisons: for instance, homologous chromosomes of the same size hybridizing with the same probes can be visualized as bands 8, 9, or 11, depending on the strain. Nevertheless, we have tried to gather together this fragmentary but valuable information by giving rough estimates of the sizes of the chromosomes identified in each publication. This very approximate method actually yielded interesting results, summarized in Table I. For all the probes except β-tubulin, only one band was present in all species; in some strains, two adjacent bands could be seen, which suggests the presence of different-sized homologues. It appears from the table that, apart from β-tubulin, none of the probes mentioned can be used for identification or taxonomic purposes. For example, the α-tubulin, the dihydrofolate reductase thy-

*See note p. 47.

midylate synthase (DHFR-TS), and the 5′-spliced leader genes have been mapped to single bands varying relatively little in size in species as distant as *L. major, L. donovani, L. braziliensis,* and *L. mexicana*. Thus, although most of the probes can detect bands of a different size in different species, nevertheless it is clear that: (1) the size variations remain within 30–200 kb between species, (2) with some probes (e.g., α-tubulin), there is as much polymorphism within as between species, and (3) there is no specific size feature for any species. Surprisingly, this seems also true for the β-tubulin probe, which has been reported as particularly useful for distinguishing *Leishmania* species, and is indeed the only one to give very polymorphic hybridization patterns. However, this appears to depend on the origin of the probe used; thus, the South American species of the complex *L. braziliensis* typically exhibit five bands carrying β-tubulin isogenes, including a small band around 280 kb (Scholler *et al.*, 1986; Dujardin *et al.*, 1989), but this was only found with the *T. brucei* probe and not with the *Leishmania* probe. Moreover, Dujardin and co-workers themselves reported five of ten isolates as "atypical" when trying to use this band pattern to identify recent *L. braziliensis* isolates.

In summary, only one probe so far (the *T. brucei* β-tubulin) might prove useful to distinguish a complex of four species from the others. An interesting alternative to gene probes might be the use of repetitive sequences. Such sequences from chromosome I were shown to hybridize differently on the karyotypes of distinct species (Samaras and Spithill, 1987), although there again caution is required because each species was represented by only one strain. Likewise, repetitive sequences have been cloned which seem specific for each of the *L. braziliensis* and *L. mexicana* complexes (A. Mendoza, S. Eresh, and D. C. Barker, personal communication).

3.5.3. Conclusions

The inconsistencies noted above, as well as the particular insight we have on the intraspecific polymorphism of *L. infantum,* lead us to believe that PFE cannot be used yet as a taxonomic tool for *Leishmania*. Marker chromosomes of a given size are unlikely to be found, because of the karyotypic plasticity that seems to be a general feature of lower eukaryotes, in particular yeasts (De Jonge *et al.*, 1986; Snell *et al.*, 1987) and protozoan parasites (Van der Ploeg *et al.*, 1984b; Kemp *et al.*, 1985; Gibson and Brost, 1986; Corcoran *et al.*, 1988; Langsley *et al.*, 1988; Sharkey *et al.*, 1988; Ponzi *et al.*, 1990). A better understanding of chromosome size polymorphisms and the precise identification of every chromosome and chromosomal band are necessary before a valid effort can be made to determine species-specific karyotypic features. Then, one cannot exclude the possibility of finding particular sequences that would show a species-specific chromosomal distribution.

Table I
Comparison of the Chromosomal Distribution of Nine Gene Probes in the Karyotypes of Different *Leishmania* Species and Strains[a]

	L. chagasi[b] *L. infantum*	*L. donovani*	*L. major*	*L. tropica*	*L. mexicana*
β-Tubulin[c]			770 + L[e]		710 + L[e]
(*Leishmania*)	650 + 770 + L[d,e]	600 + 800 + L ?[e]	500 + 800 + L ??[h]		800 + L[e]
	600 + 900 + L ??[f]	700 + 900 + L ??[e]	700 + 900 + L ??[e]		900 + L ??[e]
		500 + 600 + 900 + L ??[e]			
β-Tubulin	600 + 860 + L ??[h]	600 + 850 + L ??[h]			800 + L[g]
(*T. brucei*)	650 + 800 + L ??[e]				720 + 820 + L[e]
	650 + 850 + L ??[e]				
α-Tubulin	700 ??[i]	650–700 ??[h]	600 ?[h]		750[e]
	750–780[g]	750[f]			900[e]
rDNA	1240[e]		1500 ?[e]		1240[e]
			L ?[h]		L ?[e]
5′-SL	400 ?[f]	370 + 450 ?[e]	370[e]	400 + 500 ?[f]	400 ?[e]
		400 + 500 ?[e]	400 ?[e]		450 ?[f]
		0[e]	350 + 400[e]		
DHFR-TS	500[e]	500 ?[e]	450–550 ?[h]	500 ?[f]	500 ?[f]
					L[e]
H-region	820[e]				820[e]
hsp 70		1000 ?[e]	1000 ?[e]	1000 ?[e]	1000 ?[e]
hsp 83		1800 ?[e]	1800 ?[e]	1800 ?[e]	1800 ?[e]
gp63	700 ?[e]	700 ?[e]	700[e]		700 ?[e]

[a]The size (in kb) and number of chromosomal bands hybridizing with each probe were gathered from the following publications: Comeau *et al.* (1986), Scholler *et al.* (1986), Bishop and Miles (1987), Samaras and Spithill (1987), Button *et al.* (1989), Dujardin *et al.* (1989), Galindo and Ramirez (1989), Iovannisci and Beverley (1989), and Giannini *et al.* (1990). The sizes of the chromosomal bands are given as reported by the authors whenever applicable; if not, approximate sizes were estimated from comparisons with the yeast chromosomes or the size markers indicated on the same photographs (indicated by ? following the sizes); when the latter was not available, rough estimates were made from the few size markers indicated and with the help of similar karyotypes of strains from the same species (indicated by ?? following the sizes).

[b]Probes considered: the β-tubulin repeats from *Leishmania* and *T. brucei* are shown separately (see text); the α-tubulin, ribosomal RNA gene repeat (rDNA), and 5′-spliced leader gene (5′-SL) probes are from *Leishmania* or from *T. brucei*, both species giving similar results for these three probes; dihydrofolate reductase-thymidylate synthetase gene (DHFR-TS) and H-region are from *Leishmania;* heat shock proteins (hsp) 70 and 83 from *Drosophila* (see Samaras and Spithill, 1987); and gp63 gene is from Button *et al.* (1989) [the mapping of gp63 to different chromosomes in *L. donovani* by Maingon *et al.* (1989) and in *L. major* by Giannini *et al.* (1990), were not included since the genes isolated by these three groups appear to differ in some respects].

[c]In each box, each line corresponds to a distinct pattern from a particular study. Similar patterns located on different lines are from different studies.

[d]The figures indicate the size in kilobasepairs of the chromosomal bands seen by the probe. All sizes are approximate, and small size variations of chromosomes between strains are not mentioned if not shown or specified by the authors. $x + y + \ldots$ = different bands; x–y = single band, ranging in size from x to y kb depending on the strain; L = large chromosome ≥ 1800 kb, usually not resolved.

[e]One strain (exhibiting this pattern in the corresponding study).

[f]Two strains.

[g]Three strains.

[h]Five strains.

[i]Six strains.

[j]Eight strains.

3.6. Relative Homogeneity of the Karyotype in *Leishmania*

Apart from the great polymorphism shown above in the number and size of the chromosomal bands, a relative overall homogeneity can be noted within the whole genus. This is first visible in the molecular weight distribution and number of chromosomal bands. Grossly speaking, between 20 and 28 bands are found within the size range of the chromosomes of *S. cerevisiae,* grouped in similar

Table I
(*Continued*)

L. amazonensis	*L. braziliensis*	*L. guyanensis*	*L. panamensis*
900 + L ??[f]			
	600 + 800 + L ?[e]		
700 + 900 + L ??[e]	600 + 900 + L ?[e]		
800 + L[f]	280 + 620 + 800 + L[e]	280 + 650 + 800 + L ?[e]	280 + 650 + 800 + L ?[f]
800 + L[h]	280 + 600 + 650 + 800 + L ?[e]	280 + 600 + 650 + 800 + L ?[e]	
850 + L ?[e]	290 + 550 + 650 + 800 + L ?[j]	290 + 550 + 650 + 800 + L ?[g]	290 + 550 + 650 + 800 + L ?[g]
	but "atypical patterns"[h]		
700[f]	750[e]	780[f]	
750[h]	770[g]		
1350 ?[e]	1220[e]	>1000[e]	>1000[e]
400 ?[g]	0[e]		
450 + 500 ?[e]			
470[g]	480[e]		
820[e]	830 ?[e]		

major size classes in all species: a few well-separated small chromosomes of 200–500 kb, a cluster of a dozen often ill-resolved medium-sized chromosomes between 500 and 1000 kb, a doublet or a triplet between 1000 and 1700 kb, and a group of few large chromosomes > 1800 kb* (Figure 2; see also Figure 6 in Button *et al.*, 1989).

This schematic picture is clearly different from the karyotype of other protozoa. For instance, in the same family, *Trypanosoma* has a PFE pattern similar to *Leishmania,* but the large chromosomes extend up to 5.7 Mb, at least in *T. brucei* (Van der Ploeg *et al.*, 1989); about 100 minichromosomes of 40–150 kb have also been found in *T. brucei* (Van der Ploeg *et al.*, 1895) and *T. vivax* (Dickin and Gibson, 1989), although not in *T. cruzi* (Gibson and Miles, 1986). The insect trypanosomatid *Crithidia* has no chromosomes smaller than 700 kb (Van der Ploeg *et al.*, 1984a). The differences with other phyla appear more dramatic. In *Plasmodium,* in the haploid stage, the chromosomes number only 14 and range in size from 650 to 3000 kb (Foote and Kemp, 1989). *Eimeria* has 12 chromosomal bands ranging in size from 1.1 to over 5.7 Mb (Shirley *et al.*, 1990), *Giardia* has 8 bands ranging from 0.8 to 3.0 Mb (De Jonckheere *et al.*, 1990), *Entamoeba* has 4–9 bands ranging from 0.3 to 2.0 Mb (Valdés

*This gross picture takes into account the resolution of the banding patterns in "usual" electrophoretic conditions. Obviously, the time-consuming run of several gels at various pulse times allows the resolution at will of any size class of chromosomes.

et al., 1990), while in *Paramecium* the macronuclear DNA is only 100–800 kb (Phan *et al.*, 1989).

Of equally great interest is the constant chromosomal distribution of several genes in all the *Leishmania* species so far examined, that is, the *relative* size conservation of the chromosomes carrying these genes. This has been detailed in Section 3.5.2 (see Table I).

Thus, it is possible that the molecular karyotype polymorphisms do not affect the essential characteristics of the chromosomes in *Leishmania*. On the other hand, the chromosomal plasticity discussed in Sections 3.2. to 3.4. might be essential for certain aspects of the biology of the parasite. The characterization of such relationships awaits detailed molecular studies.

4. MOLECULAR BASIS FOR THE KARYOTYPE POLYMORPHISMS

The molecular events leading to chromosome size and number polymorphisms in *Leishmania* are still poorly understood. The hybridization patterns of single-copy DNA probes onto the karyotypes of various species and strains would tend to indicate that large-scale interchromosomal rearrangements are not involved in these polymorphisms. The most likely explanation for the common increase and decrease in size of chromosomes remains frequent DNA amplification and deletion; one example has been characterized for each of these events. These rearrangements present two interesting features: their recurrent and apparently nonrandom nature and their subtelomeric location, at least in a few chromosomes for which long-range restriction maps are available. However, the precise molecular mechanisms and the sequences involved remain to be characterized. An interesting possibility is the presence of transposable elements in the rearrangement sites. Finally, another class of karyotype modification has been discovered recently: it consists of the presence or appearance in some strains throughout the genus *Leishmania* of small multicopy linear DNAs that seem to derive from large megabase chromosomes present in all strains.

4.1. Intra- Rather Than Interchromosomal Rearrangements

We have shown in Section 3 how the sizes of homologous chromosomes (as identified by specific DNA probes) vary within and between *Leishmania* species. However, it is equally remarkable that most of the DNA sequences used as probes map on chromosomes that show relatively small size variations of 20–25% of the chromosome size between strains/species (Section 3.6). If large-scale interchromosomal rearrangements occurred, a scattering of the probes over the karyotype patterns should be observed between different strains/species. This is neither the case for 27 random DNA probes within *L. infantum* (Pagès *et al.*,

1989; unpublished data) nor for nine of the most-used gene probes reported in the literature (see references in Table I). Therefore, much chromosome size polymorphisms would seem to be due to intrachromosomal rearrangements (such as DNA amplification/deletion), as opposed to interchromosomal exchange (such as translocation or, to a lesser extent, transposition). Others have come to the same conclusion (Scholler *et al.*, 1986; Samaras and Spithill, 1987; Bishop and Miles, 1987).

4.2. DNA Amplifications and Deletions

DNA amplification/deletion has been suggested by many authors to be a cause of changes in size of homologous chromosomes. The actual demonstration of the role of amplification in these size modifications came from Iovannisci and Beverley (1989). These authors showed that an amplification of the miniexon gene array from 93 to 150 copies on one of the homologues was responsible for a 35-kb larger version of chromosome II in a variant line of *L. major*. Several different amplifications of this tandemly repeated array were found in clones of this line: they could all account for the corresponding increases in chromosome size observed. The variable size of these amplifications was attributed either to a "genetic diversity" in the line (i.e., several independent amplifications occurring simultaneously), or to a possible instability of the rearranged array, which would resolve into different sizes in different descendant cells—the newly amplified gene array could be "more prone to mitotic crossing-over, or to additional expansions or contractions."

Interestingly, Iovannisci and Beverley's work also demonstrated for the first time in *Leishmania* the occurrence on a linear chromosome of a spontaneous gene amplification during *in vitro* maintenance and not in response to an overt selection such as laboratory drug pressure. Indeed, gene amplifications have been reported at four different loci in three different *Leishmania* species, but always as extrachromosomal circular DNAs, usually during selection for drug resistance (reviewed in Beverley *et al.*, 1988) but sometimes spontaneously (Petrillo-Peixoto and Beverley, 1988, 1989). Linear gene amplification has also been described in *P. falciparum:* drug-resistant lines of this parasite were shown to have amplified a 100-kb stretch of DNA severalfold, resulting in an increase in size of chromosome V in both lines (Foote and Kemp, 1989); again, a larger amplification in one line than in another was correlated with the presence of a 200-kb larger chromosome.

However, in most cases it is not yet clear whether the size increases are in effect produced by gene amplification or by another mechanism. Actually, it is also difficult in most cases to affirm whether the size modification was primarily a reduction or an increase. A probable increase in size was demonstrated in a clone of *L. infantum* that exhibited a 20-kb larger homologue of chromosome I compared with the parental strain (see Figure 9), but the mechanism remained unknown (Bastien *et al.*, 1990). Iovannisci and Beverley (1989) could not rule out other small alterations occurring simultaneously to the miniexon amplification to explain small discrepancies (10 kb) between the size variation of the chromosome and that of the miniexon array; moreover, the enlargement of chromosome II in another *L. major* line was not the result of miniexon amplification.

The deletion of DNA sequences has, on the other hand, been invoked to explain the reduction in size of homologous chromosomes, for instance, between two clones of *L. major* (Spithill and Samaras, 1987a). A smaller homologue is also a common feature of the interclonal chromosome size polymorphism in *L. infantum* (Bastien *et al.*, 1990). More precisely, a 30-kb deletion was shown as responsible for a similar size variation of chromosome I between an *L. infantum* strain and one of its derived clones (see Figures 9 and 10): the restriction patterns of these two homologues exhibited only one size polymorphism of 30 kb in their larger restriction fragment, and the homology of the latter in the two chromosomes was confirmed by Southern analysis (Bastien *et al.*, 1990). In trypanosomes, large deletions involving a third of the chromosome size have also been reported (Van der Ploeg *et al.*, 1984c). Finally, such deletions have been well characterized in *Plasmodium*. In this organism, they are produced by chromosome breakage followed by telomeric "healing," and can account for many of the differences in size of the polymorphic chromosomes (reviewed in Foote and Kemp, 1989; Ravetch, 1989).

4.3. Recurrent Chromosomal Rearrangements

The size-variable miniexon gene amplification characterized by Iovannisci and Beverley (1989) in *L. major* was found to be a recurrent event in several lines and strains of the same species. This recalls other observations in *L. infantum* about the appearance of additional chromosomal bands in the karyotype of clones compared with parental strains; one of these extra chromosomes was found to vary in size by increments of 100 kb (1650, 1750, and 1850 kb) between different sibling clones (see Figure 8), and several clones exhibited the same size of chromosome. Moreover, when another independent strain with a very similar karyotype was cloned, identical-sized extra bands appeared in the karyotypes of the resulting clones. Finally, when one of these variant clones was subcloned, the altered chromosome disappeared in the karyotype of one of the subclones, thus reverting to the parental karyotype (Figure 8C). A similar observation was recorded in *L. amazonensis*, where a 330-kb band disappeared in one line during maintenance and reappeared in derived lines or clones (Dujardin *et al.*, 1989). These data suggest the occurrence of recurrent nonrandom events, perhaps affecting specific chromosomes. A DNA amplification was also shown to be recurrent in independently selected lines of *L. major* by Ellenberger and Beverley (1989), who suggested the presence of "hot spots" for DNA rearrangements.

4.4. Subtelomeric Location of Rearrangements

In an attempt to further elucidate the mechanisms and sequences involved in the chromosome size polymorphisms, we have constructed long-range restriction maps of five different-sized homologues of chromosome I isolated from four cloned strains of *L. infantum* (Blaineau *et al.*, 1991). These chromosomes vary in size from 290 to 350 kb. Using enzymes that cut infrequently throughout the genome, complete and partial digestion of these chromosomes followed by Southern analysis revealed a polymorphic fragment in terminal position on the chromosome in four of five cases. The size variations of this terminal fragment (from 30 to 100 kb) can completely account for the chromosome size differences observed between the clones (Figure 11). Similar macrorestriction maps for five chromosomes in different strains of *Plasmodium falciparum* have shown a great instability of the subtelomeric regions (reviewed in Foote and Kemp, 1989; Ravetch, 1989). Moreover, Southern analysis of complete digests of both total genomic and chromosome I DNA of the above *L. infantum* clones showed that the size-variable terminal region comprised a large fragment (> 20 kb) devoid of any common restriction sites (Blaineau *et al.*, 1991). It was suggested that this region might consist of long arrays of reiterated sequences. It is of particular interest that a similar "barren" repetitive DNA segment has been found in the subtelomeric regions of chromosomes of *Trypanosoma,* and that it shows size variations of up to 20 kb (Bernards *et al.*, 1983; Michels *et al.*, 1983; Aline *et al.*, 1985). Similar arrays of subtelomeric repeats, showing a remarkably nonrandom distribution of restriction sites and also variable in size between different strains, have been described in yeast (Horowitz *et al.*, 1984; Horowitz and Haber, 1985) and *Plasmodium* (Oquendo *et al.*, 1986; Pace *et al.*, 1987; Corcoran *et al.*, 1988; Vernick *et al.*, 1988).

Rearrangements in these subtelomeric regions could take place during mitosis or during meiosis. Deletions of subtelomeric fragments have been shown to occur during mitosis in *P. falciparum* and to be involved in chromosome size polymorphisms; they can result in the loss of the size-variable repeat (Corcoran *et al.*, 1988; Patarapotikul and Langsley, 1988), as well as of various genes (Corcoran *et al.*, 1986; Pologe and Ravetch, 1986, 1988; Biggs *et al.*, 1989; Cappai *et al.*, 1989). Homologous but unequal recombination between these repeats has also been invoked as an alternative mechanism generating novel-sized chromosomes in *P. falciparum* during meiosis (Corcoran *et al.*, 1988; Sinnis and Wellems, 1988). In *Leishmania,* such a mechanism is not supported by our data on the interclonal chromosomal polymorphisms: if these are due to DNA rearrangements during cloning, we have never found reciprocal size variations of both homologous chromosomes, but rather a size modification of one of the homologues only. However, no meiosis has yet been identified in *Leishmania,* and this would only concern mitotic rearrangements, which seem rare and limited if only judged from long-term *in vitro* maintenance and cloning (see Section 2.2).

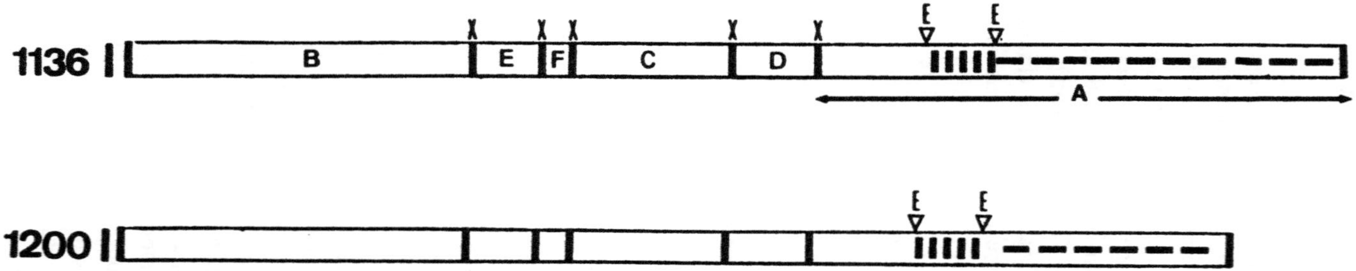

FIGURE 11. Preliminary long-range restriction maps of chromosome I homologues from two clones of *L. infantum*, LEM 1136 and 1200 (respectively, 350 and 320 kb in size). X = Xba I, E = Eco RV (fragment A only). The size polymorphism was located to a terminal Xba I fragment (A); the latter presents at least two Eco RV restriction sites that delimit a 20-kb fragment shared by both chromosomes and recognized by probe ISA132 (short vertical bars), and a size-variable fragment hybridizing with probe ISA300 (discontinuous line). The 30-kb difference in size can account for the size polymorphism shown in Figures 9 and 10 for these two homologues. For more details, see Blaineau *et al.*, 1991.

Nevertheless, there is also increasing evidence that subtelomeric repeats might be preferential sites of recombination events between homologous and even non-homologous chromosomes during mitosis, in yeast (Horowitz *et al.*, 1984; Horowitz and Haber, 1985), *Trypanosoma* (Campbell *et al.*, 1984; Shah *et al.*, 1987), and *Plasmodium* (Pace *et al.*, 1987; Dore *et al.*, 1990). In yeast, chromosome size variations could be correlated with the excision and integration of circles of repetitive subtelomeric sequences. In *T. brucei,* intertelomeric variant surface glycoprotein (VSG) gene replacement events involve recombination between the 75-base pair (bp) repeats of the barren region. In *P. berghei,* a chromosome was shown to acquire subtelomeric repeats and correspondingly increase in size during mitotic multiplication (Ponzi *et al.*, 1990). More generally, it has been suggested, for both yeast (Horowitz *et al.*, 1984) and *Plasmodium* (Corcoran *et al.*, 1988; Ponzi *et al.*, 1990) that, while the chromosomal organization is conserved in the internal regions, the terminal regions are polymorphic in length and sequence.

It is important to note that the chromosome size variations in *Leishmania,* if located near the terminal regions, cannot be due to simple telomere contraction and growth as described in *Trypanosoma* (Bernards *et al.*, 1983) or *Tetrahymena* (Larson *et al.*, 1987). Indeed, the addition of simple telomeric repeats increases the length of the chromosome by only 3–10 bp per generation, and this is insufficient to explain the large size variations of 30–100 kb observed between homologues during a maximum of 130 generations that *Leishmania* strains and clones are grown in our laboratory.

Finally, it should be emphasized that the sequences involved are not necessarily located in terminal regions of the chromosome; indeed, out of five restriction-mapped different-sized homologues of chromosome I in *L. infantum,* one was found to exhibit a 10-kb size variation in a central fragment (Blaineau *et al.*, 1991).

4.5. Interchromosomal Exchange

Few arguments presently support the hypothesis of interchromosomal (as opposed to intrachromosomal) rearrangements in *Leishmania.* It is probable that the karyotype polymorphisms between species involve more large-scale rearrangements than those within species, on account of the evolution time span. Some authors have then suggested the occurrence of translocation events during speciation as an alternative to the hypothesis of different-sized homologous chromosomes; thus, Samaras and Spithill (1987) found a 200-kb difference between the chromosomes carrying α-tubulin in *L. major* and *L. enrietti* but without being able to affirm their homology. Bishop and Miles (1987) also suggested "that translocation between chromosomes may contribute to the size polymorphisms," because of the hybridization of a chromosomal probe from one strain to two chromosomal bands in another strain; however, these bands represented a doublet, most likely due in our opinion to the presence of different-sized homologues. The duplicative transposition of a 150-kb telomeric fragment has been described in *T. brucei,* from a 2-Mb chromosome to one of 400 kb, thus yielding a novel-sized 550-kb chromosome (Van der Ploeg *et al.*, 1984c). The dispersion of β-tubulin isogenes on different chromosomes of *Leishmania* (Spithill and Samaras, 1987b), as well as the presence of shared repetitive DNA among several nonhomologous chromosomes (Samaras and Spithill, 1987; Bishop and Miles, 1987; P. Bastien, C. Blaineau, and M. Pagès, unpublished data), also support the hypothesis of interchromosomal exchange during species evolution. Nevertheless, from the available evidence, translocation events are little involved in the chromosome size polymorphisms frequently and readily observed among *Leishmania* strains of the same species.

Among other formal possibilities, one of the mechanisms that could play a role in chromosome size variations could be the insertion of foreign DNA or the transposition of DNA sequences between chromosomes. This remains very speculative, and only indirect arguments can support these hypotheses. A 30-kb circular DNA of apparently exogenous origin has been found integrated in linear chromosomes in several strains of *Leishmania* (Gajendran *et al.*, 1989) and a 6-kb viral DNA has been described in three strains of *L. guyanensis* (Tarr *et al.*, 1988; Widmer *et al.*, 1989). On the other hand, repetitive DNAs resembling mobile genetic elements have been reported in chromosomes of *Trypanosoma* (Aksoy *et al.*, 1987, 1990; Kimmel *et al.*, 1987; Murphy *et al.*, 1987) and *Crithidia* (Gabriel *et al.*, 1990). In *Leishmania,* the rearrangement site for H-region amplification was found to display properties evoking a transposon (Kamitsuka *et al.*, 1989). It is therefore plausible that such transposable elements play a part in the chromosomal rearrangements in *Leishmania.*

4.6. *De Novo* Appearance of Small Chromosomes

Recent studies have revealed a new class of karyotype modification in *Leishmania:* the recurrent and sporadic emergence of small linear DNAs (180–250 kb in size) that present features of "minichromosomes" and are present in multiple copies in the nuclear genome. These were found independently in one *L. braziliensis* strain (Scholler *et al.*, 1986), three *L. donovani* strains (Bishop and Miles, 1987; Bishop, 1990), and four *L. major* lines descendant from the same strain (Beverley and Coburn, 1990).* They were not found in a number of other strains whether of the same or of different species, nor in the original parental strains of the lines the karyotypes of which were otherwise identical

*A 50-kb linear DNA has also been found in three *L. mexicana* strains, but it has not been characterized yet and may not be present in multiple copies (S. Eresh and D. C. Barker, personal communication).

(Bishop, 1990; Beverley and Coburn, 1990; A. Mendoza and D. C. Barker, personal communication). Interestingly, they seem to all present some degree of homology (Beverley and Coburn, 1990). They also share sequence homology with one large chromosome ($\approx$ 1.5 Mb) in the respective karyotypes, suggesting that they are derived from the latter. Homologous sequences on a similar-sized chromosome seem also to be present in numerous strains that do not possess the minichromosome, including different species (Bishop and Miles, 1987; Bishop, 1990; Beverley and Coburn, 1990). It is also of particular interest that these new chromosomes present some homology with a 30-kb circular DNA (termed CD1) that is sporadically present in *L. donovani,* probably has an exogenous origin, and is integrated in linear chromosomes of varying sizes in different strains of the same or related species (Gajendran *et al.*, 1989; Stuart *et al.*, 1989). A possible relationship between the activation of copying a limited region of a large chromosome and the presence of exogenous, perhaps viral, sequences remains speculative. Beverley and Coburn (1990) attribute the formation of these minichromosomes to a "genetic lability" of the large chromosome and hypothesize the existence of an activation center from which the copying starts. The triggering factors of such a phenomenon remain unknown, although this genomic rearrangement was associated with drug resistance in three independent lines.

Interestingly, a related phenomenon was described in *Plasmodium,* following selection for drug resistance (Cowman and Lew, 1989): partial duplication of chromosome VII followed by extensive deletion yielded two homologues of a very different size (450 and 1100 kb), both distinct from the original chromosome (1300 kb). In that case, the suggested mechanism was an initial nondisjunction of the chromosome followed by secondary deletions and telomeric "healing," the resulting cell being positively selected by drug pressure.

These fascinating mechanisms of rearrangements must nevertheless be regarded within the general context of multiple chromosomal polymorphisms in strains of *Leishmania* unselected and cultivated for a short period. The multicopy minichromosomes are only rarely observed. The hybridization patterns obtained on the karyotypes with low- or single-copy DNA probes do not support the idea of small chromosomes generally being derived from large ones, even though only small portions of the genome are probed with this technique. Finally, if this mechanism (duplication plus deletion) can explain the formation of two homologous chromosomal bands instead of one, it is difficult to imagine it would occur so frequently as to concern most of the chromosomes of a strain, as suggested by the complex PFE patterns of some strains (see Section 3.4).

5. INTERPRETATION AT THE "POPULATION" LEVEL

The alleged heterogeneity of *Leishmania* strains has been implicated in the chromosome size polymorphism and in the nonstoichiometric staining intensities of the chromosomal bands. We discussed in Section 3.3 why we believe that the clonal heterogeneity, if any, does not play a direct role in these karyotype polymorphisms. However, the finding of an interclonal polymorphism leaves several questions unanswered: Are *Leishmania* strains truly heterogeneous? Is there coexistence of different versions of the same chromosome in the same cell? In the same population? Are the chromosome size polymorphisms artifactual phenomena due to artificial "triggering" conditions or are they a feature of natural populations?

The arguments given in Section 3.3 support the hypothesis of homogeneous *Leishmania* strains. Moreover, the appearance in a clone of an additional chromosomal band followed by its disappearance in some of the derived subclones (Bastien *et al.*, 1990), or vice versa (Dujardin *et al.*, 1989), strongly suggests the occurrence of chromosomal rearrangements during or shortly after the cloning procedure. However, one cannot hastily conclude that every chromosomal polymorphism proceeds from a stress imposed on the parasite population by artificial maintenance conditions. The extreme polymorphism observed between strains cultivated for a short period, as compared to the rare and limited polymorphism between sibling clones, should be a sufficient argument. Moreover, the proportion of such rearrangements among subclones of one clone appeared tenfold lower than among clones of the original parental strain (Bastien *et al.*, 1990). This implies that either the strain population is much more heterogeneous than the clonal one or it has a greater chromosomal plasticity, which seems less likely. Another indirect argument supporting the idea of multiclonal strains was the observation of the PFE pattern of an *L. infantum* strain appearing as a clear superimposition of those of two of its derived clones: one with an intensely staining additional band and one without it (Bastien *et al.*, 1990). In an attempt to reconcile both models (the mono- or the multiclonal strain), we hypothesized a strain structure that could account for the features observed in the PFE karyotypes (Bastien *et al.*, 1990): chromosomal rearrangements could occur during or shortly after cloning or isolation and become a predominant feature of the whole clone/strain population or not, depending on the time at which they occur during the initial cell divisions; alternatively, an unstable amplification prone to new or reverse rearrangements could yield several descendant subpopulations containing different-sized versions of the same chromosome. The supposed concomitant presence in the strain of closely related clones with a slightly different genome content made us choose the term "mosaic" strain, in reference to certain organizations described in higher eukaryotes. This hypothesis can particularly well explain the weaker intensity of the new distinct-sized homologues found in most clones. The stability of these different intensities over prolonged periods of *in vitro* culture could be explained by the absence of any biological advantage for the variant in culture, except perhaps to survive the alleged stress induced by cloning. An alternative explanation is the existence of aneuploidy, that is, multiple versions of the same chromosome in an individual cell, only one of which would undergo a rearrangement, for example, to yield the patterns observed in variant clones.

6. IMPLICATIONS AND OUTLOOK

6.1. Functional Significance of the Karyotype Polymorphisms

As said above, one cannot deduce from the rare occurrence of chromosomal rearrangements in laboratory conditions that the karyotype polymorphisms are not a feature of natural populations. Although *Leishmania* isolates still have to be amplified in culture before they can be examined in PFE, the strains and clones are cultivated in our laboratory in identical conditions and for a similar duration prior to their characterization. Nevertheless, very polymorphic and very similar karyotypes are found equally in strains isolated in the same area, whether at the same or at different times. In *Plasmodium,* where dramatic chromosome size polymorphisms are also a dominant feature of the PFE patterns (Kemp *et al.*, 1987), such polymorphisms were shown to occur at a similar frequency in natural populations of the parasite (Corcoran *et al.*, 1986; Langsley *et al.*, 1988; Biggs *et al.*, 1989). Therefore, they are clearly not a consequence of maintenance in artificial conditions. In clones, subtelomeric deletions have also been shown to be induced by heat stress or by cloning in trypanosomes (Bernards *et al.*, 1983). This sole mechanism may not be sufficient to explain our data. Indeed, the fact that very different chromosomes were found rearranged in clones of the same or of different *Leishmania* strains (Bastien *et al.*, 1990) does not support the idea that these rearrangements were a response to an identical stress.

Nevertheless, rearrangements do occur during *in vitro* culture and may give a selective growth advantage as shown with the amplification of the miniexon gene array in *L. major* (Iovannisci and Beverley, 1989). There is yet not much evidence in support of the conclusion that all chromosomal rearrangements involved in the karyotype polymorphisms may confer a selective advantage. In fact, much of the chromosome size variation seen in *P. falciparum* can be attributed to transcriptionally inert repetitive sequences (Van der Ploeg *et al.*, 1985; Corcoran *et al.*, 1988). Moreover, DNA deletions in *Plasmodium* can result in the complete or partial loss of specific coding sequences and sometimes in the emergence of a phenotype nonviable in natural conditions (Corcoran *et al.*, 1986, 1988; Pologe and Ravetch, 1986, 1988; Kemp *et al.*, 1987; Biggs *et al.*, 1989; Cappai *et al.*, 1989). As pointed out by Ravetch (1989), such a mechanism is therefore probably maintained for another purpose and must be among the strategies generally employed by the parasites to adapt to environmental changes. It is noteworthy that, apart from the genes mentioned above, several antigen genes are also located near chromosome ends in *Plasmodium,* and they can also be lost during subtelomeric deletions (Corcoran *et al.*, 1988). The repositioning of genes near chromosome ends is a general strategy for gene activation (Borst and Greaves, 1987) and there could be a parallel with *Trypanosoma brucei*, where chromosomal rearrangements might be involved in the regulation of differential gene expression (Van der Ploeg *et al.*, 1984a,c). Ravetch (1989) also underlines the resemblance of the rearrangement processes identified in *Plasmodium* to those occurring during the developmentally programmed fragmentation of chromosomes in ciliated protozoa (see Forney and Blackburn, 1988).

Although there is as yet no evidence for any of these mechanisms operating in *Leishmania*, it is possible that the chromosomal instability play a role in the generation of new phenotypes or in the expression of specific sequences during the life cycle of the parasite. Moreover, a population mosaic such as hypothesized above may be beneficial to the parasite by enhancing group fitness and allowing a better response to variable kinds of selective pressure encountered during the life cycle. More generally, the karyotype polymorphisms might represent a general mode of mutation in protozoan parasites, in which case it would remain to be seen why they were present at a relatively high frequency with often deleterious, or nonadvantageous, consequences.

6.2. Implications on the Ploidy of *Leishmania*

The ploidy of *Leishmania* remains an open question, mostly unexplored, despite the wide use of isoenzyme analysis for taxonomy and strain identification. Measurements of the genome size based on kinetic and chemical analysis by Leon *et al.* (1978) had already suggested diploidy for *L. donovani*. The finding of multiple-banded patterns evoking diploid heterozygotes for some isoenzymes in *Leishmania* also suggested diploidy for the loci concerned (Maazoun *et al.*, 1981; Tait, 1983; Rioux *et al.*, 1985), although other explanations were considered possible; such patterns were then found in a variety of strains using a range of different techniques (Le Blancq *et al.*, 1986; Moreno, 1989). However, certain points should be clarified before being able to draw firm conclusions from these data. For instance, although not favored by Tait (1983), gene duplications in a haploid genome could give similar isoenzyme banding patterns and explain the fixed heterozygosity observed in some species (Le Blancq *et al.*, 1987; Tibayrenc *et al.*, 1990); also, extensive posttranscriptional (reviewed in Coulson and Smith, 1990) and possibly posttranslational modifications might affect the number of isoenzymes detected. Iovannisci *et al.* (1984) then showed by biochemical genetic analysis that there exists two gene copies for adenine phosphoribosyl transferase, without formally proving allelism at this locus. Later, Beverley *et al.* (1988) and Cruz and Beverley (1990) presented the first molecular evidence for a disomy for chromosome V which carries the DHFR-TS locus in *L. major*. It should always be kept in mind that these observations concern homologous chromosomes from a population of parasites in culture and not from an individual cell. Finally, the work of Iovannisci and Beverley (1989), already mentioned in Section 3.4, clearly demonstrated the presence of two different-sized homologues of chromosome II in *L. major*. Quantitative densitometry of ethidium bromide-strained PFGE gels in the same study suggested a common ploidy for the three smallest chromosomes, which would also be indicative of a diploidy for chromosomes I and III, but such deductions are arguable (see below). We have also shown (Figure 10) that each of the four smallest chromosomes could exist in two different-sized versions in one clone of *L. infantum* (Pagès *et al.*, 1989; Blaineau *et al.*, 1991; see figure 10), and such doublets of homologous chromosomes have been found in other strains of various *Leishmania* species for chromosomes of 300–600 kb (reviewed in Section 3.4). This cumulative evidence, although incomplete, would suggest that *Leishmania* is diploid, at least for the chromosomes and loci examined. Likewise, for trypanosomes, DNA complexity measurements (Borst *et al.*, 1980) coupled with cytophotometry (Borst *et al.*, 1982), population genetic analysis of enzyme electrophoretic data (Tait, 1983), analysis of restriction site polymorphisms for three glycolytic enzyme genes (Gibson *et al.*, 1985), the observation of probable pairs of different-sized homologues (Gibson and Borst, 1986; Gibson and Miles, 1986; Engman *et al.*, 1987), and the molecular characterization of the progeny of experimental crosses in the vector (Sternberg *et al.*, 1989), all suggested that the genome was diploid.

However, in view of the fact that one is analyzing populations of cells and not individuals, the evidence in favor of a completely diploid genome remains uncertain, and the existence of aneuploidy in these organisms cannot be excluded. Spithill and Samaras (1987b) first suggested that *Leismania* could be aneuploid, but this was based on limited evidence from the greater ethidium bromide staining intensity of one chromosomal band; such deductions can be risky, since, for instance, the appearance of a larger homologue in a clone's karyotype is not always accompanied by a reduction in intensity of the smaller corresponding homologous band (Blaineau *et al.*, 1991; see also clone LEM 1284 in Figure 9). In one clone of *L. infantum* (LEM 1200 in Figure 9), the band supposed to contain both chromosome I homologues was modified in size, due to a 40-kb deletion (Bastien *et al.*, 1990); chromosome homologies were firmly established by macrorestriction mapping (Blaineau *et al.*, 1991). In a diploid model, the yield of two identical-sized versions can only be explained by a concomitant deletion on both homologues or by a complex rearrangement akin to gene conversion in one of the homologues. Alternatively, there may be only one copy of this chromosome, at least in this strain since pairs of chromosome I homologues have been demonstrated in other strains. Similarly, in *T. brucei* the analysis of translocation events (Van der Ploeg *et al.*, 1984a,c, 1989) and of the inheritance of chromosomes during genetic crosses (Le Page *et al.*, 1988) supports the idea of a haploid complement of small- and intermediate-sized chromosomes (Tait and Turner, 1990).

Conversely, the appearance of apparently supernumerary chromosomes without reduction in staining intensity of the corresponding homologous bands, as well as the weaker intensity of these extra bands in most of the clones, is suggestive of a trisomy or a polysomy for these particular chromosomes (Bastien *et al.*, 1990). An extra copy of one chromosome in a haploid genome, arising probably through partial duplication followed by DNA deletions, has been described in *Plasmodium* (Cowman and Lew, 1989). A related phenomenon might also occur in the allegedly diploid *Leishmania* (see Section 4.6), yielding several versions of particular chromosomes (polysomy) in some strains, the homologues possibly being of very different sizes. It is interesting to note that apparently true aneuploid hybrids have been obtained in some of the experimental genetic crosses in *T. brucei* but not in others (Wells *et al.*, 1987; Le Page *et al.*, 1988). Finally, an alternative hypothesis, which would fit with most PFE results, could be the coexistence of haploid and diploid (or polyploid?) cells, such as seen in certain yeasts.

Clearly, the genome structure and dynamics of trypanosomatids still leave many questions unanswered. These must be viewed within the framework of the complexity and diversity of the genetics of protozoa. For example, the chromosome number can vary greatly between related species of the same protozoan genus; in the flagellate *Euglena,* a polyploidy or a reduction in chromosome number are commonly found even in individual cells within a clonal population (Raikov, 1982). This commands a particularly open-minded attitude with regard to similar questions in *Leishmania*.

6.3. On the Presence of Genetic Exchange in *Leishmania*

The presence of genetic exchange in *Leishmania* has been the subject of constant debate among protozoologists for many years. It was hoped that the advent of PFE would help in the clarification of this debate. Unfortunately, experimental crossing experiments in the sandfly vector have as yet remained negative (Gradoni *et al.*, 1986; P. Bastien, C. Blaineau, and M. Pagès, unpublished data) and the analysis of sympatric populations is not yet available, so that only indirect arguments can be made. The presence of multiple-banded patterns for several isoenzymes, particularly in the same strain (Maazoun *et al.*, 1981; Tait, 1983; Le Blancq *et al.*, 1986; Moreno, 1989), have been considered supportive of the occurrence of genetic exchange in *Leishmania.* These isoenzymic data have been used in population genetic analyses as arguments in favor of a predominantly clonal population structure and against the presence of frequent genetic recombination (Tibayrenc *et al.*, 1990). Tibayrenc and co-workers reached similar conclusions for *Trypanosoma* and *Plasmodium*, organisms for which hybrids resulting from experimental crosses have been extensively analyzed (Jenni *et al.*, 1986; Walliker *et al.*, 1987; Wells *et al.*, 1987; Corcoran *et al.*, 1988; Le Page *et al.*, 1988; Sinnis and Wellems, 1988; Vernick *et al.*, 1988; Sternberg *et al.*, 1989). This would indicate that if mating does occur in these species between particular strains and in certain conditions, it is an uncommon and far from obligatory event. This also concurs with the analysis of Cibulskis (1988), who nevertheless emphasized the numerous pitfalls of the method. Other indirect arguments for the occurrence of genetic exchange in *Leishmania* include: the finding of a "natural" putative hybrid defined by heterozygous isoenzymic patterns and by two different-sized homologous chromosomes each of which could be inherited from one parent (Evans *et al.*, 1987); the report of multinucleate forms of *L. mexicana* in the sandfly midgut (Lainson and Shaw, 1986); and the visual demonstration of cell fusion in *L. infantum* and *L. tropica* (Lanotte and Rioux, 1990). Cell fusion remains one of the possible models for genetic exchange in trypanosomes, although not supported by recent genetic analysis (Paindavoine *et al.*, 1986; Tait and Turner, 1990). It has also been proposed as an explanation for the chromosome size and number variations between *Leishmania* strains that are identical by all other criteria (Pagès *et al.*, 1989).

It is difficult to draw any firm conclusions regarding clonality or genetic exchange from the present PFE data. Indeed, the similarity of the karyotypes of certain strains from the same geographic area (Figure 6), as well as the fact that some size classes of variable chromosomes are recurrently and independently found in different strains from this same area (P. Bastien, C. Blaineau, and M. Pagès, unpublished data), could be explained by both models: (1) a predominantly clonal evolution with frequent chromosomal rearrangements [e.g., for some trypanosome genes the frequency can be as high as 10^{-4} per generation (Van der Ploeg *et al.*, 1984c)] could lead to a progressive "karyotypic drift" from one clonal karyotype; (2) alternatively, genetic exchange within or between such strains could lead to the inheritance of distinct-sized homologues from each parent and to the spread of identical-sized chromosomes in the population; this would tend to homogenize different karyotypes and yield the above-mentioned patterns that in this respect resemble those of the parents and progeny of experimental crosses in *Trypanosoma* (Le Page *et al.*, 1988) and *Plasmodium* (Walliker *et al.*, 1987). These two hypotheses are not mutually exclusive. Also, a striking contrast persists between the extensive range of sizes of certain chromosomes among sympatric strains and the remarkable stability of these chromosome sizes in a strictly clonal reproduction (in culture), thus suggesting a diversity factor that does not exist in the laboratory.

Finally, the overall chromosomal organization in the genus *Leishmania* appears to be relatively conserved (reviewed in Section 3.6). One could argue that this overall homogeneity would not be possible in a strictly clonal evolution and that it could be maintained as allowing occasional genetic exchange between different so-called species. It is indeed intriguing that although genetic exchange naturally occurs in sympatric populations of *T. brucei* (Tait, 1983; Cibulskis, 1988), the experimental crosses mentioned above in *T. brucei* (Jenni *et al.*, 1986; Gibson, 1989) and *P. falciparum* (Walliker *et al.*, 1987) involved strains from widely separated geographical areas and that trials with *T. brucei* strains from the same area have failed (Tait *et al.*, 1985). Likewise, the only putative hybrids reported in *Leishmania* were the apparent progeny of phylogenetically very distant species found in sympatric conditions: *L. tropica*/*L. aethiopica* (Maazoun *et al.*, 1981) and *L. major*/*L. arabica* (Evans *et al.*, 1987). To conclude, and keeping in mind (1) that ploidy is still unknown and (2) that *Leishmania* populations are far from panmictic and submitted to severe bottlenecks through the life cycle, genetic exchange certainly cannot be excluded in this organism: it might take place as self-mating, which would generate less diversity and therefore be less detectable, and/or only under exceptional circumstances, still being retained by the parasite as an adaptive and evolutionary mechanism.

7. SUMMARY AND CONCLUSIONS

The advent of pulsed field electrophoresis has allowed a direct approach to the karyotype of *Leishmania*. The molecular karyotype thus obtained is a stable characteristic of a given strain, although minor modifications may occur during *in vitro* maintenance. Between 20 and 28 chromosomal bands can be resolved depending on the strain, ranging in size from approximately 250 to 2600 kb. The technique has revealed a striking degree of polymorphism in the size and number of the chromosomal bands between different strains, and this seems independent of the category (species, zymodeme, population) to which the strains belong. It appears that only certain strains originating from the same geographic area may share extensive similarities. This polymorphism can largely be accounted for by chromosome size variations, which can involve up to 25% of the chromosome length. As a result, homologous chromosomes can exist in versions of markedly different size within the same strain. When this occurs with several different chromosomes, the interpretation of PFE patterns appears difficult without prior identification of the size-variable chromosomes and of the chromosome homologies. DNA deletions and amplifications have been shown to account for some of these size modifications, but other mechanisms are probably involved; nevertheless, interchromosomal exchange does not seem to play a major role in these polymorphisms. These chromosomal rearrangements, yet in an early stage of characterization, exhibit two relevant features: they seem (1) to affect essentially the subtelomeric regions and (2) to occur in a recurrent nonrandom manner. Chromosomal rearrangements sharing the same characteristics have been identified in yeast and other protozoa such as *Trypanosoma* and *Plasmodium*. The significance of this hypervariability for the biology of the parasite remains unknown, but it can be expected that such mechanisms have been maintained for some purpose; genes specifically located near chromosome ends might benefit from rapid sequence change, alternating activation, or polymorphism of expression. The chromosomal plasticity could represent a general mode of mutation in these parasites, in parallel with genetic exchange which may be uncommon in nature. The molecular characterization of these rearrangements, the identification of each chromosome with the help of physical restriction maps and linkage maps, and the collation of such data on a number of strains and species should allow a significant progress in the understanding of the genetics of *Leishmania,* in particular as regards ploidy, generation of phenotypic diversity, and genome evolution. Finally, like other models, this is susceptible to improve our knowledge of DNA–DNA interactions and of the chromosome functional structure and dynamics.

ACKNOWLEDGMENTS. We are deeply indebted to Dr. P. D. Ready and Dr. D. F. Smith for their helpful comments and critical reading of the manuscript. We also wish to thank Dr. G. Roizès, Prof. J. A. Rioux, and Dr. R. Killick-Kendrick for their constant support, without which this work would not have been possible. We are grateful to Prof. J. A. Rioux and Dr. D. A. Evans for providing us with *Leishmania* strains, and Drs. T. W. Spithill, S. M. Beverley, D. C. Barker, and A. Mendoza for communicating unpublished data. This investigation received financial support from the UNDP/World Bank/WHO special program for research and training in tropical diseases and the INSERM (Réseau Nord-Sud).

8. REFERENCES

Aksoy, S., Lalor, T. M., Martin, J., Van der Ploeg, L. H. T., and Richards, F. F., 1987, Multiple copies of a retroposon interrupt spliced leader RNA genes in the African trypanosome, *Trypanosoma gambiense, EMBO J.* **6:**3819–3826.

Aksoy, S., Williams, S., Chang, S., and Richards, F. F., 1990, SLACS retrotransposon from *Trypanosoma brucei gambiense* is similar to mammalian LINEs, *Nucleic Acids Res.* **18:**785–792.

Aline, R. F., Jr., MacDonald, G., Brown, E., Allison, J., Myler, P., Rothwell, V., and Stuart, K., 1985, (TAA)n within sequences flanking several intrachromosomal variant surface glycoprotein genes in *Trypanosoma brucei, Nucleic Acids Res.* **13:**3161–3177.

Altschuler, M. I., and Yao, M.-G., 1985, Macronuclear DNA of *Tetrahymena thermophila* exists as defined subchromosomal-sized DNA molecules, *Nucleic Acids Res.* **13:**5817–5831.

Anonymous, 1978, Proposals for the nomenclature of salivarian trypanosomes and for the maintenance of reference collections, *Bull. WHO* **56**:467–480.

Anonymous, 1986, Recommendations, in *Leishmania. Taxonomy and Phylogeny* (J. A. Rioux, ed.), pp. 507–517, IMEEE, Montpellier, France.

Bard, E., 1989, Molecular biology of *Leishmania, Biochem. Cell Biol.* **67**:516–524.

Barker, D. C., 1987, DNA diagnosis of human leishmaniasis, *Parasitol. Today* **3**:177–184.

Barker, D. C., and Gibson, L. J., Kennedy, W. P. K., Nasser, A.A.A.A., and Williams, R. H., 1986, The potential of using recombinant DNA species-specific probes for the identification of tropical *Leishmania, Parasitology* **91**:S139–S174.

Bastien, P., Blaineau, C., Taminh, M., Rioux, J. A., Roizès, G., and Pagès, M., 1990, Interclonal variations in molecular karyotype in *Leishmania infantum* imply a "mosaic" strain structure, *Mol. Biochem. Parasitol.* **40**:53–61.

Bellis, M., Pagès, M., and Roizès, G., 1987, Electrophorèse de l'ADN en champ pulsé, *Rev. Inst. Pasteur Lyon* **20**:45–55.

Bernards, A., Michels, P. A. M., Lincke, C. R., and Borst, P., 1983, Growth of chromosome ends in multiplying trypanosomes, *Nature* **303**:592–597.

Beverley, S. M., and Coburn, C., 1990, Recurrent *de novo* appearance of small linear DNAs in *Leishmania major* and relationship to extra-chromosomal DNAs in other species, *Mol. Biochem. Parasitol.* **42**:133–142.

Beverley, S. M., Ellenberger, T. E., Iovannisci, D. M., Kapler, G. M., Petrillo-Peixoto, M. L., and Sina, B. J., 1988, Gene amplification in *Leishmania,* in *The Biology of Parasitism* (MBL Lectures in Biology 9) (P. T. Englund and A. Sher, eds.), pp. 431–448, Alan R. Liss, New York.

Biggs, B. A., Kemp, D. J., and Brown, G. V., 1989, Subtelomeric chromosome deletions in field isolates of *Plasmodium falciparum* and their relationship to loss of cytoadherence *in vitro, Proc. Natl. Acad. Sci. USA* **1986**:2428–2432.

Bishop, R. P., 1990, Extensive homologies between *Leishmania donovani* chromosomes of markedly different size, *Mol. Biochem. Parasitol.* **38**:1–12.

Bishop, R. P., and Miles, M. A., 1987, Chromosome size polymorphisms of *Leishmania donovani, Mol. Biochem. Parasitol.* **24**:263–272.

Blaineau, C., Bastien, P., Rioux, J.-A., Riozés, G., and Pagés, M., 1991, Long-range restriction maps of size-variable homologous chromosomes in *Leishmania infantum, Mol. Biochem. Parisitol.* **46**:293–302.

Borst, P., and Greaves, D. R., 1987, Programmed gene rearrangements altering gene expression, *Science* **235**:658–667.

Borst, P., Fase-Fowler, F., Frasch, A. C. C., Hoeijmakers, J. H. J., and Weijers, P. J., 1980, Characterization of DNA from *Trypanosoma brucei* and related trypanosomes by restriction endonuclease digestion, *Mol. Biochem. Parasitol.* **1**:221–246.

Borst, P., Van der Ploeg, L. H. T., Van Hoeck, J. F. M., Tas, J., and James, J., 1982, On the DNA content and ploidy of trypanosomes, *Mol. Biochem. Parasitol.* **6**:13–23.

Button, L. L., Russel, D. G., Klein, H. L., Medina-Acosta, E., Karess, R. E., and McMaster, W. R., 1989, Genes encoding the major glycoprotein in *Leishmania* are tandemly linked at a single chromosomal locus and are constitutively transcribed, *Mol. Biochem. Parasitol.* **32**:271–284.

Campbell, D. A., Van Bree, M. P., and Boothroyd, J. C., 1984, The 5′ limit of transposition and upstream barren region of trypanosome VSG gene: Tandem 76 base-pair repeats flanking $(TAA)_{90}$, *Nucleic Acids Res.* **12**:2759–2774.

Cantor, C. R., Gaal, A., and Smith, C. L., 1988, High-resolution separation and accurate size determination in pulsed-field gel electrophoresis of DNA. 3. Effect of electrical field shape, *Biochemistry* **27**:9216–9221.

Cappai, R., Van Schravendijk, M.-R., Anders, R. F., Peterson, M. G., Thomas, L. M., Cowman, A. F., and Kemp, D. J., 1989, Expression of the RESA gene in *Plasmodium falciparum* isolate FCR3 is prevented by a subtelomeric deletion, *Mol. Cell. Biol.* **9**:3584–3587.

Carle, G. F., and Olson, M. V., 1984, Separation of chromosomal DNA molecules from yeast by orthogonal-field-alternation gel electrophoresis, *Nucleic Acids Res.* **12**:5647–5664.

Carle, G. F., and Olson, M. V., 1985, An electrophoretic karyotype for yeast, *Proc. Natl. Acad. Sci. USA* **82**:3756–3760.

Carle, G. F., and Olson, M. V., 1987, Orthogonal-field-alternation gel electrophoresis, *Methods Enzymol.* **155**:468–483.

Carle, G. F., Frank, M., and Olson, M. V., 1986, Electrophoretic separations of large DNA molecules by periodic inversion of the electric field, *Science* **232**:65–68.

Chance, M. L., 1986, The nomenclature of enzyme variants with particular reference to parasitic protozoa, in *Leishmania: Taxonomy and Phylogeny* (J. A. Rioux, ed.), pp. 85–89, IMEEE, Montpellier, France.

Chang, K. P., and Hendricks, L. D., 1985, Laboratory cultivation and maintenance of *Leishmania*, in *Leishmaniasis* (K. P. Chang and R. S. Bray, eds.), pp. 213–244, Elsevier, Amsterdam.

Chu, G., Vollrath, D., and Davis, R. W., 1986, Separation of large DNA molecules by contour-clamped homogeneous electric fields, *Science* **234**:1582–1585.

Cibulskis, R. E., 1988, Origins and organization of genetic diversity in natural populations of *Trypanosoma brucei*, *Parasitology* **96**:303–322.

Comeau, A. M., Miller, S. I., and Wirth, D. F., 1986, Chromosome location of four genes in *Leishmania*, *Mol. Biochem. Parasitol.* **21**:161–169.

Corcoran, L. M., Forsyth, K. P., Bianco, A. E., Brown, G. V., and Kemp, D. J., 1986, Chromosome size polymorphisms in *Plasmodium falciparum* can involve deletions and are frequent in natural parasite populations, *Cell* **44**:87–95.

Corcoran, L. M., Thompson, J. K., Walliker, D., and Kemp, D. J., 1988, Homologous recombination within subtelomeric repeat sequences generates chromosome size polymorphisms in *P. falciparum*, *Cell* **53**:807–813.

Coulson, R. M. R., and Smith, D. F., 1990, Isolation of genes showing increased or unique expression in the infective promastigotes of *Leishmania major*, *Mol. Biochem. Parasitol.* **40**:63–76.

Cowman, A. F., and Lew, A. M., 1989, Antifolate drug selection results in duplication and rearrangement of chromosome 7 in *Plasmodium chabaudi*, *Mol. Cell. Biol.* **9**:5182–5188.

Cruz, A., and Beverley, S. M., 1990, Gene replacement in parasitic protozoa, *Nature* **348**:171–173.

De Jonckheere, J. F., Majewska, A. C., and Kasprzak, W., 1990, *Giardia* isolates from primates and rodents display the same molecular polymorphism, *Mol. Biochem. Parasitol.* **39**:23–28.

De Jonge, P., De Jongh, F. C. M., Meijers, R., Steensma, H. Y., and Scheffers, W. A., 1986, Orthogonal-field-alternation gel electrophoresis banding patterns of DNA from yeasts, *Yeast* **2**:193–204.

Dickin, S. K., and Gibson, W. C., 1989, Hybridisation with a repetitive DNA probe reveals the presence of small chromosomes in *Trypanosoma vivax*, *Mol. Biochem. Parasitol.* **33**:135–142.

Dore, E., Pace, T., Ponzi, M., Picci, L., and Frontali, C., 1990, Organization of subtelomeric repeats in *Plasmodium berghei*, *Mol. Biochem. Parasitol.* **40**:2423–2427.

Dujardin, J. C., Gajendran, N., Hamers, R., Matthijsen, G., Urjel, R., Recacoechea, M., Villarroel, G., Bermudez, H., Desjeux, P., de Doncker, S., and Le Ray, D., 1989, Leishmaniasis in the Lowlands of Bolivia: VIII Characterization and identification of Bolivian isolates by PFG karyotyping, in *Leishmaniasis. The Current Status and New Strategies for Control* (D. T. Hart, ed.), pp. 137–148, Plenum Press, New York.

Dvorak, J. A., Hall, T. E., Crane, M. S. J., Engel, J. C., McDaniel, J. P., and Uriegas, R., 1982,

Trypanosoma cruzi: Flow cytometric analysis. I. Analysis of total DNA/organism by means of mithramycin-induced fluorescence, *J. Protozool.* **29:**430–437.

Ellenberger, T. E., and Beverley, S. M., 1989, Multiple drug resistance and conservative amplification of the H region in *Leishmania major, J. Biol. Chem.* **264:**15094–15103.

Ellis, J., and Crampton, J., 1988, Characterisation of a simple, highly repetitive DNA sequence from the parasite *Leishmania donovani, Mol. Biochem. Parasitol.* **29:**9–18.

Engman, D. M., Reddy, L. V., Donelson, J. E., and Kirchhoff, L. V., 1987, *Trypanosoma cruzi* exhibits inter- and intra-strain heterogeneity in molecular karyotype and chromosomal gene location, *Mol. Biochem. Parasitol.* **22:**115–123.

Evans, D. A., Kennedy, W. P. K., Elbihari, S., Chapman, C. J., Smith, V., and Peters, W., 1987, Hybrid formation within the genus *Leishmania, Parassitologia* **29:**165–173.

Fangman, W. L., 1978, Separation of very large DNA molecules by gel electrophoresis, *Nucleic Acids Res.* **5:**653–665.

Foote, S. J., and Kemp, D. J., 1989, Chromosomes of malaria parasites, *Trends Gen.* **5:**337–342.

Forney, J. D., and Blackburn, E. H., 1988, Developmentally controlled telomere addition in wild type and mutant Paramecia, *Mol. Cell Biol.* **8:**251–258.

Frutos, R., Pagès, M., Bellis, M., Roizès, G., and Bergoin, M., 1989, Pulsed-field gel electrophoresis determination of the genome size of obligate intracellular bacteria belonging to the genera *Chlamydia, Rickettsiella,* and *Porochlamydia, J. Bacteriol.* **171:**4511–4513.

Gabriel, A., Yen, T. J., Schwartz, D. C., Smith, C. L., Boeker, J. D., Sollner-Webb, B., and Cleveland, D. W., 1990, A rapidly rearranging retrotransposon within the miniexon gene locus of *Crithidia fasciculcuta, Mol. Cell. Biol.* **10:**615–624.

Gajendran, N., Dujardin, J. C., Le Ray, D., Matthyssens, G., Muyldermans, S., and Hammers, R., 1989, Abnormally migrating chromosome identifies *Leishmania donovani* populations, in *Leishmaniasis: The Current Status and New Strategies for Control* (D. T. Hart, ed.), pp. 539–547, Plenum Press, New York.

Galindo, I., and Ramirez Ochoa, J. L., 1989, Study of *Leishmania mexicana* electrokaryotype by clamped homogeneous electric field electrophoresis, *Mol. Biochem. Parasitol.* **34:**245–252.

Gardiner, K., Laas, W., and Patterson, D., 1986, Fractionation of large mammalian DNA restriction fragments using vertical pulsed-field gradient gel electrophoresis, *Somatic Cell Mol. Genet.* **12:**185–195.

Garvey, E. P., and Santi, D. V., 1986, Stable amplified DNA in drug-resistant *Leishmania* exists as extrachromosomal circles, *Science* **233:**535–540.

Giannini, S. H., Schittini, M., Keithly, J. S., Warburton, P. W., Cantor, C. R., and Van der Ploeg, L. H. T., 1986, Karyotype analysis of *Leishmania* species and its use in classification and diagnosis, *Science* **232:**762–765.

Giannini, S. H., Curry, S. S., Tesh, R. B., and Van der Ploeg, L. H. T., 1990, Size-conserved chromosomes and stability of molecular karyotype in cloned stocks of *Leishmania major, Mol. Biochem. Parasitol.* **39:**9–22.

Gibson, W. C., 1989, Analysis of a genetic cross between *Trypanosoma brucei rhodesiense* and *T. b. brucei, Parasitology* **99:**391–402.

Gibson, W. C., and Borst, P., 1986, Size fractionation of the small chromosomes of *Trypanozoon* and *Nannomonas* trypanosomes by pulsed field gradient gel electrophoresis, *Mol. Biochem. Parasitol.* **18:**127–140.

Gibson, W. C., and Miles, M. A., 1986, The karyotype and ploidy of *Trypanosoma cruzi, EMBO J.* **5:**1299–1305.

Gibson, W. C., Osinga, K. A., Michels, P. A. M., and Borst, P., 1985, Trypanosomes of subgenus *Trypanozoon* are diploid for housekeeping genes, *Mol. Biochem. Parasitol.* **16:**231–242.

Gradoni, L., Gramiccia, M., Maroli, M., and Pozio, E., 1986, Etude expérimentale d'infections mixtes de *Phlébotomus perniciosus* Newstead, 1911 par différents zymodèmes de *Leishmania infantum* Nicolle, 1908, in *Leishmania: Taxonomy and Phylogeny* (J. A. Rioux, ed.), pp. 305–310, IMEEE, Montpellier, France.

Grimaldi, G., Jr., Tesh, R. B., and McMahon-Pratt, D., 1989, A review of the geographic distribution and epidemiology of leishmaniasis in the New World, *Am. J. Trop. Med. Hyg.* **41:**687–725.

Hightower, R. C., and Santi, D. V., 1989, Migration properties of circular DNAs using orthogonal-field-alternation gel electrophoresis, *Electrophoresis* **10:**283–290.

Hightower, R. C., Ruiz-Perez, L. M., Wong, M. L., and Santi, D. V., 188, Extrachromosomal elements in the lower eukaryote *Leishmania, J. Biol. Chem.* **263:**16970–16976.

Horowitz, H., and Haber, J. E., 1985, Identification of autonomously replicating circular subtelomeric Y′ elements in *Saccharomyces cerevisiae, Mol. Cell. Biol.* **5:**2369–2380.

Horowitz, H., Thorburn, P., and Haber, J. E., 1984, Rearrangements of highly polymorphic regions near telomeres of *Saccharomyces cerevisiae, Mol. Cell. Biol.* **4:**2509–2517.

Iovannisci, D. M., and Beverley, S. M., 1989, Structural alterations of chromosome 2 in *Leishmania major* as evidence for diploidy, including spontaneous amplification of the mini-exon array, *Mol. Biochem. Parasitol.* **34:**177–188.

Iovannisci, D. M., Goebel, D., Allen, K., Kaur, K., and Ullman, B., 1984, Genetic analysis of adenine metabolism in *Leishmania donovani* promastigotes: Evidence for diploidy at the adenine phosphoribosyl-transferase locus, *J. Biol. Chem.* **259:**14617–14623.

Jelinek, W. R., and Schmid, C. W., 1982, Repetitive sequences in eukaryotic DNA and their expression, *Annu. Rev. Biochem.* **51:**813–844.

Jenni, L., Marti, S., Schweizer, J., Betschart, B., Le Page, R. W. F., Wells, J. M., Tait, A., Paindavoine, P., Pays, E., and Steinert, M., 1986, Hybrid formation between African trypanosomes during cyclical transmission, *Nature* **322:**173–175.

Johnson, P. J., and Borst, P., 1986, Mapping of VSG genes on large expression-site chromosomes of *Trypanosoma brucei* separated by pulsed-field gradient electrophoresis, *Gene* **43:**213–220.

Kamitsuka, P. F., Iovannisci, T. E., and Beverley, S. M., 1989, Rearrangement site for H-region amplification conferring multiple drug resistance in *Leishmania* is a dispersed repetitive element [Abstract], *J. Cell Biochem.* Suppl. **13E:**107.

Kapler, G. M., Zhang, K., and Beverley, S., 1987, Sequence and S1 nuclease mapping of the 5′-region of the dihydrofolate reductase thymidylate synthase gene of *Leishmania major, Nucleic Acids Res.* **15:**3369–3383.

Kemp, D. J., Corcoran, L. M., Coppel, R. L., Stahl, H. D., Bianco, A. E., Brown, G. V., and Anders, R. F., 1985, Size variation in chromosomes from independent cultured isolates of *Plasmodium falciparum, Nature* **315:**347–350.

Kemp, D. J., Thompson, J. K., Walliker, D., and Corcoran, L. M., 1987, Molecular karyotype of *Plasmodium falciparum:* Conserved linkage groups and expendable histidine-rich protein genes, *Proc. Natl. Acad. Sci. USA* **84:**7672–7676.

Kidane, G. Z., Samaras, N., and Spithill, T. W., 1989, Cloning of developmentally regulated genes from *Leismania major* and expression following heat induction, *J. Biol. Chem.* **264:**4244–4250.

Kimmel, B. E., Ole-Moiyoi, O. K., and Young, J. R., 1987, Ingi, a 5.2-kb dispersed sequence element from *Trypanosoma brucei* that carries half of a smaller mobile element at either end and has homology with mammalian LINEs, *Mol. Cell. Biol.* **7:**1465–1475.

Kooy, R. F., Ashall, F., Van der Ploeg, M., and Overdulve, J. P., 1989, On the DNA content of *Trypanosoma cruzi, Mol. Biochem. Parasitol.* **36:**73–76.

Lainson, R., and Shaw, J. J., 1972, Leishmaniasis in the New World: Taxonomic problems, *Br. Med. Bull.* **28:**44–48.

Lainson, R., and Shaw, J. J., 1986, The genus *Leishmania* Ross, 1903. Speculations on evolution and speciation, in *Leishmania: Taxonomy and Phylogeny* (J. A. Rioux, ed.), pp. 241–245, IMEEE, Montpellier, France.

Lainson, R., and Shaw, J. J., 1987, Evolution, classification and geographical distribution, in *The Leishmaniases in Biology and Medicine*, Vol. 1 (W. Peters and R. Killick-Kendrick, eds.), pp. 1–120, Academic Press, London.

Lainson, R., and Shaw, J. J., 1990. The laboratory mix-up of *Leishmania* strains, *Trans. R. Soc. Trop. Med. Hyg.* **84**:753.

Landfear, S. M., Miller, S. I., and Wirth, D. F., 1986, Transcriptional mapping of *Leishmania enrietti* tubulin mRNAs, *Mol. Biochem. Parasitol.* **21**:235–245.

Langsley, G., Patarapotikul, J., Handunnetti, S., Khouri, E., Mendis, K. N., and David, P. H., 1988, *Plasmodium vivax:* Karyotype polymorphism of field isolates, *Exp. Parasitol.* **67**:301–306.

Lanotte, G., and Rioux, J. A., 1990, Fusion cellulaire chez *Leishmania* (Kinetoplastida, trypanosomatidae), *Comptes Rendus Acad. Sci. Paris* **310**:285–288.

Lanotte, G., Rioux, J. A., Maazoun, R., Pasteur, N., Pratlong, F., and Lepart, J., 1981, Application de la méthode numérique à la taxonomie du genre *Leishmania* Ross, 1903. A propos de 146 souches originaires de L'Ancien Monde, *Ann. Parasitol. Hum. Comp.* **56**:575–592.

Larson, D. D., Spangler, E. A., and Blackburn, E. H., 1987, Dynamics of telomere length variation in *Tetrahymena thermophila, Cell* **50**:477–483.

Le Blancq, S. M., Cibulskis, R. E., and Peters, W., 1986, *Leishmania* in the Old World: 5. Numerical analysis of isoenzyme data, *Trans. R. Soc. Trop. Med. Hyg.* **80**:517–524.

Le Blancq, S. M., Lanham, S. M., and Evans, D. A., 1987, Comparative isoenzyme profiles of Old and New World *Leishmania,* in *The Leishmaniases in Biology and Medicine* (W. Peters and R. Killick-Kendrick, eds.), pp. 543–550, Academic Press, London.

Leon, W., Fouts, D. L., and Manning, J., 1978, Sequence arrangement of the 16S and 26S rRNA genes in the pathogenic haemoflagellate *Leishmania donovani, Nucleic Acid Res.* **5**:491–504.

Le Page, R. W. F., Wells, J. M., Prospero, T. D., and Sternberg, J., 1988, Genetic analysis of hybrid *Trypanosoma* brucei, in *Molecular Genetics of Parasitic Protozoa* (M. J. Turner and D. Arnot, eds.), pp. 65–71, Cold Spring Harbor Laboratory, Cold Spring Harbor, New York.

Maazoun, R., Lanotte, G., Rioux, J. A., Pasteur, N., Killick-Kendrick, R., and Pratlong, F., 1981, Signification du polymorphisme enzymatique chez les Leishmanies, *Ann. Parasitol. Hum. Comp.* **56**:467–475.

Maigon, R. D. C., Elson, L., Hommel, M., Crampton, J. M., and Chance, M. L., 1990, Multiple distinct gp63-like genes in amastigotes of *Leishmania donovani,* in Parasites: molecular biology, drugs, and vaccine design (N. Agabian and A. Cerami, eds.), *UCLA Symp. Mol. Cell. Biol.* **131,** pp. 191–204, Wiley-Liss, New York.

Mathew, M. K., Smith, C. L., and Cantor, C. R., 1988a, High-resolution separation and accurate size determination in pulsed-field gel electrophoresis of DNA. 1. DNA size standards and effect of agarose and temperature, *Biochemistry* **27**:9204–9210.

Mathew, M. K., Smith, C. L., and Cantor, C. R., 1988b, High-resolution separation and accurate size determination in pulsed-field gel electrophoresis of DNA. 2. Effect of pulse time and electric field strength and implication for models of the separation process, *Biochemistry* **27**:9210–9216.

Mathew, M. K., Hui, C.-F., Smith, C. L., and Cantor, C. R., 1988c, High-resolution separation and accurate size determination in pulsed-field gel electrophoresis of DNA. 4. Influence of DNA topology, *Biochemistry* **27**:9222–9226.

McMahon-Pratt, D., Jaffe, C. L., Bennet, E., David, J. R., and Grimaldi, G., Jr., 1986, Studies employing monoclonal antibodies for the analysis of the genus *Leishmania* Ross, 1903, in *Leishmania: Taxonomy and Phylogeny* (J. A. Rioux, ed.), pp. 173–178, IMEEE, Montpellier, France.

Michels, P. A. M., Lin, A.-C., Bernards, A., Sloof, P., Van der Bijl, M. M. W., Schinkel, A. H., Menke, H. H., and Borst, P., 1983, Activation of the genes for variant surface glycoproteins 117 and 118 in *Trypanosoma brucei, J. Mol. Biol.* **166**:537–556.

Molyneux, D. H., and Killick-Kendrick, R., 1987, Morphology ultrastructure and life cycles, in *The Leishmaniases in Biology and Medicine,* Vol. I (W. Peters and R. Killick-Kendrick, eds.), pp. 121–176, Academic Press, London.

Moreno, G., 1989, Les complexes *Leishmania donovani* et *Leishmania infantum.* Implications taxinomiques, biogéographiques et épidémiologiques. A propos de l'analyse enzymatique de 548 souches de l'Ancien Monde, Ph.D. Thesis, (Parasitology), Université des Sciences et Techniques du Languedoc, Académie de Montpellier.

Murphy, N. B., Pays, A., Tebabi, P., Coquelet, H., Guyaux, M., Steinert, M., and Pays, E., 1987, *Trypanosoma brucei* repeated element with unusual structural and transcriptional properties, *J. Mol. Biol.* **195:**855–871.

Oquendo, P., Goman, M., Mackay, M., Langsley, G., Walliker, D., and Scaife, J., 1986, Characterization of a repetitive DNA sequence from the malaria parasite *Plasmodium falciparum, Mol. Biochem. Parasitol.* **18:**89–101.

Orbach, M. J., Vollrath, D., Davis, R. W., and Yanofsky, C., 1988, An electrophoretic karyotype of *Neurospora crassa, Mol. Cell. Biol.* **8:**1469–1473.

Pace, T., Ponzi, M., Dore, E., and Frontali, C., 1987, Telomeric motifs are present in a highly repetitive element in the *Plasmodium berghei* genome, *Mol. Biochem. Parasitol.* **24:**193–202.

Pacheco, R. S., Lopes, U. G., Morel, C. M., Grimaldi, G., Jr., and Momen, H., 1986, Schizodeme analysis of *Leishmania* isolates and comparison with some phenotypic techniques, in *Leishmania: Taxonomy and Phylogeny* (J. A. Rioux, ed.), pp. 57–65, IMEEE, Montpellier, France.

Pagès, M., Bastien, P., Veas, F., Rossi, V., Bellis, M., Wincker, P., Rioux, J. A., and Roizès, G., 1989, Chromosome size and number polymorphisms in *Leishmania infantum* suggest amplification/deletion and possible genetic exchange, *Mol. Biochem. Parasitol.* **36:**161–168.

Paindavoine, P., Zampetti-Bosseler, F., Pays, E., Schweizer, J., Guyaux, M., Jenni, L., and Steinert, M., 1986, Trypanosome hybrids generated in tse-tse flies by nuclear fusion, *EMBO J.* **5:**3631–3636.

Patarapotikul, J., and Langsley, G., 1988, Chromosome size polymorphism in *Plasmodium falciparum* can involve deletions of the subtelomeric pPFrep20 sequence, *Nucleic Acids Res.* **16:**4331–4340.

Petrillo-Peixoto, M. L., and Beverley, S. M., 1989, Amplification of a new-region of DNA in an unselected laboratory stock of *L. tarentolae:* The T region, *J. Protozool.* **36:**257–261.

Phan, H. L., Forney, J., and Blackburn, E. H., 1989, Analysis of *Paramecium* macronuclear DNA using pulsed field gel electrophoresis, *J. Protozool.* **36:**402–408.

Pologe, L. G., and Ravetch, J. V., 1986, A chromosomal rearrangement in a *P. falciparum* histidine rich protein gene is associated with the knobless phenotype, *Nature* **322:**474–477.

Pologe, L. G., and Ravetch, J. V., 1988, Large deletions result from breakage and healing of *P. falciparum* chromosomes, *Cell* **55:**869–874.

Ponzi, K. D., Janse, C. J., Dore, E., Scotti, R., Pace, T., Reterink, T. J. F., Van der Berg, F. M., and Mons, B., 1990, Generation of chromosome size polymorphism during *in vivo* mitotic multiplication of *Plasmodium berghei* involves both loss and addition of subtelomeric repeat sequences, *Mol. Biochem. Parasitol.* **41:**73–82.

Prensier, G., and Slomianny, C., 1986, The karyotype of *Plasmodium falciparum* determined by ultrastructural serial sectioning and 3-D reconstruction, *J. Parasitol.* **72:**731–736.

Raikov, I. B., 1982, The protozoan nucleus. Morphology and evolution, *Cell Biology Monographs,* Vol. 9, Springer-Verlag, Wien & New York.

Ravetch, J. V., 1989, Chromosomal polymorphisms and gene expression in *Plasmodium falciparum, Exp. Parasitol.* **68:**121–125.

Richards, T., 1989, DNA molecules observed, *Nature* **338:**461–462.

Rioux, J. A., 1986, *Leishmania: Taxonomy and Phylogeny,* IMEEE, Montpellier, France.

Rioux, J. A., Lanotte, G., Maazoun, R., and Pasteur, N., 1985, L'électrophorèse des enzymes dans le genre *Leishmania* Ross, 1903, *Parassitologia* **27**:141–156.

Rioux, J. A., Lanotte, G., Serres, E., Pratlong, F., Bastien, P., and Périéres, J., 1990, Taxonomy of *Leishmania*. Use of isoenzymes. Suggestions for a new classification, *Ann. Parasitol. hum. comp.* **65**:111–125.

Samaras, N., and Spithill, T. W., 1987, Molecular karyotype of five species of *Leishmania* and analysis of gene locations and chromosomal rearrangements, *Mol. Biochem. Parasitol.* **25**:279–291.

Scholler, J. K., Reed, S. G., and Stuart, K., 1986, Molecular karyotype of species and sub-species of *Leishmania, Mol. Biochem. Parasitol.* **20**:279–293.

Schwartz, D. C., and Cantor, C. R., 1984, Separation of yeast chromosome-sized DNAs by pulsed field gradient gel electrophoresis, *Cell* **37**:67–75.

Schwartz, D. C., and Koval, M., 1989, Conformational dynamics of individual DNA molecules during gel electrophoresis, *Nature* **338**:520–522.

Shah, J. S., Young, J. R., Kimmel, B. E., Iams, K. P., and Williams, R. O., 1987, The 5′ flanking sequence of a *Trypanosoma brucei* variable surface glycoprotein gene, *Mol. Biochem. Parasitol.* **24**:163–174.

Sharkey, A., Langsley, G., Patarapotikul, J., Mercereau-Puijalon, O., McLean, A., and Walliker, D., 1988, Chromosome size variation in malaria parasite of rodents, *Plasmodium chabaudi, Mol. Biochem. Parasitol.* **28**:47–54.

Shirley, M. W., Kemp, D. J., Pallister, J., and Prowse, S. J., 1990, A molecular karyotype of *Eimeria tenella* as revealed by contour-clamped homogeneous electric field gel electrophoresis, *Mol. Biochem. Parasitol.* **38**:169–174.

Sinnis, P., and Wellems, T. E., 1988, Long-range restriction maps of *Plasmodium falciparum* chromosomes: Crossing-over and size variation among geographically distant isolates, *Genomics* **3**:287–295.

Smith, C. L., Matsumoto, T., Niwa, O., Klco, S., Fan, J. B., Yanagida, M., and Cantor, C. R., 1987, An electrophoretic karyotype for *Schizosaccharomyces pombe* by pulsed field gel electrophoresis, *Nucleic Acids Res.* **15**:4481–4489.

Snell, R. G., and Wilkins, R. J., 1986, Separation of chromosomal molecules from *C. albicans* by pulsed field gel electrophoresis, *Nucleic Acids Res.* **14**:4401–4406.

Snell, R. G., Hermans, I. F., Wilkins, R. J., and Corner, B. E., 1987, Chromosomal variations in *Candida albicans, Nucleic Acids Res.* **5**:2369–2380.

Spithill, T. W., and Samaras, N., 1985, The molecular karyotype of *Leishmania major* and mapping of α- and β-tubulin gene families to multiple unlinked chromosomal loci, *Nucleic Acids Res.* **13**:4155–4169.

Spithill, T. W., and Samaras, N., 1987a, Chromosome size polymorphisms and mapping of tubulin gene loci in *Leishmania*, in *Molecular Strategies of Parasitic Invasion* (N. Agabian, H. Goodman, and N. Noguiera, eds.), pp. 269–278, Alan R. Liss, New York.

Spithill, T. W., and Samaras, N., 1987b, Genomic organization, chromosomal location and transcription of dispersed and repeated tubulin genes in *Leishmania major, Mol. Biochem. Parasitol.* **24**:23–37.

Steele, P. E., Carle, G. F., Kobayashi, G. S., and Medoff, G., 1989, Electrophoretic analysis of *Histoplasma capsulatum* chromosomal DNA, *Mol. Cell Biol.* **9**:983–987.

Sternberg, J., Turner, C. M. R., Wells, J. M., Ranford-Cartwright, L. C., Le Page, R. W. F., and Tait, A., 1989, Gene exchange in African trypanosomes: Frequency and allelic segregation, *Mol. Biochem. Parasitol.* **34**:269–280.

Stuart, K., Tarr, P., Aline, R. F., Jr., Smiley, B. L., Scholler, J., and Keithly, J., 1989, Small nucleic acids in *Leishmania*, in *Leishmaniasis: The Current Status and New Strategies for Control* (D. T. Hart, ed.), pp. 555–562, Plenum Press, New York.

Tait, A., 1983, Sexual processes in the kinetoplastida, *Parasitology* **86:**29–57.

Tait, A., and Turner, C. M. R., 1990, Genetic exchange in *Trypanosoma brucei, Parasitol. Today* **6:**70–75.

Tait, A., Barry, J. D., Wink, R., Sanderson, A., and Crowe, J. S., 1985, Enzyme variation in *T. brucei* ssp. II. Evidence for *T. b. rhodesiense* being a set of variants of *T. b. brucei, Parasitology* **90:**89–100.

Tarr, P. I., Aline, R. F., Smiley, B. L., Scholler, J., Keithly, J., and Stuart, K., 1988, LR1: A candidate RNA virus of *Leishmania, Proc. Natl. Acad. Sci. USA* **85:**9572–9275.

Tibayrenc, M., Kjellberg, F., and Ayala, F. J., 1990, A clonal theory of parasitic protozoa: The population structures of *Entomoeba, Giardia, Leishmania, Naegleeria, Plasmodium, Trichomonas,* and *Trypanosoma* and their medical and taxonomical consequences, *Proc. Natl. Acad. Sci. USA* **87:**2414–2418.

Valdés, J., de la Cruz Hernandez, F., Ocadiz, R., and Orozco, E., 1990, Molecular karyotype of *Entamoeba histolytica* and *E. invadens, Trans. R. Soc. Trop. Med. Hyg.* **84:**537–541.

Van der Ploeg, L. H. T., Schwartz, D. C., Cantor, C. R., and Borst, P., 1984a, Antigenic variation in *Trypanosoma brucei* analysed by electrophoretic separation of chromosome-sized DNA molecules, *Cell* **37:**77–84.

Van der Ploeg, L. H. T., Cornelissen, A. W. C. A., Barry, J. D., and Borst, P., 1984b, Chromosomes of Kinetoplastida, *EMBO J.* **3:**3109–3115.

Van der Ploeg, L. H. T., Cornelissen, A. W. C. A., Michels, P. A. M., and Borst, P., 1984c, Chromosome rearrangements in *Trypanosoma brucei Cell* **39:**213–221.

Van der Ploeg, L. H. T., Smits, M., Ponnudurai, T., Vermeulen, A., Meuwissen, J. H. E. T., and Langsley, G., 1985, Chromosome-sized DNA molecules of *Plasmodium falciparum, Science* **229:**658–661.

Van der Ploeg, L. H. T., Smith, C. L., Polvere, R. I., and Gottesdiener, K. M., 1989, Improved separation of chromosome-sized DNA from *Trypanosoma brucei* stock 427-60, *Nucleic Acids Res.* **17:**3217–3227.

Vedel, M., and Robert-Géro, M., 1987, Characterization of RNA from *Leishmania tropica* and *L. donovani* promastigotes, *Mol. Biochem. Parasitol.* **24:**81–87.

Vernick, K. D., Walliker, D., and McCutchan, T. F., 1988, Genetic hypervariability of telomere-related sequences is associated with meiosis in *Plasmodium falciparum, Nucleic Acids Res.* **16:**6973–6985.

Villalba, E., and Ramirez, J. L., 1982, Ribosomal DNA of *Leishmania brasiliensis:* Number of ribosomal copies and gene isolation, *J. Protozool.* **29:**438–441.

Vollrath, D., and Davis, R. W., 1987, Resolution of DNA molecules greater than 5 megabases by contour-clamped homogeneous electric fields, *Nucleic Acids Res.* **15:**7865–7876.

Walliker, D., Quakyi, I. A., Wellems, T. E., McCutchan, T. F., Szarfman, A., London, W. T., Corcoran, L. M., Burkot, T. R., and Carter, R., 1987, Genetic analysis of the human malaria parasite *Plasmodium falciparum, Science* **236:**1661–1666.

Wellems, T. E., Walliker, D., Smith, C. L., do Rosario, V. E., Maloy, W. L., Howard, R. J., Carter, R., and McCutchan, T. F., 1987, A histidine-rich protein gene marks a linkage group favored strongly in a genetic cross of *Plasmodium falciparum, Cell* **49:**633–642.

Wells, J. M., Prospero, T. D., Jenni, L., and Le Page, R. W. F., 1987, DNA contents and molecular karyotypes of hybrid *Trypanosoma brucei, Mol. Biochem. Parasitol.* **249:**103–116.

Widmer, G., Comeau, A. M., Furlong, D. B., Wirth, D. F., and Patterson, J. L., 1989, Characterization of a RNA virus from the parasite *Leishmania, Proc. Natl. Acad. Sci. USA* **86:**5979–5982.

Chapter 6

Intracellular Digestion of Endocytosed Proteins as a Source of Amino Acids for Protein Synthesis in *Trypanosoma cruzi*

José Luis Avila

1. INTRODUCTION AND BACKGROUND

The discovery of lysosomes, with their semipermeable membranes and formidable battery of proteolytic enzymes, gave coherence to long-standing data on the pinocytic uptake of proteins by cells. Largely through the work of Werner Straus in the early 1960s, it became clear that most cells appear to have the capacity to internalize soluble extracellular proteins by either fluid phase or adsorptive pinocytosis (Silverstein *et al.*, 1977). In most cases, the terminal events are that pinocytic vesicles migrate inward and fuse with lysosomes, so that pinocytosed proteins, and by inference other macromolecules, find themselves in contact with lysosomal enzymes within a "digestive vacuole" or "secondary lysosome" (de Duve and Wattiaux, 1966). Since the lysosomal enzymes are able to digest proteins to amino acids and dipeptides (Coffey and de Duve, 1968) and the membrane of the secondary lysosome is impermeable to peptides above molecular weight 220 Da (Ehrenreich and Cohn, 1969), it could be concluded that exogenous proteins would be retained within the vacuole until digestion was

---

José Luis Avila Instituto of Biomedicina, Apartado 4043, Caracas 1010A, Venezuela.
Subcellular Biochemistry, Volume 18: Intracellular Parasites, edited by J. L. Avila and J. R. Harris. Plenum Press, New York, 1992.

essentially complete and that the digestion products would then be transported into the cytoplasm for further catabolism or to be used for protein synthesis. This process of salvaging the cell's "building blocks" is a primary nutritional function of lysosomes. Specific lysosomal transport systems for cystine, the cationic amino acids, small neutral amino acids, and large neutral amino acids have been recently described (Gahl, 1989).

2. *TRYPANOSOMA CRUZI*

2.1. Biology of *Trypanosoma cruzi*

Trypanosoma cruzi has a complex life cycle that involves development both in a mammal and in an insect vector (various species of triatomine). The basic life cycle of *T. cruzi* is shown in Figure 1. Chagas' disease or South American trypanosomiasis is transmitted when the infected bug, in feeding, deposits feces or urine containing metacyclic trypomastigotes onto the skin; parasites invade the body through bites or by active penetration of mucous membranes. Metacyclic trypomastigotes, when inoculated into a mammal, do not multiply in the blood, but rather penetrate into cells of various tissues. Trypomastigotes, whether metacyclic or bloodstream, can survive phagocytosis by macrophages apparently because they are able to penetrate the phagocytic vacuole membrane and lie on the cytoplasmic matrix, thus evading the host cells' lysosomal responses; epimastigotes cannot do this and are killed and digested by macrophages (Nogueira and Cohn, 1976; Avila, 1983). Interestingly, it has been recently demonstrated that raising the pH of acidic cellular compartments blocks the exit of *T. cruzi* from the macrophages phagosome to the cytoplasmic matrix, suggesting the possibility that trypomastigotes release some acid hydrolase(s) inside phagosomes that is able to disrupt the membrane, allowing the parasite to escape this compartment (Ley *et al.*, 1990). Trypomastigotes can also enter nonphagocytic cells (like heart and gastrointestinal tract muscle cells) by attaching and then inducing engulfment. After penetrating host cells, trypomastigotes take about 3 hr to reorganize themselves into amastigotes, and then remain quiescent for about 35 hr before starting to divide. About nine asynchronous divisions occur that give rise to about 500 amastigotes. Doubling time is 7–14 hr, the shorter times being typical of metacyclic-induced infections. Amastigotes give rise to trypomastigote forms that find their way back into the bloodstream (Dvorak, 1976). It is generally agreed that blood trypomastigote forms do not divide in the blood, though organisms with two nuclei and/or two kinetoplasts are occasionally seen (Hoare, 1972). There are two kinds of blood trypomastigote: one is actively motile, slender, with an elongated nucleus, a subterminal kinetoplast, and a short, free flagellum, while the other is broad with slow movements, an

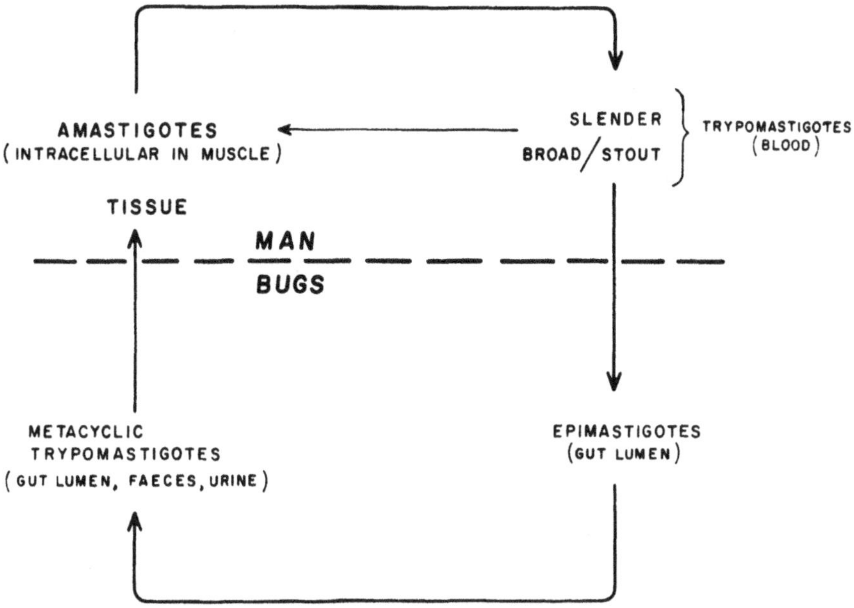

FIGURE 1. Life cycle of *Trypanosoma cruzi* in vector insects and man. Adapted from Figure 1.6 of Gutteridge and Coombs (1977).

oval nucleus, an almost terminal kinetoplast, and a long, free flagellum (Chagas, 1909). Brener and Chiari (1963) have also described a stout trypomastigote characterized by its bulky aspect, vacuolated cytoplasm, and terminal kinetoplast. Which form predominates appears to be a function of both the strain of *T. cruzi* and the age of the infection (Brener and Chiari, 1963; Cover and Gutteridge, 1978). It has been suggested that it is mainly the slender forms that invade tissues maintaining infection in mammals and that it is mainly the broad forms that are infective to insect vectors and transmit the infection (Brener, 1976).

The entire development of *T. cruzi* in the insect vector takes place within the lumen of the gut. Trypomastigotes appear to change initially in the foregut into amastigote and spheromastigote forms, which multiple by binary fission. At about 24 hr after the blood meal is taken, they have found their way to the midgut and most have changed to epimastigotes. This is probably the main form where division takes place and generally is thought to correspond to the form obtained in culture at 25 °C in liquid media. From about the fourth day on, epimastigotes are also present in the rectum. Here, some epimastigote change, either directly by gradual migration of kinetoplast backward or indirectly through a rounded spheromastigote stage, occurs to form metacyclic trypomastigotes. Recent scanning electron microscope studies have shown that metacyclics develop from the short epimastigotes attached to rectum cuticular lining, especially to the water-absorbing rectal gland (Zeledón *et al.*, 1984). *In vitro* attachment, however, does not seem necessary for metacyclogenesis, as metacyclics are generated in stationary-phase liquid cultures (Contreras *et al.*, 1985). These are voided into the feces and can initiate a new infection. Thus, there are three main environments in which *T. cruzi* occurs: within tissue cells, free in mammalian bloodstream, and in the gut lumen of insect vectors. Three different morphological forms predominate in these environments: amastigotes, trypomastigotes, and epimastigotes. The three main morphological forms are readily distinguishable on the basis of kinetoplast position, flagellum length, and extent of undulating membrane (Hoare, 1972).

2.2. Protein and Amino Acid Requirements in *Trypanosoma cruzi*

2.2.1. Proteins

T. cruzi is known to contain 43–53% protein by weight (Von Brand, 1979). It is generally accepted that *T. cruzi* need culture medium proteins for growth (O'Daly, 1975a,b; Berens *et al.*, 1976; Avila *et al.*, 1979a). In fact, this latter group has obtained continuous multiplication of mouse-virulent *T. cruzi* epi-

Table I
Effect of Several Purified Proteins on Continuous Growth of Mouse-Virulent *Trypanosoma Cruzi Epimastigotes*[a]

Proteins studied	Concentration (mg/ml)	Growth[b]
Horse heart cytochrome C	2	–
Myoglobin	1	–
Horseradish peroxidase	1	+ (42)
Lactoperoxidase	2	+ (7)
Bovine hemoglobin	3	+ (45)
Bovine liver catalase	3	+ (42)
Fetal calf serum fetuin	2	–
Egg albumin	3	–
Bovine serum albumin	2	–
γ-Casein	1	–
Bovine γ-globulin	1	–
Histone	2	–

[a]From Avila *et al.* (1979a).
[b]+, Continuous growth according to Trager (1974). The number of successive subcultures obtained for each protein is given in parentheses. –, No growth.

Table II
Biochemical Conditions Affecting the *Trypanosoma cruzi* Growth-Promoting Activity of Bovine Liver Catalase[a,b]

Experimental conditions	Percent of control *T. cruzi* growth-promoting activity
Native bovine catalase	100
Catalase subunits	86
Native catalase + 20 mM AT[c]	100
20 mM AT	96
Catalase added inside a dialysis bag	8
Mixture of L-amino acids components of catalase[d]	5
Catalase hydrolyzate	8

[a]From Avila *et al.* (1979a).
[b]Control tubes attained 20×10^6 parasites/ml after 10 days of cultivation at 28 °C, this value representing 100% control activity.
[c]AT, 3-amino-1,2,4 triazole.
[d]The amount of free L-amino acids added represents between 3–12 mg/ml bovine liver catalase. Experiments were carried out in the presence of 1 μg/ml hemin. We routinely used commercial highly purified BLC; however, occasionally we found some BLC preparations contaminated with several high-molecular-weight proteins, as determined by sodium dodecyl sulfate-polyacrylamide electrophoresis, which accounted for about 2–5% of total proteins in preparation. For elimination of contaminating proteins, phenyl-Sepharose CL-4B affinity chromatography, using 50% ethylene glycol in 10 mM $NaHCO_3$ as eluent, was performed. Extensive dialysis against 10 mM $NaHCO_3$ was immediately carried out in order to eliminate ethylene glycol present in purified protein preparation.

mastigotes from five different isolates in the presence of totally purified bovine liver catalase (BLC), bovine hemoglobin, horseradish peroxidase (HRP), and lactoperoxidase, with BLC showing optimal growth-promoting activity (Table I). Additional experiments showed that dissociation of BLC into subunits did not substantially change its growth-promoting activity. As all these active proteins possess peroxidase activity, we checked the possibility of a relationship between their activity and growth-promoting action. This was verified by isolating proteins inside dialysis bags, a condition in which any H_2O_2 released by parasites could come into contact with active proteins, whereas the parasites could not. Under these experimental conditions, there was no parasite multiplication, suggesting the importance of direct contact between active proteins and parasites for induction of cellular multiplication (Table II). Furthermore, the irreversible catalase inhibitor 3-amino-1,2,4-triazole in the 2- to 20-mM range (nontoxic doses for *T. cruzi,* but up to 100% inhibitory of catalase activity present in culture tubes) did not decrease cell division rates, suggesting that the growth-promoting effect of this protein is mediated by its nutritional capacity rather than by any enzymatic activity. Additional experiments revealed that reconstituting the exact protein composition by means of exogenous amino acids did not support parasite multiplication, suggesting the importance of primary structure of active proteins for growth-promoting activity (Table II). Our results suggest four different explanations for the strong *T. cruzi* growth-promoting activity of these four hemoproteins: (1) they provide the heme requirements of actively multiplying parasites, (2) the amino acid composition of these proteins, plus the free amino acids already present in the culture medium, exactly fulfill the nutritional requirements of the parasites (and in fact, for example, of total BLC amino acids, 40% of them are essential for *T. cruzi* metabolism), (3) there exists a certain amino acid sequence common to several of the active protein molecules capable of acting as growth factors, and (4) a combination of these factors.

We think that *T. cruzi* growth-promoting hemoproteins are active because they are being pinocytosed and processed by the very powerful proteolytic equipment of the parasites (Avila *et al.*, 1979b), thus fulfilling their heme and essential amino acid requirements, and in fact cultivation of *T. cruzi* epimastigotes in a BLC-containing medium in the absence of exogenously added amino acids actively promotes parasite multiplication (Avila *et al.*, 1983). This is the first time that a single protein (BLC) has been shown to fulfill all the essential amino acid requirements of an actively multiplying eukaryote (Avila *et al.*, 1983). Pan (1978) has been able to culture the Y strain of *T. cruzi* in a semisynthetic medium free of macromolecules but containing trypticase, as well as in a completely synthetic medium by replacing trypticase with Trager's amino acid mixture, while O'Daly *et al.* (1987) have described an enriched synthetic medium with several low-molecular-weight peptides (less than 1000 Da molecular weight). However, in the presence of these peptides, *T. cruzi* amino acids, vitamins, and nucleotide requirements increased significantly at 26 °C and 34 °C compared with fetal bovine serum-supplemented media. Also, biomass doubling time of most nutrients was longer when trypanosomes were grown on peptides instead of proteins. These results suggest that the amino acid composition of these exogenous peptides cannot fulfill all nutritional requirements for active *de novo* protein synthesis, since they are not able to substitute the essential amino acids lacking in the culture media, as probably occurs with some amino acids in protein-supplemented media.

2.2.2. Amino Acids

Trypanosoma cruzi has a pool of free amino acids, which amounts to 5.2% of total protein nitrogen; alanine accounted for 40% of these amino acids (Williamson and Desowitz, 1961; O'Daly *et al.*, 1983). Azevedo and Roitman (1980) listed 17 amino acids required for growth, namely alanine, arginine, aspartic acid, cysteine, glycine, histidine, isoleucine, hydroxyproline, leucine, lysine, methionine, phenylalanine, proline, serine, tryptophan, tyrosine, and valine, while O'Daly *et al.* (1983) have demonstrated that glutamic acid, proline, glycine, and alanine constitute 75% of the amino acid pool in *T. cruzi* epimastigotes. Additionally, these latter authors demonstrated that amino acid requirements are different in protein or fetal calf serum-derived peptide-containing media and changed according to incubation temperature. However, arginine, serine, and glycine have to be supplied to all media at all temperatures to support continuous growth (O'Daly *et al.*, 1987). Avila (1984) and O'Daly *et al.* (1987) have established that, independently of culture medium, lysine, arginine, serine, glycine, methionine, valine, isoleucine, and tryptophan are essential amino acids for *T. cruzi* epimastigote multiplication.

On the other hand, at least some amino acids can be synthesized by *T. cruzi*. Shaw *et al.* (1964) showed that radioactivity from labeled glucose appeared in alanine, aspartate, and glutamate. Interconversion of some amino acids has also been demonstrated. Hampton (1971) showed that ^{14}C-labeled alanine, aspartate, glutamate, glycine, cysteine, and threonine were recovered from epimastigotes incubated with [^{14}C]serine.

Little is known about the uptake of amino acids by *T. cruzi*. Transport mechanisms of lysine and arginine were studied by Hampton (1970, 1971). Goldberg *et al.* (1976) reported quantitative variations in uptake of arginine and lysine depending on parasite strain and culture age and qualitative differences depending on whether the parasites were percolated through a DEAE-cellulose column previous to assay or not. Chromatography resulted in increased uptake of both amino acids, modification of kinetic behavior of amino acid transport systems, and loss of the inhibitory effects reported (Goldberg *et al.*, 1976).

Bash-Lewinson and Grossowicz (1957), working with intact cells and extracts of *T. cruzi,* reported that transaminase activity of histidine pyruvate exceeded that with α-ketoglutarate, pyruvate, glutamate, and alanine. Cross *et al.* (1975) found that in *T. cruzi,* serine and histidine appear to be extensively metabolized as judged by their disappearance. The *T. cruzi* endogenous respiration rate of respiration is high (Von Brand, 1967). Zeledón (1960a) ascribes a major part of this endogenous respiration to amino acid oxidation since carbohydrate metabolism inhibitors only partially block respiration. He has shown that glutamate, glutamine, aspartate, and asparagine each stimulate oxygen consumption of *T. cruzi* culture forms two- or threefold (Zeledón, 1960b). It is interesting that Harington (1961) has shown that hemolymph of *Rhodnius prolixus* contains relatively high concentrations of histidine, leucine, phenylalanine, and proline, with proline being absolutely critical for metacyclogenesis (Contreras *et al.*, 1985).

3. THE ENDOCYTIC SYSTEM OF *TRYPANOSOMA CRUZI*

3.1. Morphological Aspects

Only those structures associated with substance ingestion processes are discussed here.

1. The *T. cruzi* cell surface is composed of a unit membrane 8–10 nm thick, covering the cell body and flagellum and lined with a flagellar pocket. On the cell membrane, there is a layer of carbohydrates easily revealed when using cytochemical methods to detect carbohydrates (DeSouza, 1984). About 12 nm below the cell membrane there is a single layer of microtubules (subpellicular microtubules) in a parallel array (De Souza, 1976; Meyer and De Souza, 1976).

2. All the Trypanosomatidae have plasma membrane invagination where the flagellum emerges from the cell, forming the flagellar pocket (Figure 2A,B). This region is characterized by the absence of subpellicular microtubules and by the presence of multiple cytoplasmic vesicles. When *T. cruzi* is fixed in the presence of substances such as ruthenium red and alcian blue-lanthanum nitrate (De Souza *et al.*, 1977, 1978b) or incubated in the presence of colloidal iron hydroxyde or cationized ferritin particles, it has been observed that these substances can penetrate inside the flagellar pocket of epimastigotes and trypomastigotes (Brown *et al.*, 1965; Vickerman, 1969; Geigy *et al.*, 1970; Jadin, 1971; Steiger, 1973; Langreth and Balber, 1975). This observation suggests that flagellum attachment to the trypomastigote cell body may act as an occluding junction which can control the nature of the substances that penetrate inside the pocket. Since in *T. cruzi* trypomastigotes, as well as in lower trypanosomatids, the incorporation of nutritive material occurs at the flagellar pocket region, we can understand why the cell needs to have control mechanisms for substances that will be ingested. It is possible that the junction is a dynamic structure that can be assembled and disassembled according to cellular needs (DeSouza, 1984). The flagellar pocket of trypanosomes is believed to represent the site of exogenous substrate predigestion in trypanosomes (Jadin, 1971).

3. Ferritin uptake in *T. brucei* (Langreth and Balber, 1975), *T. raiae* (Preston, 1969), and *Crithidia fasciculata* (Brooker, 1971) is by coated vesicles, 100 nm across. These vesicles empty into straight tubular extensions of a smooth membrane collecting system, where protein is concentrated for digestion. *Trypanosoma brucei* has been shown to take up two nutrients, low-density lipoprotein and transferrin, by receptor-mediated endocytosis (Coppens *et al.*, 1987). Interestingly, Shapiro and Webster (1989) have developed a procedure to purify a coated vesicle fraction from *T. brucei* and demonstrated that the inner coat of electron-dense material is the parasite's variant surface glycoprotein (VSG) and that these structures are involved in intracellular processing of this protein. *Trypanosoma brucei* culture forms lack the 100-nm coated vesicles of blood

FIGURE 2. (A) Thin section of an epimastigote form of *T. cruzi* showing flagellar pocket (FP), flagellum (F), and at the upper region of the parasite body two acid phosphatase-positive lysosomes. Bar = 0.5 μm. Courtesy of Dr. Antonio Bretaña. (B) Epimastigote of *T. cruzi* showing cytostome (C), kinetoplast (K), nucleus (N), and flagellar pocket (FP). Arrows indicate external lips of the cytostome. Bar = 0.5 μm. Courtesy of Dr. Antonio Bretaña.

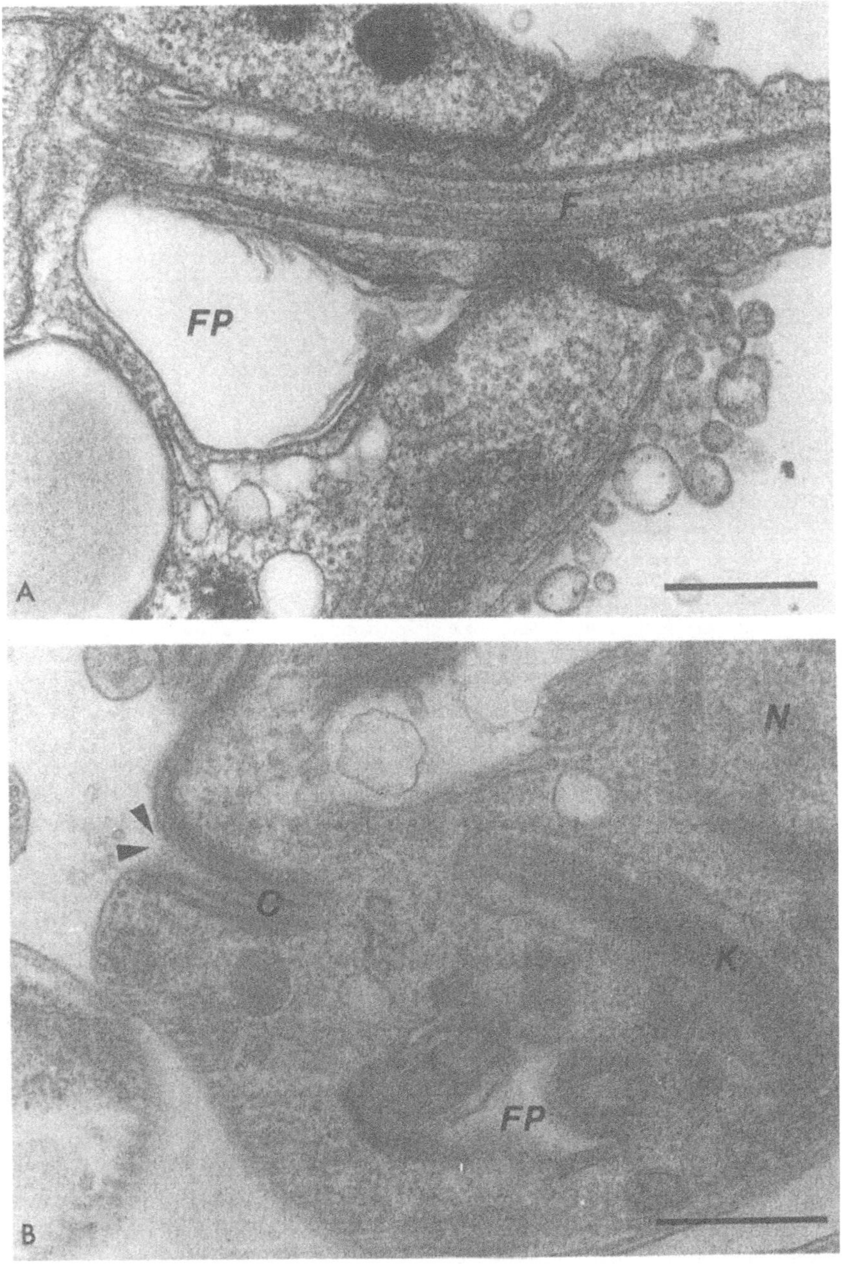
F
FP
A
N
C
K
FP
B

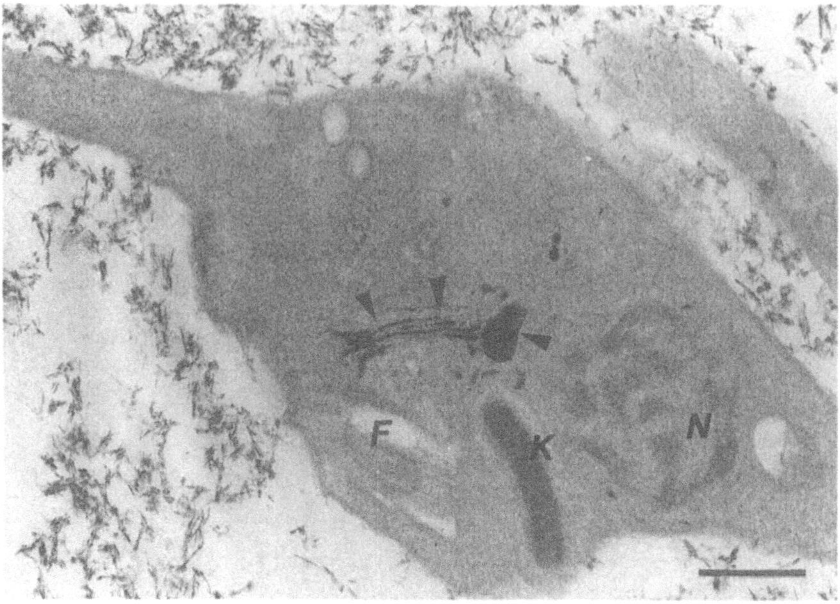

FIGURE 3. *Trypanosoma cruzi* epimastigote incubated for acid phosphatase activity. Positive staining is easily seen in the Golgi region, some close lysosomes and in a large multivesicular body. Arrows indicate acid phosphatase staining. β-Glycerophosphate was used as substrate and cerium chloride as the capture agent. Bar = 0.5 μm. Courtesy of Dr. Antonio Bretaña.

forms, as does *T. cruzi* epimastigote and trypomastigote forms (María *et al.*, 1972; Bretaña and O'Daly, 1976; De Souza *et al.*, 1978a).

4. Small (50 nm) primary lysosomes from the Golgi region fuse with digestive vacuoles (Figures 2A, 3, and 4B).

5. Steinert and Novikoff (1960) and Preston (1969) have shown in *T. mega* and *T. raiae*, respectively, incorporation of ferritin particles through a specialized surface region that was designated as the cytostome, a fact also observed by Herbert (1965a,b), who demonstrated that *T. theileri* and *T. melophagium* culture forms could take up exogenous HRP via the cytostome. This structure appears as a plasma membrane invagination that can penetrate deeply into cell cytoplasm (Milder and Deane, 1969). Cytostome cross sections show that sub-

FIGURE 4. (A) Acid phosphatase staining in *T. cruzi* epimastigote. Positive staining is seen in the flagellar pocket (FP), in a lysosome close to FP, and in the multivesicular bodies (arrows). β-Glycerophosphate was used as substrate. Courtesy of Dr. Antonio Bretaña. (B) Acid phosphatase staining in *T. cruzi* epimastigote. Positive staining is seen in several lysosomes (arrows). β-Glycerophosphate was used as substrate. Bar = 0.5 μm. Courtesy of Dr. Antonio Bretaña.

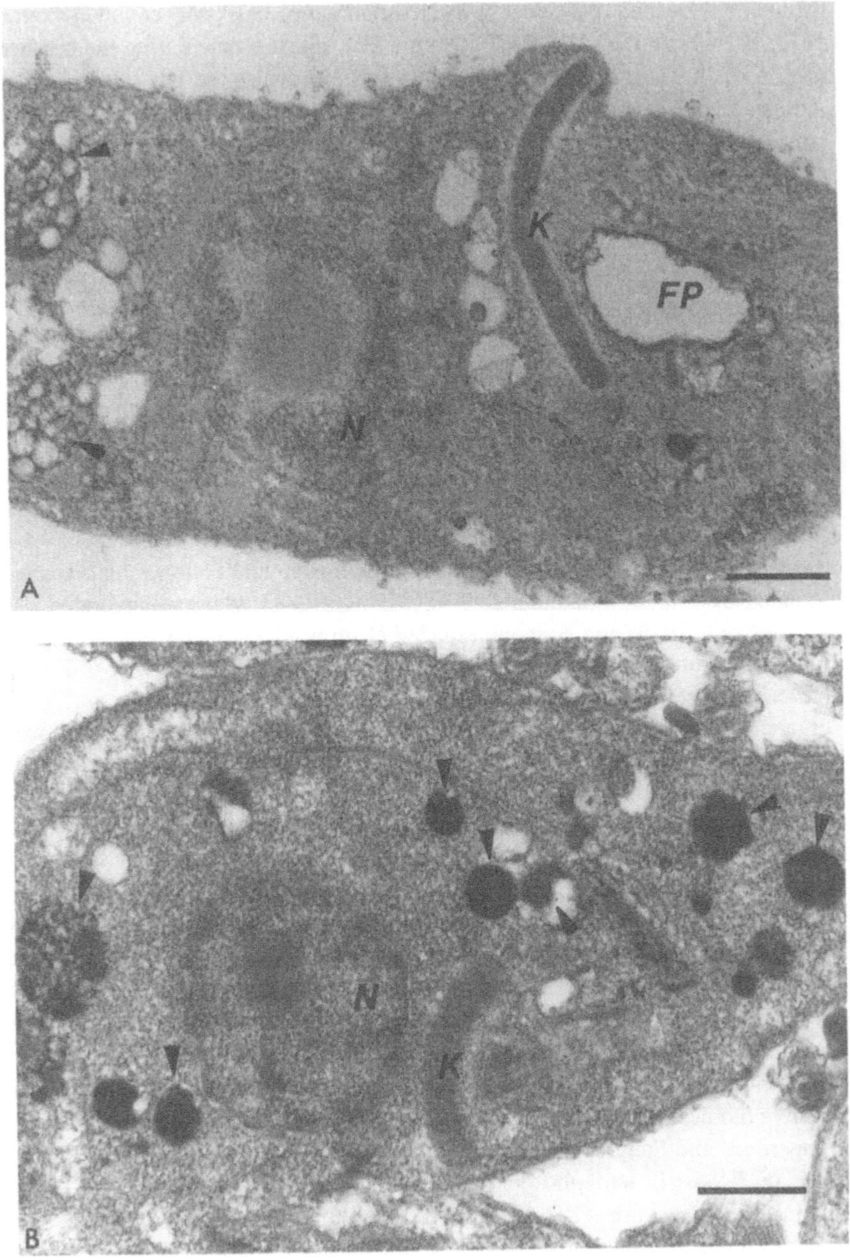
K
FP
N
A
N
K
B

pellicular microtubules remain associated with plasma membrane in this region. In *T. cruzi,* the cytostome is found only in spheromastigotes and epimastigotes (Milder and Deane, 1969; Sanabria and Aristimuño, 1970; Meyer and DeSouza, 1973; Bretaña and O'Daly, 1976) (Figure 2B). It disappears during the transformation of these forms into trypomastigotes. No cytostome has been found in African trypanosomes (Vickerman and Preston, 1976). Evidence has been obtained indicating that the plasma membrane that covers the cytostome region differs in structure and composition from the plasma membrane of other regions of the *T. cruzi* cell body. By ultrastructural cytochemistry it has been shown that the cytostome region binds more ruthenium red and concanavalin A than membranes of other regions of the protozoan. Freeze-fracture studies carried out on *T. cruzi* epimastigotes show that the cytostome appears as a particle-poor region delimited by a palisadelike row of adjacent intramembranous particles located close to the flagellar pocket (Martínez-Palomo *et al.*, 1976). During epimastigote–trypomastigote transformation, the palisadelike row of particles is gradually disintegrated, so that it is no longer observed in trypomastigotes (DeSouza *et al.*, 1978b).

6. Multivesicular bodies are membrane-bounded structures. They are located at the posterior end of the cell and their mean diameter ranges from 0.4 to 0.6 μm. High magnification examination of this structure shows that it is surrounded by a thick unit membrane that is associated with carbohydrates (DeSouza *et al.*, 1975) and contains rounded electron-lucent inclusions that are not limited by a unit membrane (Soares and DeSouza, 1988) (Figure 4A). In *T. cruzi* these structures represent about 5.65% of cell volume, and there are 16–17 organelles per cell with a mean diameter of 0.42 μm (Soares and DeSouza, 1988). These structures show acid phosphatase activity, so they may represent secondary lysosomes (Figure 4B).

From ultrastructural and biochemical studies it appears that acid hydrolases of trypanosomatid flagellates are localized in at least two different subcellular compartments, according to the parasitic stage, namely, in the flagellar pocket (acid phosphatase) and lysosomes (acid proteinase, α-mannosidase, and acid ribonuclease), at least in *T. brucei* bloodstream forms (Langreth and Balber, 1975; Steiger *et al.*, 1980; Opperdoes and Van Roy, 1982). A different subcellular localization of acid phosphatase activity, however, has been reported in cultured procyclic *T. brucei* trypomastigotes, most of this activity being associated with microsomal membranes derived from endoplasmic reticulum and digestive tubules and absent from the flagellar pocket (Langreth and Balber, 1975; Opperdoes and Steiger, 1981).

In *T. cruzi,* acid phosphatase activity has been reported in the plasma membrane (Nagakura *et al.*, 1985; Soares and DeSouza, 1988), flagellar pocket, Golgi complex (Arias-Flores *et al.*, 1986) (Figures 3 and 4A), and lysosomes (Sanabria, 1964; Vickerman and Preston, 1976; DeSouza, 1984; Arias-Flores *et*

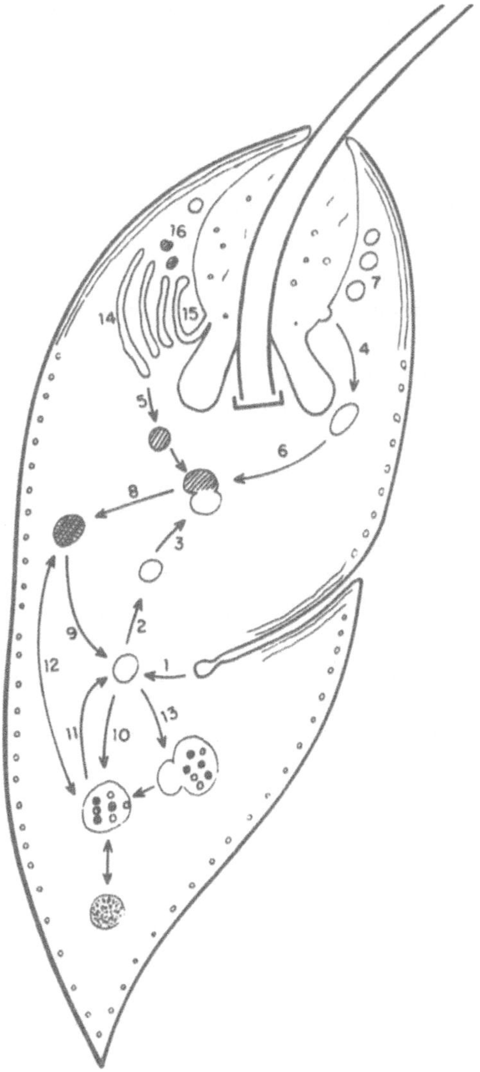

FIGURE 5. Schematic view of the endocytic system existing in *T. cruzi* epimastigotes. 1, endocytic vacuole forming from cytostome; 2 and 3, migration of endocytic vacuole in parasitic cytoplasm; 4, endocytic vacuole formed from flagellar pocket; 5, primary lysosome formation from Golgi lamellae; 3 and 6, endocytic vacuole migration toward the primary lysosome; 7, paraflagellar pocket endocytic vacuoles; 8, secondary lysosome formation; 9, secondary lysosome fusion with another new endocytic vacuole; 10, 11, and 13, endocytic vacuole fusion with multivesicular bodies; 12, secondary lysosome fusion with multivesicular body; 14, external Golgi lamellae; 15, direct discharge of Golgi materials (as acid phosphatase) in flagellar pocket; 16, primary lysosome migrating to flagellar pocket to indirectly discharge secretory materials.

al., 1986) (Figure 4B). Figure 5 presents a schematic view of the endocytic system existing in *T. cruzi* epimastigotes.

3.2. Dynamic Aspects of Endocytosis in *Trypanosoma cruzi*

The internalization of droplets of fluid from the environment was first described *in vitro* in *Amoeba* by Edwards (1925) and termed pinocytosis by Lewis (1931). Pinocytosis is one of the most important mechanisms by which cells can ingest macromolecules.

Several trypanosomatids have been shown to incorporate different protein molecules by a pinocytotic process (Steinert and Novikoff, 1960; Brown *et al.*, 1965; Herbert, 1965a,b; Preston, 1969; Geigy *et al.*, 1970; Brooker, 1971; Langreth and Balber, 1975; Bretaña and O'Daly, 1976; DeSouza *et al.*, 1978a; Avila *et al.*, 1983; Teixeira and Santana, 1989). In *T. cruzi*, pinocytic vesicles, localized in trypanosomatid cytoplasm, are always seen in association with the cytostome (Steinert and Novikoff, 1960; Bretaña and O'Daly, 1976; DeSouza *et al.*, 1978a) and the flagellar pocket region (DeSouza *et al.*, 1978a). Pinocytic vesicles containing HRP ingested by parasites fuse with each other and possibly with vesicles originated from the Golgi complex (primary lysosomes), forming large multivesicular structures (DeSouza *et al.*, 1978a). These structures have different forms. Some of them appear to contain vesicles in which peroxidase activity was not detected. Others appear uniformly electron dense, similar to those observed in mouse peritoneal macrophages (Steinman and Cohn, 1972). It was observed that even short incubation periods (about 10 min) of epimastigotes with HRP produced large multivesicular structures containing reaction products. Apparently the number of multivesicular structures is much larger in epimastigotes than in trypomastigotes. Freeze-fracture studies show that vesicle membranes and multivesicular structures formed by protein ingestion in *T. cruzi* have many fewer intramembranous particles than plasma membrane. Since they have originated from plasma membrane, they must have lost many intramembranous particles during this process (DeSouza *et al.*, 1978a).

Pinocytosis of several proteins by *T. cruzi* epimastigotes has been studied: HRP (DeSouza *et al.*, 1978a), bovine fetal serum proteins (Bretaña and O'Daly, 1976), BLC (Avila *et al.*, 1983), and specific IgG antibodies (Teixeira and Santana, 1989). These latter authors have demonstrated binding of specific human IgG antibodies to *T. cruzi* envelope with subsequent internalization by antigen receptor-mediated endocytosis (Abelha *et al.*, 1981). In this context, Cicarelli and Lopes (1989) have demonstrated a 52-kDa protein from *T. cruzi* trypomastigotes that binds human IgG. Finally, ferritin-conjugated antibodies and HRP-conjugated IgG were found inside parasite cytoplasmic vesicles (Teixeria and Santana, 1989).

4. PINOCYTOSIS IN *TRYPANOSOMA CRUZI*

4.1. Time Course and Temperature-Dependent Bovine Liver Catalase Binding to *Trypanosoma cruzi* Epimastigotes and Trypomastigotes

The time course of binding of native BLC and of ^{125}I-labeled BLC to *T. cruzi* epimastigotes and trypomastigotes at both 4 °C and 28 °C is shown in Figure 6. At 4 °C an apparent equilibrium between association and dissociation was reached after 1 hr. At 28 °C, cell-associated catalatic activity or radioactivity

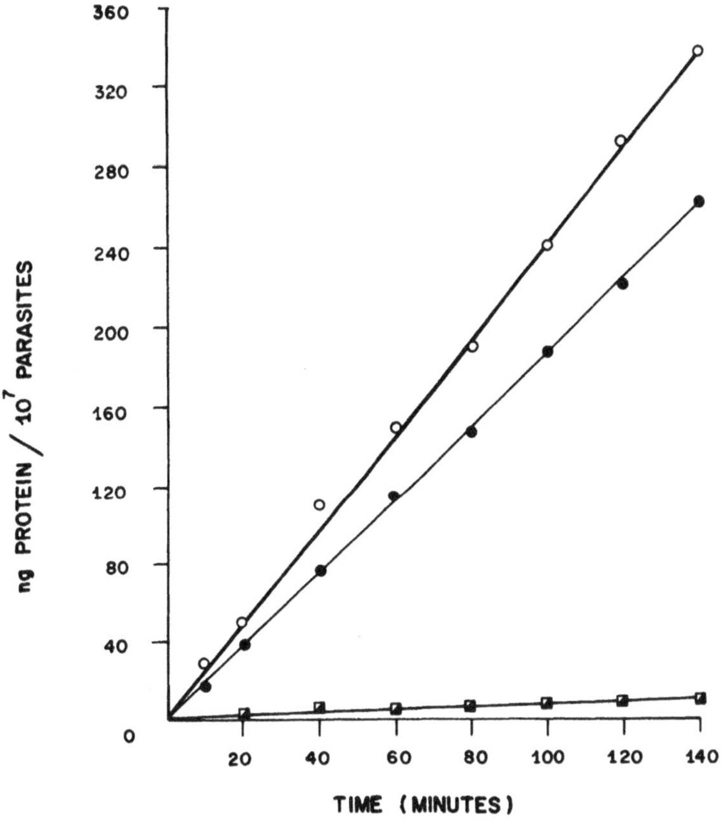

FIGURE 6. Kinetics of native and [^{125}I]-BLC uptake by *T. cruzi* epimastigotes. Uptake was quantitated at pH 6.4 using an enzymatic assay or radioactivity measurement adding 1 mg/ml BLC to incubation medium. Uptake of BLC increases linearly with the time of exposure to enzyme, indicating that pinocytosis of this marker proceeds continuously at a constant rate. The linear relation passes through the origin, showing that binding of BLC to the culture tube or to the cell surface is undetectable. ○, BLC; ●, [^{125}I]-BLC. ○, ●; BLC uptake at 28 °C; □, ■; BLC uptake at 4 °C.

increased linearly for 140 min. Trypsin treatment after incubation with iodinated BLC failed to remove cell-associated radioactivity from parasites handled at 28 °C, suggesting that these parasites are internalizing surface-bound BLC. Similar kinetic results were obtained using native HRP or [^{125}I]-HRP. This uptake probably reflects nonadsorptive pinocytosis of these macromolecules.

4.2. Concentration-Dependent Acquisition of [^{125}I]-BLC

As illustrated in Figure 7, addition of increasing concentrations of radioiodinated BLC to 1 × 10^7 organisms at 28 °C resulted in an initial stoichiometric

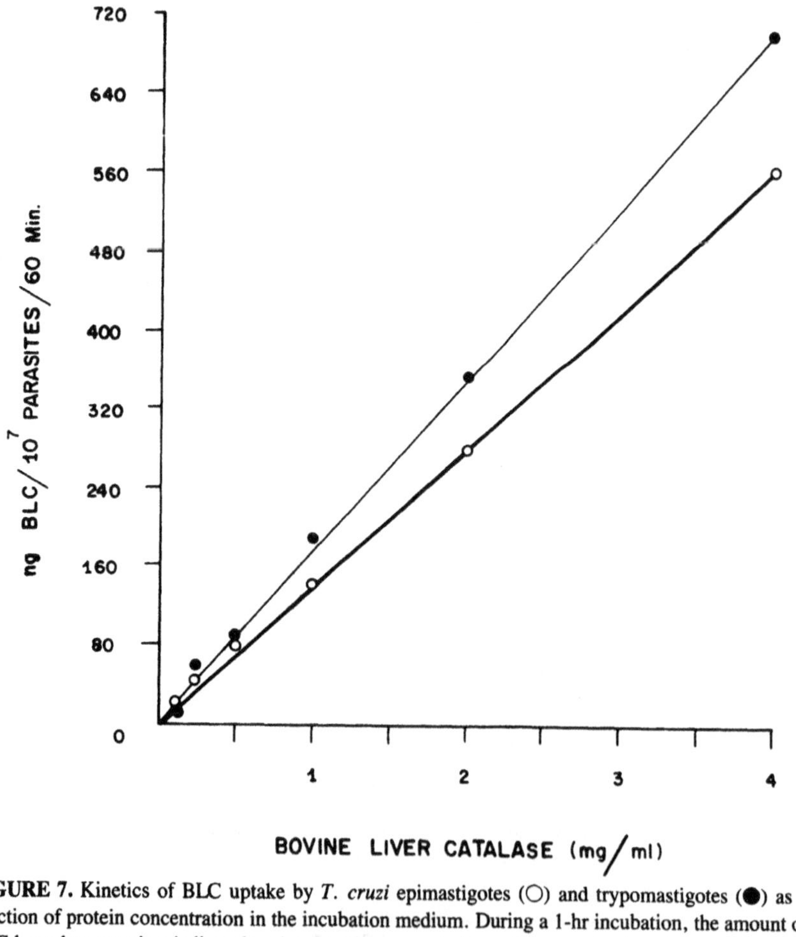

FIGURE 7. Kinetics of BLC uptake by *T. cruzi* epimastigotes (○) and trypomastigotes (●) as a function of protein concentration in the incubation medium. During a 1-hr incubation, the amount of BLC bound to parasites is linearly related to the concentration of enzyme in the culture medium, as would be expected of a material being endocytosed as a solute in pinocytic droplets. Uptake shows no evidence of saturation even at high concentration of BLC.

increase in the amount of surface-bound BLC, indicating a nonadsorptive binding component. Notably, under similar experimental conditions, *T. cruzi* trypomastigotes displayed higher endocytic activity than epimastigotes.

Taken together, the above results suggest that BLC was taken up primarily in the liquid phase and that membrane absorptive processes played a minor role. Recently, however, HDL (Prioli *et al.*, 1988) and transferrin receptors (Lima and Villalta, 1990) have been reported in *T. cruzi* epimastigotes and trypomastigotes, with the bloodstream forms being richer in these receptors. These results give evidence of the existence of adsorptive pinocytosis in these parasites, a fact already known in the case of *T. brucei* bloodstream forms (Coppens *et al.*, 1987).

4.3. Uptake of FITC-Dextran and Lucifer Yellow by *Trypanosoma cruzi* Epimastigotes and Trypomastigotes

Figure 8 shows a time course of pinocytosis of two different molecular weight substances: FITC-dextran (mol. wt. 42,000) and lucifer yellow (mol. wt. 457) by *T. cruzi* epimastigotes and trypomastigotes. As can be seen, again the pinocytic rate was higher for trypomastigotes than for epimastigotes. These results are similar to those reported by Teixeira and Santana (1989) for ^{3}H-labeled-human IgG pinocytosis. In the case of *T. cruzi* epimastigotes, we found that lucifer yellow but not FITC-dextran, was slowly exocytosed from lysosomes into the extracellular medium over a period of a few to several hours, a finding similar to that reported by Swanson *et al.* (1985) in mouse macrophages and by Buckmaster *et al.* (1987) in Chinese hamster ovary cells; namely, that retention of endocytized solutes within lysosomes is dependent on molecular size and may be limited by the rate of diffusion of molecules into shuttle vesicles.

4.4. Influence of Bovine Liver Catalase on the Uptake of Other Proteins

Table III shows that in the presence of 1 mg/ml BLC in the incubation medium, pinocytosis of other exogenous proteins was similar to that obtained in the absence of BLC, and none of those proteins altered the level of BLC uptake. Interestingly, low nonspecific protein adsorption to plastic tubes or to cells occurred as shown by the relatively close pinocytic values obtained for four different proteins (Table III and Figure 6).

4.5. Bovine Liver Catalase Uptake Rate Values Obtained in *T. cruzi* Epimastigotes

Trypanosoma cruzi epimastigote pinocytic rate was about 120 ng protein per 10^{7} cell/hr. Taking into account that in *T. cruzi* epimastigotes 1 mg of cell protein corresponds to about 10^{8} cells (Avila *et al.*, 1979b) that means a clear-

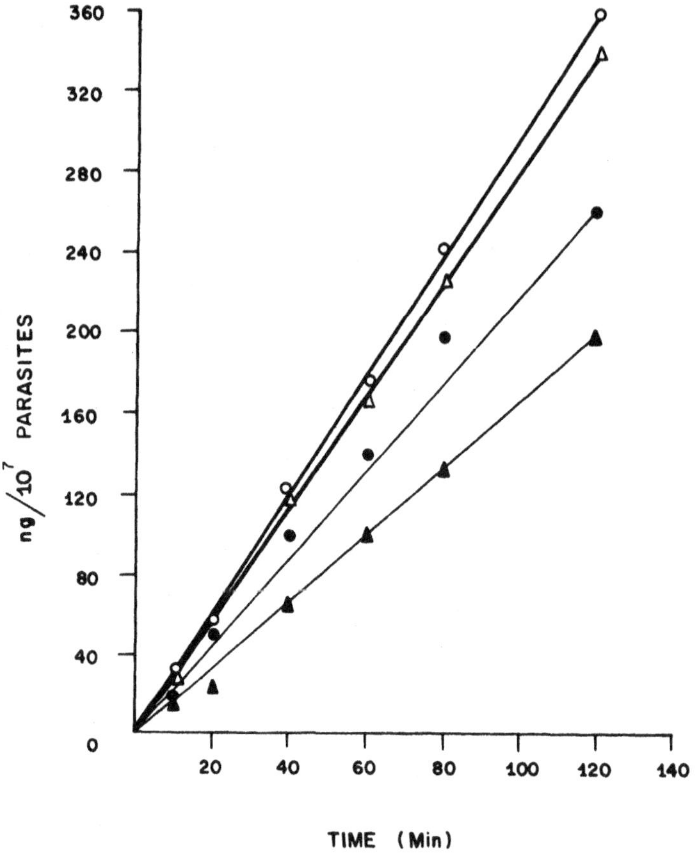

FIGURE 8. Pinocytic uptake rate of (○) FITC-dextran and (△) lucifer yellow by *T. cruzi*. Epimastigotes (●, ▲) and trypomastigotes (○, △) were incubated at 28 °C with 1 mg/ml of FITC-dextran or lucifer yellow.

ance of about 1.2 μl/mg protein per hour. This value is 12-fold higher than that reported for fluid endocytosis of polyvinylpyrrolidone (Fairlamb and Bowman, 1980) and sixfold above that of bovine serum albumin (Coppens *et al.*, 1987) by *T. brucei* bloodstream forms, indicating a striking difference in pinocytic rate due either to different experimental conditions used for each *Trypanosoma* species studied [our pinocytic conditions being similar to culture medium conditions in which *T. cruzi* epimastigotes are grown (Avila *et al.*, 1979a)] or to a higher endocytic activity of *T. cruzi* compared with *T. brucei*. The fluid pinocytosis value found for *T. cruzi* is 2.5- to 10-fold lower than that reported by Steinman *et al.* (1974) for fibroblasts and by Hammer and Rannels (1981) for rabbit pulmonary macrophages.

Table III
Influence of Bovine Liver Catalase on the Uptake of Other Soluble Proteins by *Trypanosoma cruzi* Epimastigotes[a]

Protein	Uptake of radiolabeled protein[b]		Uptake of BLC[b]
	In absence of BLC	In presence of BLC	
BSA	96 ± 21	99 ± 14	124 ± 19
HSA	94 ± 16	104 ± 19	116 ± 21
Lysozyme	115 ± 12	116 ± 17	126 ± 24

[a]Except when established, results given in this table and in others refer to experiments performed either with BLC-adapted *T. cruzi* epimastigotes or with fetal calf serum-adapted parasites that were carefully washed several times with fresh medium before being exposed to BLC or other proteins. *Trypanosoma cruzi* epimastigotes were exposed to three different radiolabeled proteins in the absence or presence of 1 mg/ml BLC. After a 1-hr exposure to 1 mg/ml of protein, parasites were washed twice with minimal essential medium and then lysed to determine the amount of bound radiolabel.
[b]Expressed as ng protein/10^7 per hr.

In terms of percent of BLC concentration in the medium, 0.0012% was internalized by 10^6 parasites/hr, a value which is about threefold lower than that reported for several fibroblast cell lines (Steinman *et al.*, 1974). In nanograms of BLC internalized per microgram of cell protein per hour at a load of 1 mg/ml BLC, the rate was 1.2 ng/μg, a value that is 12-fold higher than that reported for fibroblasts (Steinman *et al.*, 1974). As one parasite has a volume of 50 μm^3 (determined by [^{3}H]-inulin exclusion in 0.1 M acetate buffer, pH 6.4), our clearance values correspond to 2.4% cell volume per hour, a finding of the same order as that observed in mammalian cells: 3% for fibroblasts (Steinman *et al.*, 1976) and 1.7% for *T. brucei* trypomastigotes (Coppens *et al.*, 1987).

As regard the physiological significance of our *in vitro* pinocytosis data, as each trypanosome takes up 2.4% of its total volume per hour from external medium. As the BLC concentration in *T. cruzi* growth medium is 3 mg/ml and as cell division time is 55 hr (Avila *et al.*, 1979a), then these organisms would take up about 20% of their total amino acid requirements for cell duplication, 40% of total BLC amino acid composition being *T. cruzi* essential amino acids. This fact explains perhaps the 1.5-fold higher [^{3}H]leucine incorporation rate found in *T. cruzi* epimastigotes grown in free amino acid-deprived BLC-containing medium (Avila *et al.*, 1983), increased amino acid incorporation rate, and interconversion affording additional nonessential amino acids necessary for active cell multiplication.

4.6. Effect of Amino Acids and ATP on FITC-Dextran Accumulation

The time course of *T. cruzi* epimastigote accumulation of FITC-dextran was determined in three different media. One free of amino acids (basal medium) and

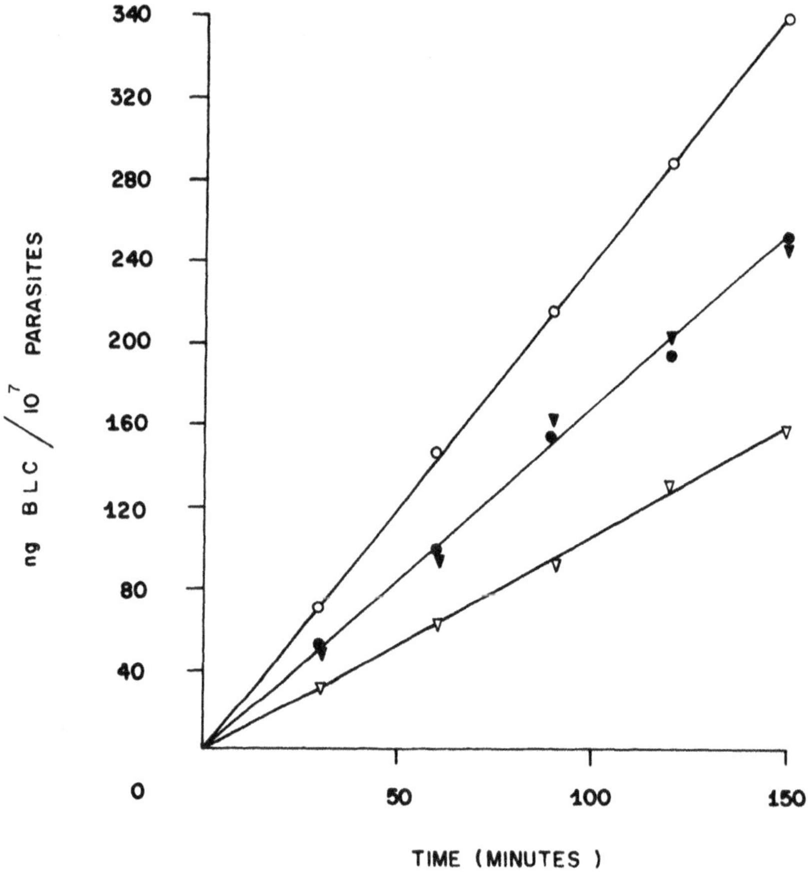

FIGURE 9. Pinocytic uptake rate of FITC-dextran by *T. cruzi* epimastigotes as a function of the composition of pinocytosis medium, ○, Amino acid-free medium; ●, MEM medium; ▼, RPMI medium; ▽, nucleotide-free medium. The exact chemical composition of fetal calf serum-containing MEM medium is as described in Avila *et al.* (1979a).

two containing essential amino acids at levels frequently employed in cell culture (1X): minimal essential medium (MEM) and RPMI. Figure 9 illustrates that, whereas there was little influence on the response to the two amino acid-containing media, omitting amino acids significantly increased net accumulation of FITC-dextran. Similar results have been reported for endocytosis of [^{14}C]sucrose by pulmonary alveolar macrophages (Besterman *et al.*, 1983). Interestingly, adding a complete vitamin mix to amino acid-free medium did not result in the lower pinocytic rate observed in presence of essential amino acids. Finally,

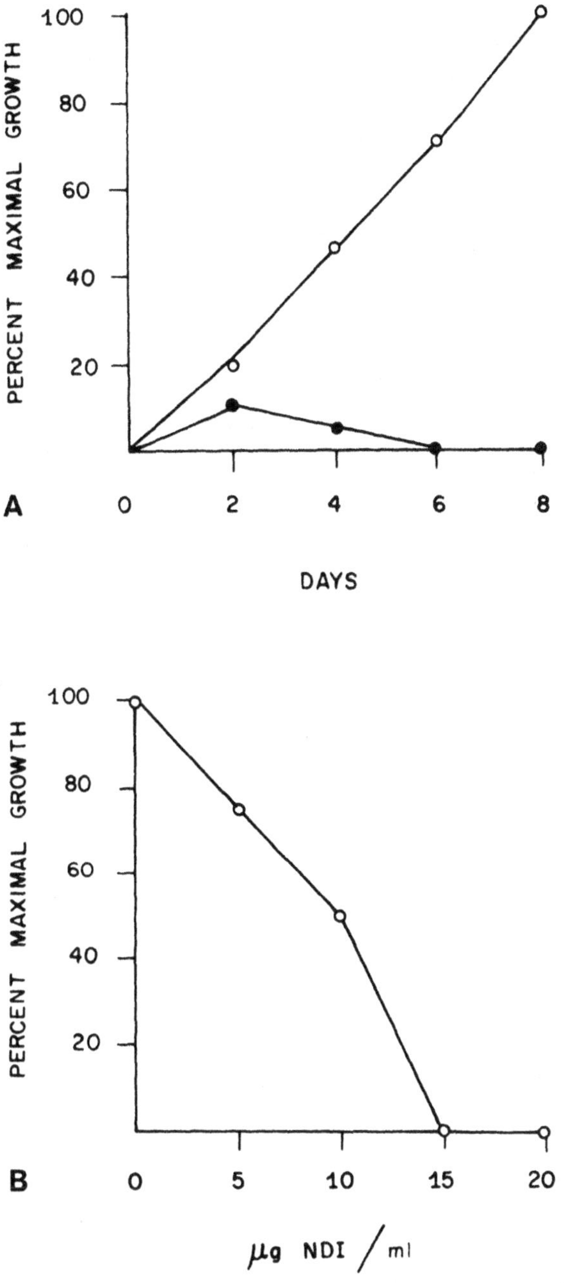

FIGURE 10. Effect of *N*-dodecyl imidazol (NDI) on *T. cruzi* epimastigote multiplication. (A) Effect of 15 μg/ml NDI on parasitic multiplication as a function of time of incubation; ○, percent control growth; ●, growth in the presence of NDI. (B) Influence of NDI concentration on parasitic multiplication.

deletion of ATP and other nucleotides present in *T. cruzi* culture medium (Avila *et al.*, 1979a) significantly lowered pinocytic rate. Similar results have been reported by Hayashi *et al.* (1973), who demonstrated that in rat liver, ATP facilitates pinocytotic engulfment of proteins into lysosomes and thereby stimulates proteolysis.

4.7. Uptake of the Lysosomotropic Drug *N*-Dodecyl Imidazol by *T. cruzi* Epimastigotes

Figure 10A shows that *N*-dodecyl imidazol (NDI), a lysosomotropic drug, in the culture medium produced a slow arrest of epimastigote multiplication, finally leading to total cell destruction. These findings are consistent with progressive rapid pinocytic uptake of NDI by *T. cruzi* epimastigotes with subsequent lysosomal accumulation until attaining a critical micellar concentration able to disrupt secondary lysosomes leading to cell death (Figure 10B), the ID_{50} being about 10 μg/ml.

Using an indirect approach, Opperdoes and Van Roy (1982) have demonstrated that in *T. brucei*-infected rats, Triton WR-1339, another lysosomotropic agent, appeared to be concentrated in parasitic lysosomes, a fact demonstrated by development of large vacuoles, reduction in α-mannosidase and acid proteinase latency, and an increase in the nonsedimentable fraction of lysosomal enzymes.

5. INTRACELLULAR DEGRADATION OF PROTEINS IN *TRYPANOSOMA CRUZI*

5.1. Proteases Present in *Trypanosoma cruzi* Epimastigotes

The intracellular breakdown of protein to amino acids involves the concerted action of endo- and exoproteinases. In *T. cruzi* epimastigotes, a multitude of proteolytic enzymes have been found. Table IV gives some characteristic features of described proteolytic activities. As shown, cell extracts of *T. cruzi* culture forms exerted hydrolytic activity on four different groups of substances commonly used as protease substrates: (1) aminoacyl-naphthylamides (BNA), substrate for amino peptidases; (2) CTN, substrate for chymotrypsinlike enzymes; (3) BAPA and BANA, substrates for trypsinlike enzymes, and azocasein and hemoglobin, substrates for a broad range of proteolytic enzymes, cathepsin D among others; and (4) NBC, substrate for cathepsin A. Of interest is the fact that hydrolysis of glycyl-l-phenylalanine-BNA and *N*-benzoyl-DL-arginine-BNA, specific substrates of cathepsins C and B, respectively, although highly

Table IV
Proteolytic Activities Demonstrated in *Trypanosoma cruzi* Epimastigotes

Substrates used[a]	Optimum pH	References
Azocasein	5.0	Itow and Camargo (1977)
BAPA	8.0	Itow and Camargo (1977)
BANA	8.0	Itow and Camargo (1977)
CTN	7.0	Itow and Camargo (1977)
Leucyl- or lysyl-BNA	7.0–8.0	Itow and Camargo (1977)
BAPA	7.0	Bongertz and Hungerer (1978)
Glycyl-l-phenylalanine-BNA	4.6	Avila *et al.* (1979b)
α-*N*-benzoyl-DL-arginine-BNA	5.8	Avila *et al.* (1979b)
L-alanyl-BNA	5.6	Avila *et al.* (1979b)
NBC-glutamyl-l-tyrosine	5.2	Avila *et al.* (1979b)
Hemoglobin	3.8	Avila *et al.* (1979b)
Hemoglobin	3.6	Steiger *et al.* (1979)
Hemoglobin	3.0	Rangel *et al.* (1981a)
Casein	6.0	Rangel *et al.* (1981b)
Azocasein	3.0–4.0	Torruella *et al.* (1981)
Endogenous	3.0–4.0	Cazzulo and Cazzulo (1982)
Endogenous	4.5	O'Daly *et al.* (1983)
Bovine serum albumin	4.0	Bontempi *et al.* (1984)
BAPA	8.0	Ashall (1990)

[a]BAPA, Benzoyl arginine-*p*-nitroanilide; BANA, Benzoyl arginine-naphthylamide; CTN, *N*-carbobenzoxy-L-tyrosine-*p*-nitrophenyl ester; BNA, β-naphthylamide; NBC, *N*-benzyloxycarbonyl-α-1.

active in parasitic extracts (Avila *et al.*, 1979b), are not activated by halogens or cysteine, as is characteristic for these cathepsins (McDonald *et al.*, 1965; Barret, 1969); therefore, they can not be considered as representing true cathepsin C and B activities.

Interestingly, Steiger *et al.* (1979) have found that trypomastigote forms of *T. cruzi* have about sevenfold higher acid proteinase-specific activity than epimastigote forms, a fact that coincides with the more rapid digestion of pinocytosed IgG (Teixeria and Santana, 1989) and of BLC and BSA (Figure 11) found in *T. cruzi* trypomastigotes compared with epimastigotes (Figures 6–8). Ashall (1990) recently demonstrated a strong cysteine-type peptidase activity in *T. cruzi*, with an optimum pH at about 8.0. Furthermore, in contrast with other proteolytic activities studied in *T. cruzi*, amastigotes and trypomastigotes contained about 16% and 14%, respectively, of the activity present in epimastigotes.

In recent experiments, we have found a neutral proteinase in living *T. cruzi* epimastigotes, active on azocasein and located on the surface of plasma membrane. Finally, the subcellular localization of acid proteinase and α-mannosidase in *T. cruzi* lysosomes has been demonstrated by Bontempi *et al.* (1989).

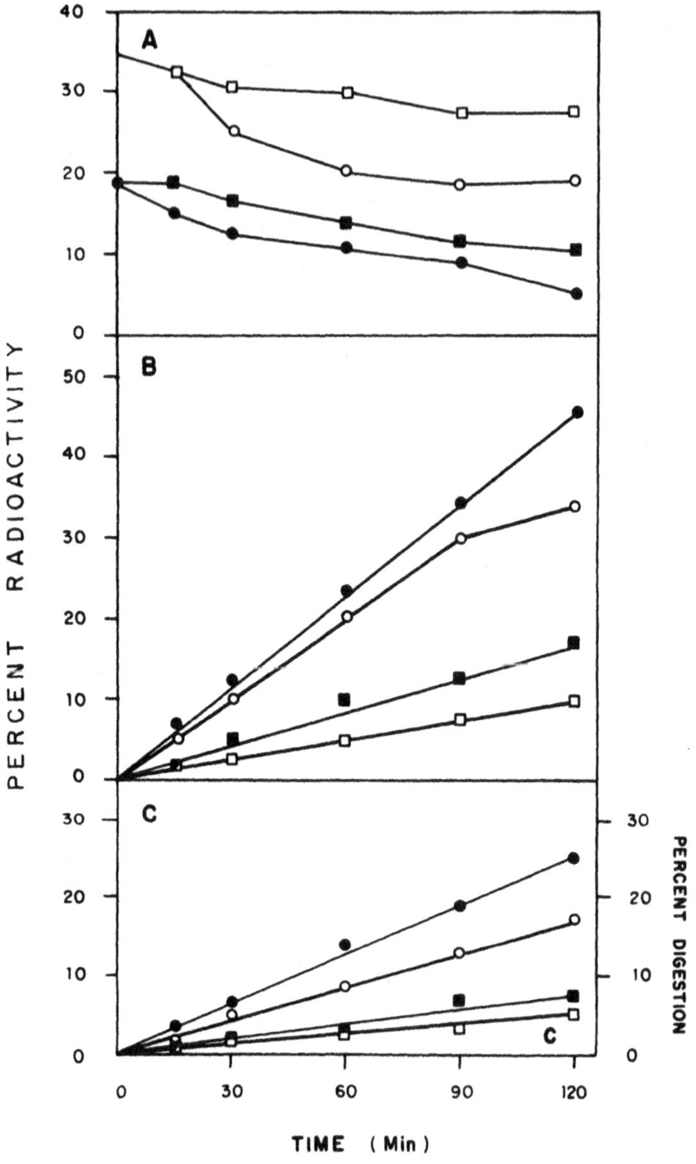

FIGURE 11. Digestion of [^{3}H]-BSA and [^{125}I]-BLC by *T. cruzi* epimastigotes. Cells were incubated for 2 hr in the presence of BSA or BLC (1 mg/ml). Cells were then washed four times with minimal essential medium and reincubated in fresh medium for periods up to 2 hr. Total radioactivity and radioactivity soluble in 8% trichloroacetic acid were measured on samples of the medium and of the cells (solubilized in 0.1 M NaOH–0.4% sodium deoxycholate) taken at several time periods of the final incubation. The increase in TCA-soluble radioactivity during incubation provided a measure of the protein digested. It was expressed as percent of the amount of exogenous protein stored by the cells (initial TCA-insoluble radioactivity). The breakdown of endogenous protein in labeled cells was measured similarly from the increase in TCA-soluble radioactivity during incubation. All experiments were performed in triplicate. (A) TCA-soluble radioactivity in the cells. (B) TCA-soluble radioactivity in the medium. (C) Net production of TCA-soluble radioactivity [A + B − (initial TCA-soluble radioactivity in cells)]. (●), BSA; (○), BLC; ■, BSA + chloroquine; □, BLC + chloroquine.

5.2. Evidence of Lysosomal Digestion of Endocytosed Proteins by *Trypanosoma cruzi* Epimastigotes

Figure 11 shows the time course of release of TCA-soluble radioactivity from *T. cruzi* epimastigotes grown in BLC but previously exposed for 2 hr to 1 mg/ml heat-denatured [^{3}H]acetylated BSA or [^{125}I]-BLC. The appearance of degradation products in the medium (Figure 11B) resulted from a combination of release of preformed products from the cells (Figure 11A) and further breakdown of stored protein. Net digestion (Figure 11C) during the experimental period amounted to some 17–25% of the total cellular radioactivity, or 35–45% of the TCA-insoluble material initially present in the cells. These values are close to those obtained for mouse peritoneal macrophages (30%) (Ohkuma *et al.*, 1986) and could be explained by the complete proteolytic equipment existing in *T. cruzi* (Table IV). This process, which for obvious reasons may be assumed to take place in lysosomes, was unaffected by 10 mM NaF but was inhibited more than 60% by 100 μM chloroquine. From Figure 11C it is evident that BSA was digested more rapidly than BLC (27% vs. 18% at 2 hr), a fact that cannot be explained by different endocytic uptake rates (Table III) but by dilution inside a lysosomal compartment of [^{125}I]-BLC with cold BLC previously pinocytosed or to a higher proteolytic sensitivity of BSA to *T. cruzi* epimastigote enzymes. Similar results were obtained if cells were incubated for 30 min after the BSA uptake period to ensure complete phagosome–lysosome fusion before digestion started to be recorded. Chloroquine inhibition thus clearly affects breakdown in lysosomes but not delivery to lysosomes. Gel filtration on Sephadex G-10 indicated that most of the TCA-soluble radioactivity of both cells and medium belonged to a material behaving like acetyl-lysine or [^{125}I]tyrosine.

When the degradation rate of pinocytosed BLC or HRP was comparatively followed by measuring loss of enzymatic activity and release of radioactivity to the cellular TCA-soluble fraction, we found that the decrease of enzymatic activity was three times faster than release of [^{125}I]iodotyrosine (Figure 12). Similar results have been reported by Steinman and Cohn (1972) in mouse peritoneal macrophages; these authors suggested that the accumulation of enzymatically inactive intermediates occurred in lysosomes.

5.3. Effect of Basic Substances, Ionophores, and Proteolytic Inhibitors on *Trypanosoma cruzi* Epimastigote Multiplication

When *T. cruzi* epimastigotes growing in BLC were exposed to a variety of basic substances and ionophores, most of which were previously found to induce macrophage vacuolation (Ohkuma and Poole, 1981) and/or to raise the intralysosomal pH (Ohkuma and Poole, 1978; Poole and Ohkuma, 1981), parasite multiplication was significantly inhibited (Tables V and VI). In intact living cells

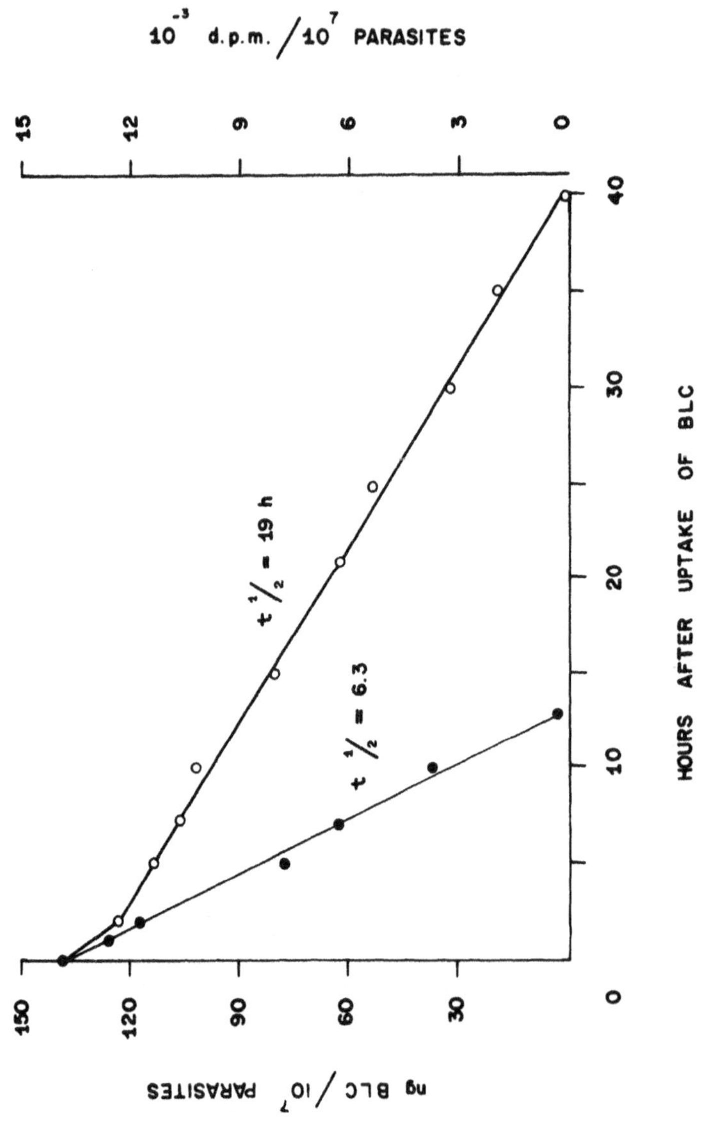
10^{-3} d.p.m./10^{7} PARASITES
ng BLC/10^{7} PARASITES
HOURS AFTER UPTAKE OF BLC
t 1/2 = 19 h
t 1/2 = 6.3

Table V
Effect of Some Basic Substances on the Multiplication of *Trypanosoma cruzi* Epimastigotes Grown in a Bovine Liver Catalase-Containing Medium

Compound	Percent control growth
None (control)[a]	100.0 ± 0.5
Chloroquine (100 μM)	15.0 ± 3.0
NH_4 Cl (10 mM)	15.2 ± 3.2
NH_4 Cl (5 mM)	28.3 ± 4.1
NH_4 Cl (1 mM)	56.1 ± 9.3
Methylamine (10 mM)	12.7 ± 2.9
Methylamine (5 mM)	19.8 ± 3.0
Methylamine (1 mM)	38.8 ± 5.2
Dimethylamine (10 mM)	17.7 ± 3.9
Trimethylamine (10 mM)	19.8 ± 4.2
Ethylamine (10 mM)	15.8 ± 5.8
Ethylamine (5 mM)	34.4 ± 7.1
Diethylamine (10 mM)	23.6 ± 5.4
Diethylamine (5 mM)	41.8 ± 9.4
Triethylamine (10 mM)	30.1 ± 5.8
Triethylamine (5 mM)	45.8 ± 8.1
n-butylamine (1 mM)	56.9 ± 7.4
Tri-*n*-butylamine (1 mM)	41.3 ± 3.2

[a]Drugs were added at time 0 to *T. cruzi* culture medium. Control growth was 17 ± 3 × 10^6 parasites/ml after 10 days of culture.

and under long exposure times, these results could be due either to direct modifications of the intralysosomal milieu or to blocking of pinocytic transit at specific point(s) from "early" to "late" prelysosomal vesicles and lysosomes, as demonstrated in Chinese hamster ovary cells (Gonzalez-Noriega *et al.*, 1980; Sullivan *et al.*, 1987).

FIGURE 12. Fate of bovine liver catalase pinocytosed by *T. cruzi* epimastigotes. Parasites cultured in 2.5% fetal calf serum-containing MEM (Avila *et al.*, 1979a) were used. Epimastigotes in log phase of growth were spun down by centrifugation at 15,000 g/min at 4 °C, washed with 50 ml cold MEM, and centrifuged again at same speed. Parasites (3 × 10^8 cells) were resuspended in 50 ml MEM containing 1 mg/ml BLC and incubated 1 hr at 28 °C under continuous shaking (150 rpm). After pinocytosis, cells were carefully washed with MEM and followed for an appropriate length of time in BLC-free MEM medium. At each point, duplicate cultures were assayed for enzyme remaining in cells, as well as enzyme in the culture medium, in order to detect enzyme being exocytosed from cells. The amount of BLC/10^7 cell diminishes exponentially until enzyme cannot be detected. The half-lives varied from 6 to 8 hr in experiments measuring catalatic activity and from 14 to 19 hr for experiments measuring TCA-insoluble [^{125}I]-BLC; (●), ng BLC/10^7 cells; (○), [^{125}I]-BLC/10^7 parasites.

Table VI
Effects of Ionophores on the Multiplication of *Trypanosoma cruzi* Epimastigotes Grown in a Bovine Liver Catalase-Containing Medium

Compound	Percent of control growth[a]
None (control)[b]	100.0 ± 5.9
Chloroquine (100 μM)	14.1 ± 2.8
Methylamine (10 mM)	18.5 ± 2.7
Nigericin (5 μg/ml)	40.8 ± 7.1
X537A (20 μg/ml)	36.9 ± 9.3
Gramicidin (2 μg/ml)	56.2 ± 4.4
2,4-dinitrophenol (1 mM)	78.2 ± 13.6
CCCP[c] (20 μM)	70.8 ± 12.5
CCCP (200 μM)	56.1 ± 10.8
CCCP (20 μM) + Valinomycin (20 μM)	48.4 ± 10.2

[a]Drugs were added at time 0 to *T. cruzi* culture medium. Control growth was 16 ± 2 × 10^6 parasites/ml after 10 days of culture.
[b]Control growth was 18 × 10^6 parasites/ml.
[c]CCCP, carbonyl cyanide *m*-chlorophenyl hydrazone.

Table VII shows the inhibitory effects on *T. cruzi* epimastigote multiplication of different proteinase inhibitors. Pepstatin (0.2 mM), antipain (0.5 mM), and leupeptin (0.5 mM) were the most effective parasitic multiplication inhibitors. Since pepstatin, phosphoramidon, antipain, leupeptin, and chymostatin are specific inhibitors of aspartic proteinases (EC 3.4.23) and metalloproteinases

Table VII
Effect of Several Proteinase Inhibitors on the Multiplication of *Trypanosoma cruzi* Epimastigotes Grown in a Bovine Liver Catalase-Containing Medium

Compounds	Percent of control growth[a]
None (control)	100.0 ± 4.8
Leupeptin (0.5 mM)	14.1 ± 2.9
Pepstatin (0.2 mM)	10.4 ± 2.3
Antipain (0.5 mM)	12.7 ± 0.9
Chymostatin (0.5 mM)	23.2 ± 3.4
E64 (1 mM)	20.1 ± 2.3
Fut 175 (1 mM)	21.8 ± 1.9
2-Arg-Leu-Phe (1 mM)	29.3 ± 4.7
3-Methyladenine (10 mM)	40.4 ± 6.2
Vanadate (10 mM)	31.2 ± 9.2
Phosphoramidon (1 mM)	38.6 ± 7.3

[a]Maximal control growth was 14 × 10^6 parasites/ml attained at 12 days postinoculation.

Table VIII
Effect of Weak Bases and Proteolysis Inhibitors on *Trypanosoma cruzi* Epimastigote *in Vivo* Protein Degradation

Compounds	Total isotope trichloroacetic acid soluble in medium	
	With fetal serum[a]	Without fetal serum[b]
None (control)	5.3 ± 1.2	10.6 ± 2.7
X537A (20 μg/ml)	5.6 ± 1.4	4.4 ± 0.7
Chloroquine (100 μM)	4.7 ± 0.3	3.6 ± 0.8
Neutral red (100 μM)	5.1 ± 0.8	3.9 ± 1.1
Tributylamine (1 mM)	5.0 ± 0.6	4.9 ± 1.7
NH_4Cl (10 mM)	4.8 ± 1.0	4.2 ± 1.3
Leupeptin (0.5 mM)	5.2 ± 1.4	4.3 ± 0.9
Pepstatin (0.2 mM)	5.7 ± 1.2	4.7 ± 0.8

[a]Parasites have been grown by at least 6 consecutive passages in 2.5% fetal calf serum in the incubation medium.
[b]Parasites have been grown by at least 6 consecutive passages in 3 mg/ml bovine liver catalase-containing culture medium.

(EC 3.4.24) but are also unspecific inhibitors of serine proteinases (EC 3.4.21) and cysteine proteinases (EC 3.4.22) (North, 1982), we see that a whole spectrum of active proteinases are needed for successful *T. cruzi* epimastigote multiplication. These results are also indirect evidence of the critical role of intracellular digestion of protein for successful *T. cruzi* epimastigote multiplication.

We must point out, however, that the *in vivo* protein degradation occurring in *T. cruzi* epimastigotes in the absence of serum (that is, using BLC as a unique protein in the culture medium) is considerably more sensitive to inhibition by cloroquine, neutral red, tributylamine, NH_4 Cl, leupeptin, and pepstatin than is the slower rate of protein degradation occurring in complete medium (Table VIII). These results are evidence of an important metabolic difference between the protein degradation rates occurring under different culture conditions. This difference could be easily explained either by an easier blocking of normal endocytic pathways as mentioned before or to the differential induction of proteolytic enzymes according to culture medium components.

5.4. Influence of pH on the *in Vitro* Hydrolysis of Bovine Liver Catalase and Bovine Serum Albumin by *Trypanosoma cruzi* Epimastigote Homogenates

When the influence of pH on the hydrolysis of [^{125}I]-BLC and [^{3}H]acetyl BSA by *T. cruzi* homogenates was investigated, two peaks were found at pH 3.5 and 6.5 (Figure 13). The contribution of a lysosomal neutral proteinase (Bongertz and Hungerer, 1978; Bontempi *et al.*, 1989) as well as the existence of nonlysosomal proteinases (Ashall, 1990) could explain the small neutral peak

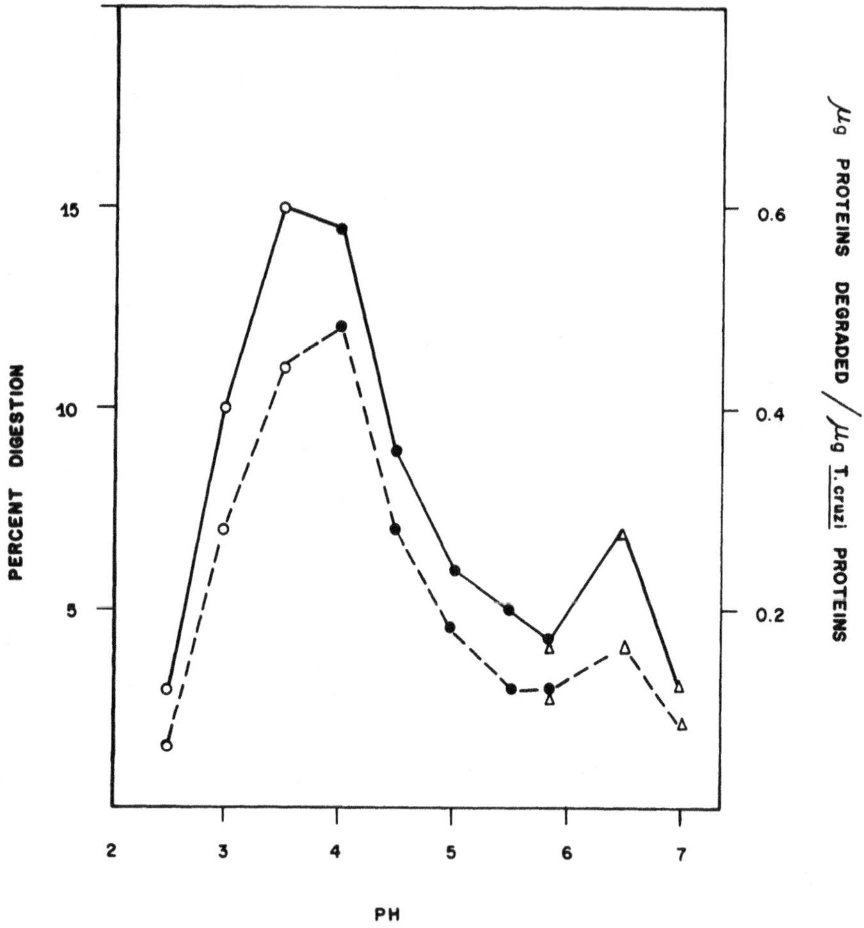

FIGURE 13. pH-Dependence of the digestion of [^{3}H]-BSA (○—○) and [^{125}I]-BLC (○---○) by *T. cruzi* epimastigote total extracts. ○, Sodium lactate buffer; ●, sodium acetate buffer; △, sodium phosphate buffer.

found in our digestion experiments. Our optimal pH for digestion of exogenous proteins by *T. cruzi* epimastigote homogenates is quite similar to that obtained for endogenous parasitic proteins (Cazzulo and Cazzulo, 1982; O'Daly *et al.*, 1983).

As most proteolytic activity in *T. cruzi* epimastigotes is carried out at acid pH, our results tend to suggest that a number of inhibitory compounds exert their effect simply by way of a pH shift. However, the possibility exists that some compounds exert additional inhibitory effects, presumably either on given cell functions or enzyme activities (Wibo and Poole, 1974). The conclusion that exogenous proteins are broken down in lysosomes and partly outside these particles is in agreement with other previous publications (Kay, 1978; Amenta and Brocher, 1981; Hershko and Ciechanover, 1982; Mortimore, 1982).

Teixeira and Santana (1989) found that while epimastigote-rich preparations usually showed a 20% decrease in the quantity of antibodies bound to their outer membrane after 12 hr of incubation at 27 °C, trypomastigote-rich preparations showed a 95% decrease under identical experimental conditions. When experiments were designed to assay free radiolabel in supernates after precipitation with 5% TCA, they found that in the case of epimastigotes the amount of free radiolabel increased from 17% at 15 min to 32% at 12 hr, while in trypomastigote-rich preparations radiolabel in TCA supernates increased from 27% at 8 hr up to 83% at 12 hr, thus suggesting that IgG is taken in, degraded by the cells, and the small peptides and amino acids formed are released into the culture medium, but also that *T. cruzi* trypomastigotes have a higher endogenous proteolytic activity than epimastigote forms.

6. EVIDENCE OF LYSOSOME-RELEASED AMINO ACIDS AS BUILDING BLOCKS FOR PROTEIN SYNTHESIS IN *TRYPANOSOMA CRUZI* EPIMASTIGOTES

6.1. Release of Lysine from Exogenous Protein Sources

The steady-state level of protein in cells is governed not only by the rate of synthesis but by the rate of degradation. There are four sources of essential amino acids available for cellular protein synthesis: intracellular free amino acids, amino acids derived from degradation of cellular protein, extracellular free amino acids, and amino acids derived from degradation of extracellular proteins. Historically, degradation of serum proteins has not been considered a significant source of amino acids for *de novo* protein synthesis, based almost solely on the work of Eagle and Piez (1960). However, new though indirect experimental evidence strongly suggests that *T. cruzi* epimastigotes, alveolar macrophages, and *Tetrahymena* can use exogenous proteins (specifically serum albumin in the case of two latter cells) as a source of essential amino acids for *de novo* protein synthesis in the presence of physiological (Avila *et al.*, 1979a; Rannels *et al.*, 1982; Grinde and Jonassen, 1987) and subphysiological levels of intracellular free amino acids (Hammer and Rannels, 1981; Avila *et al.*, 1983; Avila, 1984).

Lysine is an essential amino acid for *T. cruzi* epimastigote multiplication (Avila, 1984; O'Daly *et al.*, 1987); it accounts for about 5% of total amino acids present in BLC (Schroeder *et al.*, 1982). Having demonstrated that exogenous BLC or HRP did not alter degradation of endogenous *T. cruzi* epimastigote proteins, that lysine release from BLC did not occur in cell-free conditioned medium, and that the intracellular pool of free lysine was of stable and insignificant size compared with the size of the extracellular lysine in the presence of BLC or HRP (Avila, 1984), the increased release of lysine was attributed to the intracellular degradation of pinocytosed proteins. Therefore, to test the hypothesis that degradation of exogenous proteins provides amino acids that are reutilized for protein synthesis, we measured changes in the extracellular concentration of lysine in the presence and absence of BLC and HRP. Reutilization of lysine derived from intracellular degradation of endogenous and exogenous protein was calculated from measurements of lysine release in the absence and presence of cycloheximide. At 15 μm, cycloheximide inhibits protein synthesis in *T. cruzi* epimastigotes by 90–95%, producing only a 5% inhibition of protein degradation.

6.2. Influence of Exogenous Amino Acid Levels on the Lysine Release from Endogenous Protein Sources

In the presence of a physiological level of amino acids (1 X) (BLC absent, cycloheximide present) total lysine release (degradation) was about 2.14 nmol/10^6 cells per hour. On the other hand, in the absence of BLC and cycloheximide, net lysine release (degradation minus synthesis) occurred at a rate of 1.21 nmol/10^6 cells per hour. Therefore, lysine was recycled into the protein synthetic pathway at a rate of 0.93 nmol/10^6 cells per hour, which represents reutilization of 43% of the lysine derived from degradation of endogenous protein (Table IX). Likewise, under conditions of amino acid deprivation (0 X), 53% of the lysine derived from degradation of endogenous proteins was reutilized for protein synthesis. Thus, reutilization of lysine derived from degradation of endogenous protein was unaffected by amino acid deprivation. Therefore, endogenous protein degradation appeared insensitive to an acute depletion of free extracellular amino acids concentration, a fact that contrasts with results obtained for *Tetrahymena,* where the rate of intracellular proteolysis falls to a minimum of 3–4%/3 hr by cultivation in a medium containing free amino acids (Jonassen and Grinde, 1986). More recently, the same authors have demonstrated that among the components of the *Tetrahymena*-defined medium—amino acids, vitamins, nucleosides, acetate, and glucose—only amino acids were able to inhibit proteolysis, suggesting that regulation of autophagy by amino acids is common to a large spectrum of eukaryotic cells (Grinde and Jonassen, 1987). Among the amino acids, methionine was by far the most active. Phenylalanine, tryptophan, and leucine were also inhibitory, but to a lesser extent. Of particular interest was the fact that BSA, as a source of amino acids, showed the same effect as adding the complete and amino acid mixture, demonstrating the efficiency in uptake and degradation of exogenous proteins.

Table IX
Some Factors Affecting the Intracellular Degradation of Exogenous Protein (bovine liver catalase, BLC) by *Trypanosoma cruzi* Epimastigote as Measured by Release of Free Lysine to Incubation Medium

Amino acid concentration	BLC	15 μM cycloheximide	Lysine release[a]	Lysine release from 1 mg/ml BLC[b]
0 X	−	−	1.09 ± 0.36	
0 X	+	−	3.02 ± 0.76	1.93 ± 0.47
0 X	−	+	2.31 ± 0.81	
0 X	+	+	6.47 ± 1.38	4.16 ± 0.81
1 X	−	−	1.21 ± 0.29	
1 X	+	−	2.94 ± 0.63	1.73 ± 0.75
1 X	−	+	2.14 ± 0.56	
1 X	+	+	4.39 ± 0.82	2.25 ± 0.50

[a] nmol/10^6 per hr.
[b] *T. cruzi* epimastigotes cultivated in MEM 2.5% fetal calf serum were carefully washed twice with MEM medium alone and incubated for 4 hr at 28 °C with phosphate-buffered saline containing 1 mg/ml glucose and 5 μM lysine and additional components as specified: in the absence (0 X) or presence (1 X) of a physiological level of 13 amino acids constituents of minimal essential medium (see Avila *et al.*, 1979a); 1 mg/ml BLC or HRP; 15 μM cycloheximide. The lysine content of the media was assayed before (initial) and after (final) the 4-hr exposure to parasites and the data are expressed as the net (final minus initial) release of lysine per hr (normalized to 10^6 parasites). Lysine released from BLC degradation was calculated as the difference between lysine release in the absence (−) and presence (+) of 1 mg/ml BLC in the incubation media. Values represent the mean of four experiments; each experiment was performed in duplicate.

By measuring changes in the extracellular concentration of lysine, we found that during amino acid deprivation total protein degradation (endogenous plus exogenous) was increased 47% (6.47 at 0 X vs. 4.39 at 1 X; see Table IX). This increase resulted from a combination of invariant degradation of endogenous protein (2.31 ± 0.81 at 0 X vs. 2.14 ± 0.56 at 1 X; see Table IX) and an 85% increase in intracellular degradation of exogenous protein (4.16 at 0 X vs. 2.25 at 1 X).

6.3. Influence of Exogenous Amino Acid Levels on the Lysine Release from Exogenous Protein Sources

In order to know if degradation of exogenous protein is stimulated under amino acid deprivation conditions, we measured changes in the extracellular concentration of lysine in the presence and absence of a physiological level of free amino acids (Table IX). In the presence of BLC, total lysine release (plus cycloheximide) attributable to BLC degradation (lysine release in the presence of BLC minus lysine release in the absence of BLC) was increased 85% in the absence of amino acids (2.25 at 1 X vs. 4.16 at 0 X). Thus, unlike degradation of endogenous protein, intracellular degradation of exogenous protein appeared sensitive to amino acid deprivation. Furthermore, whereas only 23% of the lysine derived from BLC degradation was reutilized for protein synthesis in the presence of a physiological level of amino acids, reutilization efficiency increased to 54% in the absence of amino acids: 2.25 − 1.73/2.25 = 23% at 1 X; 4.16 − 1.93/4.16 = 54% at 0 X. Closely related values and percentages were obtained using HRP as a unique exogenous protein. We believe, as suggested by Besterman *et al.* (1983) that during amino acid deprivation, increased pinocytosis provides an increased alternate source of amino acids (derived from degradation of exogenous protein) for protein synthesis.

Indirect experimental evidence of a rapid amino acid utilization for *de novo* protein synthesis was obtained: (1) by incubating *T. cruzi* epimastigotes together with 3 mg/ml BLC inside a dialysis bag resuspended under strong agitation in a 1000-fold higher volume of sterile amino acid-free culture medium, and (2) by culturing *T. cruzi* epimastigotes with 3 mg/ml BLC in a sterile chamber submitted to a continuous 20 ml/hr flow of amino acid-free culture medium. Under both experimental conditions, any amino acid released extracellularly would be washed away by the 1000-fold dilution or by the continuous rapid flow of culture medium, respectively. We, however, obtained a 30-fold increase in parasite number after ten days of culture at pH 6.4. These results are evidence that under amino acid deprivation, intralysosomally released amino acids after being transported through the lysosomal membrane are rapidly used for *de novo* protein synthesis and that a minimal extracellular release of amino acids takes place. Under more physiological conditions, once amino acids are outside the lysosomal compartment they could be either utilized immediately for *de novo* protein synthesis or transported out of parasites to be taken again for protein synthesis by the same or close parasites. The increased lysine reutilization from exogenous proteins in amino acid-deprived *T. cruzi* epimastigotes could represent an adaptation mechanism of epimastigotes to changing environmental conditions existing in insect gut. That epimastigotes do not utilize endogenous proteins under amino acid deprivation is consistent with the necessity for epimastigotes to be in excellent functional order to be able to rapidly transform into metacyclic trypomastigotes to invade mammalian cells.

The present findings, along with those of Rannels *et al.* (1982) and Besterman *et al.* (1983) obtained in pulmonary macrophages and of Grinde and Jonassen (1987) in *Tetrahymena*, provide strong experimental evidence that in the absence of extracellular amino acids, intracellular degradation of exogenous protein by some cells can provide a source of essential amino acids for protein synthesis. In contrast, Eagle and Piez (1960), using HeLa cells, and Steinman *et al.* (1976), using mouse fibroblasts, concluded that degradation of exogenous protein was not a significant source of amino acids for *de novo* protein synthesis. According to our results, *T. cruzi* epimastigotes cultured in medium containing physiological levels of amino acids utilize only 23% of the lysine derived from

Table X
Effect of Deletion Alone or Combined of Several Amino Acids (At Concentrations Present in Minimal Essential Medium) and of ATP on the release of Lysine from Pinocytosed Bovine Liver Catalase (BLC) or Horseradish Peroxidase (HRP)[a]

	Lysine release in the presence of	
Substances deleted	HRP	BLC
None (control)	2.93 ± 0.86	2.36 ± 0.72
ATP	1.92 ± 0.56	1.70 ± 0.59
Methionine	3.35 ± 1.19	2.96 ± 0.79
Valine	3.40 ± 0.91	2.98 ± 0.86
Isoleucine	3.16 ± 0.79	2.82 ± 0.96
Tryptophan	3.28 ± 0.91	2.76 ± 0.62
Arginine	3.43 ± 1.05	2.89 ± 0.77
Serine	3.36 ± 0.82	3.16 ± 1.19
Glycine	3.65 ± 0.88	3.29 ± 1.42
Met + Val + Ile	4.60 ± 1.36	4.00 ± 1.54
Met + Val + Trp	4.80 ± 1.11	4.16 ± 1.47
Arg + Ser + Gly + Met	4.95 ± 1.23	4.02 ± 1.52
Arg + Ser + Gly + Val	5.00 ± 1.37	4.16 ± 1.41
Arg + Ser + Gly + Ile	4.96 ± 1.29	4.08 ± 1.76
Arg + Ser + Gly + Trp	4.99 ± 1.50	4.80 ± 1.59
Arg + Ser + Gly + Trp + Val	4.90 ± 1.72	4.90 ± 1.71

[a]Experiments were carried out with 15 μM cycloheximide in the medium. Experimental conditions were as in Table IX.

degradation of exogenous HRP or BLC for *de novo* protein synthesis. Table X shows the effects of deleting either ATP or one or several amino acids from the culture medium on lysine release from BLC or HRP by intracellular degradation. It can be seen that deleting ATP from the culture medium resulted in a 50% decrease in lysine release, suggesting that ATP could be acting either by increasing protein uptake rate as demonstrated before (Figure 9) or after hydrolysis to AMP by its intracellular incorporation and subsequently affecting several cellular functions. Table X also shows that when three or more essential amino acids for *T. cruzi* epimastigote multiplication (Avila, 1984; O'Daly *et al.*, 1987) were deleted from the culture medium there was a rapid and significant increase in lysine release. This suggests that by omitting several amino acids simultaneously there is an increased degradation of exogenously supplied proteins in order to afford actively multiplying cells with enough essential amino acids for optimal *de novo* protein synthesis. The supposition offers an explanation to the apparently contradictory results obtained by Eagle and Piez (1960) in experiments performed in culture medium that omitted six amino acids (five essentials).

Pinocytosis as a source of critical nutrients for growth and multiplication have also been demonstrated for the uptake of low-density lipoproteins (LDL) by *Trichomonas vaginalis* (Peterson and Alderete, 1984) and *T. brucei* bloodstream forms (Coppens *et al.*, 1987). It is, therefore, most likely that receptor-mediated endocytosis of LDL particles has an important function in making cholesterol available for both protozoa, as for mammalian cells (Brown and Goldstein, 1986).

7. INTRACELLULAR PROTEIN DEGRADATION IN *TRYPANOSOMA CRUZI* EPIMASTIGOTES

7.1. Enhanced Intracellular Proteolysis Provides Amino Acids for *de Novo* Protein Synthesis

The intracellular degradation of cell proteins is a well-established process that occurs in prokaryotic and eukaryotic cells. Measurements of protein degradation in cultured cells radiolabeled for short periods of time (minutes) show, in semilogarithmic plots, two different degradation kinetics with a clear discontinuity. Proteins showing the fast and the slower rate of degradation are referred to, respectively, as short- and long-lived proteins. *Trypanosoma cruzi* epimastigotes were found to degrade intracellular proteins under conditions of starvation and enrichment (incubation in MEM plus 1 mg/ml BLC or 2.5% fetal calf serum), although protein turnover was fivefold higher in starvation than in growing conditions. Notably, under this latter nutritional condition, there was no significant difference between cells cultured in fetal calf serum compared with parasites adapted to BLC-containing medium (Figure 14). Under both conditions, there is a rapid protein breakdown during the initial 30–60 min, which suggests the presence of a labile protein class that is rapidly turned over; in fact, labeling cells by 2 hr with 2.5 μCi/ml [^{3}H]leucine showed that during the initial 30-min deprivation there is a substantial radioactivity loss in the region corresponding to proteins and/or subunits with an apparent molecular weight of 40 to 100 kDa (Figure 15). Thus, proteins that turn over rapidly in *T. cruzi* are those that fall in this molecular weight range. Cells deprived for 4 hr suffer extensive protein degradation (Figure 15). This suggests that prolonged periods of carbon and energy depletion lead to generalized protein catabolism, providing amino acids for various cellular reactions.

Since metabolic inhibitors have been reported to affect overall proteolysis in different organisms (Ohkuma *et al.*, 1986), we decided to check the effect of these inhibitors on *T. cruzi* epimastigotes grown in BLC-containing medium. Figure 16 illustrates that 2,4-dinitrophenol, sodium fluoride, and iodoacetate significantly increase total proteolysis in starving cells as well as in enriched cells (results not shown). These experiments support the concept that in a parasite

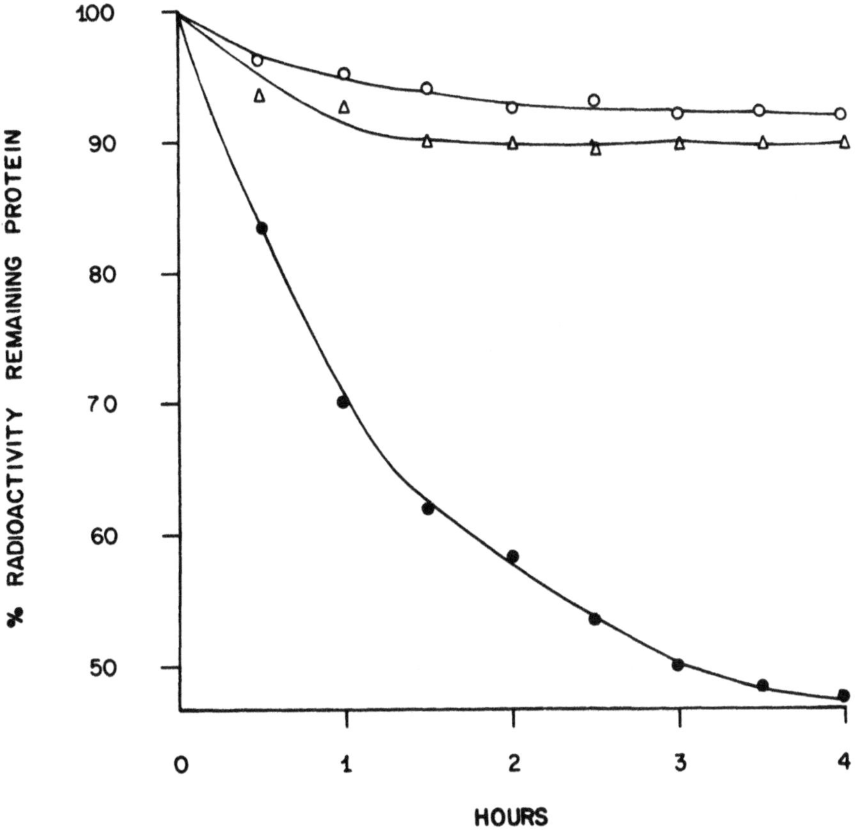

FIGURE 14. Protein turnover in *T. cruzi* epimastigotes during deprivation and growth conditions. Proteins of cells in mid-log phase were labeled with [^{3}H]leucine. After 2 hr, the cells were harvested, washed, and the degradation of labeled proteins followed over a 4-hr period under conditions of deprivation (●—●) and growth (fetal calf serum, ○—○; BLC △—△).

relying on amino acids as carbon and energy sources there is a marked increase in the rate of endogenous proteolysis and net protein catabolism during nutrient deprivation. This is consistent with similar observations in mammalian tissues (Schimke and Doyle, 1970; Goldberg and Dice, 1974). In *T. cruzi* epimastigotes there is a labile protein class or subunits with apparent molecular weight in the range 40 to 100 kDa that are degraded first. This is comparable to the situation in different animal tissues where large molecular weight proteins are degraded more rapidly than those with smaller molecular weights (Kay, 1978). Surprisingly, in contrast with results obtained in other eukaryotic cells (Poole and Wibo, 1973; Ohkuma *et al.*, 1986), metabolic inhibitors markedly increased intracellular proteolysis.

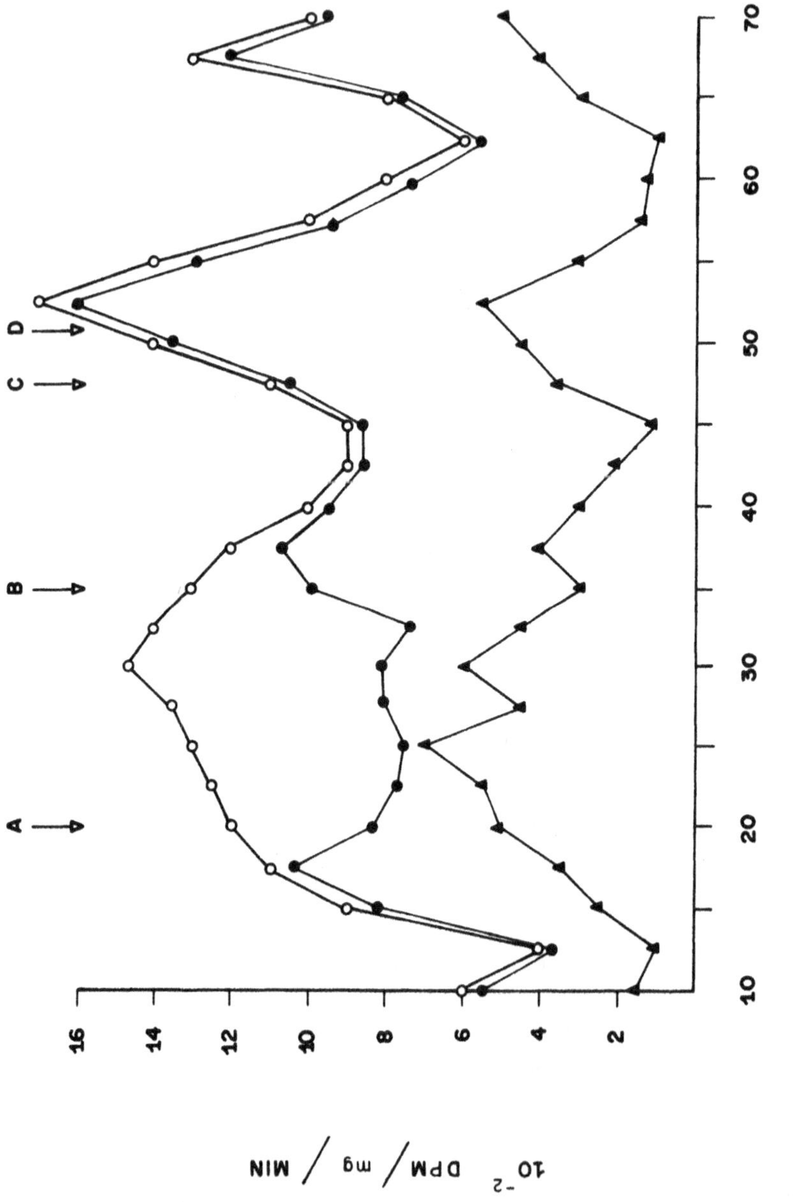

A
B
C
D
16
14
12
10
8
6
4
2
10
20
30
40
50
60
70
MM FROM ORIGIN
10⁻² DPM / mg / MIN

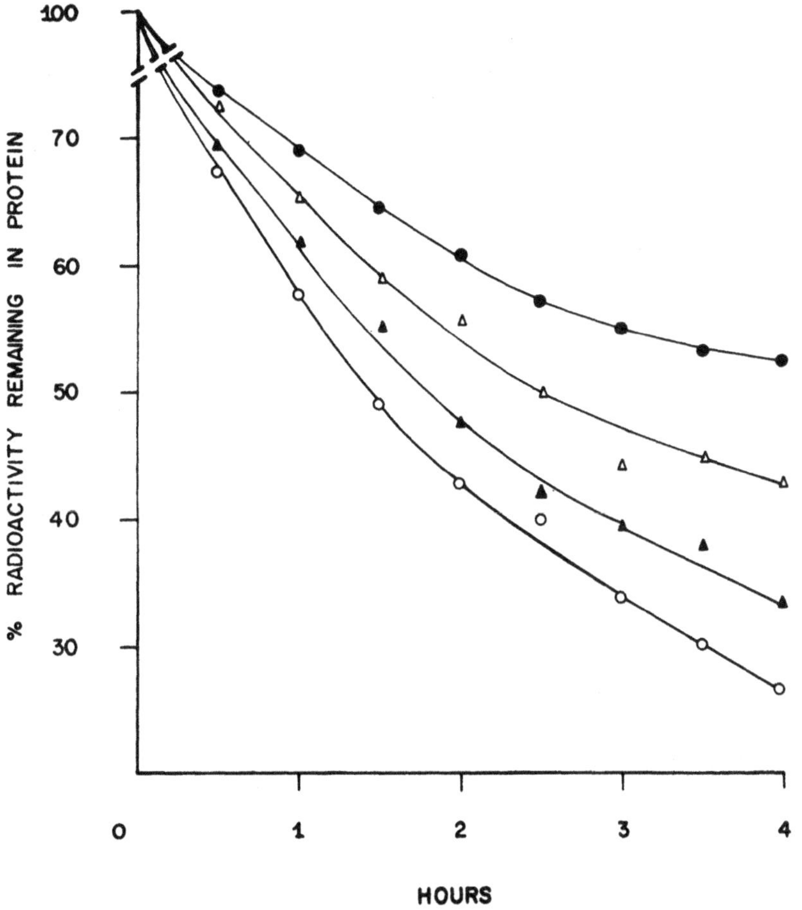

FIGURE 16. Effect of metabolic inhibitors on protein turnover in starving *T. cruzi* epimastigotes. Cellular proteins were labeled with [^{3}H]leucine and then degradation followed when treated as indicated. ●—●, Control; △—△, 2 mM sodium fluoride; ▲—▲, 2 mM iodoacetate; ○—○, 1 mM 2,4-dinitrophenol.

FIGURE 15. Differential catabolism of proteins in *T. cruzi* epimastigotes maintained by five consecutive passages in BLC-containing medium. [^{3}H]leucine (2.5 μCi/ml) was added to a mid-log-phase culture and incubated for 2 additional hours. Cells were harvested, washed, and resuspended in basal salts medium at 28 °C and samples were removed at the beginning of incubation as well as at 30 min and 4 hr later. Each sample was centrifuged and resuspended in 0.01 M phosphate buffer, pH 7.0 containing 0.02 M each of phenylmethylsulfonyl fluoride and tosyl-l-lysine chloromethyl ketone. Parasite extracts were electrophoresed in 11% (w/v) sodium dodecyl sulfate-acrylamide gels using 20 mA/slab gel. Standards used were: (A) phosphorylase b from rabbit muscle (mol. wt. 97,400); (B) alcohol dehydrogenase (mol. wt. 39,800); (C) trypsin inhibitor (mol. wt. 20,100); (D) lysozyme (mol. wt. 14,300). Labeled proteins at zero time (○—○), after 30 min (●—●), and after 4 hr starvation (▲—▲).

8. CONCLUDING REMARKS

Cultivation of *T. cruzi* epimastigotes in a BLC-containing medium, in the absence of exogenously added amino acids, actively promotes parasite multiplication (Avila *et al.*, 1983). It is the first time that a single protein (BLC) has been shown to fulfill all essential amino acid requirements of an actively multiplying eukaryote (Avila *et al.*, 1983).

This chapter has shown that (1) although the BLC uptake rate supplies only 20% of the total amino acid requirements of actively multiplying *T. cruzi* epimastigotes, 40% of BLC constitutive amino acids are essential for *T. cruzi* metabolism, a fact that assures that cells have enough of these amino acids for active protein synthesis. It is evident that amino acid interconversion pathways would be able to supply additional nonessential amino acids needed for rapid *de novo* protein synthesis. (2) *T. cruzi* epimastigotes regulate their *de novo* protein synthesis by two main mechanisms: (a) by the physiological levels of exogenous *essential* amino acids and (b) by modulating intracellular proteolysis of exogenous or endogenous proteins. The first mechanism is based in modulation of a relatively high pinocytic rate and the second on the existence of an arsenal of acid proteinases. (3) During amino acid deprivation, increased pinocytosis provides an alternate source of essential amino acids (derived from degradation of exogenous protein) for protein synthesis.

The finding that inhibitors of acid proteinase activity as well as basic substances that increased parasitic intralysosomal pH strongly affected *T. cruzi* epimastigote multiplication suggests that proteolysis plays a role in parasitic nutrition and underscores the potential of proteinases as targets for chemotherapeutic attack. Proteinase inhibitors are likely to have antitrypanosomal activity and it should be possible to design pro-drugs to be activated by these enzymes. It is also possible to study specificity of *T. cruzi* proteinases to find selective parasitic inhibitors. In this sense it is interesting to note that recently a novel lytic peptide has been shown to inhibit *T. cruzi* trypomastigote and amastigote multiplication *in vitro* (Jaynes *et al.*, 1988).

ACKNOWLEDGMENTS. This work is a product of fruitful collaborations with many colleagues. In particular, we thank María Argelia Polegre, Angela Avila, Antonio Bretaña, Roberto Pérez-Kepp, María Fernanda Correa, Hernán Carrasco, and Juan Ernesto Lüdert. I thank Dr. R. Firestone for the generous gift of *N*-dodecyl imidazol, María Eugenia Pinardi for help with idiomatic English, Alfonso Sierra for graphic art assistance, and Francisco Rivero for logistic support. This work was supported by Consejo Nacional de Investigaciones Cientificas y Tecnológicas (CONICIT) and Fundación Polar. I thank Prof. J. Convit for his continued interest and support.

9. REFERENCES

Abelha, J., Azevedo, M. O., and Teixeira, A. R. L., 1981, *Trypanosoma cruzi:* Antigen-receptor-mediated endocytosis of antibody, *Mem. Inst. Oswaldo Cruz* **76:**213–225.

Amenta, J. S., and Brocher, S. C., 1981, Minireview: Mechanisms of protein turnover in cultured cells, *Life Sci.* **28:**1195–1208.

Arias-Flores, M., Bretaña, A., Avila, J. L., and Contreras-Bretaña, M., 1986, Localización citoquímica de la actividad de fosfatasa acida en diversos miembros de la familia Trypanosomatidae, *Second Venezuelan Meeting on Electron Microscopy,* Caracas, pp. 44–46.

Ashall, F., 1990, Characterization of an alkaline peptidase of *Trypanosoma cruzi* and other trypanosomatids, *Mol. Biochem. Parasitol.* **38:**77–88.

Avila, J. L., 1983, New rational approaches to Chagas' disease chemotherapy, *Interciencia* **8:**405–417.

Avila, J. L., 1984, *Trypanosoma cruzi: Importancia de la Proteolisis Intralisosomial en la Nutrición del Parasito,* Universidad Central de Venezuela, Facultad de Medicina, Caracas, pp. 1–29.

Avila, J. L., Bretaña, A., Casanova, M. A., Avila, A., and Rodriguez, F., 1979a, *Trypanosoma cruzi:* Defined medium for continuous cultivation of virulent parasites, *Exp. Parasitol.* **48:**27–35.

Avila, J. L., Casanova, M. A., Avila, A., and Bretaña, A., 1979b, Acid and neutral hydrolases in *Trypanosoma cruzi.* Characterization and assay, *J. Protozool.* **26:**304–311.

Avila, J. L., Pérez-Kepp, R., and Bretaña, A., 1983, A minimal medium for the cultivation of infective *Trypanosoma cruzi* epimastigotes, *J. Gen. Microbiol.* **129:**285–291.

Azevedo, H. P., and Roitman, I., 1980, Requirements of amino acids and vitamins by *Trypanosoma cruzi,* VII Reuniao Anual de Pesquisa Basica em Doenca de Chagas, Caxambu, MG, Brazil.

Barrett, A. J., 1969, Properties of lysosomal enzymes, in *Lysosomes in Biology and Pathology,* Vol. 2, (J. T. Dingle and H. B. Fell, eds.), pp. 245–312, North Holland Publishing Co., Amsterdam.

Bash-Lewinson, D., and Grossowicz, N., 1957, Transaminases of *Trypanosoma cruzi, Bull. Res. Council Israel* **6E:**91–92.

Berens, L. R., Brun, R., and Krassner, S. M., 1976, A simple monophasic medium for axenic culture of hemoflagellates, *J. Parasitol.* **63:**360–365.

Besterman, J. M., Airhart, J. A., Low, R. B., and Rannels, D. E., 1983, Pinocytosis and intracellular degradation of exogenous protein: Modulation by amino acids, *J. Cell Biol.* **96:**1586–1591.

Bongertz, V., and Hungerer, K. D., 1978, *Trypanosoma cruzi:* Isolation and characterization of a protease, *Exp. Parasitol.* **45:**8–18.

Bontempi, E., Franke de Cazzulo, B. M., Ruiz, A. M., and Cazzulo, J. J., 1984, Purification and some properties of an acidic protease from epimastigotes of *Trypanosoma cruzi, Comp. Biochem. Physiol.* **77B:**599–604.

Bontempi, E., Martínez, J., and Cazzulo, J. J., 1989, Subcellular localization of a cysteine proteinase from *Trypanosoma cruzi, Mol. Biochem. Parasitol.* **33:**33–42.

Brener, Z., 1976, Significance of morphological variation of bloodstream forms, in *American Trypanosomiasis Research,* Scientific Publications 318, pp. 127–131, Pan American Health Organization, Washington, D.C.

Brener, Z., and Chiari, E., 1963, Variacoes morfologicas observadas em diferentes amostras de *Trypanosoma cruzi, Rev. Inst. Med. Trop. S. Paulo* **6:**220–224.

Bretaña, A., and O'Daly, J. A., 1976, Uptake of fetal proteins by *Trypanosoma cruzi.* Immunofluorescence and ultrastructural studies, *Int. J. Parasitol.* **6:**379–386.

Brooker, B. E., 1971, The fine structure of *Crithidia fasciculata* with special reference to the organelles involved in the ingestion and digestion of protein, *Zeits. Zellforsch. Mikrosc. Anat.* **116:**532–563.

Brown, M. S., and Goldstein, J. L., 1986, A receptor-mediated pathway for cholesterol homeostasis, *Science* **232:**34–47.

Brown, K. N., Armstrong, J. A., and Valentine, R. C., 1965, The ingestion of protein molecules by blood forms of *Trypanosoma rhodesiense, Exp. Cell Res.* **39:**129–135.

Buckmaster, M. J., Lo Braico, D., Ferris, A. L., and Storrie, B., 1987, Retention of pinocytized solute by CHO cell lysosomes correlates with molecular weight, *Cell Biol. Int. Rep.* **11:**501–507.

Cazzulo, J. J., and Franke de Cazzulo, B. M., 1982, Proteolytic activity on endogenous substrates in cell-free extracts of *Trypanosoma cruzi, Experientia (Basel)* **38:**1135–1137.

Chagas, C., 1909, Nova tripanosomiase humana. Estudos sobre a morfologia eo ciclo evolutivo do *Schizotrypanum cruzi* n. gen. n. sp., agente etiologico de nova entidade morbida do homen, *Mem. Inst. Oswaldo Cruz* **1:**159–218.

Cicarelli, R. M. B., and Lopes, J. D., 1989, Characterization of a protein from *Trypanosoma cruzi* trypomastigotes that cleaves nonimmune IgG bound through its Fab fragment, *J. Immunol.* **142:**1685–1690.

Coffey, J. W., and de Duve, C., 1968, Digestive activity of lysosomes. I. The digestion of proteins by extracts of rat liver lysosomes, *J. Biol. Chem.* **243:**3255–3263.

Contreras, V. T., Salles, J. M., Thomas, N., Morel, C. M., and Goldenberg, S., 1985, *In vitro* differentiation of *Trypanosoma cruzi* under chemically defined conditions, *Mol. Biochem. Parasitol.* **16:**315–327.

Coppens, I., Opperdoes, F. R., Courtoy, P. J., and Baudhuin, P., 1987, Receptor-mediated endocytosis in the bloodstream form of *Trypanosoma brucei. J. Protozool.* **34:**465–473.

Cover, B., and Gutteridge, W. E., 1978, The course of infection of *Trypanosoma cruzi in* Γ-irradiated rats, *Trans. R. Soc. Trop. Med. Hyg.* **72:**596–601.

Cross, G. A., Klein, R. A., and Baker, J. R., 1975, *Trypanosoma cruzi* growth, amino acid utilization and drug action in a defined medium, *Ann. Trop. Med. Parasitol.* **69:**513–519.

de Duve, C., and Wattiaux, R., 1966, Functions of lysosomes, *Annu. Rev. Physiol.* **28:**435–492.

DeSouza, W., 1976, Associations membrane-microtubule chez *Trypanosoma cruzi, J. Microscop. Biol. Cell.* **25:**189–190.

DeSouza, W., 1984, Cell biology of *Trypanosoma cruzi, Int. Rev. Cytol.* **86:**197–285.

DeSouza, W., Grynberg, N., and Nery-Guimaraes, F., 1975, Aspectos ultraestructuais de forma epimastigota de *Trypanosoma cruzi* em medio LIT, *Rev. Soc. Bras. Med. Trop.* **9:**143–156.

DeSouza, W., Arguello, C., Martínez-Palomo, A., Trissl, D., Gonzalez-Robles, A., and Chiari, E. 1977, Surface charge of *Trypanosoma cruzi.* Binding of cationized ferritin and measurement of cellular electrophoretic mobility, *J. Protozool.* **24:**411–415.

DeSouza, W., deCarvalho, T. U., Benchimol, M., and Chiari, E., 1978a, *Trypanosoma cruzi:* Ultrastructural, cytochemical and freeze-fracture studies of protein uptake, *Exp. Parasitol.* **45:**101–115.

DeSouza, W., Martínez-Palomo, A., and Gonzalez-Robles, A., 1978b, The cell surface of *Trypanosoma cruzi:* Cytochemistry and freeze fracture, *J. Cell Sci.* **33:**285–299.

Dvorak, J. A., 1976, New *in vitro* approach to quantitation of *Trypanosoma cruzi*–vertebrate cell interactions, in *American Trypanosomiasis Research,* Scientific Publication No. *318,* pp. 109–120, Pan American Health Organization. Washington, D.C.

Eagle, H., and Piez, K. A., 1960, The utilization of proteins by cultured human cells, *J. Biol. Chem.* **235:**1095–1097.

Edwards, J. C., 1925, Formation of food-cups in *Amoeba* induced by chemicals, *Biol. Bull.* **48:**236–241.

Ehrenreich, B. A., and Cohn, Z. A., 1969, The fate of peptides pinocytosed by macrophages *in vitro, J. Exp. Med.* **129:**227–243.

Fairlamb, A. H., and Bowman, I. B. R., 1980, *Trypanosoma brucei:* Maintenance of concentrated suspensions of bloodstream trypomastigotes *in vitro* using continuous analysis for measurement of endocytosis, *Exp. Parasitol.* **49:**366–380.

Gahl, W. A., 1989, Lysosomal membrane transport in cellular nutrition, *Annu. Rev. Nutr.* **9:**39–61.

Geigy, R., Steiger, R., and Hecker, H., 1970, Beiträge zur pinocytose von *Trypanosoma (Trypanozoon) brucei,* Plimmer and Bradford, 1899, *Acta Trop.* **27:**221–227.

Goldberg, A. L., and Dice, J. F., 1974, Intracellular protein degradation in mammalian and bacterial cells, *Annu. Rev. Biochem.* **43:**835–869.

Goldberg, S. S., Silva Pereira, A. A., Chiari, E., Mares-Guia, M., and Gazzinelli, G., 1976, Comparative kinetics of arginine and lysine transport by epimastigotes and trypomastigotes from two strains of *Trypanosoma cruzi, J. Protozool.* **23:**179–186.

Gonzalez-Noriega, A., Grubb, J. H., Talkad, V., and Sly, W. S., 1980, Chloroquine inhibits lysosomal enzyme pinocytosis and enhances lysosomal enzyme secretion by impairing receptor recycling, *J. Cell Biol.* **85:**839–852.

Grinde, B., and Jonassen, T. O., 1987, Methionine is a regulator of starvation-induced proteolysis in *Tetrahymena, Exp. Cell. Res.* **173:**496–503.

Gutteridge, W. E., and Coombs, G. H., 1977, *Biochemistry of Parasitic Protozoa,* MacMillan Press, London.

Hammer, J. A., and Rannels, D. E., 1981, Protein turnover in pulmonary macrophages. Utilization of amino acids derived from protein degradation, *Biochem. J.* **198:**53–65.

Hampton, H. r., 1970, Lysine transport in the culture form of *Trypanosoma cruzi:* Kinetics and inhibition of uptake by structural analogues, *Int. J. Biochem.* **1:**706–714.

Hampton, J. R., 1971, Arginine transport in the culture form of *Trypanosoma cruzi, J. Protozool.* **18:**701–703.

Harington, J. S., 1961, Studies on the amino acids of *Rhodnius prolixus.* I. Analysis of the haemolymph, *Parasitology* **51:**309–318.

Hayashi, M., Hiroi, Y., and Natori, Y., 1973, Effect of ATP on protein degradation in rat liver lysosomes, *Nature* **242:**163–166.

Herbert, I. V., 1965a, Cytochemistry of *in vitro* cultured *Trypanosoma theileri, Exp. Parasitol.* **16:**348–362.

Herbert, I. V., 1965b, Cytoplasmic inclusions and organelles of *in vitro* cultured *Trypanosoma theileri* and *Trypanosoma melophagium, Exp. Parasitol.* **17:**24–40.

Hershko, A., and Ciechanover, A., 1982, Mechanisms of intracellular protein breakdown, *Annu. Rev. Biochem.* **51:**335–364.

Hoare, C. A., 1972, *The Trypanosomes of Mammals. A Zoological Monograph,* Blackwell Scientific Publications, Oxford.

Itow, S., and Camargo, E. P., 1977, Proteolytic activities in cell extracts of *Trypanosoma cruzi, J. Protozool.* **24:**591–595.

Jadin, J. M., 1971, Cytologie et cytophysiologie des Trypanosomatidae, *Acta Zool. Path. Antverpiensia* **53:**1–168.

Jaynes, J. M., Burton, C. A., Barr, S. B., Jeffers, G. W., Julian, G. R., White, K. L., Enright, F. M., Klei, T. R., and Laine, R. A., 1988, *In vitro* cytocidal effect of novel lytic peptides on *Plasmodium falciparum* and *Trypanosoma cruzi, FASEB J.* **2:**2878–2883.

Jonassen, T. O., and Grinde, B., 1986, Proteolytic response to nutritional step-down in *Tetrahymena, Exp. Cell. Res.* **163:**165–174.

Kay, W. W., 1978, Intracellular protein degradation, *Biochem. Soc. Trans.* **6:**789–797.

Langreth, S. G., and Balber, A. E., 1975, Protein uptake and digestion in bloodstream and culture forms of *Trypanosoma brucei, J. Protozool.* **22:**40–53.

Lewis, W. H., 1931, Pinocytosis, *Bull. Johns Hopkins Hosp.* **49:**17–36.

Ley, V., Robbins, E. S., Nussenzweig, V., and Andrews, N. W., 1990, The exit of *Trypanosoma cruzi* from the phagosome in inhibited by raising the pH of acidic compartments, *J. Exp. Med.* **171**:401–413.

Lima, M. F., and Villalta, F., 1990, *Trypanosoma cruzi* receptors for human transferrin and their role, *Mol. Biochem. Parasitol.* **38**:245–252.

María, T. A., Tafuri, W., and Brener, Z., 1972, The fine structure of different bloodstream forms of *Trypanosoma cruzi*, *Ann. Trop. Med. Parasitol.* **66**:423–431.

Martínez-Palomo, A., DeSouza, W., and Gonzalez-Robles, A., 1976, Topographical differences in the distribution of surface coat components and intramembrane particles. A cytochemical and freeze-fracture study in culture forms of *Trypanosoma cruzi*, *J. Cell Biol.* **69**:507–513.

McDonald, J. K., Reilly, T. J., and Ellis, S., 1965, A chloride-activated dipeptidyl-β-naphthyl-amidase of the pituitary gland, *Life Sci.* **4**:1665–1670.

Meyer, H., and DeSouza, W., 1973, On the fine structure of *Trypanosoma cruzi* in tissue cultures of pigment epitelium from the chicken embryo. Uptake of melanin granules by the parasite, *J. Protozool.* **20**:590–593.

Meyer, H., and DeSouza, W., 1976, Electron microscopic study of *Trypanosoma cruzi* periplast in tissue cultures. I. Number and arrangement of the peripheral microtubules in the various forms of the parasite life cycle, *J. Protozool.* **23**:385–390.

Milder, R., and Deane, M. P., 1969, The cytostome of *Trypanosoma cruzi* and *T. conorhini*, *J. Protozool.* **16**:730–737.

Mortimore, G. E., 1982, Mechanisms of cellular protein catabolism, *Nutr. Rev.* **40**:1–12.

Nagakura, K., Tachibana, H., and Kaneda, Y., 1985, Alteration of the cell surface acid phosphatase concomitant with the morphological transformation in *Trypanosoma cruzi*, *Comp. Biochem. Physiol.* **81B**:815–817.

Nogueira, N., and Cohn, Z., 1976, *Trypanosoma cruzi:* Mechanism of entry and intracellular fate in mammalian cells, *J. Exp. Med.* **143**:1402–1420.

North, M. J., 1982, Comparative biochemistry of the proteinases of eucaryotic microorganisms, *Microbiol. Rev.* **46**:308–340.

O'Daly, J. A., 1975a, A new liquid medium for *Trypanosoma (Schizotrypanum) cruzi*, *J. Protozool.* **22**:265–270.

O'Daly, J. A., 1975b, Serum proteins promoting [^{3}H]-thymidine uptake by *Tryhpanosoma (Schizotrypanum) cruzi* (Chagas) *in vitro*, *J. Protozool.* **22**:555–559.

O'Daly, J. A., Serrano, L. E., and Rodríguez, M. B., 1983, Free amino acid pool and proteolytic enzymes in *Trypanosoma cruzi* cultured *in vitro*, *Int. J. Parasitol.* **13**:433–440.

O'Daly, J. A., Rodríguez, M. B., and Garlin, G., 1987, *Trypanosoma cruzi:* Growth requirements at different temperatures in fetal bovine serum or peptide supplemented media, *Exp. Parasitol.* **64**:78–87.

Ohkuma, S., and Poole, B., 1978, Fluorescence probe measurement of the intralysosomal pH in living cells and the perturbation of pH by various agents, *Proc. Natl. Acad. Sci. USA* **75**:3327–3331.

Ohkuma, S., and Poole, B., 1981, Cytoplasmic vacuolation of mouse peritoneal macrophages and the uptake into lysosomes of weakly basic substances, *J. Cell Biol.* **90**:656–664.

Ohkuma, S., Chudzik, J., and Poole, B., 1986, The effects of basic substances and acidic ionophores on the digestion of exogenous and endogenous proteins in mouse peritoneal macrophages, *J. Cell Biol.* **102**:959–966.

Opperdoes, F. R., and Steiger, R. F., 1981, Localization of hydrolases in cultured procyclic trypomastigotes of *Trypanosoma brucei*, *Mol. Biochem. Parasitol.* **4**:311–323.

Opperdoes, F. R., and Van Roy, J., 1982, Involvement of lysosomes in the uptake of macromolecular material by bloodstream forms of *T. brucei*, *Mol. Biochem. Parasitol.* **6**:181–190.

Pan, S. C., 1978, *Trypanosoma cruzi:* Cultivation in macromolecule-free semisynthetic and synthetic media, *Exp. Parasitol.* **46**:108–112.

Peterson, K. M., and Alderete, J. F., 1984, Selective acquisition of plasma proteins by *Trichomonas vaqinalis* and human lipoproteins as a growth requirement for this species, *Mol. Biochem. Parasitol.* **12**:37–48.

Poole, B., and Ohkuma, S., 1981, Effect of weak bases on the intralysosomal pH in mouse peritoneal macrophages, *J. Cell Biol.* **90**:665–669.

Poole, B., and Wibo, M., 1973, Protein degradation in cultured cells: The effect of fresh medium, fluoride, and iodoacetate on the digestion of cellular protein of rat fibroblasts, *J. Biol. Chem.* **248**:6221–6226.

Preston, T. M., 1969, The form and function of the cytostomecytopharynx of the culture forms of the elasmobranch haemoflagellate *Trypanosoma raiae* Laveran and Mesnil, *J. Protozool.* **16**:320–333.

Prioli, R. P., Rosenberg, I., Shivakumar, S., and Pereira, M. E. A., 1988, Specific binding of human plasma high density lipoprotein (cruzin) to *Trypanosoma cruzi, Mol. Biochem. Parasitol.* **28**:257–264.

Rangel, H. A., Araujo, P. M. F., Repka, D., and Costa, M. G., 1981a, *Trypanosoma cruzi:* Isolation and characterization of a proteinase, *Exp. Parasitol.* **52**:199–209.

Rangel, H. A., Aranjo, P. M. F., and Camargo, I. J. B., Bonfitto, M., Repka, D., Sakurda, J. K., and Atta, A. M., 1981b, Detection of a proteinase common to epimastigote, trypomastigote and amastigote of different strains of *Trypanosoma cruzi, Tropenmed. Parasitol.* **32**:87–92.

Rannels, D. E., Low, R. B., Youdale, T., Volkin, E., and Longmore, J., 1982, The use of radioisotopes in quantitative studies of lung metabolism, *Fed. Proc.* **41**:2833–2839.

Sanabria, A., 1964, The ultrastructure of *Trypanosoma cruzi.* II. Crithidial and leishmanial forms, *Exp. Parasitol.* **15**:125–137.

Sanabria, A., and Aristimuño, J., 1970, Nuevos estudios acerca de la ultraestructura del *Trypanosoma cruzi* en el miocardio del ratón, *Acta Cient. Venez.* **21**:107–118.

Schmike, R. T., and Doyle, D., 1970, Control of enzyme levels in animal tissues, *Annu. Rev. Biochem.* **39**:929–976.

Schroeder, W. A., Shelton, J. R., Shelton, J. B., Robberson, B., Apell, G., Fang, R. S., and Bonaventura, J., 1982, The complete amino acid sequence of bovine liver catalase and the partial sequence of bovine erythrocyte catalase, *Arch. Biochem. Biophys.* **214**:397–421.

Shapiro, S. Z., and Webster, P., 1989, Coated vesicles from the protozoan parasite *Trypanosoma brucei:* Purification and characterization, *J. Protozool.* **36**:344–349.

Shaw, J. J., Voller, A., and Bryant, C., 1964, Intermediary carbohydrate metabolism in four species of Trypanosomatidae, *Ann. Trop. Med. Parasitol.* **58**:17–24.

Silverstein, S. C., Steinman, R. M., and Cohn, Z. A., 1977, Endocytosis, *Annu. Rev. Biochem.* **46**:669–722.

Soares, M. J., and DeSouza, W., 1988, Cytoplasmic organelles of trypanosomatids: A cytochemical and stereological study, *J. Submicrosc. Cytol. Pathol.* **20**:349–361.

Steiger, R. F., 1973, On the ultrastructure of *Trypanosoma* (*Trypanozoon*) *brucei* in the course of its life cycle and some related aspects, *Acta Trop.* **30**:63–168.

Steiger, R. F., Van Hoof, F., Bontemps, J., Nyssens-Jadin, M., and Druetz, J-E., 1979, Acid hydrolases of trypanosomatid flagellates, *Acta Trop.* **36**:335–341.

Steiger, R. F., Opperdoes, F. R., and Bontemps, J., 1980, Subcellular fractionation of *Trypanosoma brucei* bloodstream forms with special reference to hydrolases, *Eur. J. Biochem.* **105**:163–175.

Steinert, M., and Novikoff, A. B., 1960, The existence of a cytostome and the occurrence of pinocytosis in the trypanosome *Trypanosoma mega, J. Biophys. Biochem. Cytol.* **8**:563–569.

Steinman, R. M., and Cohn, Z. A., 1972, The interaction of soluble horseradish peroxidase with mouse peritoneal macrophages *in vitro, J. Cell Biol.* **55**:186–204.

Steinman, R. M., Silver, J. M., and Cohn, Z. A., 1974, Pinocytosis in fibroblasts. Quantitative studies *in vitro, J. Cell Biol.* **63**:949–969.

Steinman, R. M., Brodie, S. E., and Cohn, Z. A., 1976, Membrane flow during pinocytosis. A stereological analysis, *J. Cell Biol.* **68**:665–687.

Sullivan, P. C., Ferris, A. L., and Storrie, B., 1987, Effects of temperature, pH elevators, and energy production inhibitors on horseradish peroxidase transport through endocytic vesicles, *J. Cell Physiol.* **131**:58–63.

Swanson, J., Yirinic, B. D., and Silverstein, S. C., 1985, Phorbol esters and horseradish peroxidase stimulate pinocytosis and redirect the flow of pinocytosis fluid in macrophages, *J. Cell Biol.* **100**:851–859.

Teixeira, A. R. L., and Santana, J. M., 1989, *Trypanosoma cruzi:* Endocytosis and degradation of specific antibodies by parasite forms, *Am. J. Trop. Med. Hyg.* **40**:165–170.

Torruella, M., Franke de Cazzulo, B. M., Engel, J. C., Ruiz, A. M., Segura, E. L., and Cazzulo, J. J., 1981, *Trypanosoma cruzi* and *trypanosoma rangeli:* Glutamate dehydrogenases and proteolytic activities, *Comp. Biochem. Physiol.* **70B**:463–468.

Trager, W., 1974, Nutrition and biosynthetic capabilities of flagellates: Problems of *in vitro* cultivation and differentiation, in *Trypanosomiasis and Leishmaniasis,* CIBA Foundation Symposium, Vol. 20, pp. 225–245, Elsevier, Amsterdam.

Vickerman, K., 1969, On the surface coat and flagellar adhesion in trypanosomes, *J. Cell. Sci.* **5**:163–193.

Vickerman, K., and Preston, T. M., 1976, Comparative cell biology of the kinetoplastid flagellates, in *Biology of the Kinetoplastida,* Vol. 1 (W. H. R. Lumsden and D. A. Evans, eds.), pp. 35–129, Academic Press, London and New York.

Von Brand, T., 1967, Metabolism of *Trypanosoma cruzi.* A review of recent developments, in *Medicina Tropical* (A. Anselmi, ed.), pp. 261–275, Editorial Fournier, Mexico.

Von Brand, T., 1979, *Biochemistry and Physiology of Endoparasites,* Elsevier/North Holland Biomedical Press, Amsterdam.

Wibo, M., and Poole, B., 1974, Protein degradation in cultured cells. II. The uptake of chloroquine by rat fibroblasts and the inhibition of cellular protein degradation and cathepsin B_1, *J. Cell Biol.* **63**:430–440.

Williamson, J., and Desowitz, R. S., 1961, The chemical composition of trypanosomes. I. Protein, amino acid and sugar analysis, *Exp. Parasitol.* **11**:161–175.

Zeledón, R., 1960a, Comparative physiological studies on four species of hemoflagellates in culture. IV. Effect of metabolic inhibitors on the respiration, *Rev. Biol. Trop.* **8**:181–195.

Zeledón, R., 1960b, Comparative physiological studies on four species of hemoflagellates in culture. II. Effect of carbohydrates and relate substances and some amino compounds on the respiration, *J. Parasitol.* **46**:541–551.

Zeledón, R., Bolaños, R., and Rojas, M., 1984, Scanning electron microscopy of the final phase of the life cycle of *Trypanosoma cruzi* in the insect vector, *Acta Trop. (Basel)* **41**:39–43.

Chapter 7

Energy Metabolism in *Trypanosoma cruzi*

Juan José Cazzulo

1. INTRODUCTION

The American trypanosomiasis, Chagas' disease, affects about 20 million people in Central and South America. The causative agent is a parasitic flagellate, *Trypanosoma cruzi,* which has a complex life cycle, involving a replicative form, the amastigote, and a nonreplicative form, the bloodstream trypomastigote, in the mammalian host. There is also a replicative form, the epimastigote, and a nonreplicative form, the infective metacyclic trypomastigote, in the insect vector (Brener, 1973). Although at present all these developmental stages can be obtained in axenic culture (Contreras *et al.*, 1985) or in mammalian cell culture (Andrews and Colli, 1982), or from infected mammals (Gutteridge *et al.*, 1978), most metabolic and enzymological studies so far have been performed with cultured epimastigotes, presumably identical to those naturally present in the insect vector. Comparative studies using epimastigotes, amastigotes, and bloodstream trypomastigotes (Gutteridge and Rogerson, 1979) suggest that, at variance with African trypanosomes (Opperdoes, 1987), the different developmental stages of *T. cruzi* have qualitatively similar metabolic characteristics.

Juan José Cazzulo Instituto de Investigaciones Bioquimicas "Luis F. Leloir," Fundación Campomar—CONICET—Falcultad de Ciencias Exactas y Naturales, Universidad de Buenos Aires, 1405 Buenos Aires, Argentina.
Subcellular Biochemistry, Volume 18: Intracellular Parasites, edited by J. L. Avila and J. R. Harris. Plenum Press, New York, 1992.

This chapter deals with the energy metabolism in *T. cruzi* and covers two main, closely related, aspects: carbohydrate catabolism, in particular the aerobic fermentation of glucose, and protein and amino acid catabolism. The relative weight of both metabolic pathways in fulfilling the energetic requirements of the parasite seems to vary both with the parasite developmental stage and with the nutrient availability in the medium. Since some fairly recent reviews have dealt with these subjects (Gutteridge and Rogerson, 1979; Cannata and Cazzulo, 1984a; Cazzulo, 1984), emphasis here will be put on the more recent findings and concepts and on the possible metabolic interrelationships.

2. TRANSPORT AND UTILIZATION OF CARBOHYDRATE, PROTEINS, AND AMINO ACIDS

It has been known for many years that epimastigotes of *T. cruzi* are able to take up and consume both carbohydrates and amino acids, excreting them into the medium as some reduced glucose catabolites, chiefly succinate, and NH_3, respectively (von Brand, 1979). Nevertheless, there has been some controversy over which is the most important and natural nutrient for the parasite (Urbina and Azavache, 1984). Some fairly well-established facts at present are: (1) epimastigotes are able to consume both nutrients, although glucose is preferred over amino acids and consumed first when both are present (Cannata and Cazzulo, 1984a; Cazzulo *et al.*, 1985). This happens despite the fact that the natural environment of the epimastigote, the triatomine's gut, is likely to be rich in proteins and amino acids and poor in carbohydrate. (2) Despite some claims, no culture medium completely devoid of glucose seems to have been successfully used for growth of the parasite, at least for more than a few doublings. (3) Different developmental stages of the parasite seem to prefer one type of nutrient or the other; thus, axenic-cultured amastigotelike forms have an exclusively glycolytic catabolism (Engel *et al.*, 1987), whereas epimastigotes are able to use both glucose (which is the preferred substrate during the exponential phase of growth) and amino acids (which are better consumed near the stationary phase) (Cazzulo *et al.*, 1985; Engel *et al.*, 1987). Metacyclic trypomastigotes have recently been reported as having a nonglycolytic catabolism, deriving their energy exclusively from proteins and amino acids (Urbina *et al.*, 1989).

Both glucose and amino acids are actively taken up from the culture medium by the epimastigotes (see Cannata and Cazzulo, 1984a; Cazzulo, 1984, for review); proteins are incorporated into the parasites by pinocytosis, which seems to take place mostly at the level of the flagellar pocket (DeSouza, 1984), and they seem to be able to cover the energy requirements of the parasite (Avila *et al.*, 1979a, 1983) (see also Chapter 6). *Trypanosoma cruzi,* like the other Trypanosomatids, does not seem to be able to synthesize and store any reserve polysaccharide, and therefore carbohydrate needs to be taken up continuously for utilization (see Cannata and Cazzulo, 1984a, for review). The presence of fructose 1,6-*bis*-phosphatase in the parasite, recently reported (Adroher *et al.*, 1987), suggests the possibility of the operation of a gluconeogenic pathway. Protein and amino acids, on the other hand, are likely to be an energy reserve for the parasite, which at least under some circumstances, such as the epimastigote–metacyclic trypomastigote differentiation, seems to be able to use as fuel a significant proportion of its own protein (Rodriguez *et al.*, 1990). There is a large amino acid pool, almost half of which is L-alanine (Williamson and Desowitz, 1961), which can be oxidized by the parasite; indeed, it is necessary to starve epimastigotes for 24 hr in order to detect a clear stimulation of respiration by glucose or amino acids (Sylvester and Krassner, 1976).

Glucose is consumed at a fairly high rate [0.7 $\mu mol/min^{-1}$ per g (wet weight)$^{-1}$] by the epimastigotes; this rate is nevertheless about fivefold lower than that shown by other Trypanosomatids, such as *Crithidia fasciculata* or *Trypanosoma brucei* (Cazzulo *et al.*, 1988a).

3. CARBOHYDRATE CATABOLISM: THE AEROBIC FERMENTATION OF GLUCOSE

3.1. End Products of Glucose Catabolism

From the early studies of Chang (1948) and Ryley (1956), it is known that *T. cruzi* epimastigotes and bloodstream trypomastigotes produce organic acids during glucose catabolism. These studies, as well as those of Bowman *et al.* (1963) and Cáceres and Fernandes (1976), consistently indicated succinate as a major catabolite, although they differed to some extent with reference to other minor products. This question has been reexamined recently using two different approaches: (1) specific enzymatic assay of catabolites in culture medium and cell suspension supernatants; (2) the noninvasive technique of ^{13}C-labeled nuclear magnetic resonance (NMR), which allows one to follow glucose catabolism by living cells. Both techniques indicate that succinate is the major catabolite, L-alanine being the second quantitatively important product (Cazzulo *et al.*, 1985; Engel *et al.*, 1987; Frydman *et al.*, 1990); only a little of the enzymatically detected acetate (Cazzulo *et al.*, 1985; Engel *et al.*, 1987) actually originates from glucose (Frydman *et al.*, 1990), most of it probably arising from lipid and amino acid catabolism. The NMR evidence suggests that there are two different alanine pools with different access to deuterium from the medium; one of them is excreted, whereas the other is retained within the cells (Frydman *et al.*, 1990). These experiments, performed with the Tulahuén strain (Tul 0 stock) of the parasite, have given no indication of other catabolites reported by earlier authors, such as lactate, pyruvate, and malate (see Cannata and Cazzulo, 1984a, for review); this discrepancy might be attributed either to the use, in the earlier work, of less specific assay methods or simply to strain differences. Glycerol and ethanol, which are produced by other Trypanosomatids, are not produced at all by *T. cruzi* epimastigotes (Cazzulo *et al.*, 1988a). Comparison of two developmental forms of the parasite indicated that axenic cultured amastigotelike forms and epimastigotes of the same stock presented only quantitative differences in the end products excreted (Engel *et al.*, 1987).

3.2. Enzyme Pathways for Glucose Catabolism

Evidence obtained in several laboratories indicates that both glycolysis and the pentose phosphate shunt are operative in epimastigotes of *T. cruzi* (see Cannata and Cazzulo, 1984a, for review); the balance between the two pathways seems to depend on the strain of the parasite (Mancilla and Náquira, 1964). Glycolysis in *T. cruzi* presents two distinctive characters compared with the mammalian host: (1) most of the glycolytic enzymes are placed in a microbody, which for this reason was named "glycosome" by Opperdoes and Borst (1977), who discovered it in *T. brucei;* and (2) hexokinase and phosphofructokinase, key regulatory enzymes in most glycolytic systems, from bacteria to mammals, are little or not affected by most common effectors (Urbina and Crespo, 1984; Racagni and Machado de Domenech, 1983; Aguilar and Urbina, 1986; Taylor and Gutteridge, 1986b). The only control found so far for a glycolytic enzyme in Trypanosomatids is the strong activation of pyruvate kinase by very low concentrations of fructose 2,6-*bis*-phosphate (van Schaftingen *et al.*, 1985). The pyruvate kinase from *T. cruzi* epimastigotes (Juan *et al.*, 1976) is strongly activated by fructose 2,6-*bis*-phosphate, which is able to counteract (at micromolar concentrations) the inhibitory effects of millimolar concentrations of P_1 and ATP (Cazzulo *et al.*, 1986b). Millimolar concentrations of fructose 1,6-*bis*-phosphate are also able to active pyruvinate kinase (Juan *et al.*, 1976; Cazzulo *et al.*, 1989b). As a result of the almost complete lack of inhibitory controls, the operation of glycolysis is not influenced appreciably by the presence or absence of oxygen; this is seen as a complete lack of the Pasteur effect (Cazzulo *et al.*, 1988a).

The early demonstration of the aerobic fermentation of glucose posed the question of whether the tricarboxylic acid cycle and the respiratory chain were operative in *T. cruzi*. The present evidence (reviewed in Cannata and Cazzulo, 1984a) indicates that both pathways are actually functional, and all the cycle enzymes, with the only exception of α-oxoglutarate dehydrogenase, have been detected in cell-free extracts. Citrate synthase is a typically eukaryotic enzyme, with the usual response to effectors (Juan *et al.*, 1977). Figure 1 summarizes our present knowledge of these pathways, together with some enzymes involved in catabolite production.

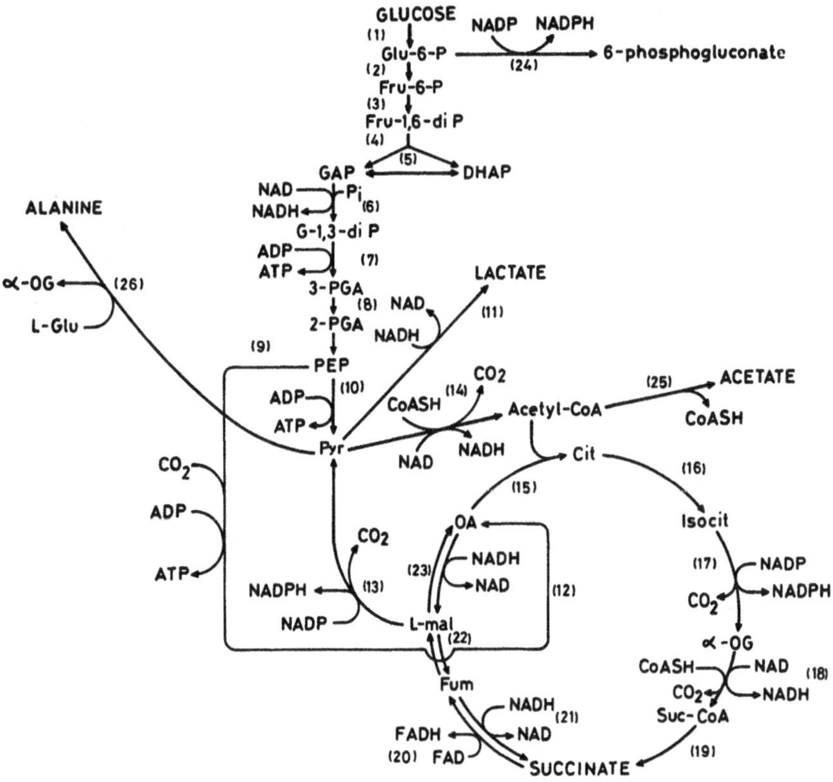

FIGURE 1. Enzyme pathways for glucose catabolism in *Trypanosoma cruzi*. Abbreviations: Glu-6-P, glucose-6-phosphate; Fru-6-P, fructose-6-phosphate; Fru-1,6-diP, fructose 1,6-*bis*-phosphate; GAP, glyceraldehyde-3-phosphate; DHAP, dihydroxyacetone phosphate; G-1,3-diP, 1,3-diphosphoglycerate; 3-PGA, 3-phosphoglycerate; 2-PGA, 2-phosphoglycerate; PEP, phosphoenolpyruvate; Pyr, pyruvate; OA, oxaloacetate; L-mal, L-malate; Fum, fumarate; Suc-CoA, succinyl-CoA; α-OG, α-oxoglutarate; Isocit, isocitrate; Cit, citrate; L-glu, L-glutamate. The enzymes involved are the following: (1) hexokinase; (2) hexose phosphate isomerase; (3) phosphofructokinase; (4) aldolase; (5) triosephosphate isomerase; (6) glyceraldehyde-3-phosphate dehydrogenase; (7) phosphoglycerokinase; (8) phosphoglyceromutase; (9) enolase; (10) pyruvate kinase; (11) α-hydroxyacid dehydrogenase; (12) phosphoenolpyruvate carboxykinase; (13) malic enzynme; (14) pyruvate dehydrogenase; (15) citrate synthase; (16) aconitase; (17) isocitrate dehydrogenase; (18) α-oxoglutarate dehydrogenase; (19) succinate thiokinase; (20) succinate dehydrogenase; (21) fumarate reductase; (22) fumarase; (23) malate dehydrogenase; (24) glucose-6-phosphate dehydrogenase; (25) acetyl-CoA hydrolase; (26) alanine aminotransferase. Enzyme 18 has not been detected yet in *T. cruzi*.

Recent studies have focused on the variation of the levels of some glycolytic and Krebs cycle enzymes during differentiation of axenic amastigotelike forms to epimastigotes (Engel *et al.*, 1987) and of epimastigotes to metacyclic trypomastigotes (Adroher *et al.*, 1988, 1990). The decreasing levels of glycolytic enzymes and increasing levels of key cycle enzymes during these transitions support the idea that the amastigotelike forms are exclusively glycolytic, but the epimastigotes have a mixed metabolism and the metacyclic trypomastigotes are almost exclusively oxidative, conducting little glycolysis if at all.

3.3. Systems for Reoxidation of Glycolytic NADH

In order to remain functional, all glycolytic systems must have an efficient system for the reoxidation of the NADH generated in the reaction of glyceraldehyde-3-phosphate dehydrogenase. This system is usually the respiratory chain under aerobic conditions and several possible fermentative pathways under anaerobic conditions; among the latter, lactate production by skeletal muscle or ethanol production by yeast are the best known. *Trypanosoma cruzi* has several possible ways for NADH reoxidation, all of which seem to be operative under aerobic conditions. These are, in addition to reoxidation in the respiratory chain, production of succinate and probably also of L-alanine. Production of lactate, reported by some early workers (Ryley, 1956; Cáceres and Fernandes, 1976), and probably mediated by a low lactate dehydrogenase activity (Ryley, 1956; Raw, 1959; Gutteridge and Rogerson, 1979; Taylor *et al.*, 1980), which may be due to one of the isoenzymes of α-hydroxyacid dehydrogenase (Coronel *et al.*, 1981), seems to be of little or no importance, even under anaerobic conditions (Frydman *et al.*, 1990). Ethanol or other alcohols have not been found as products of glucose catabolism by *T. cruzi* (Frydman *et al.*, 1990; Cazzulo *et al.*, 1988a). The "alcohol dehydrogenase," acting best with *n*-propanol as substrate and NADP as coenzyme and proposed as a marker enzyme for taxonomic purposes (Miles *et al.*, 1980), has been recently shown to be an NADP-linked aldehyde reductase, able to use a number of aldehydes as substrates, which is not likely to be involved in the reoxidation of reduced coenzymes (Arauzo and Cazzulo, 1989). This leaves us with two, and perhaps three, main pathways for NADH reoxidation in *T. cruzi*, which are analyzed below.

3.3.1. Succinate Production

The production of succinate was shown by Bowman *et al.* (1963) to depend on CO_2 fixation. Using $^{14}CO_2$, both under aerobic and anaerobic conditions, radioactivity appeared in the carboxyl groups of succinate, whereas radioactivity from [^{14}C]glucose appeared, in anaerobisis, exclusively in the methylene groups. These experiments are consistent with CO_2 fixation on a C3 monocarboxylic acid arising from glycolysis, either phosphoenolpyruvate (PEP) or pyruvate, to give a C4 dicarboxylic acid (Bowman *et al.*, 1963). The primary CO_2 fixation product was initially supposed to be either oxaloacetate or L-malate; the former might be produced by several enzymes, among them pyruvate carboxylase and PEP-carboxykinase (PEPCK), and the latter by malic enzyme.

Crithidia fasciculata contains all three enzymes (Cannata *et al.*, 1980; Cazzulo *et al.*, 1980), whereas *T. cruzi* contains only PEPCK and malic enzyme (Cataldi de Flombaum *et al.*, 1977). Recent [^{13}C]-NMR evidence (De los Santos *et al.*, 1985) has shown that the CO_2-fixing enzyme in *C. fasciculata* is unmistakably PEPCK, despite the presence of a low pyruvate carboxylase activity. The labeling pattern in similar experiments performed with *T. cruzi* is also compatible with this possibility (Frydman *et al.*, 1990) as is the fact that in both Trypanosomatids the relative proportion of succinate and the other major product (ethanol for *C. fasciculata*, L-alanine for *T. cruzi*) changes with the addition of bicarbonate to the incubation medium (De los Santos *et al.*, 1985; Frydman *et al.*, 1990). Moreover, recently Urbina *et al.* (1989) have shown *in vivo*, in [^{13}C]-NMR experiments, that addition to the cells of an inhibitor of PEPCK, 3-mercaptopicolinic acid, results in a decrease in succinate production with a concomitant increase in L-alanine production. The PEPCK was shown to be ADP-linked, as those from yeast and bacteria, and at variance with the PEPCK of higher animals, which is specific for GDP or IDP (Cataldi de Flombaum *et al.*, 1977), and has been recently purified and studied by Urbina (1987), who showed a strong inhibition of oxaloacetate decarboxylation by ATP, which might be relevant for metabolic regulation *in vivo*.

For succinate production, the oxaloacetate synthesized by the PEPCK reaction would be reduced to L-malate by malate dehydrogenase, L-malate would be dehydrated to fumarate by fumarase (Baernstein, 1953), and fumarate would be finally reduced to succinate by a fumarate reductase (Boveris *et al.*, 1986). Reduction to L-malate would be enough for the reoxidation of all glycolytic NADH, if both PEP molecules were converted into succinate, with no energy loss, since one ATP is synthesized by both pyruvate kinase and PEPCK. The reason for succinate production and excretion is not clear yet, but has been consistently found by all research workers studying this subject, whereas L-malate itself has been only seldom reported (Cáceres and Fernandes, 1976).

3.3.2. L-Alanine Production

We have recently proposed (Cazzulo *et al.*, 1988a) that the production of L-alanine might be linked to the reoxidation of glycolytic NADH. The reduced coenzyme can be reoxidized in the reaction of the NAD-linked glutamate dehydrogenase, leading to L-glutamate synthesis; the latter will then react with the pyruvate produced by glycolysis, by the transamination reaction catalyzed by alanine aminotransferase present in the parasite (Barros and Caldas, 1983). L-Glutamate and α-oxoglutarate would be required in catalytic amounts, whereas NH_3 would be consumed. It is noteworthy that excretion of NH_3 into the medium is low or negligible during glucose consumption (Cazzulo *et al.*, 1985, 1988a; Engel *et al.*, 1987); this may be due to actual lack of production or to preferential

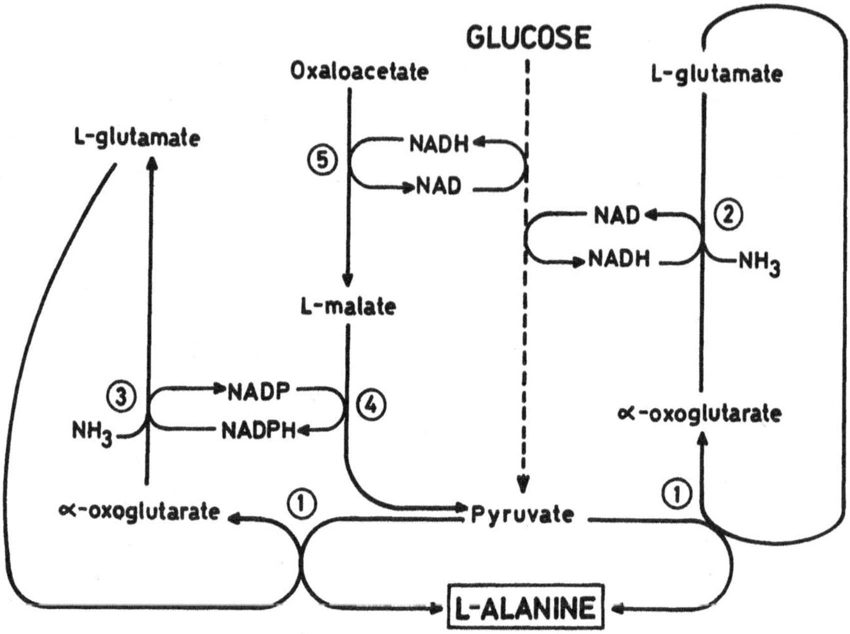

FIGURE 2. Possible participation of the glutamate dehydrogenases in the reoxidation of glycolytic NADH in *Trypanosoma cruzi*. The enzymes indicated are: 1, alanine aminotransferase; 2, NAD-linked glutamate dehydrogenase; 3, NADP-linked glutamate dehydrogenase; 4, malic enzyme; 5, malate dehydrogenase.

use for transamination. Production of L-alanine or succinate would be equivalent in terms of NADH oxidation. The subcellular localization of the enzymes required for this pathway has not been determined yet; in the case of alanine aminotransferase, it is known that there are, in most *T. cruzi* stocks, at least two isoenzymes (Miles *et al.*, 1978). The existence of cytosolic and mitochondrial isoenzymes is therefore possible, and the results of Taylor and Gutteridge (1986a), although not conclusive, support this possibility.

L-Alanine might also be obtained in the cytosol from L-malate (synthesized by the reduction of oxaloacetate inside the glycosome) through the concerted action of malic enzyme, the NADP-linked glutamate dehydrogenase and alanine aminotransferase; this pathway (Figure 2) would convert L-malate and NH_3 into L-alanine and CO_2 and would have similar requirements as the one involving the NAD-linked glutamate dehydrogenase, plus catalytic amounts of NADP. The observed decrease in L-alanine production in the presence of added $NaHCO_3$ might be due to product inhibition of the decarboxylation reaction catalyzed by malic enzyme. This latter pathway has the advantage that it does not require the existence of a cytosolic pool of NADH arising from glycolysis, even if the pathway occurs in the soluble fraction; malic enzyme II and the NADP-linked glutamate dehydrogenase are known to be cytosolic (Cannata *et al.*, 1980; Cannata and Cazzulo, 1984a,b). The pathway involving the NADP-linked glutamate dehydrogenase may be essential in the case of *C. fasciculata*, which has a very low pyruvate kinase activity (Cazzulo *et al.*, 1989b) and must nevertheless obtain pyruvate to convert it into ethanol, with concomitant NADH oxidation (Cazzulo *et al.*, 1985, 1988a).

3.3.3. Respiratory Chain

At least part of the glycolytic NADH in *T. cruzi* is reoxidized through a respiratory chain consisting of cytochromes *b*, c_{558}, and two alternative terminal oxidases, cytochrome oxidase ($a + a_3$) and cytochrome *o* (Stoppani *et al.*, 1980; Gutteridge, 1981). Cytochrome *d* seems also to be present (Affranchino *et al.*, 1986). Evidence obtained by different groups (Carneiro and Caldas, 1982; Affranchino *et al.*, 1986) suggests the presence of at least three different terminal oxidases in *T. cruzi* epimastigotes, namely, one sensitive to cyanide and azide, a second sensitive to cyanide but insensitive to azide, and a third insensitive to both inhibitors. The first, cytochrome oxidase, seems to be the most important terminal oxidase in the parasite; the second would be cytochrome *o* and the third might be cytochrome *d* (Affranchino *et al.*, 1986). Recent studies indicate the existence of wide variations in the cytochrome contents of different parasite stocks (Schwartz de Tarlovsky *et al.*, 1985), and also in different developmental stages; Engel *et al.* (1987) showed that the extracellular amastigotelike forms completely lacked spectrophotometrically detectable cytochromes, whereas these were acquired on differentiation to epimastigotes. Intracellular (tissue) amastigotes, as well as bloodstream trypomastigotes, on the other hand, contain a full complement of cytochromes (Gutteridge and Rogerson, 1979). The amounts of cytochromes *b* and $a + a_3$ present in epimastigotes are significantly lower than those found in mammalian mitochondria (Boiso *et al.*, 1979a), and the resulting relative inefficiency of the respiratory chain might be the reason for the existence of the aerobic fermentation, as a complementary pathway for NADH oxidation. Mitochondrial particles obtained from culture epimastigotes and bloodstream trypomastigotes and tissue amastigotes are able to oxidize NADH, succinate, and glycerol 3-phosphate (Gutteridge and Rogerson, 1979). Cyanide and antimycin A are inhibitors of both the endogenous respiration of epimastigotes (Stoppani *et al.*, 1980) and the consumption of oxygen in the presence of glucose (Gutteridge and Rogerson, 1979). Antimycin A inhibited growth of the epimastigotes and considerably increased production of succinate from labeled glucose or CO_2, with a decrease in alanine production (Boiso *et al.*, 1979b). These findings suggest the importance of the respiratory chain for both NADH oxidation and ATP synthesis, through coupled oxidative phosphorylation (see below).

The transport of glycolytic NADH from the glycosome, where it is produced, to the mitochondrion for oxidation in the respiratory chain may be mediated by L-malate, synthesized inside the glycosome, and oxidized in the mitochondrion (see Section 3.4). Mitochondrial oxidation of NADH produced in the cytoplasm might be accomplished through the shuttle system α-oxoisocaproate–α-hydroxyisocaproate; isoenzyme II of α-hydroxyacid dehydrogenase and leucine aminotransferase, essential to obtain α-oxoisocaproate from L-leucine, have been reported as present both in the cytosol and in the mitochondrion (Montamat *et al.*, 1987). The importance of such a cytosol–mitochondrion shuttle system in Trypanosomatids is still a matter of speculation; moreover, Taylor and Gutteridge (1986a) have reported that α-hydroxyacid dehydrogenase is present in the glycosome and in the cytosol of *T. cruzi* epimastigotes, whereas the transaminases are mostly cytosolic, with a small particulate, presumably mitochondrial, component. The possible action of a glycerol 3-phosphate shuttle system, like the one present in mammalian cells and also present in *C. fasciculata* (Bacchi *et al.*, 1968), seems to be ruled out by the absence in *T. cruzi* of an NAD-linked glycerol 3-phosphate dehydrogenase (Cazzulo *et al.*, 1988a), present in the glycosome of *T. brucei* and *Crithidia luciliae* (Opperdoes *et al.*, 1977).

3.4. Subcellular Compartmentation and Regulation of Glucose Catabolism

As described above, inhibitory controls usually present in most organisms and responsible for the Pasteur effect, are lacking in Trypanosomatids. The organisms seem to be adapted to consume as much glucose as is able to enter the cell, and the only apparently important control point is the activation of the cytosolic pyruvate kinase by fructose 2, 6- (or 1,6-) *bis*-phosphate (Cazzulo *et al.*, 1989b). Even if lack of controls is useful for the parasite, which has no obvious reason to save its host's glucose, some kind of regulatory mechanism is necessary to prevent futile cycles, which might interfere with the aerobic fermentation process. In this sense, the reaction catalyzed by malic enzyme must be controlled to avoid recycling of C4 dicarboxylic acid, synthesized by PEPCK from the C3 PEP, back to the C3 pyruvate. The present evidence suggests that subcellular compartmentation is the major factor involved in the regulation of glycolysis in Trypanosomatids. First, the high rate of glucose consumption seems to be due to the very high concentration of both glycolytic enzymes and intermediates inside the glycosome; this has been particularly well studied in *T. brucei* (Opperdoes, 1987). Glycosomes are

FIGURE 3. Compartmentation of enzymes involved in glucose catabolism in *Trypanosoma cruzi*. Abbreviations are the same as in Figure 1, with the addition of ALA for L-alanine; some of the reactions, shown in Figure 1, have been omitted here for the sake of clarity.

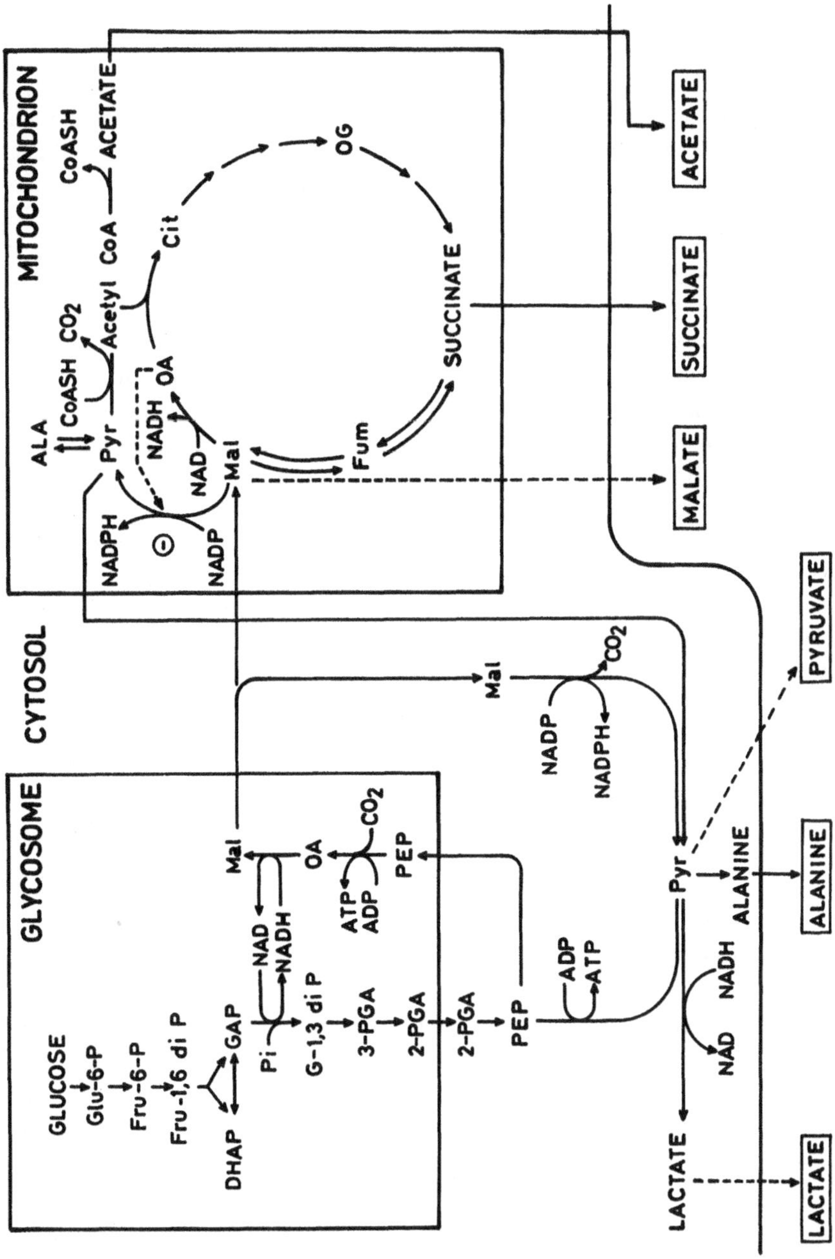
MITOCHONDRION
CYTOSOL
GLYCOSOME
ALA
Pyr
CoASH
CO_2
Acetyl CoA
CoASH
ACETATE
Cit
OG
SUCCINATE
Fum
Mal
OA
NAD
NADH
NADPH
NADP
Mal
NADP
NADPH
CO_2
GLUCOSE
Glu-6-P
Fru-6-P
Fru-1,6 di P
DHAP
GAP
Pi
NAD
NADH
G-1,3 di P
3-PGA
2-PGA
2-PGA
PEP
Mal
OA
PEP
ATP
ADP
CO_2
ADP
ATP
Pyr
ALANINE
NADH
NAD
LACTATE
ACETATE
SUCCINATE
MALATE
PYRUVATE
ALANINE
LACTATE

certainly present in *T. cruzi* (Taylor *et al.*, 1980; Cannata *et al.*, 1982) and have recently been purified and characterized by Taylor and Gutteridge (1987), who showed that hexokinase, phosphoglucose isomerase, phosphofructokinase, aldolase, triosephosphate isomerase, and glyceraldehyde 3-phosphate dehydrogenase are glycosomal enzymes, whereas phosphoglycerate kinase, phosphoglycerate mutase, enolase, and pyruvate kinase are cytosolic. Second, the possible futile cycle between PEPCK and malic enzyme seems to be prevented by the fact that the former is glycosomal (Cannata *et al.*, 1982), whereas the latter has two isoenzymes, one mitochondrial and the other cytosolic (Cazzulo *et al.*, 1980; Cannata *et al.*, 1982). Malate dehydrogenase also has two different isoenzymes, one glycosomal and one mitochondrial (Cannata and Cazzulo, 1984b). The subcellular distribution of PEPCK and the malate dehydrogenases is identical to that demonstrated for the procyclic trypomastigotes of *T. brucei* (Opperdoes, 1987). Figure 3 summarizes the data on subcellular compartmentation of enzymes in *T. cruzi*, as well as the interconnections between glycolysis and the tricarboxylic acid cycle. Phosphoenol pyruvate is produced in the cytosol, since Taylor *et al.* (1980) showed this location for enolase, and it can then follow two different pathways: (1) it can be taken up by the glycosome and carboxylated to oxaloacetate by PEPCK; or (2) it can be converted into pyruvate by the cytosolic pyruvate kinase. Glycosomal oxalocetate will be converted into L-malate by the glycosomal malate dehydrogenase, whereas cytosolic pyruvate will be transaminated to L-alanine. Pyruvate can also be obtained in the cytosol by the reaction of malic enzyme II or in the mitochondrion by the reaction of malic enzyme I; this raises the possibility of two different pools of pyruvate, which might explain the presence of two different L-alanine pools, recently detected by [^{13}C]-NMR (Frydman *et al.*, 1990). Production of L-malate inside the glycosome will leave this organelle in redox balance, if both molecules of PEP are taken up by the glycosome and converted into oxaloacetate and afterward into L-malate. L-Malate will probably be transferred then to the mitochondrion via cytosol; part of it may be decarboxylated to pyruvate by malic enzyme II, producing NADPH for cytosolic biosynthetic processes. When L-malate reaches the mitochondrial matrix, its fate will depend on whether the Krebs cycle is actively operating or not. Under anaerobic conditions, or if for some reason the operation of the cycle is low, some oxaloacetate obtained by the malate dehydrogenase reaction will accumulate and it will be able to inhibit the highly sensitive (K_i for oxaloacetate 9μM) malic enzyme I (Cannata *et al.*, 1979), thus blocking L-malate decarboxylation and diverting most of the L-malate to succinate via fumarate. On the other hand, if the cycle is working actively, the oxaloacetate concentration will be kept low by the citrate synthase reaction, malic enzyme I will be active, and part of the L-malate will be decarboxylated to pyruvate and enter the cycle as acetyl-CoA to be oxidized. Activity of malic enzyme I during operation of the tricarboxylic acid cycle can be inferred from the results of Stoppani *et al.* (1980), who showed that during oxidation of [^{14}C]-acetate by the epimastigotes, a substantial amount of labeled L-alanine is formed. Labeled L-alanine can be obtained in these circumstances only from pyruvate arising from a Krebs cycle intermediate, like L-malate.

The present evidence suggests a central role for L-malate in the aerobic fermentation process and a key position of malic enzyme in the regulation of the pathway.

4. OXIDATIVE PHOSPHORYLATION

4.1. Phosphorylation *in Vivo* and in Submitochondrial Particles

Oxidation of substrates in the respiratory chain of *T. cruzi* is coupled to phosphorylation, as was first shown by Stoppani and Boiso (1973), though the evidence was indirect. In fact, these authors showed that the intracellular ATP content was decreased by about 50% *in vivo* by the addition of several well-known inhibitors of oxidative phosphorylation (*N,N'*-dicyclohexylcarbodiimide, or DCCD) and the respiratory chain (cyanide) and also by uncouplers (carbonyl cyanide *m*-chlorophenylhydrazone, or CCP). Only recently direct evidence of oxidative phosphorylation in this parasite has been obtained (Affranchino *et al.*, 1985). Submitochondrial particles obtained by differential centrifugation after parasite disruption by grinding with glass powder in a mortar, were shown to be able to oxidize substrates such as L-malate, glycerol-3-phosphate, succinate and ascorbate plus tetramethyl-*p*-phenylendiamine (TMPD). Evidence of respiratory control, with values in the range of 2.8–2.0, was found and suggested the operation of coupling sites II and III; respiration and phosphorylation were uncoupled by CCP or sonication. NADH oxidation was not controlled by ADP nor inhibited by antimycin or cyanide, thus suggesting the possible presence of a special NADH dehydrogenase in the outer mitochondrial membrane, as found in other organisms (Affranchino *et al.*, 1985). The apparent absence of phosphorylation site I might be a common property of Trypanosomatids, since it was previously reported for two species of Crithidia (Kusel and Storey, 1972; Toner and Weber, 1972; Opperdoes *et al.*, 1976).

4.2. Mitochondrial ATPase

As in all other systems known, either prokaryotic or eukaryotic, oxidative phosphorylation is mediated in *T. cruzi* by a mitochondrial oligomycin-sensitive Mg^{2+}-ATPase. The enzyme was detected in epimastigotes (De Sastre and Stoppani, 1973; Frasch *et al.*, 1978b), solubilized from the mitochondrial membranes (Frasch *et al.*, 1978a), and purified to electrophoretic homogeneity (Cataldi de Flombaum *et al.*, 1979). The ATPase is made up of subunits similar to those found in other systems and is inhibited by oligomycin (in its membrane-bound form) or efrapeptin, aurovertin, and citreoviridin (both membrane-bound and soluble forms) (Cataldi de Flombaum and Stoppani, 1981), and also by organo-mercurial compounds, in particular by fluorescein mercuric acetate (Cataldi de Flombaum *et al.*, 1979).

5. PROTEIN AND AMINO ACID CATABOLISM

5.1. Production of NH_3

Proteins and amino acids are consumed by epimastigotes in culture, with production of ammonia (von Brand, 1979), which is liberated into the medium, raising again the pH lowered by the excretion of organic acids (Cáceres and Fernandes, 1976). *Trypanosoma cruzi,* at variance with other Trypanosomatids, is strictly ammonotelic (Yoshida and Camargo, 1978) and lacks all the enzymes of the urea cycle (Camargo *et al.*, 1978).

5.2. Proteolytic Enzymes

Epimastigotes of *T. cruzi* have been known for many years to have a considerable intracellular amino acid pool, about half of which is made up by L-alanine (Williamson and Desowitz, 1961); proline, glycine, and glutamic acid are also abundant (O'Daly *et al.*, 1983). These free amino acids, which can be used for the synthesis of proteins or some derived substances or for energy production, are likely to be produced by proteolysis of exogenous or endogenous proteins or to be taken up from the culture medium (see also Chapter 6). In the case of L-alanine, a substantial proportion can originate in glucose catabolism, as discussed above.

A number of proteolytic activities have been described in *T. cruzi* epimastigotes, starting with the work of Itow and Camargo (1977) and Avila *et al.* (1979b); it is difficult to tell whether all the activities described belong to different enzymes, since these early studies were performed with crude extracts (see Cazzulo, 1984, for review). Three proteinases, all belonging to the group of the cysteine proteinases, have been purified to homogeneity from epimastigotes. One of them has a high molecular weight (about 200 kDa) and hydrolyzes efficiently benzoyl-arginine-*p*-nitroanilide (BAPA) (Bongertz and Hungerer, 1978). This proteinase might be the same enzyme recently reported and studied in further detail by Ashall (1990); the other two have molecular weight values of about 60 kDa (Rangel *et al.*, 1981a,b; Bontempi *et al.*, 1984) and might be the same enzyme, except for some differences in substrate and inhibitor specificity (see Cazzulo, 1984, for review). The study of the proteinases in trypanosomes and leishmanias as well as in other protozoal parasites, such as plasmodia, and in helminths, such as the schistosomes, has received considerable interest in recent years, since, in addition to their obvious participation in the nutrition of the parasite at the expense of the host, they may be relevant to other aspects of the host–parasite relationship. In the case of *T. cruzi,* proteinases might be involved in penetration of the trypomastigote into the host cell (Piras *et al.*, 1985), as well as in evasion from the immune response of the host (Krettli *et al.*, 1980; Teixeira and Santana, 1989).

The 60-kDa cysteine proteinase has been studied in detail recently; it is a high-mannose-type glycoprotein (Cazzulo *et al.*, 1989a, 1990a) placed in the lysosomes (Bontempi *et al.*, 1989), which presents in the sequenced parts of the molecule high homology with cathepsin L and papain (Cazzulo *et al.*, 1989a; Eakin *et al.*, 1990), as well as with the cysteine proteinase recently cloned and sequenced in *T. brucei* (Mottram *et al.*, 1989). The enzyme, for which the name "cruzipain" has been proposed (Cazzulo *et al.*, 1990b), seems to be the major cysteine proteinase of the parasite, present not only in epimastigotes, but also in cell-culture trypomastigotes and amastigotes (Campetella *et al.*, 1990). The levels of the proteinase, as detected both by enzyme activity and immunoreactivity in Western blots, are considerably lower in amastigotes and trypomastigotes than in epimastigotes, which suggests that the expression of the enzyme is developmentally regulated (Campetella *et al.*, 1990). Recent self-proteolysis experiments (U. Hellman, C. Wernstedt, and J. J. Cazzulo, unpublished data) indicate that the enzyme possesses a 25-kDa proteolysis-resistant, glycosylated C-terminal extension, which may have some similarity with that recently described for the *T. brucei* enzyme (Mottram *et al.*, 1989).

In addition to cysteine proteinases, which are by far the best studied in this parasite, recent evidence has been presented (by DNA sequencing) for the presence of a serine proteinase (Sakanari *et al.*, 1989); there also is a recent report of a metalloproteinase, which might have some resemblance to the *Leishmania* gp63 (Greig and Ashall, 1990). There is no evidence for the possible presence of aspartic proteinases in *T. cruzi,* and high-molecular-weight multicatalytic proteinases have not been reported yet in the parasite, although their presence seems likely, considering their universal distribution (Dahlmann *et al.*, 1989).

5.3. Amino Acid Catabolism

Trypanosoma cruzi epimastigotes are able to oxidize a number of amino acids, presumably through the tricarboxylic acid cycle (Mancilla *et al.*, 1967; Sylvester and Krassner, 1976). The catabolic pathways for amino acids such as leucine and proline are probably similar to those present in other organisms, but so far as I know, they have not been examined in detail and none of the enzymes involved has been demonstrated. The only exception would be serine hydroxymethyltransferase, which is involved in the serine–glycine interconversion, which has been characterized by Nosei and Avila (1985). On the other hand, the enzymes responsible for the exchange and disposal of amino nitrogen have been studied in much greater detail.

The enzymes involved in the liberation of the amino group of the different amino acids as ammonia are the transaminases, many of which transfer the amino group to α-oxoglutarate to give L-glutamate, and the glutamate dehydrogenases, which liberate the $-NH_2$ group as NH_3. The contribution of amino acid oxidases to the production of ammonia is in general of little importance, and none of these enzymes has been reported so far in Trypanosomatids. Since the early studies of Bash-Lewinson and Grossowicz (1957) and Zeledón (1960), it is known that *T. cruzi* contains transaminase activities. A number of amino acids were shown to be able to transaminate with α-oxoglutarate or pyruvate to yield L-glutamate or L-alanine, respectively (Zeledón, 1960). The enzymes responsible, aspartate aminotransferase (Cazzulo *et al.*, 1977) and alanine aminotransferase (Barros and Caldas, 1983) have been partially purified from epimastigotes and characterized; they have molecular weights in the vicinity of 100 kDa and seem not to differ significantly from their mammalian counterparts. Both enzymes have at least two isoenzymes in most *T. cruzi* isolates and are among the six enzyme systems proposed by Miles *et al.* (1978, 1980) for the classification of isolates of the parasite into zymodemes. As noted above, leucine aminotransferase has also been detected in *T. cruzi* epimastigotes (Montamat *et al.*, 1987).

Trypanosoma cruzi epimastigotes contain two different glutamate dehydrogenases, one NADP-linked and one NAD-linked (Cazzulo *et al.*, 1979; Walter and Ebert, 1979). The parasite therefore resembles bacteria, fungi, and plants in this respect and clearly differs from higher animals, which possess only one glutamate dehydrogenase able to act both with NAD and NADP. Both enzymes have been purified to homogeneity (Juan *et al.*, 1978; Walter and Ebert, 1979) and their properties studied in some detail (see Cazzulo, 1984, for review). The regulatory properties of the NAD-linked enzyme have been studied by Urbina and Azavache (1984). Recently, the NADP-linked enzyme has been shown to be very similar to the enzyme from *Escherichia coli* in terms of amino acid composition and of *N*-terminal sequence; 33 amino acids at the *N*-terminus presented 65% identity with the *E. coli* enzyme, including a 14-amino-acid *N*-terminal extension, which is not present in the enzyme from *Neurospora crassa* (Cazzulo *et al.*, 1988b). The role both enzymes play in *T. cruzi* metabolism is still doubtful; there is some evidence suggesting that the NADP-linked glutamate dehydrogenase might be biosynthetic (Carneiro and Caldas, 1983), whereas the NAD-linked enzyme might have a catabolic role, since its activity increases in the late exponential phase of growth, when amino acids are certainly being consumed and NH_3 is produced (Cazzulo *et al.*, 1985) and is strongly inhibited by ATP

(Cazzulo *et al.*, 1979; Walter and Ebert, 1979; Urbina and Azavache, 1984) and acetyl-CoA (Urbina and Azavache, 1984).

6. RELATIONSHIP BETWEEN CARBOHYDRATE AND AMINO ACID CATABOLISM

As stated in Section 1, there seems to be a delicate balance between the metabolic pathways for energy production from either carbohydrates or amino acids. Oxidation of the carbon chains of the latter compounds requires the function of the tricarboxylic acid cycle (Sylvester and Krassner, 1976). The question of which catabolic pathway is more important to the parasite has been raised and is still a matter of speculation. The oxidation in the cycle of intermediates of the cycle itself, such as oxaloacetate and α-oxoglutarate, arising from amino acid catabolism, requires their previous conversion into acetyl-CoA. This can be accomplished through the action of malic enzyme on L-malate or through the action of an oxaloacetate decarboxylase on oxaloacetate; the latter role can be fulfilled by the concerted action of PEPCK and pyruvate kinase just as well. Urbina and Azavache (1984) and Urbina (1987) suggest that the preferential use of glucose before amino acids by *T. cruzi* epimastigotes is due to the tight regulation of NAD-linked glutamate dehydrogenase and PEPCK and the lack of regulation of the glycolytic pathway; the former enzymes would remain inhibited during active glycolysis by the high ATP concentration, whereas they would become fully active after glucose is exhausted and the ATP contents decreases. However, the experimental evidence of PEPCK participation in succinate excretion mentioned above indicates that under conditions of active glycolysis the enzyme is acting as an auxiliary to this pathway, despite the inhibition of the carboxylation reaction by ATP reported by Urbina (1987). According to this author, amino acid catabolism would be mediated by transport of L-malate from the mitochondrion to the glycosome, formation of oxaloacetate by malate dehydrogenase, decarboxylation of this intermediate to PEP by PEPCK, transport of PEP to the cytosol, formation of pyruvate by pyruvate kinase, and transport of pyruvate into the mitochondrion where it would be oxidized to acetyl-CoA and incorporated into the cycle through the citrate synthase reaction. In support of this proposal, Urbina *et al.* (1989) report inhibition of proline oxidation by 3-mercaptopicolinic acid. Although this pathway is feasible, since the reaction catalyzed by PEPCK is freely reversible, the possibility that mitochondrial malic enzyme I is involved instead of PEPCK is also likely. In fact, in this way the conversion of Krebs cycle intermediates into acetyl-CoA would take place in only one subcellular compartment, the mitochondrion, instead of requiring three (mitochondrion, glycosome, cytosol, and mitochondrion again), as in Urbina and Azavache's proposal. As pointed out above, the strong inhibition of the malic enzyme I by oxaloacetate suggests that it will be active only when the concentration of this intermediate is kept low by citrate synthase and the cycle is fully active.

There can be little doubt that PEPCK in *T. cruzi* and other Trypanosomatids has a catabolic function, as suggested by its lack of the glucose repression characteristic of PEPCKs having a gluconeogenic role (Cazzulo *et al.*, 1985). This applies to any of its proposed roles in the parasite's catabolism.

Although more experimental evidence will be required to solve the remaining doubts about the relative relevance of carbohydrate and amino acid catabolism in *T. cruzi,* it seems reasonable to assume that both have similar importance and that one or the other will be preferred according to substrate availability and also to the enzyme makeup of the different developmental forms of the parasite.

ACKNOWLEDGMENTS. I am indebted to Professor J. J. B. Cannata for helpful discussions. The work performed at the author's laboratory and reviewed here was aided by grants from the Consejo Nacional de Investigaciones Cientificas y Técnicas (CONICET) and the Secretaria de Ciencia y Técnica (SECYT) de la República Argentina; the UNDP/World Bank/WHO Special Programme for Research and Training in Tropical Diseases (TDR); and the Swedish Agency for Research Cooperation with Developing Countries (SAREC). The author is a member of the Carrera del Investigador Cientifico from CONICET.

7. REFERENCES

Adroher, F. J., Osuna, A., and Lupiáñez, J. A., 1987, Fructose 1,6-bisphosphatase activity in two *Trypanosoma cruzi* morphological forms, *J. Parasitol.* **73**:438–441.

Adroher, F. J., Osuna, A., and Lupiáñez, J. A., 1988, Differential energetic metabolism during *Trypanosoma cruzi* differentiation. 1. Citrate synthase, NADP-isocitrate dehydrogenase, and succinate dehydrogenase, *Arch. Biochem. Biophys.* **267**:252–261.

Adroher, F. J., Osuna, A., and Lupiáñez, J. A., 1990, Differential energetic metabolism during *Trypanosoma cruzi* differentiation. II. Hexokinase, phosphofructokinase, and pyruvate kinase, *Mol. Cell. Biochem.* **94**:71–80.

Affranchino, J. L., Tarlovsky, M. N. S. de, and Stoppani, A. O. M., 1985, Respiratory control in mitochondria from *Trypanosoma cruzi, Mol. Biochem. Parasitol.* **16**:289–298.

Affranchino, J. L., Schwarcz de Tarlovsky, M. S., and Stoppani, A. O. M., 1986, Terminal oxidases in the Trypanosomatid *Trypanosoma cruzi, Comp. Biochem. Physiol.* **85**:381–388.

Aguilar, Z., and Urbina, J. A., 1986, The phosphofructokinase of *Trypanosoma* (*Schizotrypanum*) *cruzi:* Purification and kinetic mechanism, *Mol. Biochem. Parasitol.* **21**:103–111.

Andrews, N. W., and Colli, W., 1982, Adhesion and interiorization of *Trypanosoma cruzi* in mammalian cells, *J. Protozool.* **29**:264–269.

Arauzo, S., and Cazzulo, J. J., 1989, The NADP-linked aldehyde reductase from *Trypanosoma cruzi.* Subcellular localization and some properties, *FEMS Microbiol. Lett.* **58**:283–286.

Ashall, F., 1990, Characterization of an alkaline peptidase of *Trypanosoma cruzi* and other Trypanosomatids, *Mol. Biochem. Parasitol.* **38:**77–88.

Avila, J. L., Bretaña, A., Casanova, M. A., Avila, A., and Rodriguez, F., 1979a, *Trypanosoma cruzi:* Defined medium for continuous cultivation of virulent parasites, *Exp. Parasitol.* **48:**27–35.

Avila, J. L., Casanova, M. A., Avila, A., and Bretaña, A., 1979b, Acid and neutral hydrolases in *Trypanosoma cruzi.* Characterization and assay, *J. Protozool.* **26:**304–311.

Avila, J. L., Pérez-Kepp, R., and Bretaña, A., 1983, A minimal medium for the cultivation of infective *Trypanosoma cruzi* epimastigotes, *J. Gen. Microbiol.* **129:**285–291.

Bacchi, C. J., Hutner, S. H., Ciaccio, E. I., and Marcus, S. M., 1968, O_2-polarographic studies on soluble and mitochondrial enzymes of *Crithidia fasciculata;* Glycerol phosphate enzymes, *J. Protozool.* **15:**576–584.

Baernstein, H. D., 1953, Malic dehydrogenase and related enzymes in the culture form of *Trypanosoma cruzi, Exp. Parasitol.* **2:**380–396.

Barros, E. G., and Caldas, R. A., 1983, Partial purification and characterization of glutamic-pyruvic transaminase from *Trypanosoma cruzi, Comp. Biochem. Physiol.* **74B:**449–452.

Bash-Lewinson, D., and Grossowicz, N., 1957, Transaminases of *Trypanosoma cruzi, Bull. Res. Counc. Israel* **6E:**91–92.

Boiso, J. F. de, Docampo, R., and Stoppani, A. O. M., 1979a, Citocromos de la forma epimastigote del *Trypanosoma cruzi, Rev. Arg. Microbiol.* **11:**22–36.

Boiso, J. F. de, Docampo, R., Stoppani, A. O. M., and Frasch, A. C. C., 1979b, Efecto de la antimicina A sobre la respiración y el crecimiento del *Trypanosoma cruzi, Medicina (Buenos Aires)* **39:**362–368.

Bongertz, V., and Hungerer, K. D., 1978, *Trypanosoma cruzi:* Isolation and characterization of a protease, *Exp. Parasitol.* **45:**8–18.

Bontempi, E., Franke de Cazzulo, B. M., Ruiz, A. M., and Cazzulo, J. J., 1984, Purification and some properties of an acidic protease from epimastigotes of *Trypanosoma cruzi, Comp. Biochem. Physiol.* **77B:**599–604.

Bontempi, E., Martinez, J., and Cazzulo, J. J., 1989, Subcellular localization of a cysteine proteinase from *Trypanosoma cruzi, Mol. Biochem. Parasitol.* **33:**43–47.

Boveris, A., Hertig, C. M., and Turrens, J. F., 1986, Fumarate reductase and other mitochondrial activities in *Trypanosoma cruzi, Mol. Biochem. Parasitol.* **19:**163–169.

Bowman, I. B. R., Tobie, E. J., and von Brand, Th., 1963, CO_2 fixation studies with the culture form of *Trypanosoma cruzi, Comp. Biochem. Physiol.* **9:**105–114.

Brener, Z., 1973, Biology of *Trypanosoma cruzi, Annu. Rev. Microbiol.* **27:**347–382.

Cáceres, O., and Fernandes, J. F., 1976, Glucose metabolism, growth and differentiation of *Trypanosoma cruzi, Rev. Brasil. Biol.* **36:**397–410.

Camargo, E. P., Coelho, J. A., Moraes, G., and Figueiredo, E. N., 1978, *Trypanosoma* spp., *Leishmania* spp. and *Leptomonas* spp.: Enzymes of ornithine-arginine metabolism, *Exp. Parasitol.* **46:**141–144.

Campetella, O., Martinez, J., and Cazzulo, J. J., 1990, A major cysteine proteinase is developmentally regulated in *Trypanosoma cruzi, FEMS Microbiol. Lett.* **67:**145–150.

Cannata, J. J. B., and Cazzulo, J. J., 1984a, The aerobic fermentation of glucose by *Trypanosoma cruzi, Comp. Biochem. Physiol.* **79B:**297–308.

Cannata, J. J. B., and Cazzulo, J. J., 1984b, Glycosomal and mitochondrial malate dehydrogenases in epimastigotes of *Trypanosoma cruzi, Mol. Biochem. Parasitol.* **11:**37–49.

Cannata, J. J. B., Frasch, A. C. C., Cataldi de Flombaum, M. A., Segura, E. L., and Cazzulo, J. J., 1979, Two forms of malic enzyme with different regulatory properties in *Trypanosoma cruzi, Biochem. J.* **184:**409–419.

Cannata, J. J. B., Docampo, R., Valle, E., Franke de Cazzulo, B. M., and Cazzulo, J. J., 1980,

Distribución intracelular de las enzimas fijadoras de CO_2 en *Trypanosoma cruzi* y *Crithidia fasciculata, Medicina (Buenos Aires)* **40**(Suppl. 1):143–153.

Cannata, J. J. B., Valle, E., Docampo, R., and Cazzulo, J. J., 1982, Subcellular localization of phosphoenolpyruvate carboxykinase in the trypanosomatids *Trypanosoma cruzi* and *Crithidia fasciculata, Mol. Biochem. Parasitol.* **6**:151–160.

Carneiro, M., and Caldas, R. A., 1982, Evidence for three respiratory terminals in *Trypanosoma cruzi* epimastigotes, *Acta Trop.* **39**:41–49.

Carneiro, V. T. C., and Caldas, R. A., 1983, Regulatory studies of L-glutamate dehydrogenase from *Trypanosoma cruzi* epimastigotes, *Comp. Biochem. Physiol.* **75B**:61–64.

Cataldi de Flombaum, M. A., and Stoppani, A. O. M., 1981, Influence of efrapeptin, aurovertin and citreoviridin on the mitochondrial adenosine triphosphatase from *Trypanosoma cruzi, Mol. Biochem. Parasitol.* **3**:143–155.

Cataldi de Flombaum, M. A., Cannata, J. J. B., Cazzulo, J. J., and Segura, E. L., 1977, CO_2-fixing enzymes in *Trypanosoma cruzi, Comp. Biochem. Physiol.* **58B**:67–69.

Cataldi de Flombaum, M. A., Frasch, A. C. C., and Stoppani, A. O. M., 1979, Adenosine triphosphatase from *Trypanosoma cruzi:* Purification and properties, *Comp. Biochem. Physiol.* **65B**:103–109.

Cazzulo, J. J., 1984, Protein and amino acid catabolism in *Trypanosoma cruzi, Comp. Biochem. Physiol.* **79B**:309–320.

Cazzulo, J. J., Juan, S. M., and Segura, E. L., 1977, Glutamate dehydrogenase and aspartate aminotransferase in *Trypanosoma cruzi, Comp. Biochem. Physiol.* **56B**:301–303.

Cazzulo, J. J., Franke de Cazzulo, B. M., Higa, A. I., and Segura, E. L., 1979, NAD-linked glutamate dehydrogenase in *Trypanosoma cruzi, Comp. Biochem. Physiol.* **64B**:129–131.

Cazzulo, J. J., Valle, E., Docampo, R., and Cannata, J. J. B., 1980, Intracellular distribution of CO_2-fixing enzymes in *Trypanosoma cruzi* and *Crithidia fasciculata, J. Gen. Microbiol.* **117**:271–274.

Cazzulo, J. J., Franke de Cazzulo, B. M., Engel, J. C., and Cannata, J. J. B., 1985, End products and enzyme levels of aerobic glucose fermentation in Trypanosomatids, *Mol. Biochem. Parasitol.* **16**:329–343.

Cazzulo, J. J., Arauzo, S., Franke de Cazzulo, B. M., and Cannata, J. J. B., 1988a, On the production of glycerol and L-alanine during the aerobic fermentation of glucose by Trypanosomatids, *FEMS Microbiol. Lett.* **51**:187–192.

Cazzulo, J. J., Nowicki, C., Santomé, J. A., Wernstedt, C., and Hellman, U., 1988b, Amino acid composition and *N*-terminal sequence of the NADP-linked glutamate dehydrogenase from *Trypanosoma cruzi, FEMS Microbiol. Lett.* **56**:215–220.

Cazzulo, J. J., Couso, R., Raimondi, A., Wernstedt, C., and Hellman, U., 1989a, Further characterization and partial amino acid sequence of a cysteine proteinase from *Trypanosoma cruzi, Mol. Biochem. Parasitol.* **33**:33–41.

Cazzulo, J. J., Cazzulo Franke, M. C., and Franke de Cazzulo, B. M., 1989b, On the regulatory properties of the pyruvate kinase from *Trypanosoma cruzi* epimastigotes, *FEMS Microbiol. Lett.* **59**:259–264.

Cazzulo, J. J., Hellman, U., Couso, R., and Parodi, A. J. A., 1990a, Amino acid and carbohydrate composition of a lysosomal cysteine proteinase from *Trypanosoma cruzi.* Absence of phosphorylated mannose residues, *Mol. Biochem. Parasitol.* **38**:41–48.

Cazzulo, J. J., Cazzulo Franke, M. C., Martinez, J., and Franke de Cazzulo, B. M., 1990b, Some kinetic properties of a cysteine proteinase (Cruzipain) from *Trypanosoma cruzi, Biochim. Biophys. Acta* **1037**:186–191.

Chang, S. L., 1948, Studies on hemoflagellates. IV. Observations concerning some biochemical activities in culture, and respiration of three species of Leishmanias and *Trypanosoma cruzi, J. Infect. Dis.* **82**:109–121.

Contreras, V. T., Salles, J. M., Thomas, N., Morel, C. M., and Goldenberg, S., 1985, *In vitro* differentiation of *Trypanosoma cruzi* under chemically defined conditions, *Mol. Biochem. Parasitol.* **16**:315–327.

Coronel, C. E., Rovai, L. E., Gerez de Burgos, N. M., Burgos, C., and Blanco, A., 1981, Properties of α-hydroxyacid dehydrogenase isoenzymes from *Trypanosoma cruzi*, *Mol. Biochem. Parasitol.* **4**:29–38.

Dahlmann, B., Kopp, F., Kuehn, L., Niedel, B., Pfeifer, G., Hegerl, R., and Baumeister, W., 1989, The multicatalytic proteinase (prosome) is ubiquitous from eukaryotes to archaebacteria, *FEBS Lett.* **251**:125–131.

De los Santos, C., Buldain, G., Frydman, B., Cannata, J. J. B., and Cazzulo, J. J., 1985, Carbon-13 nuclear magnetic resonance analysis of (1-^{13}C)-glucose metabolism in *Crithidia fasciculata* Evidence of CO_2 fixation by phosphoenolpyruvate carboxykinase, *Eur. J. Biochem.* **149**:421–429.

De Sastre, M. B. R., and Stoppani, A. O. M., 1973, Demonstration of Mg^{2+}-activated adenosine triphosphatase in *Trypanosoma cruzi*, *FEBS Lett.* **31**:137–142.

DeSouza, W., 1984, Cell biology of *Trypanosoma cruzi*, *Int. Rev. Cytol.* **86**:197–285.

Eakin, A. E., Bouvier, J., Sakanari, J. A., Craik, C. S., and McKerrow, J. H., 1990, Amplification and sequencing of genomic DNA fragments encoding cysteine proteinases from protozoan parasites, *Mol. Biochem. Parasitol.* **39**:1–8.

Engel, J. C., Franke de Cazzulo, B. M., Stoppani, A. O. M., Cannata, J. J. B., and Cazzulo, J. J., 1987, Aerobic glucose fermentation by *Trypanosoma cruzi* axenic culture amastigote-like forms during growth and differentiation to epimastigotes, *Mol. Biochem. Parasitol.* **25**:1–10.

Frasch, A. C. C., Segura, E. L., Cazzulo, J. J., and Stoppani, A. O. M., 1978a, Adenosine triphosphatase activities in *Trypanosoma cruzi*, *Comp. Biochem. Physiol.* **60B**:271–275.

Frasch, A. C. C., Cazzulo, J. J., and Stoppani, A. O. M., 1978b, Solubilization and some properties of the Mg^{2+}-activated adenosine triphosphatase from *Trypanosoma cruzi*, *Comp. Biochem. Physiol.* **61B**:207–212.

Frydman, B., de los Santos, C., Cannata, J. J. B., and Cazzulo, J. J., 1990, Carbon-13 nuclear magnetic resonance analysis of [1-^{13}C]glucose metabolism in *Trypanosoma cruzi*. Evidence of the presence of two alanine pools and of two CO_2 fixation reactions, *Eur. J. Biochem.* **192**:363–368.

Grieg, S., and Ashall, F., 1990, Electrophoretic detection of *Trypanosoma cruzi* peptidases, *Mol. Biochem. Parasitol.* **39**:31–38.

Gutteridge, W. E., 1981, *Trypanosoma cruzi:* Recent biochemical advances, *Trans. R. Soc. Trop. Med. Hyg.* **75**:484–492.

Gutteridge, W. E., and Rogerson, G. W., 1979, Biochemical aspects of the biology of *Trypanosoma cruzi*, in *Biology of the Kinetoplastida*, Vol. 2 (W. H. Lumsden and D. A. Evans, eds.), pp. 619–652, Academic Press, London.

Gutteridge, W. E., Cover, B., and Gaborak, M., 1978, Isolation of blood and intracellular forms of *Trypanosoma cruzi* and preliminary study of their metabolism, *Parasitology* **76**:159–176.

Itow, S., and Camargo, E. P., 1977, Proteolytic activities in cell extracts of *Trypanosoma cruzi*, *J. Protozool.* **24**:591–595.

Juan, S. M., Cazzulo, J. J., and Segura, E. L., 1976, The pyruvate kinase from *Trypanosoma cruzi*, *Acta Physiol. Latinoam.* **26**:424–426.

Juan, S. M., Cazzulo, J. J., and Segura, E. L., 1977, The citrate synthase from *Trypanosoma cruzi*, *J. Parasitol.* **63**:921–922.

Juan, S. M., Segura, E. L., and Cazzulo, J. J., 1978, Purification and some properties of the NADP-linked glutamate dehydrogenase from *Trypanosoma cruzi*, *Int. J. Biochem.* **9**:395–400.

Krettli, A., Thomas, N., and Eisen, H., 1980, Escape mechanisms of *Trypanosoma cruzi* from the host immune system, in *Les Colloques de l'INSERM. Cancer Immunology and Parasite Immunology*, Vol. 97 (L. Israel, P. Lagrange, and J. C. Salomon, eds.), pp. 553–558, INSERM, Paris.

Kusel, J. P., and Storey, B. T., 1972, Evidence for the presence of two phosphorylation sites in mitochondria isolated from the trypanosomatid hemoflagellate *Crithidia fasciculata*, *Biochem. Biophys. Res. Commun.* **46**:501–507.

Mancilla, R., and Náquira, C., 1964, Comparative metabolism of ^{14}C-glucose in two strains of *Trypanosoma cruzi*, *J. Protozool.* **11**:509–513.

Mancilla, R., Náquira, C., and Lanas, C., 1967, Protein biosynthesis in Trypanosomatidae. II. The metabolic fate of DL-leucine-1-^{14}C in *Trypanosoma cruzi*, *Exp. Parasitol.* **21**:154–159.

Miles, M. A., Souza, A., Póvoa, M., Shaw, J. J., Lainson, R., and Toyé, P. J., 1978, Isozymic heterogeneity of *Trypanosoma cruzi* in the first autochtonous patients with Chagas' disease in Amazonian Brazil, Nature **272**:819–821.

Miles, M. A., Lanham, S. M., de Souza, A. A., and Póvoa, M., 1980, Further enzymic characters of *Trypanosoma cruzi* and their evaluation for strain identification, *Trans. R. Soc. Trop. Med. Hyg.* **74**:221–237.

Montamat, E. E., Arauzo, S. S., and Blanco, A., 1987, Subcellular localization of leucine aminotransferase and α-hydroxyacid dehydrogenase in *Trypanosoma cruzi*, *Mol. Biochem. Parasitol.* **22**:185–193.

Mottram, J. C., North, M. J., Barry, J. D., and Coombs, G. H., 1989, A cysteine proteinase cDNA from *Trypanosoma brucei* predicts an enzyme with an unusual C-terminal extension, *FEBS Lett.* **258**:211–215.

Nosei, C., and Avila, J. L., 1985, Serine hydroxymethyltransferase activity in *Trypanosoma cruzi*, *Trypanosoma rangeli* and American *Leishmania* spp., *Comp. Biochem. Physiol.* **81**:701–704.

O'Daly, J. A., Serrano, L. E., and Rodriguez, M. B., 1983, Free amino acid pool and proteolytic enzymes in *Trypanosoma cruzi* cultured *in vitro*, *Int. J. Parasitol.* **13**:433–440.

Opperdoes, F. R., 1987, Compartmentation of carbohydrate metabolism in Trypanosomes, *Annu. Rev. Microbiol.* **41**:127–151.

Opperdoes, F. R., and Borst, P., 1977, Localization of nine glycolytic enzymes in a microbody-like organelle in *Trypanosoma brucei:* The glycosome, *FEBS Lett.* **80**:360–364.

Opperdoes, F. R., de Rijke, D., and Borst, P., 1976, Reactions involved in energy transfer in trypanosomes. I. Characterization of the mitochondrial adenine nucleotide translocator and the ATPase of *Crithidia luciliae*, *Comp. Biochem. Physiol.* **54B**:7–12.

Opperdoes, F. R., Borst, P., Bakker, S., and Leene, W., 1977, Localization of glycerol-3-phosphate oxidase in the mitochondrion and particulate NAD^+-linked glycerol-3-phosphate dehydrogenase in the microbodies of the bloodstream form of *Trypanosoma brucei*, *Eur. J. Biochem.* **76**:29–39.

Piras, M. M., Henriquez, D., and Piras, R., 1985, The effect of proteolytic enzymes and protease inhibitors on the interaction *Trypanosoma cruzi*–fibroblasts, *Mol. Biochem. Parasitol.* **14**:151–163.

Racagni, G. E., and Machado de Domenech, E. E., 1983, Characterization of *Trypanosoma cruzi* hexokinase, *Mol. Biochem. Parasitol.* **9**:181–188.

Rangel, H. A., Araújo, P. M. F., Repka, D., and Costa, M. G., 1981a, *Trypanosoma cruzi:* Isolation and characterization of a proteinase, *Exp. Parasitol.* **52**:199–209.

Rangel, H. A., Araújo, P. M. F., Camargo, I. J. B., Bonfitto, M., Repka, D., Sakurada, J., and Atta, A. M., 1981b, Detection of a proteinase common to epimastigote, trypomastigote and amastigote of different strains of *Trypanosoma cruzi*, *Tropenmed. Parasit.* **32**:87–92.

Raw, I., 1959, Some aspects of carbohydrate metabolism of cultural forms of *Trypanosoma cruzi*, *Rev. Inst. Med. Trop. Sao Paulo* **1**:192–194.

Rodriguez, F., Ramirez, J. L., and Rangel-Aldao, R., 1990, Gene expression of tubulin during the differentiation of *Trypanosoma cruzi*, Abstracts, VI PAABS Congress, Sao Paulo, Brazil, p. 70.

Ryley, J. F., 1956, Studies on the metabolism of Protozoa. 7. Comparative carbohydrate metabolism of eleven species of Trypanosome, *Biochem. J.* **62**:215–222.

Sakanari, J. A., Staunton, C. E., Eakin, A. E., Craik, C. S., and McKerrow, J. H., 1989, Serine proteases from nematode and protozoan parasites: Isolation of sequence homologs using generic molecular probes, *Proc. Natl. Acad. Sci. USA* **86:**4863–4867.

Schwarcz de Tarlovsky, M. S., Affranchino, J. L., Stoppani, A. O. M., Isola, E. L. D., Lammel, E. M., and Gonzalez Cappa, S. M., 1985, Citocromos en diferentes estadios; cepas y poblaciones de *Trypanosoma cruzi, Rev. Arg. Microbiol.* **17:**121–130.

van Schaftingen, E., Opperdoes, F. R., and Hers, H.-G., 1985, Stimulation of *Trypanosoma brucei* pyruvate kinase by fructose 2,6-bisphosphate, *Eur. J. Biochem.* **153:**403–406.

Stoppani, A. O. M., and Boiso, J. F. de, 1973, Oxidative phosphorylation in *Trypanosoma cruzi, Experientia* **29:**1494–1496.

Stoppani, A. O. M., Docampo, R., Boiso, J. F. de, and Frasch, A. C. C., 1980, Effect of inhibitors of electron transport and oxidative phosphorylation on *Trypanosoma cruzi* respiration and growth, *Mol. Biochem. Parasitol.* **2:**3–21.

Sylvester, D., and Krassner, S. M., 1976, Proline metabolism in *Trypanosoma cruzi* epimastigotes, *Comp. Biochem. Physiol.* **55B:**443–447.

Taylor, M. B., and Gutteridge, W. E., 1986a, The occurrence and distribution of α-hydroxy-acid dehydrogenase in some members of the order Kinetoplastida, *FEBS Lett.* **199:**237–241.

Taylor, M. B., and Gutteridge, W. E., 1986b, The regulation of phosphofructokinase in epimastigote *Trypanosoma cruzi, FEBS Lett.* **201:**262–266.

Taylor, M. B., and Gutteridge, W. E., 1987, *Trypanosoma cruzi:* Subcellular distribution of glycolytic and some related enzymes in epimastigotes, *Exp. Parasitol.* **63:**84–97.

Taylor, M. B., Berghausen, H., Heyworth, P., Messenger, N., Rees, L. J., and Gutteridge, W. E., 1980, Subcellular localisation of some glycolytic enzymes in parasitic flagellated protozoa, *Int. J. Biochem.* **11:**117–120.

Teixeira, A. R. L., and Santana, J. M., 1989, *Trypanosoma cruzi:* Endocytosis and degradation of specific antibodies by parasite forms, *Am. J. Trop. Med. Hyg.* **40:**165–170.

Toner, J. J., and Weber, M. M., 1972, Respiratory control in mitochondria from *Crithidia fasciculata, Biochim. Biophys. Res. Commun.* **46:**652–660.

Urbina, J. A., 1987, The phosphoenolpyruvate carboxykinase of *Trypanosoma* (*Schizotrypanum*) *cruzi* epimastigotes: Molecular, kinetic, and regulatory properties, *Arch. Biochem. Biophys.* **258:**186–195.

Urbina, J. A., and Azavache, V., 1984, Regulation of energy metabolism in *Trypanosoma* (*Schizotrypanum*) *cruzi* epimastigotes. II. NAD^+-dependent glutamate dehydrogenase, *Mol. Biochem. Parasitol.* **11:**241–255.

Urbina, J. A., and Crespo, A., 1984, Regulation of energy metabolism in *Trypanosoma* (*Schizotrypanum*) *cruzi* epimastigotes. I. Hexokinase and phosphofructokinase, *Mol. Biochem. Parasitol.* **11:**225–239.

Urbina, J. A., Osorno, C. E., and Rojas, A., 1989, The central role of phosphoenol pyruvate carboxykinase in the energy metabolism of *Trypanosoma cruzi*. Studies using 3-mercaptopicolinic acid, *Mem. Inst. Oswaldo Cruz* **84**(Suppl. II):59.

Von Brand, Th., 1979, *Biochemistry and Physiology of Endoparasites,* Elsevier/North Holland, Amsterdam.

Walter, R. D., and Ebert, F., 1979, Evidence for NADH- and NADPH-linked glutamate dehydrogenases in *Trypanosoma cruzi* epimastigotes, *J. Protozool.* 26:653–656.

Williamson, J., and Desowitz, R. S., 1961, The chemical composition of trypanosomes. I. Protein, amino acid and sugar analysis, *Exp. Parasitol.* **11:**161–175.

Yoshida, N., and Camargo, E. P., 1978, Ureotelism and ammonotelism in Trypanosomatids, *J. Bacteriol.* **136:**1184–1186.

Zeledón, R., 1960, Comparative physiological studies on four species of hemoflagellates in culture. V. Transaminases, *Rev. Brasil. Biol.* **20:**309–414.

Chapter 8

Malarial Lipids

An Overview

Henri J. Vial and Marie Laure Ancelin

1. LIPIDS: WHAT FOR?

The process of erythrocyte invasion by the malaria-causing protozoan, *Plasmodium,* is associated with the sudden appearance of a membrane-delimited foreign body and the growth of the intracellular parasite is associated with a dramatic increase in the total membrane area. This heterotrophic organism installs structures to converse with the host erythrocyte, producing appropriate changes by remote control in the erythrocytic host membrane to assure its own nutrition, maturation, and survival against the host immune system. *Plasmodium* has proved to be a useful and rich model because it concentrates complex biochemistry and very intricate mechanisms to escape host immune responses.

Growth of the parasite within the host erythrocyte and the subsequent formation of up to 6–20 merozoites require a large increase in total membrane. The invasive stage of malaria, the ovoid merozoite, measures about 1–2 μm across, is surrounded by a complex of three membranes (pellicle), and contains a nucleus, mitochondria, endoplasmic reticulum, a cytostome, and apical organelles called rhoptries and micronemes. The latter organelles probably contain mem-

Henri J. Vial and Marie Laure Ancelin Membrane Interactions, CNRS URA 530, 34095 Montpellier, France.
Subcellular Biochemistry, Volume 18: Intracellular Parasites, edited by J. L. Avila and J. R. Harris. Plenum Press, New York, 1992.

brane-active substances that are released during the invasion process and act on the host cell membrane to induce endocytosis. This process gives rise to a parasitophorous vacuolar membrane (PVM) that surrounds the intracellular parasite (Aikawa and Miller, 1983; Aikawa, 1988; Mitchell and Bannister, 1988). The PVM is generally thought to originate from the plasma membrane of the erythrocyte, but recent data indicate that parasitic material derived from the apical organelles, including lipidic material, could be added when the membrane of red cells expands inward (Mikkelsen *et al.*, 1988; Mitchell and Bannister, 1988; Bannister and Mitchell, 1989).

The PVM and plasma membrane are retained throughout the entire intraerythrocytic developmental cycle and consequently some substances must be physically added to the PVM to allow its further expansion. Another consequence is that the cytosolic compartment of the intraerythrocytic parasite is separated from the blood plasma by the erythrocytic membrane of the host and by the two very closely apposed membranes, an outer PVM and the parasite plasma membrane. Hence, the individual properties of each membrane are probably of crucial importance when the parasite has to incorporate supplies of ions or organic precursors or eliminate catabolic products. During parasite maturation, complex membranous material accumulates in the future merozoites, suggesting that molecules capable of assembling into membranes are formed and stored at the apex of the merozoite (Aikawa and Miller, 1983; Bannister and Mitchell, 1986; Aikawa, 1988; Mitchell and Bannister, 1988).

This dramatic increase in membrane is associated with a considerable increase in total lipid content in the malarial-infected erythrocyte, irrespective of the *Plasmodium species*. These changes were reviewed in great detail by Holz (1977) and also by Sherman (1979, 1984). Since then, the worldwide resurgence of malaria has stimulated a redoubling of efforts, orchestrated by the World Health Organization and other groups, to devise novel strategies of transmission control and to find new treatments for drug-resistant *P. falciparum* malaria and its potentially lethal complications. Here we present the lipids and the malarial parasite. This chapter predominantly deals with *Plasmodium* invading mature mammalian erythrocytes because of the considerable and recent advances in this field, especially concerning the biosynthetic metabolic pathway on which our laboratory has concentrated much effort. In other cases, studies on lipid metabolism of *Plasmodium*-infected cells are complicated by the presence of a lipid metabolism of the nucleated bird erythrocyte or of the internal membranes of rodent reticulocyte. We will make frequent reference to *P. knowlesi,* a malaria parasite of *Macaca* monkeys that can infect man. We have never observed a significant reticulocyte response with this parasite, as reported in other similar systems (Weatherall *et al.*, 1983; Boonpucknavig *et al.*, 1984). This primate model is of particular interest because of its close similarities to the human malaria parasite. We will nevertheless refer to other *Plasmodium* species, though not exhaustively.

We will first describe lipid molecules before investigating the scene of the crime and examining the identity and amounts of lipids in *Plasmodium*. Then, we will focus on the means the parasites use to obtain them, and we will see that the parasitic villain possesses a bewildering variety of metabolic pathways providing its lipids. This metabolism appears to be quite essential to the life of the parasite and we will show examples of what can be done to disorganize it. Last, we examine the subtle changes induced in the lipid organization of the host erythrocyte after *Plasmodium* infection, which has been widely investigated but has led to contradictory results.

2. STRUCTURE OF LIPIDS

Lipids constitute about 50% of the mass of most animal membranes. The most striking feature of lipids is their enormous diversity and complexity with slightly different charge, packing, and hydrophibicity. The three major types of lipids in biological membranes are cholesterol, phospholipids (PL) (which are the most abundant), and glycolipids (Figure 1). All are amphipathic, that is, they have a hydrophobic nonpolar part and a hydrophilic end. Certainly, the major role of these amphipathic molecules is to form the bilayer matrix with which the proteins interact. These lipids are arranged in a continuous bilayer 4 to 5 nm thick, which regulates the activity of membrane-associated enzymes. On the other hand, the triacylglycerols (TAG), although they are not membrane components, constitute the most abundant storage fuel and are known to undergo active metabolism in eukaryotic cells. Recently, lipid and lipid-derived metabolites have been recognized as second messengers in signal transduction (Gennis, 1989).

Phospholipids are the most abundant cell membrane lipids, particularly glycerophospholipids. As illustrated in Figure 1, the hydrophobic moiety of these latter lipids is derived from long-chain fatty acid (FA) molecules, which are esterified via a glycerol molecule and phosphate ester to a hydrophilic base, characteristic of the PL class. The PL composition of cellular membranes varies considerably with respect to the head group, the most common containing a quaternary ammonium (phosphatidylcholine, or PC), a free amino group (phosphatidylethanolamine, or PE), the L-amino acid serine (phosphatidylserine, or PS), or the sugar inositol (phosphatidylinositol, or PI). The volume of the polar head determines the ability to form stable bilayer, for example, a stable bilayer is impossible with PE. The polar head group also determines what global charge the PL can carry. Both PC and PE, which carry a negative charge on the phos-

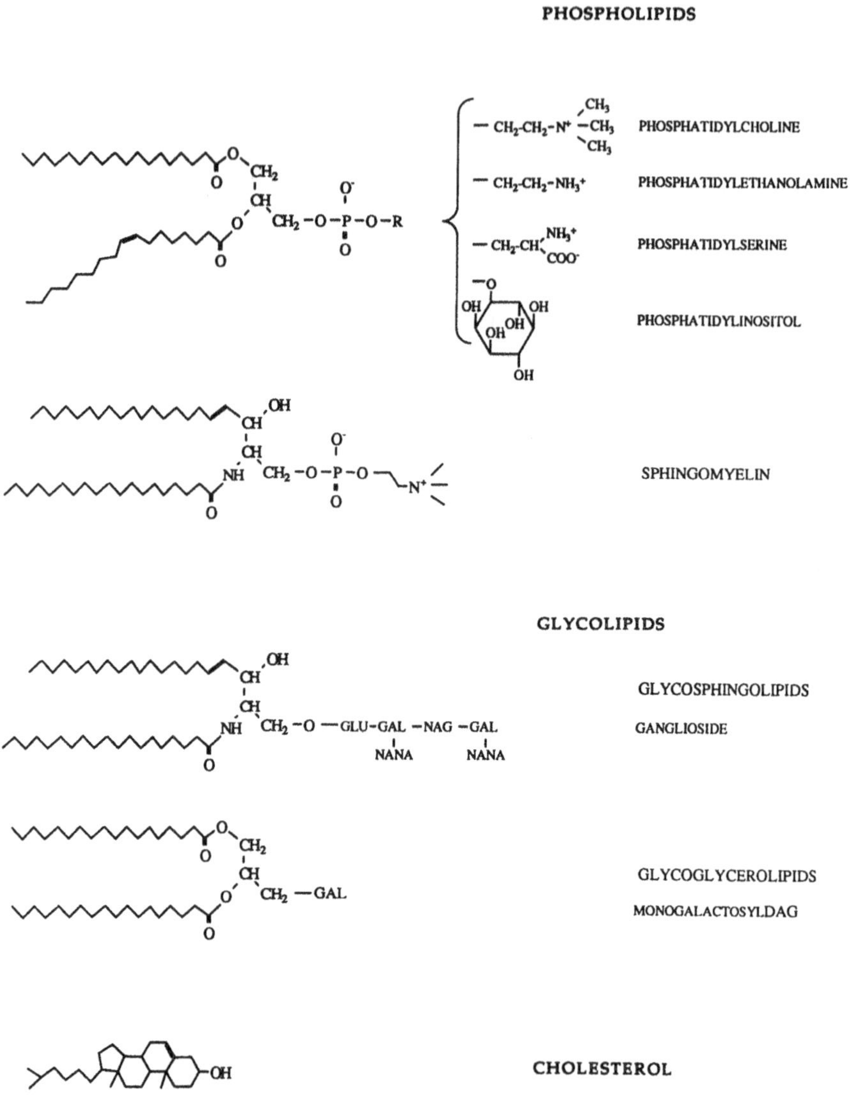

FIGURE 1. The structure of membrane lipids.

phate and a positive charge on the amine, are zwitterionic but electrically neutral. On the other hand, PS and PI exhibit an overall negative charge at neutral pH. The passage of the polar head through the hydrophobic membrane interior is an energetically unfavorable reaction; consequently, the transmembrane movement of PL (flip-flop) does not occur spontaneously. In addition to the variability in the polar head region, the FA chains of PL are subjected to considerable variations in each lipid class, the length of the commonly found FA varies from as few as 12 carbons to as many as 26, while the number of double bonds per FA commonly ranges from none to as many as six. The most commonly found isomers are *cis* isomers, causing a bend of about 30° in the hydrocarbon chain. Evidently, such a *cis*-unsaturated molecule cannot pack so tightly as a saturated hydrocarbon, resulting in a more fluid membrane. This example illustrates how minor modifications of membrane components can change the detailed properties of a membrane without changing the overall structure. The distribution of FA in PL is specific to the PL and membrane type. However, unsaturated FA is esterified predominantly at position 2 of the glycerol and saturated FA occurs mostly at position 1. There is a further variation in glycerophospholipids to which the hydrocarbon chain is attached through an ether linkage (alkyl-PL), which may be unsaturated (plasmalogens) (not illustrated in Figure 1). Sphingomyelin (SM), another type of PL, is condensed from a sphingosine long-chain base with a FA-derivative and phosphorylcholine as the polar substituent.

Glycolipids probably occur in all animal cells. They are distinguished by a polar head group consisting of one or more sugar residues. Glycosphingolipids and gangliosides are related to SM, having a short carbohydrate chain in place of the phosphate and base moities. The gangliosides have a negative charge due to their sialic acids and have important functions, especially on the cell surface.

Cholesterol represents the large class of sterol lipids and is found predominantly in the plasma membrane of mammalian cells (Yeagle, 1987; Lange *et al.*, 1989). This compact and rigid molecule is quite different from other lipids, having a large planar apolar steroid region and one hydroxyl group on the A ring, constituting the polar region. Last, there are a number of other lipid components found in membranes, such as free FA and lyso-PL, monoacyl-, diacyl-, and triacylglycerol polyisoprenoid (Gennis, 1989).

3. LIPIDS OF *PLASMODIUM* AND THE SURROUNDING PLASMA

It has been clearly established that the lipid content of the erythrocyte increases considerably after *Plasmodium* infection and that there are many differences between the lipid composition of malarial parasites and that of the original host erythrocyte. Taken as a whole, considerable increases are observed in total PL as well as in neutral lipids, diacylglycerols (DAG), and TAG. The casual increase in cholesterol content remains controversial and will be discussed (Section 3-2).

We will first briefly describe the many changes occurring in the lipids of the plasma bathing the infected erythrocyte, followed by a detailed description of the lipids in the infected cells. Last, we will describe recent experiments revealing the quite distinct lipid composition of the parasite and the host plasma membrane, as well as subtle changes in the molecular species of the PL of the latter.

3.1. Plasma Lipids during Malarial Infection

The overall plasma lipid content usually shows a net increase in the malarial-infected animal. As a general rule, there is a two- to five-fold increase in the TAG and FA content (Holz, 1977). In absolute amounts of unesterified FA, the changes are relatively small and not of great significance compared with their abundant turnover in plasma. Nevertheless, alterations in their relative proportions have been shown (Angus *et al.*, 1971b). Total cholesterol may increase or decrease slightly but there is a surprisingly large increase in free cholesterol occurring at the expense of the cholesterol ester (Angus *et al.*, 1971a; Holz, 1977; McClean *et al.*, 1976a; Majumdar *et al.*, 1981; Ononghbu and Onyeneke, 1983). The consequence of increases in the free cholesterol/esterified cholesterol ratio for the erythrocytes is discussed below.

The plasmatic PL level can increase, but by less than 100%. In monkey and man, PC represents about 75% of the total PL, followed by SM and PE. Phosphatidylserine and phosphatidylinositol are quasi-absent from plasma (Angus *et al.*, 1971a; McClean *et al.*, 1976a; Ononghbu and Onyeneke, 1983; personal observations). Plasma lipids are present as lipoprotein complexes: FA and lyso-PL are bound to plasma albumin, whereas the other PL form more complex structures subdivided according to their density into four major classes. Chylomicrons and very-low-density lipoproteins (VLDL) contain large quantities of TAG. Low-density lipoproteins (LDL) carry the largest portion of the total blood cholesterol, and high-density lipoproteins (HDL) most of the remainder. Phospholipids are distributed mainly between HDL, LDL, and to a lesser extent VLDL (Jackson *et al.*, 1976).

In mice, the TAG-rich lipoproteins, that is, chylomicrons and VLDL, increase during early malarial infection and correlate with the parasitemia. This is followed by quantitative changes in the distribution of cholesterol-rich lipoproteins, involving an increase in LDL and a decrease in HDL (Dei-Cas *et al.*, 1986). This variation is also observed in the *P. knowlesi*-infected *Macaca* (Sainte-Marie and Vial, unpublished results) and in man during *P. falciparum, P. malariae, P. vivax,* and *P. ovale* infection (Maurois *et al.*, 1979; Dei Cas *et al.*, 1986). This serious dyslipoproteinemia could originate from malarial infection, which quickly induces lipolysis of adipose tissue as a result of stimulation of adenylcyclase activity (Eisenberg *et al.*, 1975). Mobilized FAs are released into the serum and transformed into TAG by the liver, resulting in steatosis (Angus *et al.*, 1971d; Maurois *et al.*, 1981). This is quickly followed by an increase in VLDL secretion. Moreover, a decrease in the lipoprotein lipase activity of adipose tissue intensifies the accumulation of chylomicrons and

VLDL in plasma (Al Rifai *et al.*, 1982; Maurois *et al.*, 1980), whereas an inhibition of serum lecithin-cholesterol acyl transferase leads to an increase in the free cholesterol/cholesterol ester ratio and a decrease in HDL (Maurois *et al.*, 1985). This severe dyslipoproteinemia probably has important physiological effects (Dei-Cas *et al.*, 1986).

3.2. What Is the Role of Cholesterol?

Most studies measuring the cholesterol content of *Plasmodium*-infected erythrocytes have reported an increased concentration of this sterol following parasitization. However, in rodents the results are complicated by the presence of reticulocytes, which appear at the same time as the parasites (reticulocytes contain internal membranes that can account for the cholesterol increases) (Holz *et al.*, 1977). In mature erythrocytes, the cholesterol increase appears to be random in certain monkeys (Angus *et al.*, 1971c; Majumdar *et al.*, 1981; Vial *et al.*, 1984a).

There are several possible sources of cholesterol in infected cells. First, it could be synthesized from acetate via a pathway involving the regulatory enzyme 3-hydroxy-3-methyl glutaryl-coenzyme A reductase. However evidence now seems to suggest that the malarial parasite is incapable of *de novo* synthesis of sterols, as measured by incorporation of [^{14}C]acetate, and $^{3}H_2O$, whatever the parasite developmental stage (Trigg, 1968; Cenedella *et al.*, 1969; Holz, 1977; Vial *et al.*, 1984a). This is consistent with the nonsignificant activity of the regulatory enzyme hydroxymethylglutaryl-CoA reductase in *Plasmodium* (Vial *et al.*, 1984a). Second, it is well known that erythrocyte cholesterol is in dynamic equilibrium with the cholesterol of plasma lipoproteins, so that exchanges could occur between the parasite, the erythrocyte, and plasma. We have in fact shown that *P. knowlesi*= and *P. falciparum*-infected erythrocytes exchange cholesterol with plasma at the same rate and level as normal erythrocytes. The parasitized cells do not show preferential uptake of free cholesterol from plasma lipoproteins (Vial *et al.*, 1984a). We have described the great changes in plasma cholesterol that result in large increases in free cholesterol at the expense of the cholesterol ester. This free and ready exchange of cholesterol between plasma lipoproteins and the erythrocyte is probably the origin of the increase.

It follows that the cholesterol increase of the erythrocyte *in vivo* is not specific to the infected erythrocyte but also involves uninfected erythrocytes, that is, it only reflects the elevated plasmatic free cholesterol. This is corroborated by the notably greater amount of cholesterol in nonparasitized erythrocytes of infected monkeys (Angus *et al.*, 1971c). In the case of *P. gallinaceum*, which infests chickens, the PL concentration of red cells varies with the parasitemia, whereas the cholesterol content remains high even after parasite clearance (Seed and Kreier, 1972). Last, the formation of the acanthocyte, which is an irreversible deformation of the discocyte (related to an increase in the cholesterol/PL ratio) during *P. berghei* infection confirms that the cholesterol increase is not restricted to the infected erythrocyte (Dei Cas *et al.*, 1986).

Thus, an increase in plasma-free cholesterol induces an increase in the cholesterol of all erythrocytes (infected or not), which raises the question as to where this cholesterol overloading occurs in infected erythrocytes. Rock *et al.* (1971b), using the rather imprecise saponin lysis method (Beaumelle et al., 1987), previously found a very low level of cholesterol in parasite membranes. This was confirmed in the *P. chabaudi*–mouse system (Wunderlich *et al.*, unpublished data; Table I). This indicates that the intracellular parasite does not possess cholesterol, and when the increase in cholesterol occurs, it is probably confined to the erythrocyte membrane, which is capable of absorbing such an

Table I
Phospholipid Composition of the Plasma Membranes of Normal and Infected Erythrocytes and Parasites[a]

	Plasma membrane of normal erythrocyte	Plasma membrane of infected erythrocyte	Free parasite	Plasma membrane of erythrocyte	Plasma membrane of infected erythrocyte	Whole infected erythrocyte
PC	4.6 ± 0.5	48.0 ± 2.3	59.4 ± 1.7	39.3 ± 3.0	40.2 ± 3.2	49.3 ± 1.6
(16:0/18:2)				(40.2)	(49.0)	(48.7)
(16:0/16:0)				(7.0)	(5.1)	(3.3)
(16:0/18:1)				(30.6)	(28.1)	(25.6)
(18:0/20:4)				(5.0)	(1.6)	(2.4)
PE	26.8 ± 0.8	25.5 ± 1.7	24.4 ± 1.2	26.3 ± 2.3	25.3 ± 2.6	35.2 ± 0.7
(16:0/18:2)				(23.3)	(43.0)	(41.3)
(16:0/18:1)				(14.4)	(20.4)	(21.2)
(16:0/20:4)				(20.7)	(5.7)	(5.5)
(18:0/20:4)				(9.9)	(2.1)	(2.1)
(18:1/20:4)				(7.2)	(1.6)	(2.6)
SM	12.5 ± 0.5	11.4 ± 1.4	3.9 ± 0.6	16.2 ± 1.2	16.7 ± 1.5	3.6 ± 0.3
PS	9.6 ± 0.4	10.7 ± 2.3	5.8 ± 0.5	18.2 ± 2.0	18.1 ± 1.7	10.2 ± 1.15
PI	3.1 ± 0.2	3.8 ± 0.7	5.6 ± 0.9			
Lyso PC	1.8 ± 0.2	0.5 ± 0.3	1.0 ± 0.4			
Cholesterol /PL (mole/mole)	0.95	0.94	0.09	1.02		0.25

[a]The phospholipid composition is expressed as mol% of total phospholipids. The mouse erythrocytes, infected or not by the trophozoite stage of *P. chabaudi*, were fractionated in free parasites and host cell membranes by glycerol-enhanced osmotic shock (Wunderlich *et al.*, 1987). The erythrocyte membrane of normal and *P. knowlesi*-infected cell (mature stages) were isolated on Affi-gel beads (Van der Shaft *et al.*, 1987; Vial *et al.*, 1989b). Infected cells correspond to 100% trophozoite-infected cells. The diacyl species composition of PC and PE, obtained by reverse HPLC and GLC, are shown in brakets, and are expressed as the mol% of the total molecular species. In this case, values were obtained from normal or 20–30% infected-erythrocytes at different stages of development of the parasite. Only the main molecular species or main alterations are shown (see Simoes *et al.*, 1990).

increase (see also Section 7). However, note here that feeding with a cholesterol-rich diet decreases the chloroquine sensitivity in *P. berghei* and *P. chabaudi* infections (Ginsburg *et al.*, 1988). The causes of this phenomenon need further investigation.

3.3. Neutral Lipids and PL in Infected Erythrocytes

In the normal erythrocyte, FA and the neutral lipids are at a very low level, and TAG is barely detectable. After infection by *P. knowlesi,* the level of DAG rose by 9- to 13-fold while that of FA and TAG increased 2- to 7-fold (Angus *et al.*, 1971c; Laser *et al.*, 1975; Majumdar *et al.*, 1981; Beaumelle and Vial, 1986). Linoleic acid, which represents less than 3% of the FA fraction, was dramatically increased to 10% for FA and 18–22% for DAG and TAG. This occurs particularly at the expense of myristic acid, which is abundant in these lipids in the original cells (Beaumelle and Vial, 1986).

One of the main characteristics of malarial infection is the considerable increase in PLs, which are already abundant in the original cells. Their levels rise by about 50% from the ring stage to 500–700% in the late form, just before *P. knowlesi* merozoite release (Angus *et al.*, 1971c; Majumbar *et al.*, 1981; Beaumelle and Vial, 1986; Vial *et al.*, 1984a; Van der Schaft *et al.*, 1987). There are clear differences in the composition of individual PLs distinguished by their polar heads between the whole infected erythrocyte and the original erythrocyte. The latter contains four major types of PL, that is, PC (30–45%), PE (25–40%), and SM and PS (10–20% each). Phosphatidylinositol and other PLs such as phosphatidic acid (PA), cardiolipin, and lyso-PL are present in trace amounts ($< 3\%$). After infection, the proportion of the two main PLs is reinforced, since PC and PE together account for 75–85% of the parasitic PL. The PI is greatly increased after infection, both in terms of its percentage and much more in terms of its absolute amount, bearing in mind the global increase in PL. On the other hand, SM and PS decrease sharply (Rock *et al.*, 1971b; De Zeeuw *et al.*, 1972; McClean *et al.*, 1976a; Van der Schaft *et al.*, 1987; Moll *et al.*, 1990a).

The considerable increase in PL is associated with a comparable increase in the amount of FA bound to the glycerol backbone. It is again striking that the overall FA composition of the erythrocyte is changed after infection. Data in the literature agree with the amount of the saturated FA palmitic acid that is increased at the expense of stearic acid in the *P. knowlesi* system. In terms of polyunsaturated acids, there is an increase in both oleic and linoleic acid, whereas arachidonic acid decreases sharply (Angus *et al.*, 1971c; Rock, 1971a; Beaumelle and Vial, 1986). It should be noted that a decreased proportion of FA does not usually indicate a decrease in the total amount of FA, due to the large increase in the total lipid content. The particular attention paid to the hemolytic role of *cis*-vaccenic acid (Holz, 1977; Holz *et al.*, 1977) appears to have been unjustified (Maurois *et al.*, 1981; Beaumelle and Vial, 1986). Extrapolation of these data to other *Plasmodium* species (see for review Holz, 1977; Stocker *et al.*, 1987) would be quite risky, especially because of the differences in FA metabolism between Plasmodium species.

Studies of changes in the FA profile of different lipid classes (including the monitoring of plasmalogens) have been performed only on the duck parasite (Beach *et al.*, 1977) and by our group during *P. knowlesi* infection. The FA compositions of choline-containing PLs (PC and SM) are not as altered in their FA pattern as other PLs (PE, PS, PI, and lyso-PC). The rise in linoleic acid is recovered in each PL except PC. The unsaturation index is globally unaltered, but is altered for each individual PL except PC and SM (Beaumelle and Vial, 1986). *Plasmodium knowlesi*–intraerythrocytic development is associated with a decrease of more than 50% in the relative amounts of both 1-alkyl-2 acyl- and 1-alkenyl-2-acyl-PE, which included the plasma membrane of trophozoite-infected erythrocytes. But the level of 1-alkenyl-2-acyl is increased into PI (Beaumelle and Vial, 1986; Simoes *et al.*, 1990).

3.4. Glycolipids in Infected Erythrocytes

There are some lipid molecules (the glycolipids) that show a most striking and consistent asymmetrical distribution in the plasma membranes of animal cells. They are found only in the outer half of the bilayer and their sugar groups are exposed at the cell surface. The oligosaccharides that form the head groups of glycolipid molecules can be complex. This, together with their strategic position on the cell surface, suggests that they may play a part in intercellular communication, but this function has not yet been demonstrated. Glycolipids occurring in animal tissues are derived from both sphingosine (glycosphingolipids) or glycerol (glycoglycerolipids), whereas in bacteria and plants almost all glycolipids are derived from glycerol (Slomiany *et al.*, 1987).

Presently, very little information is available concerning glycolipids in malaria-infected erythrocytes, although these minor lipids could have crucial functions in the antigenic and contact properties of infected cells. Their existence in *Plasmodium* is now established (Sherwood *et al.*, 1986), and the absence of ^{35}S or ^{32}P incorporation into glycolipids probably rules out the possibility of the presence of sulfolipids or phosphosphingolipids. The lack of negative charge in the [^{3}H]glucosamine compounds would suggest that there are no gangliosides. This also agrees with a previous study revealing that *P. knowlesi* did not contain or synthesize sialic acids (Schauer *et al.*, 1984).

Concerning the biosynthesis of glycolipids in *Plasmodium*, it was notably shown that there are stage-specific differences in glycolipid synthesis in the sexual and asexual erythrocytic stages and vector stages of the parasite life cycle. For the asexual erythrocytic stage, plasmodial glycolipid synthesis occurs concomitantly with glycoprotein synthesis and both increase during schizogony, with the most intensive synthesis at the schizont stage (that corresponds to a time of maximum membrane and glycoprotein synthesis), suggesting that these compounds may be membrane components (Sherwood *et al.*, 1986). Nevertheless, the precise glycolipid structure, cellular location, antigenicity, and biosynthesis as well as their role in pathogenesis remain to be further explored.

3.5. Lipids of Various Fractions of the Infected Erythrocyte

The lipid ccomposition of the infected erythrocyte reflects the separate contribution of the erythrocyte and the intracellular parasite itself, which is composed of different membranes separating various compartments. There is no significant reticulocytosis in mammalian models and it is now possible to concentrate infected erythrocytes to almost 100%. The following questions have recently received partial responses:

1. What is the composition of the pure parasite itself?
2. Does the lipid composition of plasmodial membranes differ from that of the host erythrocytic membrane, considering that this membrane can be modified after infection?
3. Are these changes specific to the infected erythrocyte, since lipids of the plasma also show great alterations during malarial infection?

Comparing the PL composition of the isolated *Plasmodium* with that of the original erythrocyte yields the same observations as above, that is, a larger proportion of PC (45–50%), PE, and PI (which reaches 4–11%). On the other hand, SM is present at only 4% and PS becomes nearly nonexistent (Rock *et al.*, 1971b; De Zeeuw *et al.*, 1972; McClean *et al.*, 1976b). Similar observations are obtained by extrapolating the parasite composition from the amounts found in the whole infected erythrocyte minus the PL present in the host erythrocyte membrane (Van der Schaft *et al.*, 1987). This general observation can also be roughly applied to other malarial parasites, notably *P. lophurae* and *P. berghei* (Holz, 1977).

Nevertheless, it is more difficult to compare the PL composition of the isolated free parasite and the erythrocyte host membrane, since the composition of the latter can be changed after infection (see Section 7). Wunderlich *et al.* (1987) developed a new procedure for concomitant isolation of viable parasites and host cell plasma membranes from *P. chabaudi*-infected erythrocytes with very low cross contamination. A collaborative work has shown that the absence of cholesterol in the purified parasite is striking. The low level or the absence of SM and PS in *Plasmodium* is confirmed (Wunderlich *et al.*, unpublished data; Table I).

More importantly, isolation of the plasma membrane gave highly interesting results (Table I). The molar ratio of cholesterol/PL was not changed when plasma membrane originated from mouse erythrocytes that had been infected by *P. chabaudi,* indicating the same stoichiometry between these two lipids in normal and infected erythrocytes. In the same way, the PL composition of the erythrocyte plasma membrane, when referring to the PL class, is not changed after *P. chabaudi* infection, indicating that there is no modification in lipid composition of the host membrane with respect to PL content, class distribution, and cholesterol content. This well corroborates other studies restricted to PL distribution in *P. knowlesi* (Joshi *et al.*, 1987; Van der Schaft *et al.*, 1987) and *P. falciparum* (Joshi and Gupta, 1988) using Affi-gel beads to isolate the erythrocyte plasma membrane.

More subtle analysis, however, by reversed high performance liquid chromatography, complemented by gas liquid chromatography, revealed striking changes in the molecular species of the constitutive PLs, that is, in their constitutive FAs (Simoes *et al.*, 1990; Table I). The PC molecular species composition of the isolated host erythrocyte membrane is identical to that of the *P. knowlesi*-infected cell. However, this composition differs from that of unparasitized erythrocytes that had been recovered from the same batch of infected monkey blood: there is a remarkable increase in 16 : 0/18 : 2-PC, which seems to be compensated for by a decrease in 16 : 0/16 : 0-, 16 : 0/18 : 1-, and 18 : 0/20 : 4-PC. Marked differences are also observed for PE (the two lipid classes, PC and PE, exhibit quite distinct species composition in erythrocytes) of the plasma membrane of unparasitized and *P. knowlesi*-parasitized erythrocytes. There is a marked increase in 16 : 0/18 : 2- and 16 : 0/18 : 1-PE while species containing 20 : 4, that is 16 : 0/20 : 4, 18 : 0/20 : 4, and 18 : 1/20 : 4 are severely decreased. Moreover, the fact that unparasitized erythrocytes from infected monkeys exhibit some differences when compared to erythrocytes from never-infected monkeys is intriguing. This may be correlated with the serum protein-mediated FA alteration of uninfected red cells from malaria-infected blood (Beaumelle and Vial, 1988c). Generally, these results point out that an intense dynamic of PL must operate in the infected cells (see Section 7). Also, since the different molecular species may be associated with specific lipid packing, it may be conceivable that by changing the molecular species composition of the host cell membrane, the parasite manipulates its physiological properties in order to create the conditions that will allow the uptake of nutrients and assure its survival.

It has recently been shown that there is a decrease in SM content, a decline in the cholesterol/PL ratio, and a reduced exchange rate of cholesterol in the host membrane of *P. falciparum*-infected cells (Maguire and Sherman, 1990). These results conflict with those showing an unaltered PL composition in *P. falciparum* (Schwartz *et al.*, 1987; Joshi and Gupta, 1988) and *P. knowlesi* (Joshi *et al.*,

1987; Van der Schaft *et al.*, 1987) and also unmodified exchangeability of cholesterol in 30–60% parasitized *P. knowlesi*-infected erythrocytes (Vial *et al.*, 1984a). This discrepancy could result from the fact that *P. falciparum* cultures often include outdated blood and also from the numerous subsequent treatments of the cells for the synchronization, concentration, and plasma membrane isolation. As an example, we have found that only a portion of erythrocyte cholesterol was exchangeable with plasma cholesterol in a 30–60% parasitized *P. knowlesi*-infected suspension (Vial *et al.*, 1984a) due to an equilibrium between cholesterol in the membrane and in the plasma (Lange and D'Alessandro, 1977). Thus, the observed changes could well result from modified properties of old cells after numerous treatments, including brief exposure to very low ionic strength iso-osmotic buffer, which has been shown to make all of the cell cholesterol available for exchange (Lange *et al.*, 1981).

However, although in our opinion this loss of cholesterol probably does not occur *in vivo* in the isolated host membrane, it probably reveals modified properties of the infected erythrocytes. Besides, the loss of SM and cholesterol is not so intriguing due to the special interaction of these two lipids and their covariance in subcellular fractions (Yeagle, 1985, 1987; Lange *et al.*, 1989).

4. BIOSYNTHETIC METABOLIC PATHWAYS OF GLYCEROLIPIDS

Immediately after the invasion process, the intracellular parasite must procure or synthesize many compounds, not only to ensure its survival but also to allow its rapid multiplication, which requires enormous energy. Any lipid metabolism consideration of the malarial parasite must begin with an examination of the host erythrocyte and its plasma environment, since their lipid composition and lipid metabolism can govern the availability of lipid precursors for the parasite. The erythrocytes that have received the most attention are (1) mature mammalian erythrocytes invaded by *P. knowlesi* or *P. falciparum;* (2) erythrocytes or reticulocytes of rats and mice invaded by *P. berghei*, and (3) mature, nucleated duck erythrocytes invaded by *P. lophurae*. Due to the presence of mitochondria and sparsely distributed rough endoplasmic reticulum, or of a nucleus, the lipid metabolism of each erythrocyte-type has special characteristics. For example, rat reticulocytes probably retain some capacity for lipid synthesis as well as for lipid exchange. Mature nucleated avian erythrocytes can apparently carry out *de novo* PL, but not FA synthesis (Holz, 1977; Sherman, 1984). Here we will mainly deal with mature mammalian erythrocytes, which most closely resemble human malaria and for which recent, nonambiguous data are available.

It would be reasonable to consider the possibility that the essential PLs are imported in the form of diacylglycerol (DAG), TAG, or as complex PL. Results concerning this possibility are presented in Section 4.6. Indeed, there is an intense capacity for transbilayer movement of PL in the plasma membrane of the infected erythrocyte (see Section 6.3), as well as uptake of vesicular PL, which is followed by considerable dynamic movements of these molecules within infected erythrocytes (see Section 7). However, the plasmatic concentrations of PS, PI, and PE are very low, and no assurance and *a fortiori* no quantification have been obtained concerning the procurement of preformed lipid molecules from serum lipoproteins *in situ*.

The mature erythrocyte only has a very low turnover of its membrane lipids and is devoid of any lipid biosynthetic activity (Van Deenen and De Gier, 1975). In the following section, we will discuss the biosynthetic capacity of *Plasmodium*, which operates at the expense of precursors originating from the plasma.

Glycerophospholipid biosyntheses follow biochemical pathways that involve the activated form of either PA (i.e., CDP-diacylglycerol) or of a polar head (e.g., CDP-choline). The synthesis begins with two molecules of FA being transferred from a donor (usually a fatty acyl-CoA) to a molecule of glycerophosphate forming the most primitive key intermediate PL, PA. In subsequent steps, the metabolic pathways diverge depending on the organism and the type of PL. In prokaryotes, a phosphorylated nucleotide, CMP, is attached to PA, and then another molecule of *sn*-glycerol-3-phosphate or the amino acid serine is added. In this last case, PS is synthesized and finally a molecule of CO_2 is eliminated from the serine, yielding PE. In mammalian cells, only PI (and trace levels of phosphatidylglycerol and cardiolipin) appear to be synthesized from CDP-DAG. Base exchange is believed to be the principal route for PS biosynthesis. On the other hand, most of the PA can be converted into the intermediate DAG by the enzyme phosphatidate phosphohydrolase. The DAG species can then be used for TAG synthesis or, more frequently, lead to the biosynthesis of PC or PE after the condensation of DAG with the activated form of the polar head, that is, CDP-choline or CDP-ethanolamine (Bell and Coleman, 1980; Kennedy, 1986; Yeagle, 1987).

There is no net synthesis of major PL and other lipids in mature mammalian erythrocytes. Their turnover is low and can be attributed to two main processes: (1) the exchange of PL between plasma lipoproteins and membrane, and (2) the acylation of lyso-PL originating from surrounding plasma and possibly from membrane. The relative contribution of these two mechanisms appears to be of the same order of magnitude (Nelson, 1972; Van Deenen and de Gier, 1975). Therefore, the large increase in PL content observed during intracellular maturation of *Plasmodium* is related to the specific metabolism of the parasite.

Lipid metabolism in the *Plasmodium*-infected mature mammalian erythrocytes was first studied by the groups of Rock and Angus. Our group has carried out detailed studies on this metabolism (see for review Vial *et al.*, 1990) in which incubations resembled physiological conditions as closely as possible (Vial *et al.*, 1989a).

We have never observed incorporation of choline or FA into SM to any appreciable extent, which probably indicates that this PL is not synthesized by *Plasmodium*. the abundant synthesis of DAG and TAG (Vial *et al.*, 1982b, 1989a) is in agreement with the presence of relatively large amounts of these lipids in the cells. Let us note that DAG are no longer considered to be ordinary intermediates of glycerolipid synthesis and degradation since a second messenger function has been attributed to them (Nishizuka, 1988; Gennis, 1989).

The biosynthesis of glycerolipids is far greater in the schizont stage, as observed in synchronous *P. falciparum* (Vial *et al.*, 1982a) and *P. knowlesi* (unpublished results) infected erythrocytes. The problem becomes crucial with PS and TAG, whose syntheses occur essentially when the parasite is mature, and also with PI, whose synthesis is practically nonexistent in the ring stage. No function has yet been attributed to TAG, which is known to be an energy store in mammalian organisms. External TAG does not exchange with the parasite (Vial *et al.*, 1982b) and the FA part of endogenous TAG is not used by the parasite at any time in the whole cycle of *P. knowlesi* (Vial *et al.*, unpublished results). We have identified significant pathways of PL biosynthesis not previously detected or even invalidated (Rock 1971b), the major objective of our work having been to quantify each metabolic pathway. Thus, the scheme that illustrates the biosynthetic ability of *Plasmodium* is now nearly complete (Figure 2).

4.1. Fatty Acid Sources

It is now well accepted that no route for FA biosynthesis exists in *Plasmodium*-invaded mature mammalian erythrocytes or any FA elongation or desaturation processes (Holz, 1977; personal observations). Thus, the only possible source for FA is plasma lipids. Plasma FA and lyso-PL both serve as sources of FA required for the parasite and they act by a competitive mechanism. In infected erythrocytes, lyso-PLs are not key intermediates in the esterification of FA to PL but instead induce a very fast release of the FA portion (Vial *et al.*, 1982b, 1989a). This implies the existence of a powerful lysophospholipase system in contrast to the normal erythrocyte in which lyso-PC behaves as an acyl acceptor and strongly promotes FA incorporation (Van Deenen and De Gier, 1975).

There is still controversy as to whether the cellular permeation of long-chain FA occurs by passive diffusion or energy-independent facilitated diffusion. The rapid filtration method failed to determine the transport mechanism of FA into infected erythrocytes (high adsorption of radioactive FA to cells that are infected or not, no effect of phloretin) (Beaumelle, 1987) although it succeeded in hepatocytes (Stremmel and Berk, 1986). Fatty acid binding proteins (FABP) are assumed to play an important role in the intracellular transport (and possibly

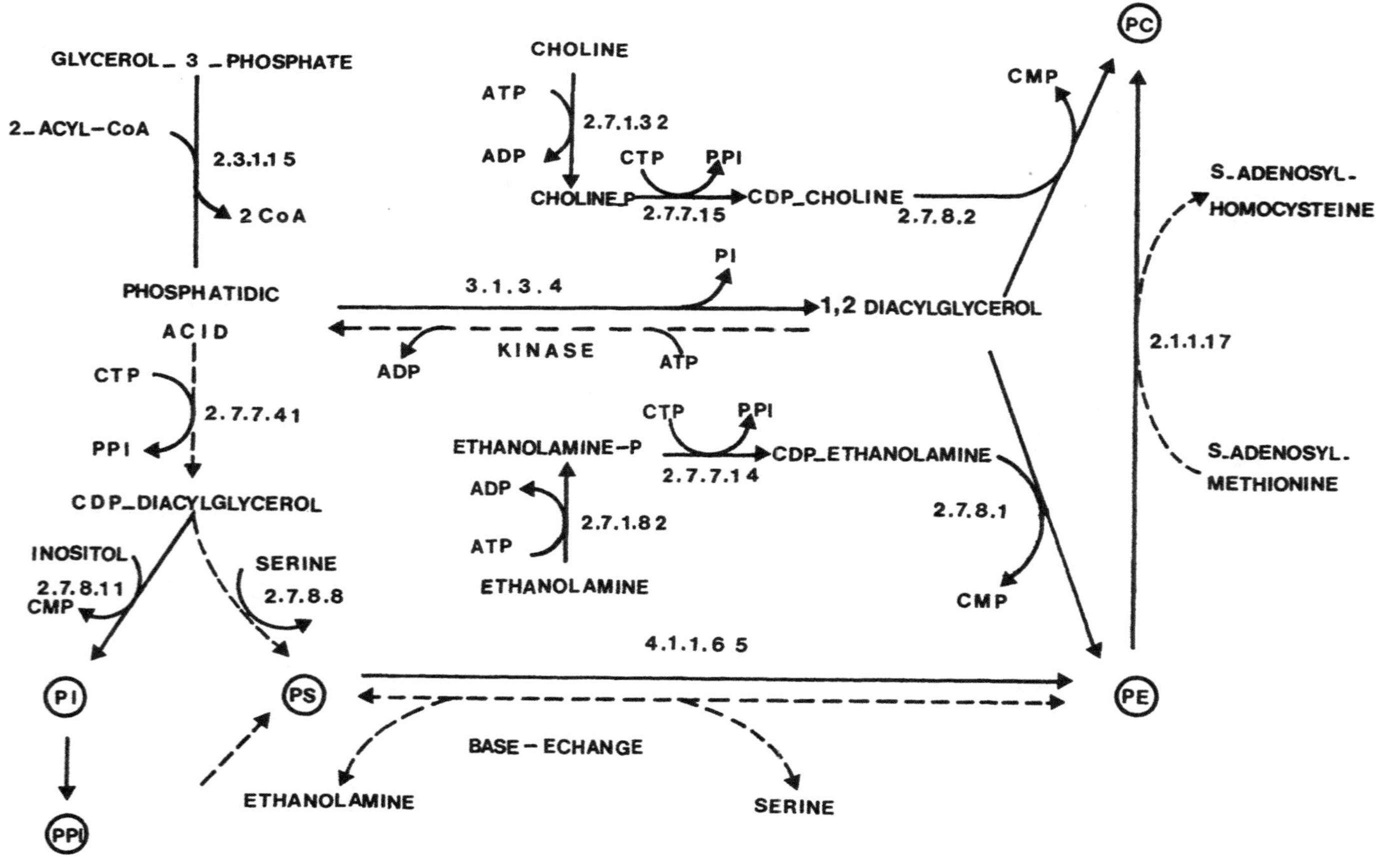

FIGURE 2. Phospholipid biosynthesis in *Plasmodium*-infected erythrocytes. Dotted lines correspond to undemonstrated pathways.

storage) of FA, comparable with the extracellular function of albumin, and could play a crucial role in the cellular lipid metabolism of many eukaryotic cells, especially in cells showing a very active lipid metabolism (Glatz and Veerkamp, 1985). However, we could not detect FABP in *P. knowlesi* homogenates by coelution of labeled FA from a gel-filtration column (Beaumelle, 1987). Nevertheless, a *P. falciparum* protein related to the heat shock protein (hsp 75) that can firmly bind FA (not exchangeable under our conditions, according to Guidon and Hightower, 1986), has been reported (Bianco *et al.*, 1986; Ardeshir *et al.*, 1987).

The activation of FA to acyl-CoA thioesters by acyl-CoA synthetase (EC 6.2.1.3) is required for most cellular reactions involving FA. This enzyme, which metabolizes FA, is probably the first controllable step involving FA. *P. knowlesi* has a high acyl-CoA synthetase activity, about 20 times more active than that of control erythrocytes. The parasite has at least two distinct enzymatic sites, and therefore probably two enzymes for FA activation, one specific for saturated FA and the other one for polyunsaturated species. However it does not seem to possess an arachidonic-specific acyl-CoA synthetase as found in other cells (Beaumelle and Vial, 1988a).

The rate of FA incorporation into the erythrocyte is increased at least 20-fold after *P. knowlesi* infection. Comparable incorporations occur for palmitic, stearic, oleic, linoleic, or arachidonic acid. However, large amounts of the saturated acids are recovered in an unesterified form inside the cells (unpublished results). Linoleic acid is incorporated most intensively into the parasitic PL, which confirms the great importance of this FA (Rock *et al.*, 1971a; Vial *et al.*, 1982b; and unpublished results; Mikkelsen *et al.*, 1988). The positioning of FA studied by phospholipase A_2 (PLA_2)-mediated selective hydrolysis shows that the saturated FAs, palmitate and stearate and the polyunsaturated linoleate and arachidonate are essentially attached to the sn-1 and sn-2 positions on the glycerol backbone, respectively, whereas oleate may occupy both positions (Beaumelle, 1987; Vial *et al.*, unpublished results).

4.2. Biosynthesis of PC

Three pathways have been characterized for PC biosynthesis in *Plasmodium*. The *de novo* pathway (Kennedy pathway), which involves the condensation of DAG with CDP-choline, is the best known (see Figure 3). Choline first enters the infected erythrocytes by a transport-mediated process, whose main kinetic characteristics are the same as in normal cells except for a considerable increase in the V_m (Ancelin *et al.*, 1991). Choline is then activated for its entry into the biosynthetic pathway by choline kinase (EC 2.7.1.32) that is located in the parasitic cytosol (Ancelin and Vial, 1986a). This enzyme does not appear to be capable of catalyzing the phosphorylation of ethanolamine (Ancelin and Vial, 1986b; Ancelin *et al.*, 1986). Further activation to CDP-choline is carried out by

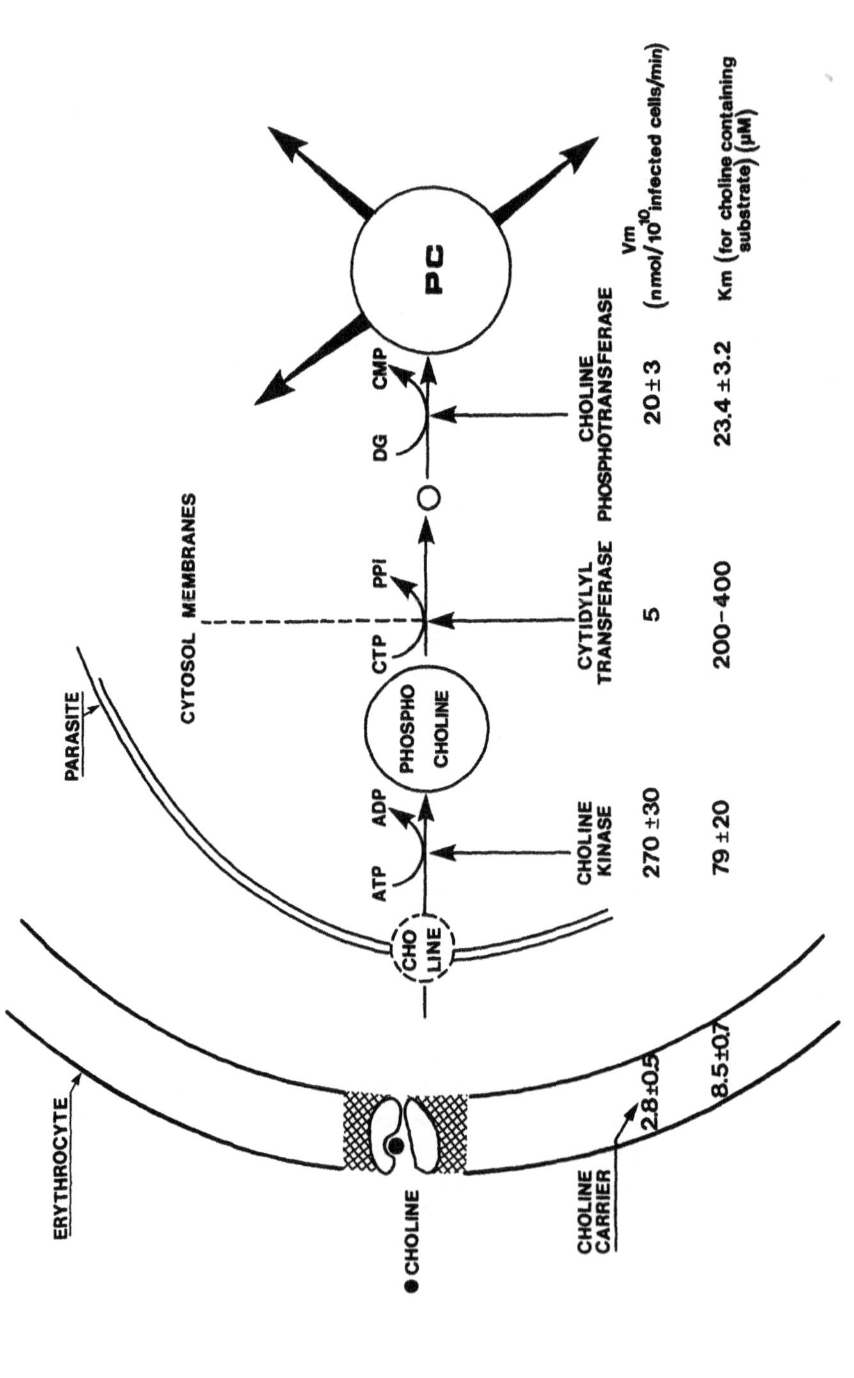
ERYTHROCYTE
CHOLINE
CHOLINE CARRIER
2.8±0.5
8.5±0.7
PARASITE
CHO LINE
ATP
ADP
PHOSPHO CHOLINE
CYTOSOL
MEMBRANES
CTP
PPi
DG
CMP
PC
CHOLINE KINASE
CYTIDYLYL TRANSFERASE
CHOLINE PHOSPHOTRANSFERASE
270 ±30
5
20±3
Vm (nmol/10^10 infected cells/min)
79 ±20
200–400
23.4 ±3.2
Km (for choline containing substrate) (µM)

choline-phosphate cytidylyltransferase (EC 2.7.7.15), which is distributed in both the cytosolic and membrane fractions of *Plasmodium* and is a rate-limiting step in *de novo* PC biosynthesis (Ancelin and Vial, 1989). In other cells, cytidylyltransferase appears to be active primarily in its membrane-bound dephosphorylated form (Vance and Pelech, 1984). Indeed, it has been demonstrated that *de novo* PC biosynthesis may be reversibly regulated *in vitro* by phosphorylation–dephosphorylation of the regulatory enzyme, cytidylyltransferase, in a cyclic AMP-dependent manner (Sanghera and Vance, 1989). Protein kinase C has also been implicated in the regulation of this enzymatic activity in response to the effects of phorbol esters on PC metabolism in various cells (Pelech and Vance, 1989). Recently the phosphorylation of cytidylyltransferase was also demonstrated *in vivo* in HeLa cells (Watkins and Kent, 1990).

To complete the Kennedy pathway, synthesis of PC is achieved through the reaction of CDP-choline with DAG catalyzed by CDP-choline : 1,2 DAG choline phosphotransferase (EC 2.7.8.2). This parasitic membrane-bound enzyme shows high selectivity for molecular species of diacylglycerol and is probably responsible for the various molecular species of parasitic PL (Vial *et al.*, 1984c).

Phosphatidylcholine biosynthesis can also occur by stepwise addition of three methyl groups (probably from S-adenosylmethionine) on the nitrogen of PE (Vial *et al.*, 1982b). In eukaryotic cells, it is only a minor pathway in most tissues, except liver (Vance and Ridgway, 1988). However in *Plasmodium*, PC originating from the methylation of PE could represent a large proportion of the total PC pool (see Section 4.6).

The third pathway of PC biosynthesis to be considered here is the reacylation of external lyso-PC. As seen above, a major part of lyso-PC is readily deacylated after its entry into the cell. However, radioactive [^{14}C]palmitoyl lyso-PC shows a specific incorporation pattern in PC that differs from those of glycerol and oleate (Vial *et al.*, 1989a). The high recovery in PC could indicate a preferential acylation of part of lyso-PC or of the glycerophosphorylcholine released by the lysophospholipase.

4.3. Biosynthesis of PE

The major pathway for PE biosynthesis involves another PL, that is, PS with the participation of PS decarboxylase, which converts the serine head group into the ethanolamine head group. This reaction is plethoric in erythrocytes infected by *P. knowlesi* (Vial *et al.*, 1982b) and *P. falciparum* (Vial *et al.*,

FIGURE 3. Metabolic fate of choline inside *P. knowlesi*-infected erythrocytes. The circles are proportional to the pool size of each metabolite; dotted circle refers to the (low) content in choline. Dark arrows indicate the gradual increase in PC, which accumulates in the parasite membranes.

1982a), giving rise to PE, which in turn can be methylated to PC. In normal cells, the decarboxylation reaction is believed to take place in mitochondria.

Although the rate of PE formation by PS decarboxylation is enough to satisfy most of the parasite requirements for this PL, a *de novo* biosynthesis pathway analogous to that described for PC also operates in *Plasmodium*. The enzyme ethanolamine kinase (EC 2.7.1.82), which catalyzes the first step of this pathway, is quite distinct from the enzyme that catalyzes the reaction involving choline discussed above (Ancelin and Vial, 1986b; Ancelin *et al.*, 1986). The final step in the pathway involves membrane-bound ethanolamine transferase (EC 2.7.8.1), whose substrate dependence for the molecular species of DAG is distinct from cholinephosphotransferase (Vial *et al.*, 1984c). Last, the reacylation of lyso-PE has not been investigated, but this could basically involve the same reaction as described for lyso-PC.

4.4. Biosynthesis of PS

Phosphatidylserine is present only at a very low level in *Plasmodium* but is a key intermediate from which PE can be derived by decarboxylation. In turn, PE also serves as a precursor for PC through subsequent methylation. Taking into account these conversions into PE and PC, PS is the PL that is the most abundantly biosynthesized by the infected erythrocyte (Vial *et al.*, 1982b, 1989a; see Section 4.6).

In animal tissues, the biosynthesis of PS is thought to occur solely by the enzyme-catalyzed, energy-independent interchange of L-serine with the head group of PE or PC. In *Plasmodium*, the amplitude of the phenomenon relative to its quantitative transformation into PE and PC precludes such a possibility, since a molecule cannot originate from the one it produces. Hence, the synthesis of PS probably takes place via the CDP–DAG linked reaction, first discovered in *Escherichia coli* (Raetz and Dowhan, 1990). Further investigations are needed to characterize this metabolic pathway.

4.5. Biosynthesis of PI

Phosphatidylinositol experiences the highest relative increase after *Plasmodium* infection. It is probably synthesized by a set of reactions using the CDP–DAG pathway (Vial *et al.*, 1982b). The metabolism of PI is the most interesting since it involves the phosphorylated derivatives PI-bisphosphate (PIP_2) from which the powerful second messengers, DAG and inositol 1,4,5-triphosphate (IP_3), are released in response to the activation of a phospholipase C. DAG activates protein kinase C and IP_3 releases intracellular calcium (Nishizuka, 1988). We have shown a *de novo* synthesis of PIP_2 in *P. knowlesi*. This biosynthesis has only been observed at the schizont stage. Furthermore, the calcium ionophore ionomycin induces a release of neosynthesized IP_3, probably by activating an intracellular phospholipase C (Vial *et al.*, unpublished results). The natural stimulus for the activation of phospholipase C is unknown but the polyphosphoinositide cycle probably has important physiological functions in the biology of the parasite. Phosphoinositide hydrolysis at the extraflaggelation step during zygote formation has recently been shown in *P. falciparum* malaria (Jett *et al.*, 1989). The fact that phorbol esters that are potent activators of protein kinase C and stimulate gametocyte formation at 30 nM (Trager and Gill, 1989) points to a potential role of the polyphosphoinositide cycle in the biology of *Plasmodium*.

4.6. Can These Metabolic Pathways Provide All the PL Needed by the Parasite?

As discussed below (see Section 7), parasites are capable of incorporating and metabolizing exogenous PL (Van der Shaft *et al.*, 1987; Moll *et al.*, 1988), such as PC, PS, or PE. Thus, the question arises whether the mere uptake of preformed PL molecules from plasma can account for a part of the parasite PL requirements. First, the existence of such an uptake *in situ* remains to be explored. Second, this appears quite unlikely, especially for PE (representing less than 7% of the total plasmatic PL) and above all for PI and PS, which are almost absent in plasma (around 2% and $< 0.1\%$ of total plasma PL, respectively) (Angus *et al.*, 1971a; Stocker *et al.*, 1987), and which must therefore mainly be provided to the parasite by its own biosynthetic pathways.

In our studies, we have consistently quantified each metabolic pathway, especially when it had not previously been detected. Table II shows the PL requirements of *P. knowlesi* for one erythrocytic cycle, as deduced from the lipid composition at the schizont stage. It is likely that *P. falciparum* has the same PL requirements.

The minimal amounts that could originate from parasite biosynthetic pathways during one cycle of *P. knowlesi* (24 hr) or *P. falciparum*, (48 hr) development have been deduced from experiments carried out in our laboratory over the last eight years using the appropriate radioactive precursors at physiological concentrations. The calculations, especially for the incorporation of ethanolamine into PC and serine into PE and PC, are complicated because of the isotopic equilibrium time of the endogenous pools of PE and PS. This could be critical for *P. knowlesi*, for which experiments were usually carried out for 3 hr. Nevertheless, *P. falciparum* experiments lasted over 15 hr (i.e., the dynamic equilibrium was clearly reached) and showed comparable results. Last, the rates of incorporation of some precursors are largely dependent on the parasite stage (Vial *et al.*, 1982a). Indeed, the rates of incorporation of palmitate into total PL and of choline into PC increase linearly with the maturation of the parasite. In contrast,

Table II
Comparison of Overall Phospholipid Increase to the Measured Syntheses During One Erythrocytic Cycle of *P. knowlesi* and *P. falciparum*[a]

	PI	PS	PE	PC
Theoretical increase in PL				
[nmol/10^{10} infected cells per cycle)	750	250	5100	5700
Inositol	**75 ± 30**			
	50 (R)			
	460 (S)			
Choline				**480 ± 80**
				770 ± 150
				590 (R)
				1830 (S)
Ethanolamine			**1600 ± 700**	**460 ± 130**
			1700 ± 700	*230 ± 90*
Serine		**450 ± 170**	**1400 ± 340**	**160 ± 15**
		630 ± 170	*1200 ± 470*	*80 ± 15*
		570	*800*	*30 (R)*
		1550	*4400*	*150 (S)*
Glycerol 3-P			**2100**	**1900**
Glycerol	**190 ± 20*b***		**870 ± 160**	**650 ± 50**
Theoretical fatty acid requirements[b]	330	100	2000	2300
Palmitate	**200[b]**		**1200**	**800**
Oleate	**750 ± 220[b]**		**2210 ± 440**	**1100 ± 140**
			2330	*1400*
Palmitoyl-lysoPC	**280 ± 70[b]**		**1120 ± 80**	**2800 ± 1300**

[a]The theoretical increase in individual PL due to *Plasmodium* is based on a 350% increase in PL at the schizont stage. For FA theoretical needs, the calculation is based on the same assumption and in supposing that the FA account for 20% of total FA. Other values indicate the amount of PL biosynthesized from the various precursors and are in bold characters for *P. knowlesi* and italics for *P. falciparum*. Data expressed in nmol/10^{10} infected cells represent average obtained with ring and schizont forms. R and S stand for data from either ring or schizont stage, respectively. The precursors were at physiological concentrations, i.e., 40 μM inositol and choline, 20 μM ethanolamine, 140 μM serine, 160 μM palmitate and oleate, 70 μM LPC. Glycerol and glycerol-3-P were at 1–1.5 mM.

[b]These fractions were composed of PS + PI.

the incorporation of FA into TAG, the transformation of PS into PE, and the methylation step of PE show a marked increase during the second half of the cycle. Moreover, the rate of inositol incorporation into PI is nearly tenfold higher when the parasite is mature than in the ring stage, although this could be related to new permeability pathways for the precursor induced at the end of the schizogony (Ginsburg *et al.*, 1986).

Regardless of the species, the incorporation of glycerol or FA into the various PLs roughly reflects the PL composition of the infected cell, that is, the main incorporation is into PE and PC (80–90%) and the remainder into PS and PI (10%). Twenty to twenty-five percent of radioactive serine incorporated into PL is recovered in PS, while labeling of PE and PC accounts for 70% and 5–10%, respectively, indicating a very high PS decarboxylase activity.

Generally speaking, regardless of the parasite species, incorporation of the various precursors into each PL represents a large part of the parasite requirements: 10% for PI; 20% for PC synthesis from choline, ethanolamine, and serine; 60% for PE from ethanolamine and serine; while the requirements for PS synthesis and FA incorporations are largely covered. However, keeping in mind the increased syntheses with parasite maturation and comparing the values at the schizont stage (only available for *P. falciparum*), the incorporations of inositol and choline into PI and PC can reach 60 and 40% of the parasite requirements, respectively.

Let us now examine the proportion of each pathway for PC or PE synthesis. Phosphatidylcholine molecules originate from either *de novo* synthesis or methylation of PE, each accounting for 40–45% of the total biosynthesized PC in *P. knowlesi*, whereas in *P. falciparum*, *de novo* synthesis appears to dominate (about ⅔) over PE methylation. Phosphatidylethanolamine molecules result from *de novo* synthesis and from serine via PS decarboxylation, nearly equally, for both species. At the schizont stage, however, the PE portion synthesized from PS decarboxylation may well increase and surpass the fraction that is synthesized *de novo* (⅔ *vs* ⅓).

4.7. Equilibrium between the Different Pathways for PL Biosyntheses and PL Compartmentalization

Faced with all of these possibilities for synthesizing its main PL (see simplified scheme, Figure 4), the parasite is able to maintain an equilibrium between these different ways of synthesis. There is a reciprocal interaction between *de novo* PC biosynthesis and methylation of *de novo* PE. Indeed, an increase in *de novo* PC synthesis resulted in a decrease in the methylation of *de novo* PE (Ancelin and Vial, 1989). In turn, an increase in *de novo* synthesis of PE and of its methylated product, PC, has led to a concomitant alteration of the *de novo* PC synthesis (Ancelin *et al.*, 1991).

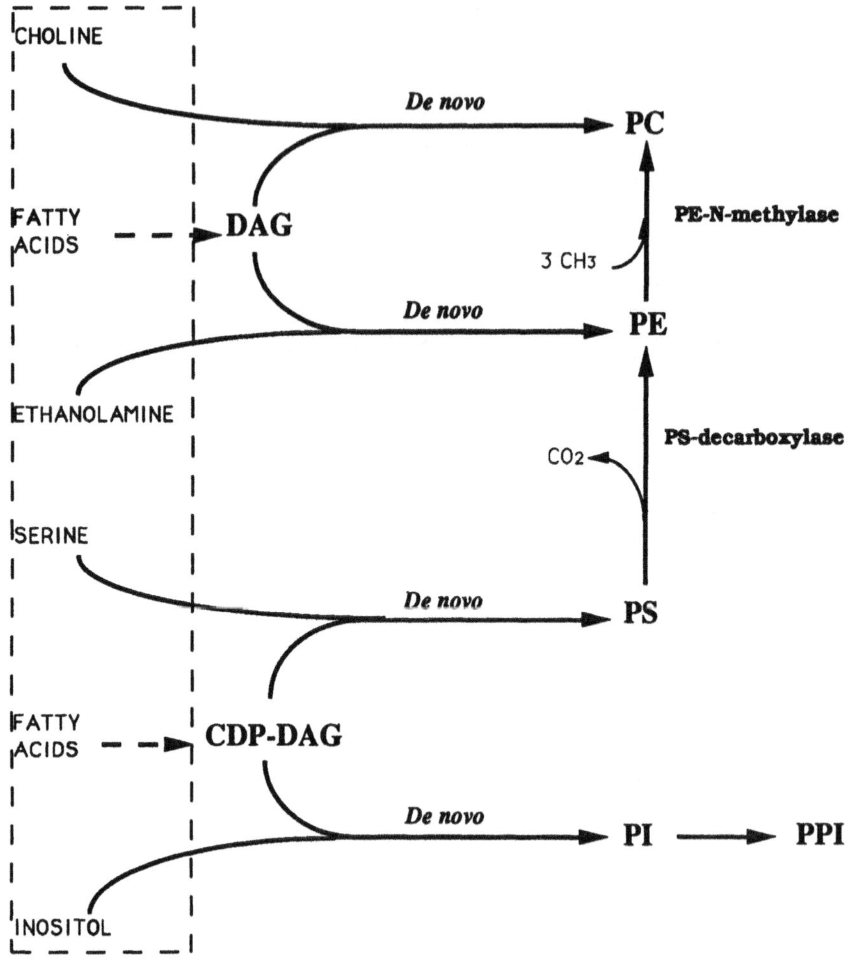

FIGURE 4. Simplified scheme of phospholipid metabolism in *Plasmodium*-infected erythrocytes. PL precursors (polar head groups and fatty acids) drawn from plasma are in the dotted frame.

Such an equilibrium also seems to occur for PE synthesis from serine and ethanolamine. An increase in external ethanolamine, resulting in an increase in *de novo* biosynthesized PE, was accompanied by a decrease of the PE formed from PS decarboxylation. At higher ethanolamine concentrations, a diminution of the PC synthesized from serine or ethanolamine also occurred. In turn, an increase in the PE formed from serine after PS decarboxylation provoked a diminution of the biosynthesized *de novo* PE (unpublished results). By contrast, there seems to be no such interference between *de novo* PC biosynthesis and serine metabolism into PL, especially into PC (Ancelin, personal observations).

Taken together, this indicates coordinate regulation of the parasite PL synthetic pathways, whose mechanism remains to be explored in *Plasmodium*-infected erythrocytes. In hepatocytes, the inhibition of the methylation of PE into PC with a concomitant increase in *de novo* PC synthesis is clearly related to an activation of the rate-limiting step, that is, choline phosphate cytidylyltransferase (Vance and Ridgway 1988). In *Plasmodium,* this enzyme, which is also a limiting step (Ancelin and Vial, 1989), could play such a role.

Last, the possibility (or not) of interregulation between different modes of PL biosynthesis brings up the question of a possible compartmentalization of PL in the infected erythrocyte. In liver, it has already been demonstrated that depending on its way of synthesis, PE methylation has different sensitivity to 3-deazaadenosine (DZA). Methylation of the *de novo* PE is inhibited by DZA, whereas methylation of the PE formed by PS decarboxylation is DZA-insensitive. Phosphatidylcholine, resulting from the methylation of this latter PE, would be specifically involved in lipoprotein secretion (Vance and Ridgway, 1988). In *Plasmodium,* the lack of interference between choline and serine metabolisms could indicate that the PL synthesized from these pathways could have different localizations depending on their mode of synthesis.

Moreover, in normal eukaryotic cells, *de novo* PL biosyntheses are essentially at the endoplasmic level, whereas PS decarboxylase is mitochondrial. Phosphatidylethanolamine-*N*-methylase is located in plasma membranes, except in liver where it takes place in microsome rough endoplasmic reticulum (Vance and Ridgway, 1988). In *Plasmodium*-infected erythrocytes, such information is missing, mainly because of the lack of any fractionation procedure. Nevertheless, regarding their modes of synthesis, the idea has not bee discarded that there can be various metabolic pools for the same PL, which could play different roles in parasite biology, notably in the construction of the various membranes of the infected erythrocyte, including cytoplasmic vesicles or membrane structures in the erythrocyte, PVM, or parasitic or intraparasitic membranes.

4.8. Adjustment of the Lipid Composition

Glycerolipids provide the basic structure of the parasite membrane since it does not contain appreciable levels of cholesterol and SM. They are an extremely diverse group of molecules, with a variety of head groups (class distribution) and FA (molecular species). Proper cell function depends to a large extent on the physical and physiological properties of the membranes formed from these lipids (Lynch and Thompson, 1984; Spector and Yorek, 1986). Concerning the head groups, it is worth noting that the parasitic PS decarboxylase and PE methylase activities allow head group remodeling, considering the required end products. Currently, the Ca^{2+}-dependent polar head group exchanges have not yet been explored.

For molecular species, naturally occurring PLs typically have a saturated or monounsaturated FA at the sn-1 position and an unsaturated FA at the sn-2 position. The fine adjustment of the composition of the fatty acyl group is usually controlled by enzymes that are compartmentalized in cells (Stubbs and Smith, 1984). Elongation and desaturation of precursor FAs do not occur at an appreciable level in *Plasmodium* (Rock, 1971a; personal observations). The deacylation–reacylation reaction is a means of membrane lipid regeneration but its most important function may be in modulating membrane function. There is at present little evidence of modifications of the membrane FA composition by deacylation and reacylation of existing membrane phosphoglycerides. We have never observed a release of the FA part of intact PL molecules (PC, PS, PE) nor a transacylation reaction (Beaumelle, 1987; Moll *et al.*, 1988) occurring in the parasitized cell. Furthermore, it appears that fluorescent lipids are not deacylated when they are introduced into the infected erythrocyte, although these PL molecules reach the intracellular parasite (Haldar *et al.*, 1989). We found no intracellular PLA_2 activity in homogenates of *P. knowlesi*-infected erythrocytes using palmitoyl-[^{14}C]oleyl-PC and varying the conditions of ions and pH (unpublished results). However, Ginsburg's group (Krugliak *et al.*, 1987), using a fluorescent PL, detected an acid CA^{2+}-independent phospholipase activity in *P. falciparum* cytosol but the results were not quantified. Hence, FAs are probably initially introduced into PL during *de novo* synthesis and are further strictly controlled by choline- and ethanolamine-phosphotransferase (Vial *et al.*, 1984c).

Thus, the parasite probably does not possess the ability to retailor lipid molecular species. By contrast the transport of lipids and their dynamics could constitute a tactical mechanism for modulating membrane properties (see Section 7).

4.9. Conclusion

Altogether these data show that under physiological conditions the incorporation of the various precursors is probably sufficient to cover most parasite requirements and they further illustrate the bewildering versatility for the synthesis of PL, notably for PE and PC, thus placing the parasite midway between prokaryotic and eukaryotic organisms.

To summarize, we can try to place *Plasmodium* on the animal evolution scale with some reference points along the chain (Table III). Bacterial plasma membranes are often composed of one main type of PL and contain no cholesterol (Raetz and Dowhan, 1990). On the other hand, the membranes of most eukaryotic cells contain not only large amounts of cholesterol, but also a variety

Table III
Some Comparative Characteristics of Lipids of *Plasmodium*, a Lower Eukaryote, with Prokaryotes and Multicellular Organisms

	Eukaryotic cells	
Prokaryotes (Bacteria)	Plasmodium	Multicellular Organisms
No cholesterol	No cholesterol	Presence of cholesterol
One main type of PL	At least 4 types of PL	Variety of PL
PE synthesis from PS decarboxylation	Both systems	*De novo* synthesis of PE and PC

of PL. *Plasmodium* is situated at an intermediate position, with at least four types of PL, but probably very little, if any, cholesterol. Its PL biosynthesis pathways resemble those of yeasts (Carman and Henry, 1989).

Concerning biosynthetic pathways, the protozoon *Plasmodium* also appears midway between the prokaryote, whose PE synthesis results exclusively from the decarboxylation of PS, and multicellular organisms that synthesize PE only by the *de novo* Kennedy pathway. Thus, this single-cell eukaryote *Plasmodium* occupies an intermediate position. Consequently, it exhibits a great capacity in the synthesis of its PL, that appears to be as intricate as in many multicellular organisms.

5. PHOSPHOLIPID METABOLISM: AN ORIGINAL TARGET FOR NEW MALARIA CHEMOTHERAPY

One of the fundamental goals in the study of *Plasmodium* biochemistry involves the discovery of metabolic pathways that might be beneficial in the design of drugs capable of exterminating parasites without injuring the host. Phospholipid metabolism is quite essential to the life of the parasite and we here consider what might be done to disorganize it.

We have discussed above the extent and the specificity of PL metabolism in *Plasmodium*-infected erythrocytes. This metabolism is plethoric, and is carried out by the parasitic machinery in host erythrocytes devoid of PL metabolism. We have seen that this murderous parasite uses a variety of complex metabolic pathways. However, to ensure its development the parasite depends on the host for supplies of the polar head groups and FA molecules that it cannot synthesize. It uses these molecules to build all of the membranes required for its intra-erythrocytic multiplication. Hence, interference in this vital parasite metabolism would probably provoke the death of the parasite. This could be the basis for new chemotherapies.

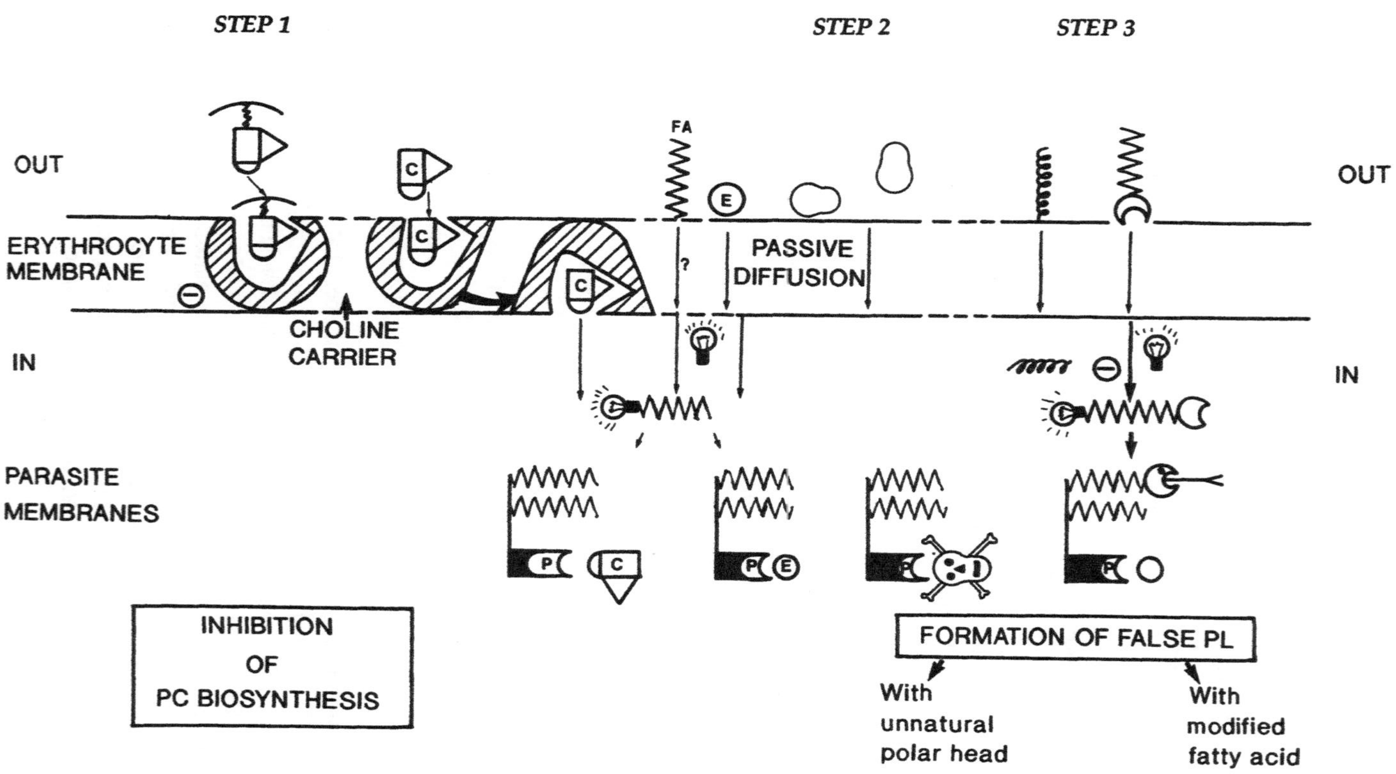

FIGURE 5. Phospholipid metabolism as an ideal target for original chemotherapy. C, choline; E, ethanolamine; FA, fatty acid.

Many steps within PL metabolism are potential targets for such chemotherapy. The access of precursor molecules (such as choline) to the intracellular parasite could be blocked (step 1, Figure 5), which includes blocking the transport system if there is one. Alternatively, false precursors could be added to deceive the parasite. After their incorporation into the parasite PL, the false metabolites would alter the lipid composition of the membrane and their physicochemical properties, leading to parasite death. Small analogues of polar head groups retaining a free hydroxyl group (step 2) or unnatural FA (step 3) could be included in this series. Also, metabolic pathways essential to PL biosynthesis could be inhibited, such as the incorporation of FA into PL, the *de novo* biosynthesis of PC, the methylation step converting PE into PC, or the pathway for PS biosynthesis. Blockage of this last pathway would be ideal since it is absent from the host, but such molecules have yet to be discovered.

We have already shown that three types of compounds that interfere with the supply of plasmatic precursors (FA or polar heads) or with their metabolism, can kill the parasite *in vitro* by altering parasite PL metabolism. A first example involves inhibition of FA acylation. We showed an effective inhibition of *P. falciparum* growth by unnatural FA with 50% inhibitory concentrations (IC_{50}) of 7 to 90 μM. Their action was clearly correlated with the inhibition of acyl-CoA synthetase, the first enzyme that metabolizes FA, that is, acyl-CoA synthetase (Beaumelle and Vial, 1988b). Second, small molecules, analogues of choline or ethanolamine, that retain a free hydroxyl group, have IC_{50} of 50 μM at best. They are incorporated by the parasite and then transformed into unnatural PL, which we have identified by mass spectrometry in the case of D-amino-butanol used as an analogue. It accumulates at the expense of the natural form (PE), and probably both of these factors (decrease of natural PL and accumulation of false PL) lead to parasite death. We have studied the influence of the length of the chain between the hydroxyl group and the nitrogen, and the role of steric hindrance by nitrogen substitution. The distance between the nitrogen and hydroxyl group was found to be a crucial factor and cannot be longer than 2 methylene groups for these small molecules. Substitutions of nitrogen at the beta position were far less effective than at the alpha position. We also defined some rules concerning nitrogen substitution (Vial *et al.*, 1984b).

Larger molecules possessing one or two quaternary ammoniums are potential analogues of choline. Their IC_{50} was found to be much better between 0.7 and 10 μM (Ancelin *et al.*, 1985) and they do not induce the formation of false metabolites. These compounds block the entry of choline into the infected erythrocyte, causing a specific decrease in the biosynthesis of PC (Ancelin and Vial, 1986c). From these first results, and considering that PC is (with PE) the major PL of the infected erythrocyte, a strategy including the chemical synthesis of a range of potential inhibitors of choline metabolism was developed. Based on the very detailed studies of the intracellular dynamics of choline (see Figure 3), and especially on the determination of choline transport as a limiting step of the *de novo* pathway (Ancelin and Vial, 1989), we focused particularly on choline transport inhibitors. Last, and in contrast to PE, PC synthesis requires a precursor (choline) whose entry in infected cells remains well controlled throughout the parasite cycle and involves a carrier-mediated process (Ancelin *et al.*, 1991), which can be a very accessible pharmacological target.

Our results concerning 60 polar head analogues allows the study of different parameters, such as the importance of the nitrogen and hydroxyl groups, length of the chain that separates them, and the role of steric hindrance by *N* substitutions. Finally, this provides information on the nature of the pharmacological target, the choline transporter of *P. falciparum*-infected erythrocytes. The choline carrier would possess an anionic site capable of accommodating positively charged polar head groups, with a hydrophobic domain (whose length is approximatively that spanned by 10–12 methylene groups), adjacent to the anionic site. The anionic site of the carrier appears to be capable of accommodating small polar head groups (Ancelin *et al.*, unpublished results).

We have shown a very close correlation between the capacities of 35 compounds to inhibit *P. falciparum* growth and their ability to inhibit the synthesis of specific PL (Figure 6). Compounds possessed the same antiplasmodial activity whether they were assayed on chloroquine-sensitive or polypharmacoresistant *P. falciparum* strains. Measurement of the effects of 15 of these compounds on a lymphoblastoid cell line demonstrated the total absence of correlation between inhibition of parasite growth and inhibition of lymphoblastoid cell growth (Figure 6). This demonstrates that structural requirements to inhibit PL metabolism are very specific to the infected erythrocyte and that this metabolism constitutes a very realistic target for new malaria chemotherapy. Basically, it further illustrates that knowledge concerning metabolism specific to the parasite can provide a wide range of targets for research in a rational drug design program (see also Chapter 13).

Another possible chemotherapy approach could be based on the particular properties of the erythrocyte membrane after infection. Indeed, a modified acylated pancreatic PLA_2 (Lau-Ampa) was synthesized by the Utrecht group and used in the presence of *Plasmodium*-infected erythrocytes (Moll *et al.*, 1990b; see Section 6.2). This acylated PLA_2 could selectively attack and destroy infected erythrocytes, due to the modification in the lateral surface pressure caused by malarial infection. Hence, this could lead to a chemotherapy concept based on the selective recognition of infected red blood cells by enzymes or even drugs.

One more approach to malarial chemotherapy indirectly involves lipids, since changes in the degree of saturation of the FA constituting membrane lipids may influence the susceptibility of membrane components to oxidative damage. Malarial parasites are destroyed within intact host erythrocyte membranes by

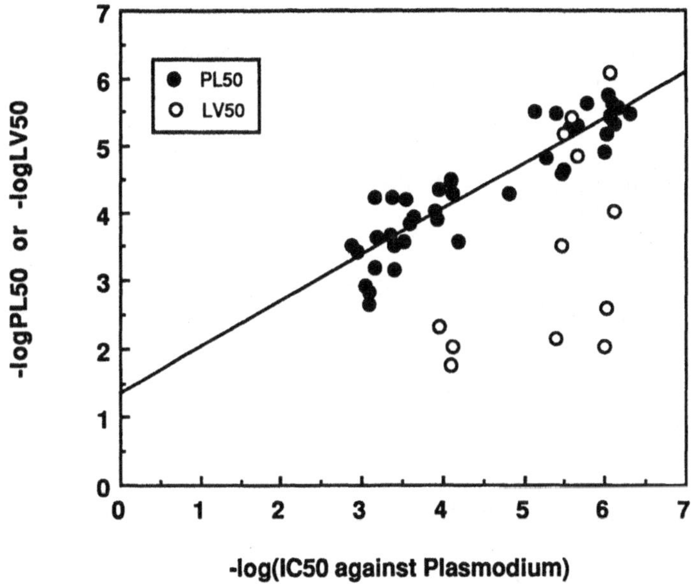

FIGURE 6. Correlation between the inhibition of PL metabolism and the inhibition of parasite growth. PL_{50} and IC_{50} correspond to the concentration producing 50% inhibition of PL synthesis and parasite growth, respectively. For comparison, we also reported the LV_{50} (concentration producing 50% inhibition of the viability of a lymphoblastoid cell line) of some compounds tested on the growth of S.A.R. lymphoblastoid cells (open symbols).

oxygen radical-producing drugs, which implies a difference in the susceptibility of erythrocytic and parasitic membranes to oxidative damage. It also appears that an increase in spontaneous lipid peroxidation occurs in response to increased parasite load and maturation (Buffinton *et al.*, 1986; Golender and Chevio, 1989).

Based on these facts, Levander *et al.*, (1989, 1990) offer a new approach to the problem of drug-resistant malaria. They fed mice with polyunsatured FA, of the ω-3 series, for example, and then the mice were subsequently infected by *P. berghei* or *P. yoelii*. This probably leads to a massive incorporation of this FA into lipids of the infected cells, modifying the membrane properties of the parasite, and probably also increases spontaneous lipid peroxidation in the infected erythrocytes. In the absence of protection, that is, in the case of vitamin E deficiency (vitamin E is the most powerful lipid-soluble antioxidant), the parasite cannot withstand the peroxidative reaction and the mice were saved. Thus, factors such as differences in the amount and composition of constituent lipids are also likely to be critical in determining the relative susceptibilities of the different membranes to peroxidative damage.

6. STRUCTURAL ORGANIZATION OF LIPIDS IN THE HOST CELL MEMBRANE

6.1. Functional Modifications of the Host Cell Membrane May Involve Changes in Lipids

During intraerythrocytic development of the malarial parasite, there are several disturbances involving the host cell membrane. These disturbances can be either morphological (Aikawa and Miller, 1983; Allred *et al.*, 1986; Aikawa, 1988) or biochemical, the latter involving the appearance of parasite proteins, the expression of new antigenic determinants (Howard, 1988b; Hommel and Semoff, 1988), and increases in permeability (Homewood and Neame, 1974; Kutner *et al.*, 1985; Ginsburg and Stein, 1987; Cabantchik, 1989; Ancelin *et al.*, 1991).

It is also apparent that intracellular development of the parasite causes a stage-dependent alteration of the physical properties of the host cell membrane. Such experiments included merrocyanin 540, a fluidity-sensing fluorescent dye (Sherman and Greenan, 1984), fluorescence polarization microscopy with FA derivatives (Howard and Sawyer, 1980), or specific plasma membrane fluorescent probes (Deguercy *et al.*, 1986; Koppaka *et al.*, 1989), electron spin resonance of FA (Butler *et al.*, 1984; Forte and Roca, 1989) or PL (Tarashi *et al.*, 1986) derivatives, or of hydrocarbon chains (Allred *et al.*, 1983). All of these experiments indicated that upon infection, erythrocyte plasma membranes show a lower degree of molecular order and increase in the probe motions, that is, higher fluidity and lower viscosity. These differences in membrane fluidity probably reflect biochemical modifications in the protein and lipid composition of the membrane and/or altered relationships with the underlying cytoskeleton. Disruption or reorganization of the erythrocyte cytoskeleton during the formation of PVM in the invasion process and also subcellular trafficking (Atkinson *et al.*, 1987; Taylor *et al.*, 1987; Howard, 1988a,b; Hommel and Semoff, 1988; Mitchell and Bannister, 1988; Bannister and Mitchell, 1989) may contribute to disturbances of the host erythrocyte membrane.

These changes in fluidity might be explained by changes in the cholesterol–PL ratio (see above) since cholesterol is known to impose conformational restriction on the movement of membrane PL hydrocarbon chains, thereby stabilizing membrane fluidity (Yeagle, 1985, 1987). However, the host cell membrane also becomes disordered in *P. falciparum*-infected erythrocytes (Sherman and Greenan, 1984; Tarashi *et al.*, 1986) although cholesterol changes probably do not occur, since the modifying cause (i.e., a change in external free cholesterol content) is suppressed (see Section 3.2).

Neither class distribution nor probably asymmetric distribution of PL is changed in the host membrane (see Section 6.2). In contrast, we have shown an enhancement in the flip rate of spin-labeled PL occurring at an early stage in parasite development (see below). The striking changes in the molecular species of the constitutive PC of the host cell membrane, with a marked increase in 16:0, 18:2-PC, and PE (Simoes *et al.*, 1990; see Table I) could be the basis of such changes. These changes in the molecular species of PL (i.e., in the FA joined in pairs on the glycerol backbone) affect the physicochemical properties of the hydrocarbon region of the bilayer. They are therefore particularly consistent with experiments involving probe molecules at different depths in the membrane, which show that the largest modifications take place in the center of the lipid bilayer (Howard and Sawyer, 1980; Allred *et al.*, 1983).

6.2. Asymmetric Distribution of the PL

The protein molecules embedded in membranes or attached to them, or even those that span the thickness of the membrane, invariably have a fixed asymmetric orientation. The smaller PL molecules, which make up the structural matrix of the membrane, have also been found to have an asymmetric distribution, although this asymmetry is partial rather than absolute. The erythrocyte membrane has been well studied by a variety of methods establishing the striking asymmetry in PL distribution. The external leaflet of the membrane consists primarily of the bulk of PC and SM. The inner leaflet of the erythrocyte membrane is enriched in anionic PLs, with most of the PE and all of the PS (Op den Kamp, 1979; Schwartz *et al.*, 1985a).

There is no doubt that *Plasmodium*-infected erythrocytes show substantial changes in their membrane structure, functional properties, and antigenicity. Such modifications may also occur in uninfected erythrocytes from infected hosts, although information concerning these aspects appears to be scarce (see Beaumelle and Vial, 1988c). Such changes probably also apply to the lipid organization. Structural alterations in membrane lipid asymmetry might be of physiological importance in regulating blood coagulation by giving the cell a negative surface charge, thereby allowing the interaction of various coagulation factors. Exposure of PS at the outer surface also enhances adhesion of the erythrocyte with endothelium cells or increases its recognition by mononuclear phagocytes (Schwartz *et al.*, 1985b; Zwaal and Bevers, 1986; Zwall *et al.*, 1989). Migration of the amino-PL, especially that of anionic PS, to the outer surface could have dramatic consequences for the erythrocyte whether infected or not, since it could induce the major pathophysiological effects of malarial infection, namely hemostatic plug formation, interaction of parasitized cells with endothelium, and destruction of infected or uninfected erythrocytes by macrophages (Weatherall *et al.*, 1983; Boonpucknavig *et al.*, 1984; Howard, 1988a; Hommel and Semoff, 1988).

At the present time, highly contradictory results concerning both infected and uninfected erythrocytes have been published. These studies were initiated by Gupta's group using PLA_2 (*Naja naja* snake venom) and the reagent 2,4,6 trinitrobenzene sulfonic acid (TNBS), which labels free amino groups. They found an inverse accessibility of the major PL to these probes, with only 25% of PC and around 50% of the PE present in the outer surface of the *P. knowlesi*-infected erythrocyte (parasitemia > 90%, early ring). Under these conditions, all PS remained internal (Gupta and Mishra, 1981). In a subsequent work, using bee venom and pancreatic PLA_2, they noted enhanced accessibility of both amino-PL (i.e., including PS) to the enzyme, which depended on the developmental stage of the intracellular parasite but not on parasitemia (Joshi *et al.*, 1987). This group also found that nonparasitized erythrocytes, either from infected monkeys or from the same monkeys 30 days after chloroquine cure, also contained significantly larger amounts of PE and PS in the outer leaflet of their membrane (Gupta *et al.*, 1982). These surprising observations were later restricted to chronically infected or splenectomized monkeys (Joshi *et al.*, 1986). Very recently, an abnormal degradation of PE and PS by PLA_2 was also observed in cultured *P. falciparum*-infected erythrocytes (not recovered in uninfected erythrocytes of the suspension), which appeared to be related to the developmental stage of the parasite (Joshi and Gupta, 1988). A similar alteration of the transbilayer distribution of PL in *P. falciparum* infection was also found by Schwartz *et al.* (1987), although only minor changes in PS distribution were observed and these were found only with chemical (TNBS) but not by enzymatic methods.

On the other hand, our studies failed to show any major abnormality in the transbilayer distribution of the PL in the plasma membrane of erythrocytes from *P. knowlesi*-infected or chloroquine-cured monkeys. Three independent methods, that is, treatment of intact cells with two different types of phospholipases, fluorescamine labeling of amino PL and a PC transfer-mediated exchange procedure, showed that the outer layer of the membrane from uninfected erythrocytes of infected monkeys and of erythrocytes from cured monkeys contain about 82, 70, and 20% of SM, PC, and PE, respectively. No PS could be found in this outer monolayer. Concerning the infected erythrocyte, PLA_2 digestion showed no degradation of either PI, PS, and PA, indicating that these PLs remain exclusively in the inner monolayer of the erythrocyte. We observed no plateau in the course of PE and PC hydrolysis, precluding conclusive data for these PLs using this method. On the other hand, 82% of the SM could be localized in the outer monolayer with sphingomyelinase, and fluorescamine labeling demonstrated that the asymmetric distribution of PE is probably not changed at any stage of parasite development (Van der Schaft *et al.*, 1987).

Very recently, a palmitoyl amidinated pancreatic PLA_2 (Pal-AMPA) was shown to attack the outer layer of normal erythrocytes extremely rapidly due to its superior membrane-penetrating capacities (Van der Wiele *et al.*, 1988). This modified enzyme resulted in a clear plateau of PC and PE hydrolysis that was reached in a matter of minutes in infected erythrocytes also. It thereby avoids the problem of cell lysis and possible flip-flop of PL (see Section 6.3). Using this tool, we confirmed that around 15% of the PE in the host cell membrane appeared to be accessible for Pal-AMPA, whereas virtually no PS was hydrolyzed. Thus, the plasma membrane of *P. knowlesi*-infected erythrocytes has a normal asymmetric transbilayer PL distribution (Moll *et al.*, 1990a).

When considering the above controversies, it should first be understood that in studies on PL distribution between the two leaflets of the lipid bilayer in the host cell plasma membrane, it is essential to correlate the percentage of PL accessible to the chemical or enzymatic probe with the total amount of PL present in that membrane. Consequently, the results must take into account the large amount of intracellular PL that is contributed by the parasite. Gupta's group has isolated the erythrocytic plasma membrane from the contaminating parasite by either lysing the cells with high concentrations of the detergent saponin or by using hypoosmotic medium. However, such a protocol is not suitable for achieving the proposed goal (Beaumelle *et al.*, 1987). Furthermore, multiple centrifugation, especially in the presence of detergent, may lead to selective loss of PL or lyso-PL, or even part of the membrane. Moreover, Schwartz *et al.* (1987) found an increase of 110% in the PL content of *P. falciparum*-infected erythrocyte (after correction for parasitemia) but did not correct their results for the presence of intracellular PL. In any case, regarding the normal susceptibility of uninfected cells to the treatment, it is obvious that the degradation of over 40% of the total PE by PLA_2 in schizont-infected erythrocytes (parasitemia 12–18%) comprises a fraction that considerably exceeds the entire PE content of the host cell plasma membrane. It was also surprising that "to minimize hemolysis" they had to use half the amount of enzyme for the treatment of uninfected erythrocytes compared with that of infected erythrocytes, although the latter are basically more fragile.

The experiments of Gupta's group were preceded by multiple density gradient centrifugation, each with intense cell washings in a saline medium. Incubations were then run in a buffer containing 100 mM KCl. This probably led to severe starvation of the parasite, particularly of glucose, the only energy source of *Plasmodium* [the intracellular level satisfies the parasite requirements for only a matter of minutes (Zolg *et al.*, 1984)], and probably had dramatic effects on cell function and viability of the parasite. Furthermore, incubations in buffers containing 100 mM KCl (Joshi *et al.*, 1987; Joshi and Gupta, 1988) or 130 mM KCl (Schwartz *et al.*, 1987) are known to inhibit the Na^+/K^+ ATPase (Skou, 1988) and therefore may give rise to cell swelling, particularly as the plasma membrane of the infected cell shows considerably increased permeability (Sherman, 1979, 1988; Ginsburg and Stein, 1987; Cabantchik, 1989). Cell swelling (Woodward and Zwaal, 1972) may explain why the pancreatic PLA_2 was able to attack its substrates in the infected cells, whereas it failed to do so in our experiments (Moll *et al.*, unpublished data). The use of 10 mM $CaCl_2$ for the porcine pancreatic PLA_2 probably has dramatic consequences, especially if the Ca^{2+}/Mg^{2+} ATPase cannot function normally, which is probable during energy starvation. All of these conditions can lead to an increase in Ca^{2+} concentration in the erythrocyte cytoplasm. Such an increase is known to inhibit the PE and PS translocation system (Zachowski *et al.*, 1986), which may already be largely impaired because of energy deprivation. Clearly, such conditions may affect the transbilayer PL distribution.

It should be added that the observed changes in membrane PL organization have often been attributed to the same cause, that is, an increase in the erythrocyte cytoplasmic Ca^{2+} concentration (Schwartz *et al.*, 1987; Joshi *et al.*, 1987; Joshi and Gupta, 1988). Although the Ca^{2+} content of the infected erythrocytes increases 20- to 40-fold at schizogony, this may be rather speculative, because this increase is localized in the parasite compartment (Leida *et al.*, 1981; Tanabe *et al.*, 1982; Krungkrai and Yuthavong, 1983).

We tried to carefully avoid these drawbacks in our experiments, which involved the use of RPMI 1640 for preparing cells and an isotonic medium with glucose for incubations. Our method is based on the fact that the PL composition (according to the polar heads) of the plasma membrane does not change after infection and uses SM as internal standard, making the expression of the extent of chemical or enzymatic labeling easier (Van der Schaft *et al.*, 1987). In addition, we always performed dynamic studies as a function of time and reagent concentration, since reaching a plateau in the hydrolysis profile is a prerequisite to obtain conclusive results (Op den Kamp, 1979). Besides, the apparent absence of PS in the exofacial bilayer leaflet of the *P. falciparum*-parasitized erythrocyte was recently confirmed by the noninvasive, highly specific and sensitive prothrombinase assay (Moll *et al.*, 1990a). The method is based on the fact that the conversion of prothrombin into thrombin is critically dependent on the presence of membrane-bound PS, allowing detection when less than 1% of the PS present in the erythrocyte is located in the outer layer of the plasma membrane of the infected cell.

This statement is reinforced by our inability to find a modified distribution of PL in uninfected erythrocytes as reported elsewhere (Gupta *et al.*,, 1982), although these changes were later explained by the absence of spleen function (Joshi *et al.*, 1986) (all our monkeys were splenectomized).

Therefore, we should be cautious about attributing functional consequences to dubious modifications of the asymmetric distribution of PL in the parasitized erythrocyte. Furthermore, properties of sequestration of infected erythrocytes in the venous capillaries and cytoadherence of the infected erythrocytes could originate from protein modification of the erythrocyte membrane (Hommel and Semoff, 1988; Howard, 1988a; Miller, 1989).

6.3. Transbilayer Mobility of PL

The membrane is by no means static, each monolayer being a two-dimensional fluid with rapid lateral diffusion of the lipid molecules. In the second dimension, however, the mobility of the lipids is severely restricted. The motion of a lipid molecule from one monolayer to the other, or flip-flop, occurs at a low rate. Transbilayer PL organization in the erythrocyte membrane seems to be maintained by interactions of the amino-PL, especially the anionic PS, with the underlying cytoskeletal protein and also by an ATP-driven specific amino-PL translocase (Tilley *et al.*, 1986; Williamsom *et al.*, 1987; Devaux, 1988).

Using electron spin resonance, we have shown a very large increase in the flip movement of SM, PC, PE, and PS in the erythrocyte membrane after *P. knowlesi* infection, which occurs as early as the ring stage (Beaumelle *et al.*, 1988). A rapid transbilayer PL movement (flip) of the infected erythrocyte membrane was confirmed by Haldar *et al.* (1989) in studies using NBD-PC.

However, considerably enhanced flip-flop of PL in the host cell plasma membrane should have led to increased availability of PL during the 1-hr PLA_2-mediated PL hydrolysis, which did not seem to be the case (Van der Schaft *et al.*, 1987), unless PL hydrolysis itself stops enhanced transbilayer motility. The fast uptake of PLs into the intracellular parasite, as soon as they are introduced into the erythrocytic membrane (Van der Schaft *et al.*, 1987; Moll *et al.*, 1988) offers another explanation: spin-labeled PL does not reflect the flip-flop movement of endogenous PLs because their short chain at position 2 makes them closely resemble lyso-PL, which is known to be transferred to the parasite very quickly (Vial *et al.*, 1982b, 1989a).

7. DYNAMICS OF LIPIDS WITHIN INFECTED CELLS

Although it is well known that cholesterol can readily move from one membrane to another (Yeagle, 1985), the entire cholesterol content of the infected erythrocyte is found in the host cell membrane (see Section 3.2). However, the membrane components that contribute to the thermodynamic partitioning of cholesterol among membranes have yet to be determined. The fact that the total SM content does not change and remains associated with the erythrocyte membrane after infection could help in confining the cholesterol to the host meembrane (Yeagle, 1985, 1987).

In the presence of external vesicles containing PC, PE, or PS, *P. knowlesi*-

infected cells have the capacity to directly accumulate PL molecules, in contrast to uninfected cells, which show only a low nonspecific adherence of such vesicles. When external vesicles contain a large excess of PC (compared with the PC in the erythrocyte membrane), infected cells have the capacity to incorporate intact PC molecules (i.e., without any metabolism) from the medium, again in contrast to uninfected cells. In the presence of a PC-specific transfer protein, the external PC accumulates to a greater extent and surpasses the erythrocyte PC present in the plasma membrane, confirming that the newly introduced PL accumulated in the parasite. The rate and amplitude of the exchange increase with the maturation of the parasite (Van der Schaft *et al.*, 1987; Moll *et al.*, 1988). A similar intracellular translocation to the parasite was recently observed when *P. falciparum*-infected cells were incubated with fluorescent NBD-PC, reaching approximately 20% of the PC in the host membrane (Haldar *et al.*, 1989).

These experiments were extended to PS and PE, given that erythrocyte membrane PL represented a 10- to 20-fold excess over vesicle PL. Similarly, infected cells had the ability to accumulate labeled PL, a process that could be accelerated in the presence of nonspecific lipid transfer proteins. Indeed, this led to PL depletion from the medium after 6–8 hr incubation, whereas in the presence of a nonspecific lipid transfer protein, complete uptake was achieved within 30 min. We also observed a modification of the polar groups of the newly introduced PS and PE, which were transformed into PE and PC, respectively, showing that these lipids were incorporated into and metabolized by the parasite. The extent and rate of their transformation appeared to depend on the way in which the PLs were introduced into the infected cells (in the absence or presence of the transfer protein). This probably reflects a compartmentalization of PL inside the infected cells (Moll *et al.*, 1988).

These experiments show that there are rapid movements of PL between the erythrocytic plasma membrane and the intracellular parasite, at least when additional PLs are introduced into the host membrane. In the absence of transfer proteins, PL vesicles could fuse with the external membrane or enter by endocytosis. A PC-specific transfer protein catalyzes a one-to-one exchange process, whereas the nonspecific lipid transfer protein may mediate a net transfer of exogenous PL. The rapid translocation of PL from the outer monolayer of the plasma membrane to the inner monolayer (flip movement) does not necessarily indicate an enhanced transbilayer flip-flop of endogenous PL since a disequilibrium may provide the driving force for the inward movement of the exogenous molecule.

In any case, there must be some kind of exchanges or transport systems between the host erythrocyte membrane and the parasite. This latter possibility is largely corroborated by the intense trafficking of PL in *P. knowlesi*-infected erythrocytes, which makes the PL molecular species of the erythrocyte membrane very close to that of the internal parasite (Simoes *et al.*, 1990; Table I). Localization of intracellularly biosynthesized fluorescent PL within the merozoite near its apical end (within the rhoptries?) is probably followed by their incorporation into the host membrane and into the PVM during invasion (Mikkelsen *et al.*, 1988).

Phospholipids, like membrane proteins but unlike cholesterol, cannot spontaneously partition between membrane compartments and their intracellular movement must therefore be a mediated process. Vesicle-mediated PL transport is intuitively attractive, but there has been little evidence to support it. That there is a transport from the parasite membrane through the erythrocyte cytoplasm to the erythrocyte membrane, involving organelles, is generally accepted and potential pathways have been described (Howard, 1988b; Aikawa, 1988). The reverse route of subcellular trafficking, with internalization by endocytosis and fusion of small vesicles with the PVM (Rodriguez and Jungery, 1986; Burns and Pollack, 1988), requires a more detailed description. The movement of PL through a region of continuity (possibly a "dynamic continuity") between the erythrocyte membrane and the intracellular membrane via the PVM should be explored. Lipid dynamics could also receive important contributions from membranes intensely recycled from phagosomal vacuoles or cytostomes (Bannister and Mitchell, 1986) [which are more frequently found in the trophozoite than in the ring stage (Langreth *et al.*, 1978), that is, when infected cell membrane needs are high]. Another major mechanism for PL transport could involve a role of lipid exchange proteins. However, their role and specificity in PL movement and membrane biogenesis remains to be established.

At the moment, transfer mechanisms are not sufficiently well established to allow conclusions regarding the direct uptake of PL molecules from plasmatic lipoproteins or the extent of such a movement in terms of net transfer. This mechanism could help microsomial PS transfer to mitochondria, where decarboxylation usually takes place (Bell and Coleman, 1980; Kennedy, 1986; Bishop and Bell, 1988) or stimulate the intense movement of membranes at the schizont stage (Mitchell and Bannister, 1988). Whatever the nature of the mechanism, it probably contributes to the great changes in the PL molecular species of the erythrocyte membrane. It should be noted, however, that the uptake of PL from plasma can in any case account for the total requirements of the parasite. This is particularly true of PE, PS, and PI, which are almost absent in plasma after infection. The significance of lipid dynamics for membrane biogenesis turnover and recycling has not been established, since the final localization of PLs and related enzymes within the parasite are presently difficult to determine.

8. SUMMARY AND PERSPECTIVES

The asexual development of *Plasmodium* within the mature mammalian erythrocyte is accompanied by intense membrane biogenesis, notably to ensure the increase in the size of the parasite itself and also of the parasitophorous vacuolar membranes. A considerable increase in the content of most lipids except cholesterol (namely, phospholipids, neutral lipids, and fatty acids) occurs. The phospholipid composition and the constitutive fatty acids of the parasite differ markedly from the original host cell membrane. Particularly notable is the absence of cholesterol and sphingomyelin in the parasite membranes. How can the parasite obtain such a quantity of new lipid molecules in a host cell that is totally devoid of any lipid biosynthetic activity? Like the normal erythrocyte, the infected cell remains unable to synthesize cholesterol or fatty acids. By contrast, it exhibits an intense biosynthesis of neutral lipids and a bewildering variety of modes of phospholipid biosynthesis. Phosphatidylcholine is synthesized by a *de novo* pathway and also by methylation of phosphatidylethanolamine, which itself originates from *de novo* biosynthesis or from decarboxylation of phosphatidylserine. Hence, interference with this intense and specific phospholipid metabolism suggests a wide range of potent tools for new malaria chemotherapy. Indeed, compounds that interfere with the entry of plasmatic precursors (fatty acids or polar heads) or with their metabolisms are lethal to the parasite. Last, we focus on the structural modifications of the host cell membrane with respect to lipids, including higher fluidity and increased transbilayer mobility of phospholipids. Possible modification of the asymmetric distribution of phospholipids in the original host membrane is discussed in light of the various methods used and their limits. The intense dynamics of lipids in the infected erythrocytes can be partly accounted for by the capacity of infected cells to take up and metabolize exogenous vesicles of phospholipids.

Thus, the basic pathways of PL metabolism become increasingly well characterized. There is no existing data concerning the intraparasitic compartmentalization and specificity of localization of lipids, or of the PL biosynthetic enzymes, mainly because the fractionation procedure is still missing. Numerous enzymes have been well characterized, but none have been purified. If this problem could be solved, it would probably enable intracellular localization via specific antibodies. Enzymes conducting the biosynthesis of PS have not been explored. Elucidation of the pathways of intermembrane movement and the mechanisms by which selectivity is imposed for this process remain unsolved by malariologists and biochemists alike. In this respect, the existence of specific flippase proteins in the host plasma membrane, as well as specific PL carriers inside the infected erythrocyte, has not been proven. The possibilities of obtaining viable mutants using only specified PL biosynthetic pathways would allow the dissection of lipid transport pathways. However, in the case of PS biosynthesis, obtaining such mutants by serine auxotrophy seems difficult, due to the possible availability of this precursor through parasitic hemoglobin digestion. The isolation and molecular analysis of the genes encoding the enzymes of phospholipid biosynthesis has not yet been initiated.

ACKNOWLEDGMENTS. The authors are especially grateful to B. Beaumelle, G. Moll, M. Thuet, F. Wunderlich, J. Philippot, A. Bienvenüe, J. Op den Kamp, B. Roelofsen, and L. L. M. Van Deenen, who contributed to the studies reviewed here. Our study was partly supported by grants from the UNDP/World Bank/WHO Special Program for Research and Training in Tropical Diseases (Grant T16-181-M2-15B), the Institut National de la Santé et de la Recherche Médicale (CRE 910616), the Commission of the European Communities (T S2-0180-T), and the "Centre National de la Recherche Scientifique."

9. REFERENCES

Aikawa, M., 1988, Morphological changes in erythrocytes induced by malarial parasites, *Biol. Cell* **64:**173–181.

Aikawa, M., and Miller, L. H., 1983, Structural alteration of the erythrocyte membrane during malarial parasite invasion and intraerythrocytic development, in *Malaria and the Red Cell* (Ciba Symposium 94), pp. 45–63, Pitman, London.

Allred, D. R., Sterling, C. R., and Morse, P. D., 1983, Increased fluidity of *Plasmodium berghei*-infected mouse red blood cell membrane detected by electron spin resonance spectroscopy, *Mol. Biochem. Parasitol.* **7:**27–39.

Allred, D. R., Gruenberg, J. E., and Sherman, I. W., 1986, Dynamic rearrangements of erythrocyte membrane internal architecture induced by infection with *Plasmodium falciparum, J. Cell Sci.* **81:**1–16.

Al Rifai, W. A., Maurois, P., Fruchart, J. C., and Charet, P., 1982, A lipose tissue lipoprotein lipase activity in *Plasmodium chabaudi* and *Plasmodium vinckei* rodent malaria, *Biochimie* **64:**127–132.

Ancelin, M. L., and Vial, H. J., 1986a, Choline kinase activity in *Plasmodium*-infected erythrocytes: Characterisation and utilization as useful parasite specific marker in malarial fractionation studies, *Biochim. Biophys. Acta* **875:**52–58.

Ancelin, M. L., and Vial, H. J., 1986b, Several lines of evidence demonstrating that *Plasmodium falciparum,* a parasitic organism, has distinct enzymes for the phosphorylation of choline and ethanolamine, *FEBS Lett.* **202:**217–223.

Ancelin, M. L., and Vial, H. J., 1986c, Quaternary ammonium compounds efficiently inhibit *Plasmodium falciparum* growth *in vitro* by impairment of choline transport, *Antimicrob. Agents Chemother.* **29:**814–820.

Ancelin, M. L., and Vial, H. J., 1989, Regulation of phosphatidylcholine biosynthesis in *Plasmodium*-infected erythrocytes, *Biochim. Biophys, Acta* **1001:**82–89.

Ancelin, M. L., Vial, H. J., and Philippot, J. R., 1985, Inhibitors of choline transport into *Plasmodium*-infected erythrocytes are effective antiplasmodial compounds *in vitro, Biochem. Pharmacol.* **34:**4068–4071.

Ancelin, M. L., Vial, H. J., and Philippot, J. R., 1986, Characterization of choline and ethanolamine kinase activities in *Plasmodium*-infected erythrocytes, in *Enzymes of Lipid Metabolism II* (L. Freysz, H. Dreyfus, R. Massarelli, and S. Gatt, eds.), pp. 59–64, Plenum Press, New York.

Ancelin, M. L., Parant, M., Thuet, M. J., Philippot, J. R., and Vial, H. J., 1991, Increased permeability to choline in simian erythrocytes after *Plasmodium knowlesi* infection, *Biochem. J.* **273:**701–709.

Angus, M. G. N., Fletcher, K. A., and Maegraith, B. G., 1971a, Studies on the lipids of *Plas-*

modium knowlesi-infected rhesus monkeys (*Macaca mulatta*). I: Changes in serum lipids, *Ann. Trop. Med. Parasitol.* **65:**135–154.

Angus, M. G. N., Fletcher, K. A., and Maegraith, B. G., 1971b, Studies on the lipids of *Plasmodium knowlesi*-infected rhesus monkeys (*Macaca mulatta*). II: Changes in serum non-esterified fatty acids, *Ann. Trop. Med. Parasitol.* **65:**155–167.

Angus, M. G. N., Fletcher, K. A., and Maegraith, B. G., 1971c, Studies on the lipids of *Plasmodium knowlesi*-infected rhesus monkeys (*Macaca mulatta*). IV: Changes in erythrocytic lipids, *Ann. Trop. Med. Parasitol.* **65:**429–439.

Angus, M. G. N., Fletcher, K. A., and Maegraith, B. G., 1971d, Studies on the lipids of *Plasmodium knowlesi*-infected rhesus monkeys (*Macaca mulatta*). III: Changes in liver lipids, *Ann. Trop. Med. Parasitol.* **65:**419–427.

Ardeshir, F., Flint, J. F., Richman, S. J., and Reese, R. T., 1987, A 75 Kd merozoite surface protein of *Plasmodium falciparum* which is related to the 70 Kd heat-shock proteins, *EMBO J.* **6:**493–499.

Atkinson, C. T., Aikawa, M., Perry, G., Fujino, T., Bennett, V., Davidson, E. A., and Howard, R. J., 1987, Ultrastructural localization of erythrocyte cytoskeletal and integral membrane proteins in *Plasmodium falciparum*-infected erythrocytes, *Eur. J. Cell Biol.* **45:**192–199.

Bannister, L. H., and Mitchell, G. H., 1986, Lipidic vacuoles in *Plasmodium knowlesi* erythrocytic schizonts, *J. Protozool.* **33:**271–275.

Bannister, I. H., and Mitchell, C. M., 1989, The fine structure of secretion by *Plasmodium knowlesi* merozoites during red cell invasion, *J. Protozool.* **36:**275–283.

Beach, D. H., Sherman, I. W., and Holz, G. G., 1977, Lipids of *Plasmodium lophurae*, and of erythrocytes and plasmas of normal and *P. lophurae*-infected Pekin Ducklings, *J. Parasitol.* **63:**62–75.

Beaumelle, B. D., 1987, Metabolisme des acides gras et dynamique des phospholipides dand l'érythrocyte infecté par le parasite du paludisme. Thesis, Université des Sciences et Techniques du Languedoc, Montpellier.

Beaumelle, B. D., and Vial, H. J., 1986, Modification of the fatty acid composition of phospholipids and neutral lipids after infection of the simian erythrocyte by *Plasmodium knowlesi*, *Biochim. Biophys. Acta* **877:**262–270.

Beaumelle, B. D., and Vial, H. J., 1988a, Acyl-CoA synthetase activity in *Plasmodium knowlesi*-infected erythrocytes displays peculiar substrate specificities, *Biochim. Biophys. Acta* **958:**1–9.

Beaumelle, B. D., and Vial, H. J., 1988b, Correlation of the efficiency of fatty acid derivatives in suppressing *Plasmodium falciparum* growth in culture with their inhibitory effect on acyl-CoA synthetase activity, *Mol. Biochem. Parasitol.* **28:**39–42.

Beaumelle, B. D., and Vial, H. J., 1988c, Uninfected red cells from malaria-infected blood: Alteration of fatty acid composition involving a seric protein: An *in vivo* and *in vitro* study, *In Vitro Cell. Develop. Biol.* **24:**711–718.

Beaumelle, B. D., Vial, H. J., and Philippot, J. R., 1987, Reevaluation, using marker enzymes, of the ability of saponin and ammonium chloride to free *Plasmodium* from infected erythrocytes, *J. Parasitol.* **73:**743–748.

Beaumelle, B. D., Vial, H. J., and Bienvenüe, A., 1988, Enhanced transbilayer mobility of phospholipids in malaria-infected monkey erythrocytes. A spin label study, *J. Cell Physiol.* **135:**94–100.

Bell, R. M., and Coleman, R., 1980, Enzymes of glycerolipid synthesis in eukaryotes, *Annu. Rev. Biochem.* **49:**459–487.

Bianco, A. E., Favarolo, J. M., Burkot, T. R., Culvenor, J. G., Crewther, P. E., Brown, G. V., Anders, R. F., Coppel, R. L., and Kemp, D. J., 1986, A repetitive antigen of *Plasmodium falciparum* that is homologous to heat shock protein 70 of *Drosophila melanogaster*, *Proc. Natl. Acad. Sci. USA* **83:**8713–8717.

Bishop, W., and Bell, R. M., 1988, Assembly of phospholipids into cellular membranes. Biosynthesis, transmembrane movement and intracellular translocation, *Annu. Rev. Cell. Biol.* **1988:**579–610.

Boonpucknavig, V., Srichaikul, T., and Punyagupta, T., 1984, Clinical pathology, in *Antimalarial Drugs I* (W. Peters and W. H. Richards, eds.), pp. 127–176, Springer-Verlag, Berlin.

Buffinton, G. D., Hunt, N., Cowden, W., and Clark, A., 1986, Malaria: A role for reactive oxygen species in parasite killing and host pathology, in *Free Radicals, Cell Damage and Disease* (C. Rice-Evans, ed.), pp. 201–220, Richelieu Press, London.

Burns, E. R., and Pollack, S., 1988, *P. falciparum* infected erythrocytes are capable of endocytosis, *In Vitro Cell. Develop. Biol.* **24:**481–486.

Butler, K. W., Deslauriers, R., and Smith, I. C. P., 1984, *Plasmodium berghei:* Electron spin resonance and lipid analysis of infected mouse erythrocyte membranes, *Exp. Parasitol.* **57:**178–184.

Cabantchik, Z. I., 1989, Altered membrane transport of malaria-infected erythrocytes: A possible pharmacological target, *Blood* **74:**1464–1471.

Carman, G., and Henry, S., 1989, Phospholipid biosynthesis in yeast, *Annu. Rev. Biochem.* **1989:**635–639.

Cenedella, R. J., Jarrell, J. J., and Saxe, L. H., 1969, Lipid synthesis *in vivo* from 1-^{14}C-oleic acid and 6-^{3}H-glucose by intraerythrocytic *Plasmodium berghei, Mil. Med.* **134:**1045–1055.

Deguercy, A., Schrevel, J. Duportail, G., Laustriat, G., and Kuhry, J. G., 1986, Membrane fluidity changes in *P. berghei*-infected erythrocytes, investigated with a specific plasma membrane fluorescent probe, *Biochem. Int.* **12:**21–31.

Dei-Cas, E., Maurois, P., and Vernes, A., 1986, Physiopathologie du Paludisme, *Médecine/Sciences* **2:**322–330.

Devaux, P. F., 1988, Phospholipid flippases, *FEBS Lett.* **234:**8–12.

De Zeeuw, R. A., Wisbeck, J., Rock, R. C., and McCormick, G. I., 1972, Composition of phospholipids in *Plasmodium knowlesi* membranes and in host rhesus erythrocyte membranes, *Proc. Helminthol. Soc. Wash.* **39:**412–418.

Eisenberg, S., Levy, R. I., Paoletti, R., and Kritchevsky, D., 1975, Lipoprotein metabolism during experimental rodent malaria, *Adv. Lipid Res.* **13:**1–89.

Forte, T., and Roca, L., 1989, Thermal properties of red blood cells infected by malaria parasites, *Exp. Parasitol.* **69:**272–280.

Gennis, R. B., 1989, *Biomembranes, Molecular Structure and Function,* Springer-Verlag, New York.

Ginsburg, H., and Stein, D., 1987, New permeability pathways induced by the malarial parasite in the membrane of its host erythrocyte: Potential routes for targeting of drugs into infected cells, *Biosci. Rep.* **7:**455–463.

Ginsburg, H., Kutner, S., Zangwil, M., and Cabantchik, Z. I., 1986, Selectivity of pores induced in host erythrocyte membrane by *Plasmodium falciparum*. Effect of parasite maturation, *Biochim. Biophys. Acta* **861:**194–196.

Ginsburg, H., Landau, I., and Baccam, D., 1988, Effect of cholesterol-rich diet on the susceptibility of rodent malarial parasites to chloroquine chemotherapy, *Life Sci.* **42:**7–10.

Glatz, J. F. C., and Veerkamp, J. H., 1985, Intracellular fatty acid-binding proteins, *Int. J. Biochem.* **17:**13–22.

Golender, J., and Chevio, M., 1989, Oxidant stress and malaria: Host–parasite interrelationships in normal and abnormal erythrocytes, *Semin. Hematol.* **26:**313–325.

Guidon, P. T., and Hightower, L. E., 1986, The 73 kilodalton heat shock cognate protein purified from rat brain contains nonesterified palmitic and stearic acids, *J. Cell. Physiol.* **128:**239–245.

Gupta, C. M., and Mishra, G. C., 1981, Transbilayer phospholipids asymmetry in *Plasmodium knowlesi*-infected host cell membrane, *Science* **212:**1047–1049.

Gupta, C. M., Alam, A., Mathur, P. N., and Dutta, G. P., 1982, A new look at nonparasitized red cells of malaria-infected monkeys, *Nature* **299**:259–261.

Haldar, K., DeAmorim, A. F., and Cross, G. A. M., 1989, Transport of fluorescent phospholipid analogues from the erythrocyte membrane to the parasite in *Plasmodium falciparum* infected erythrocytes, *J. Cell Biol.* **108**:2183–2192.

Holz, G. G., 1977, Lipids and the malarial parasite, *Bull. WHO* **55**:237–248.

Holz, G. G., Beach, D. H., and Sherman, I. W., 1977, Octadecenoic fatty acids and their association with hemolysis in malaria, *J. Protozool.* **24**:566–574.

Homewood, C. A., and Neame, K. D., 1974, Malaria and the permeability of the host erythrocyte, *Nature* **252**:718–719.

Hommel, M., and Semoff, S., 1988, Expression and function of erythrocyte-associated surface antigens in malaria, *Biol. Cell* **64**:183–203.

Howard, R. J., 1988a, Malarial proteins at the membrane of *Plasmodium falciparum*-infected erythrocytes and their involvement in cytoadherence to endothelial cells, *Prog. Allergy* **41**:98–147.

Howard, R. J., 1988b, *Plasmodium falciparum* proteins at the host erythrocyte membrane: Their biological and immunogical significance and novel parasite organelles which deliver them to the cell surface, in *The Biology of Parasitism* (P. P. Englund and C. A. Sher, eds.), pp. 111–145, Alan R. Liss, New York

Howard, R. J., and Sawyer, W. H., 1980, Changes in the membrane microviscosity of mouse red blood cells infected with *Plasmodium berghei* detected using *n*-(9-anthroloxy) fatty acid fluorescent probes, *Parasitology* **80**:331–342.

Jackson, R. L., Morrisett, J. D., and Gotto, A. M., 1976, Lipoprotein structure and metabolism, *Physiol. Rev.* **56**:259–316.

Jett, M., Martin, S., and Schneider, I., 1989, Phosphoinositide hydrolysis in *P. falciparum* malaria, *FASEB J.* **3**:A371.

Joshi, P., and Gupta, C. M., 1988, Abnormal membrane organization in *P. falciparum* human erythrocytes, *Br. J. Haematol.* **68**:255–259.

Joshi, P., Alam, A., Chandra, R., Puri, S. K., and Gupta, C. M., 1986, Possible basis for membrane changes in nonparasitized erythrocytes of malaria-infected animals, *Biochim. Biophys. Acta* **862**:220–222.

Joshi, P., Dutta, G. P., and Gupta, C. M., 1987, An intracellular simian malarial parasite (*Plasmodium knowlesi*) induces stage-dependent alterations in membrane phospholipid organization of its host erythrocyte, *Biochem. J.* **246**:103–108.

Kennedy, E. P., 1986, The biosynthesis of phospholipids, in *Lipids and Membranes, Past, Present and Future* (Op den Kamp *et al.*, eds.), pp. 171–207, Elsevier, Amsterdam.

Koppaka, V., Sharma, R., and Lala, A. K., 1989, Fluorescence studies on erythrocyte membrane isolated from *Plasmodium berghei*-infected mice, *Mol. Cell. Biochem.* **91**:167–172.

Krugliak, M., Waldman, Z., and Ginsburg, H., 1987, Gentamicin and amikacin repress the growth of *Plasmodium falciparum* in culture, probably by inhibiting a parasite acid phospholipase, *Life Sci.* **40**:1253–1257.

Krungkrai, J., and Yuthavong, Y., 1983, Enhanced Ca^{2+} uptake by mouse erythrocytes in malarial (*Plasmodium berghei*) infection, *Mol. Biochem. Parasitol.* **7**:227–235.

Kutner, S., Breuer, W. V., Ginnsburg, H., Aley, S. B., and Cabantchik, Z. I., 1985, Characterization of permeation pathways in the plasma membrane of human erythrocytes infected with early stages of *Plasmodium falciparum*. Association with parasite development, *J. Cell Physiol.* **125**:521–527.

Lange, Y., and D'Alessandro, J. S., 1977, Characterization of mechanisms for transfer of cholesterol between human erythrocytes and plasma, *Biochemistry* **16**:4339–4343.

Lange, Y., Dolde, J., and Steck, T. L., 1981, The rate of transmembrane movement of cholesterol in the human erythrocyte, *J. Biol. Chem.* **256**:5321–5323.

Lange, Y., Swaisgood, M. H., Ramos, B. V., and Steck, T. L., 1989, Plasma membranes contain half the phospholipid and 90% of the cholesterol and sphingomyelin in cultured human fibroblasts, *J. Biol. Chem.* **264**:3786–3793.

Langreth, S. G., Jensen, J. B., Reese, R. T., and Trager, W., 1978, Fine structure of human malaria *in vitro, J. Protozol.* **25**:443–452.

Laser, H., Klein, R. A., Kemp, P., Lander, D., and Miller, N. G. A., 1975, Changes in the neutral lipid content of erythrocytes parasitized by *Plasmodium knowlesi, Proc. Brit. Soc. Parasitol.* **71**:5–7.

Leida, M. N., Mahoney, J. R., and Eaton, J. W., 1981, Intraerythrocytic calcium metabolism, *Biochem. Biophys. Res. Commun.* **103**:402–406.

Levander, O. A., Ager, A., Morris, V., and May, R., 1989, Menhaden-fish oil in a vitamin E-deficient diet: Protection against chloroquine-resistant malaria in mice, *Am. J. Clin. Nutr.* **50**:1237–1239.

Levander, O. A., Ager, A., Morris, V., and May, R., 1990, *Plasmodium yoelii:* Comparative antimalarial activities of dietary fish oils and fish oil concentrates in vitamin E-deficient mice, *Exp. Parasitol.* **70**:323–329.

Lynch, D. V., and Thompson, G. A., 1984, Retailored lipid molecular species: A tactical mechanism for modulating membrane properties, *Trends Biochem. Sci.* **9**:442–445.

Maguire, P. A., and Sherman, I. W. S., 1990, Phospholipid composition, cholesterol content and cholesterol exchange in *Plasmodium falciparum*-infected red cells, *Mol. Biochem. Parasitol.* **38**:105–112.

Majumdar, R. S., Chandanani, R. E., Ganduli, N. K., Chakarvarti, R. N., and Mahajan, R. C., 1981, Red cell and plasma lipids in monkeys infected with *Plasmodium knowlesi, Ind. J. Med. Res.* **73**:50–54.

Maurois, P., Vernes, A., Charet, P., Nouvelot, A., Becquet, R., and Giard, R., 1979, Changes in serum lipoproteins during malaria therapy with *Plasmodium vivax, Ann. Trop. Med. Parasitol.* **73**:491–493.

Maurois, P., Charet, P., Nouvelot, A., Fruchart, J. C., Vernes, A., and Biguet, J., 1980, Kinetic study of serum lipoproteins, total cholesterol and triacylglycerides in various models of experimental rodent malaria, *Ann. Trop. Med. Parasitol.* **74**:17–28.

Maurois, P., Charet, P., Fournet, B., and Fruchart, J. C., 1981, Metabolism of lipoproteins in rodent malaria: Relationship between lipolysis, steatosis and increased biosynthesis of VLDL, *Ann. Parasitol. (Paris)* **56**:9–19.

Maurois, P., Pessah, M., Briche, I., and Alcindor, L. G., 1985, Alterations of lecithin-cholesterol acyltransferase activity during *Plasmodium chabaudi* rodent malaria, *Biochimie* **67**:227–239.

McClean, S., Kabat, A., Sampugna, J., Purdy, W. C., and McCormick, G., 1976a, Analysis of the effect of malaria on lipid composition of Rhesus plasma, *Anal. Chim. Acta* **86**:255–261.

McClean, S., Purdy, W. C., Kabat, A., Sampugna, J., DeZeeuw, R., and McCormick, G., 1976b, Analysis of the phospholipid composition of *Plasmodium knowlesi* and Rhesus erythrocyte membranes, *Anal. Chim. Acta* **82**:175–185.

Mikkelsen, R. B., Kamber, M., Wadwa, K. S., Lin, P. S., and Schmidt-Ullrich, R., 1988, The role of lipids in *Plasmodium falciparum* invasion of erythrocytes: A coordinated biochemical and microscopic analysis, *Proc. Natl. Acad. Sci. USA* **85**:5956–5960.

Miller, L. H., 1989, Malaria. Binding of infected red cells, *Nature* **341**:18.

Mitchell, G. H., and Bannister, L. H., 1988, Malarial parasite invasion. Interaction with red cell membrane, *CRC Crit. Rev. Oncol./Haematol.* **8**:255–310.

Moll, G. N., Vial, H. J., Ancelin, M. L., Op den Kamp, J. A. F., Roelofsen, B., and Van Deenen, L. L., 1988, Phospholipid uptake by *Plasmodium knowlesi* infected erythrocytes, *FEBS Lett.* **232**:341–346.

Moll, G. N., Vial, H. J., Bevers, E. M., Ancelin, M. L., Roelofsen, B., Comfurius, P., Slotboom, A. J., Zwaal, R. F. A., Op den Kamp, J. A. F., and Van Deenen, L. L., 1990a, Phospholipid asymmetry in the plasma membrane of malaria infected erythrocytes, *Biochem. Cell Biol.* **68:**579–585.

Moll, G. N., Vial, H. J., Ancelin, M. L., Op den Kamp, J. A. F., Roelofsen, B., Slotboom, A. J., and Van Deenen, L. L. M., 1990b, Selective elimination of malaria-infected erythrocytes by a modified phospholipase A_2, *Biochim. Biophys. Acta* **1024:**189–192.

Nelson, G. J., 1972, Lipid composition and metabolism of erythrocytes, in *Blood Lipids and Lipoproteins* (G. J., Nelson, ed.), pp. 317–386, John Wiley and Sons, New York.

Nishizuka, Y., 1988, The molecular heterogeneity of protein kinase C and its implication for cellular regulation, *Nature* **234:**661–665.

Ononghbu, I. C., and Onyeneke, E. C., 1983, Plasma lipid changes in human malaria, *Tropenmed. Parasit.* **34:**193–196.

Op den Kamp, J., 1979, Lipid asymmetry in membranes, *Annu. Rev. Biochem.* **48:**47–71.

Pelech, S. L., and Vance, D. E., 1989, Signal transduction via phosphatidylcholine cycles, *Trends Biochem. Sci.* **14:**28–30.

Raetz, C. R. H., and Dowhan, W., 1990, Biosynthesis and function of phospholipids in *Escherichia coli, J. Biol. Chem.* **265:**1235–1238.

Rock, R. C., 1971a, Incorporation of ^{14}C-labelled fatty acids into lipids of rhesus erythrocytes and *Plasmodium knowlesi in vitro, Comp. Biochem. Physiol.* **40B:**893–906.

Rock, R. C., 1971b, Incorporation of ^{14}C-labelled non-lipid precursors into lipids of *Plasmodium knowlesi in vitro, Comp. Biochem. Physiol.* **40B:**657–669.

Rock, R. C., Standefer, J., and Little, W., 1971a, Incorporation of ^{32}P-orthophosphate into membrane phospholipids of *Plasmodium knowlesi* and host erythrocytes of *Macaca mulata, Comp. Biochem. Physiol.* **40B:**543–561.

Rock, R. C., Standefer, J. C., Cook, R. T., Little, W., and Sprintz, H., 1971b, Lipid composition of *Plasmodium knowlesi* membranes: Comparison of parasites and microsomal subfractions with host rhesus erythrocyte membranes, *Comp. Biochem. Physiol.* **38B:**425–437.

Rodriguez, M. H., and Jungery, M., 1986, A protein on *Plasmodium falciparum*-infected erythrocytes functions as a transferrin receptor, *Nature* **324:**388–391.

Sanghera, J. S., and Vance, D. E., 1989, CTP:phosphocholine cytidylyltransferase is a substrate for cAMP-dependent protein kinase *in vitro, J. Biol. Chem.* **264:**1215–1223.

Schauer, R., Wember, M., and Howard, R. J., 1984, Malaria parasites do not contain or synthesize sialic acids, *Hoppe Seylers Z. Physiol. Chem.* **365:**185–194.

Schwartz, R. S., Chiu, D., and Lubin, B., 1985a, Plasma membrane phospholipid organization in human erythrocytes, *Curr. Top. Hematol.* **5:**63–112.

Schwartz, R. S., Tanaka, Y., Fidler, I. J., Chiu, D. T., Lubin, B., and Schroit, A. J., 1985b, Increased adherence of sickled and phosphatidylserine-enriched human erythrocytes of cultured human peripheral blood monocytes, *J. Clin. Invest.* **75:**1965–1972.

Schwartz, R. S., Olson, J. A., Raventos-Suarez, C., Yee, M., Heath, R. H., Lubin, B., and Nagel, R. L., 1987, Altered plasma membrane phospholipid organization in *Plasmodium falciparum*-infected human erythrocytes, *Blood* **69:**401–407.

Seed, T. M., and Kreier, J. P., 1972, *Plasmodium gallinaceum:* Erythrocyte membrane alteration and associated plasma changes induced by experimental infection, *Proc. Helminthol. Soc.* **39:**387–411.

Sherman, I. W., 1979, Biochemistry of *Plasmodium* (malarial parasites), *Microbiol. Rev.* **43:**453–495.

Sherman, I. W., 1984, Metabolism, in *Antimalarial Drugs,* vol. I. (W. Peters and W. H. Richards, eds.), pp. 31–71, Springer-Verlag, Berlin.

Sherman, I. W., 1988, Mechanisms of molecular trafficking in malaria, *Parasitology* **96:**857–881.

Sherman, I. W., and Greenan, R. T., 1984, Altered red cell membrane fluidity during schizogonic development of malarial parasites (*Plasmodium falciparum* and *P. lophurae*), *Trans. Roy. Soc. Trop. Hyg.* **78:**641–644.

Sherwood, J. A., Spitalnik, S. L., Aley, S. B., Quakyi, I. A., and Howard, R. J., 1986, *Plasmodium falciparum* and *P. knowlesi:* Initial identification and characterization of malaria synthesized glycolipids, *Exp. Parasitol.* **62:**127–141.

Simoes, A. P., Moll, G. N., Vial, H. J., Beaumelle, B., Roelofsen, B., and Op Den Kamp, J., 1990, *Plasmodium knowlesi* induces alterations in phosphatidylcholine and phosphatidylethanolamine molecular species composition of parasitized monkey erythrocytes, *Biochim. Biophys. Acta* **1022:**135–145.

Skou, J. C., 1988, Overview: The Na, K-pump, *Methods Enzymol.* **156:**1–25.

Slomiany, B. L., Murty, L. N., Liau, Y. H., and Slomiany, A., 1987, Animal glycoglycerolipids, *Prog. Lipid Res.* **26:**29–51.

Spector, A. A., and Yorek, M. A., 1986, Membrane lipid composition and cellular function, *J. Lipid Res.* **26:**1013–1025.

Stocker, R., Cowden, W. B., Tellam, R. L., Weidemann, M. J., and Hunt, N. H., 1987, Lipids from *Plasmodium vinckei*-infected erythrocytes and their susceptibility to oxidative damage, *Lipids* **22:**51–57.

Stremmel, W., and Berk, P. D., 1986, Hepatocellular influx of [^{14}C]oleate reflects membrane transport rather than intracellular metabolism or binding, *Proc. Natl. Acad. Sci. USA* **83:**3086–3090.

Stubbs, C. D., and Smith, A. D., 1984, The modification of mammalian membrane polyunsaturated fatty acid composition in relation to membrane fluidity and function, *Biochim. Biophys. Acta* **779:**89–137.

Tanabe, K., Mikkelsen, R. B., and Wallach, D. F. H., 1982, Calcium transport of *Plasmodium chabaudi*-infected erythrocytes, *J. Cell Biol.* **93:**680–684.

Taraschi, T. F., Parashar, A., Hooks, M., and Rubin, H., 1986, Perturbation of red cell membrane structure during intracellular maturation of *Plasmodium falciparum, Science* **232:**102–104.

Taylor, D. W., Para, M., and Stearns, M. E., 1987, *Plasmodium falciparum:* Fine structural changes in the cytoskeletons of infected erythrocytes, *Exp. Parasitol.* **64:**178–176.

Tilley, L., Cribier, S., Roelofsen, B., Op den Kamp, J. A. F., and Van Deenen, L. L. M., 1986, ATP-dependent translocation of amino phospholipids across the human erythrocyte membrane, *FEBS Lett.* **194:**21–27.

Trager, W., and Gill, G. S., 1989, *Plasmodium falciparum* gametocyte formation *in vitro:* Its stimulation by phorbol diesters and by 8-bromo cyclic adenosine monophosphate, *J. Protozool.* **36:**451–454.

Trigg, P. I., 1968, Sterol metabolism of *Plasmodium knowlesi in vitro, Ann. Trop. Med. Parasitol.* **62:**481–487.

Vance, D. E., and Pelech, S. L., 1984, Enzyme translocation in the regulation of phosphatidylcholine biosynthesis, *Trends Biochem. Sci.* **9:**17–20.

Vance, D. E., and Ridgway, D., 1988, The methylation of phosphatidylethanolamine, *Prog. Lipid Res.* **27:**61–79.

Van Deenen, L. L. M., and De Gier, J., 1975, Lipids of the red cell membrane, in *The Red Blood Cell* (G. Surgenor, ed.), pp. 147–211, Academic Press, New York.

Van Der Schaft, P. H., Beaumelle, B., Vial, H., Roelofsen, B., Op Den Kamp, J. A. F., and Van Deenen, L. L. M., 1987, Phospholipid organization in monkey erythrocytes upon *Plasmodium knowlesi* infection, *Biochim. Biophys, Acta* **901:**1–14.

Van der Wiele, F., Atsma, W., Roelofsen, B., Van Linde, M., Van Binsbergen, J., Radvanyi, F., Raykova, D., Slotboom, A. J., and De Haas, G. H., 1988, Site-specific epsilon-NH_2 monoacylation of pancreatic phospholipase A_2.2. Transformation of soluble phospholipase A_2 into a highly penetrating "membrane-bound" form, *Biochemistry* **27:**1688–1694.

Vial, H., Thuet, M., and Philippot, J., 1982a, Phospholipid biosynthesis in synchronous *Plasmodium falciparum* cultures, *J. Protozool.* **29:**258–263.

Vial, H., Thuet, M., Broussal, J., and Philippot, J., 1982b, Phospholipid biosynthesis by *Plasmodium knowlesi*-infected erythrocytes: The incorporation of phospholipid precursors and the identification of previously undetected metabolic pathways, *J. Parasitol.* **68:**379–391.

Vial, H. J., Philippot, J. R., and Wallach, D. F. M., 1984a, A reevaluation of the status of cholesterol in erythrocytes infected by *Plasmodium knowlesi* and *Plasmodium falciparum*, *Mol. Biochem. Parasitol.* **13:**53–65.

Vial, H. J., Thuet, M. J., Ancelin, M. L., Philippot, J. R., and Chavis, C., 1984b, Phospholipid metabolism as a new target for malaria chemotherapy. Mechanism of action of D-2-amino-1-butanol, *Biochem. Pharmacol.* **33:**2761–2770.

Vial, H. J., Thuet, M. J., and Philippot, J. R., 1984c, Cholinephosphotransferase and ethanolaminephosphotransferase activities in *P. knowlesi* infected erythrocytes. Their use as parasite specific markers, *Biochim. Biophys. Acta* **795:**372–383.

Vial, H. J., Ancelin, M. L., Thuet, M. J., and Philippot, J. R., 1989a, Phospholipid metabolism in *Plasmodium knowlesi*-infected erythrocytes: Guidelines for further studies on radioactive precursor incorporation, *Parasitology* **98:**351–357.

Vial, H. J., Van der Schaft, P., Beaumelle, B. D., Thuet, M. J., and Op Den Kamp, J., 1989b, Improved isolation of *Plasmodium knowlesi*-infected erythrocyte host–cell membrane on polycationic beads, *Parasitol. Res.* **75:**419–421.

Vial, H. J., Ancelin, M. L., Philippot, J. R., and Thuet, M. J., 1990, Biosynthesis and dynamics of lipids in *Plasmodium*-infected mature mammalian erythrocytes, *Blood Cells* **16:**531–555.

Watkins, J. D., and Kent, C., 1990, Phosphorylation of CTP:phosphocholine cytidylyltransferase *in vivo*. Lack of phorbol ester treatment in HeLa Cells, *J. Biol. Chem.* **265:**2190–2197.

Weatherall, D. J., Abdalla, S., and Pippard, M. J., 1983, The anemia of *Plasmodium falciparum* malaria, in *Malaria and the Red Cell* (Ciba Foundation Symposium), pp. 74–94, Pitman, London.

Williamsom, P., Antia, R., and Schlegel, R. A., 1987, Maintenance of membrane phospholipid asymmetry. Lipid-cytoskeletal interactions or lipid pump? *FEBS Lett.* **219:**316–320.

Woodward, C. B., and Zwaal, R. F. A., 1972, The lytic behavior of pure phospholipases A_2 and C towards osmotically swollen erythrocytes and resealed ghosts, *Biochim. Biophys. Acta* **274:**272–278.

Wunderlich, F., Helwig, M., Schillinger, G., Vial, H., Philippot, J., and Speth, V., 1987, Isolation and characterization of parasites and host cell ghosts from erythrocytes infected with *Plasmodium chaubaudi*, *Mol. Biochem. Parasitol.* **23:**103–115.

Yeagle, P. L., 1985, Cholesterol and the cell membrane, *Biochim. Biophys. Acta* **822:**267–287.

Yeagle, P., 1987, *The Membrane of the Cell*, Academic Press, New York.

Zachowski, A., Favre, E., Hervé, P., and Devaux, P., 1986, Outside–inside translocation of aminophospholipids in the human erythrocyte membrane is mediated by a specific enzyme, *Biochemistry* **25:**2585–2590.

Zolg, J. W., Macleod, A. J., Scaife, J. G., and Beaudoin, R. L., 1984, The accumulation of lactic acid and its influence on the growth of *Plasmodium falciparum* in synchronized cultures, *In Vitro Cell. Develop. Biol.* **20:**205–215.

Zwaal, R. F. A., and Bevers, E. M., 1986, Structural and functional aspects of the platelet plasma membrane, in *Lipids and Membranes: Past, Present and Future* (J. A. Op den Kamp, B. Roelofsen, and K. W. Wirtz, eds.), pp. 231–258, Elsevier, Amsterdam.

Zwall, R. F. A., Bevers, E., Comfurius, P., Rosing, J., Tilly, R., and Verhallen, P., 1989, Loss of membrane phospholipid asymmetry during activation of blood platelets and sickled red cells; mechanisms and physiological significance, *Mol. Cell. Biochem.* **91:**23–31.

Chapter 9

Genetic Analysis of Malaria Parasites

Brian Fenton and David Walliker

1. INTRODUCTION

Malaria is the disease caused by protozoan parasites of the genus *Plasmodium*. There are many species parasitizing such organisms as rodents, reptiles, birds, and mammals (Garnham, 1966). The parasites are transmitted from one host to another by an anopheline mosquito vector. Although the vertebrate host for each species of *Plasmodium* is specific, the anopheline vector need not be. For example, *Plasmodium falciparum*, the most pathogenic species infecting man, can be transmitted by different species of mosquito and yet only develops naturally within man.

The diseases caused by the human malarias are a heavy burden on the tropical regions of the world. There are estimated to be 200 million cases in Africa alone and as many as 2 million people may die as a result.

The first genetic study of malaria parasites was attempted by crossing two strains of *P. gallinaceum*, a malaria parasite of birds (Greenberg and Trembley, 1954). Many basic genetic principles were established using the rodent malaria species *P. chabaudi*, *P. berghei*, *P. vinckei* and *P. yoelii*. These rodent models were found to be particularly suitable for genetic analysis for three main reasons:

Brian Fenton Cell Cycle Genetics Group, Department of Biochemistry, Medical Sciences Institute, University of Dundee, Dundee DD1 4HN, Scotland. **David Walliker** Institute of Cell, Animal, and Population Biology, University of Edinburgh, Edinburgh EH9 3JN, Scotland.
Subcellular Biochemistry, Volume 18: Intracellular Parasites, edited by J. L. Avila and J. R. Harris. Plenum Press, New York, 1992.

(1) they could be readily maintained in laboratory animals, (2) relatively large numbers of crosses and progeny could be made, and (3) they exhibited variation in features such as drug response, antigens, virulence, and enzymes that could be used for genetic analysis. It was thought likely that knowledge gained from studying the inheritance of these characters would have important implications for human malaria species. The contribution of the rodent malarias to the understanding of the genetic composition of malaria parasites has been documented by Walliker (1983a,b) and Beale and Walliker (1988).

In 1987 the first genetic analysis of the human malaria parasite *P. falciparum* was described (Walliker *et al.*, 1987). A number of technical advances made this possible, in particular the development of an *in vitro* culture system for the asexual blood forms of this species (Haynes *et al.*, 1976; Trager and Jensen, 1976). This was followed by a technique for *in vitro* growth and successful maturation of gametocyte and gamete stages infective to mosquitoes (Campbell *et al.*, 1982; Ponnudurai *et al.*, 1982). Mosquitoes that have fed on these gametocytes produce the infective sporozoite forms in their salivary glands. The life cycle can be completed by infecting chimpanzees, from which new *in vitro* cultures of the resulting blood forms can be established. This chapter will concentrate on the results obtained using *P. falciparum*.

2. LIFE CYCLE

2.1. The Malaria Life Cycle and Its Relationship to Genetic Studies

The life cycle of the human malaria parasite *P. falciparum* is presented in Figure 1. The parasite is haploid with the exception of the zygote stage within the mosquito. The first stage to invade the human body is the sporozoite. This is injected from the salivary glands of an infected mosquito as it takes a blood meal. The sporozoite invades a liver parenchyma cell where it undergoes exo-erythrocytic growth and mitotic divisions to form, after 7 days, some 30,000 merozoites. Merozoites invade erythrocytes in the blood stream. Once inside, the merozoite develops over a 48-hr period from ring stage to trophozoite and finally to a schizont. The schizont is composed of up to 24 merozoites which are released into the blood stream and invade further erythrocytes. This process is the asexual cycle. In later stages of infection gametocytes are formed. It has been demonstrated *in vitro* that a proportion of mature schizonts produce merozoites, which invade erythrocytes to form only gametocytes (Bruce *et al.*, 1990). Gametocytes form gametes after ingestion by a female mosquito. Both male and female gametes are formed from a malaria clone originating from a single blood form (Downs, 1947). As the organism is haploid, this suggests that differentiation of male and female gametes cannot depend on chromosomal segregation, as

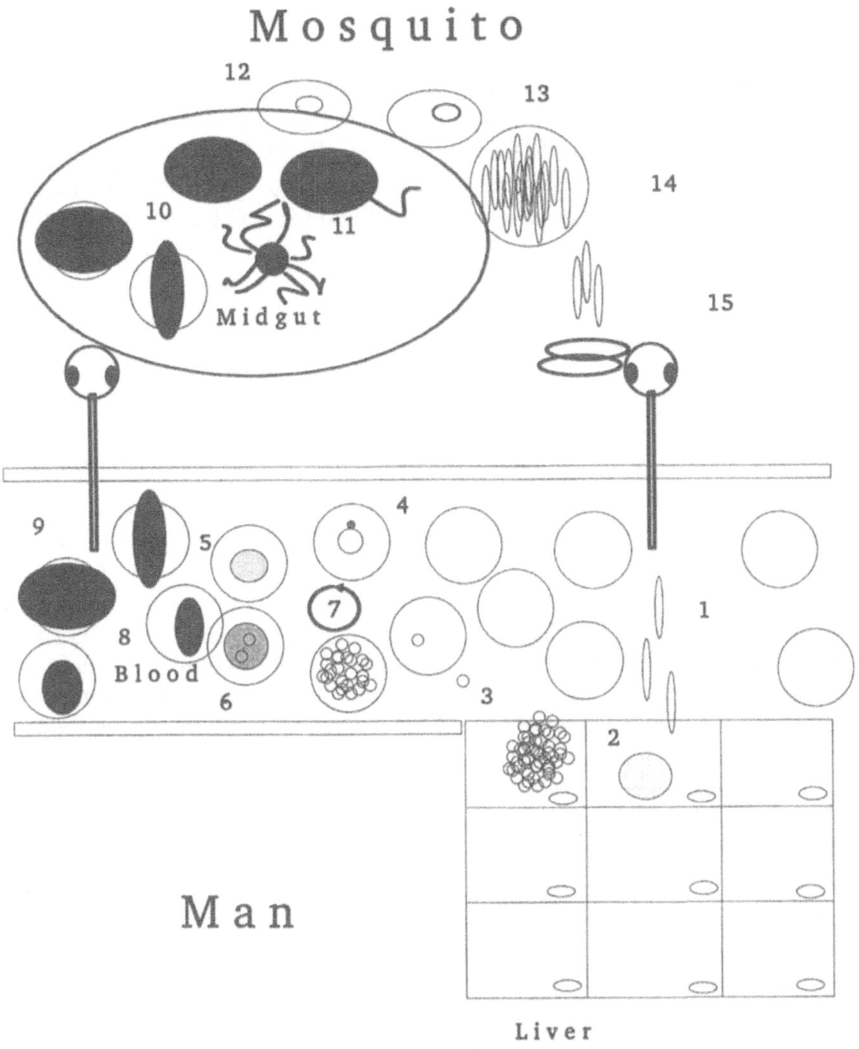

FIGURE 1. The life cycle of *Plasmodium falciparum*. 1, Sporozoite invades liver parenchyma cell; 2, exoerythrocytic multiplication; 3, merozoite release and erythrocyte invasion; 4, ring stage; 5, trophozoite stage; 6, schizont stage; 7, the asexual cycle; 8, parasitaemia increases; 9, gametocytes form in blood; 10, gametocytes are ingested and form gametes; 11, Fertilization to form zygote; 12, Ookinete penetrates midgut wall; 13, sporogony; 14, Mature sporozoites migrate to and invade the salivary glands (15).

in higher eukaryotes. Within the mosquito midgut the gametocytes are released from their erythrocytes. The male gametocytes undergo exflagellation, releasing eight motile microgametes. These fuse with and fertilize the female macrogametes. This fusion forms the diploid zygote. Zygotes develop into ookinetes, which penetrate the mosquito midgut wall. On the hemocoel side of the wall, the ookinetes form oocysts. The parasite is believed to undergo meiosis within a few hours of zygote formation (Sinden and Hartley, 1985). After approximately 10 days, sporozoites have developed within the oocyst. These are released and migrate to the salivary glands, to be injected into the next person on whom the mosquito feeds.

3. GENETIC ANALYSIS

3.1. Introduction

The complex nature of the malaria parasite's life cycle presents difficulties in making a conventional type of genetic analysis by crossing experiments. The procedure used in making a *P. falciparum* cross is outlined in Figure 2 (based on Walliker *et al.*, 1987).

Cloned asexual parasite lines are first established by limiting dilution or by micromanipulation of blood forms in *in vitro* culture (Rosario, 1981; Trager *et al.*, 1981). This process gives genetically pure populations of haploid parasites. Culture conditions are then changed to allow the formation of gametocytes. These are mixed in equal numbers and fed to mosquitoes. Within the mosquito, gametes form and undergo fertilization. Since each parasite clone produces both male and female gametes, both cross and self-fertilization events can take place. After the parasites have matured to sporozoites in the mosquitoes, they are allowed to feed directly on a mammalian host; alternatively, the sporozoites can be dissected from the mosquitoes and inoculated directly. The sporozoites then enter the mammalian liver. Approximately 10 days later, blood forms may be detected in erythrocytes. Samples of blood are then removed and asexual parasites established in culture. If crossing between the parent clones has occurred in the mosquitoes, these cultures will consist of a mixture of both parental-type and recombinant-type progeny. To characterize resultant parasites in detail, individual progeny clones are established and examined for the characters that distinguished the parent clones. The presence of recombinant-type parasites, that is, parasite clones with nonparental combinations of markers, indicates that crossing between the parent clones has taken place.

The presence of recombinant parasites can be tested without cloning if there are differences in drug response in the parents. In the case of the 3D7 and HB3 cross (Walliker *et al.*, 1987), pyrimethamine treatment can be used on the un-

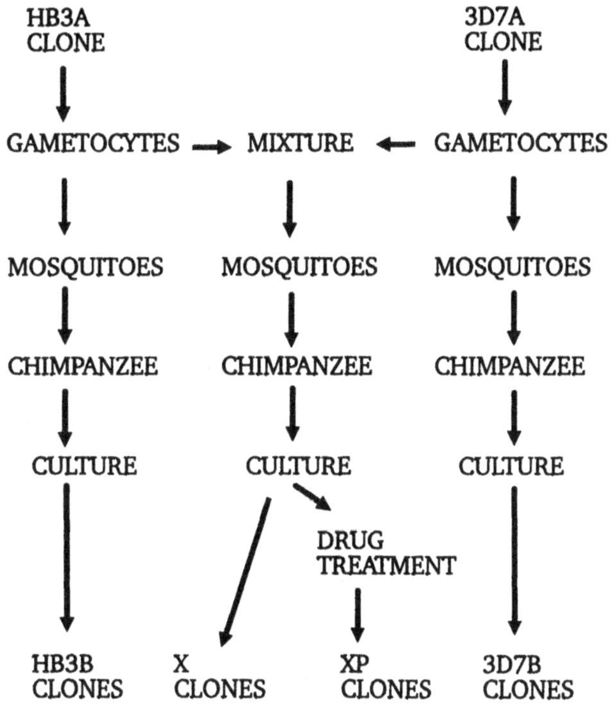

FIGURE 2. The procedure used in constructing a cross between two clones (3D7 and HB3) of the human malaria parasite *Plasmodium falciparum*.

cloned progeny because the HB3 parent was resistant to this drug; the treatment therefore eliminates 3D7 parental parasite genotypes from the mixture of uncloned progeny. 3D7 markers still present in the drug-treated progeny must therefore result from the presence of recombinant type parasites.

At the same time as gametocytes are mixed to perform the cross, parental gametocytes are also fed separately to mosquitoes. The result of these infections allows the infectivity of the individual parental gametes to the mosquito to be determined.

3.2. Inheritance of Specific Characters

3.2.1. Principles

One of the first procedures in examining the genetics of an organism is to isolate genetically distinct individuals. In eukaryotic organisms that are much used for genetic analysis such as yeast and *Drosophila*, this is relatively easy as

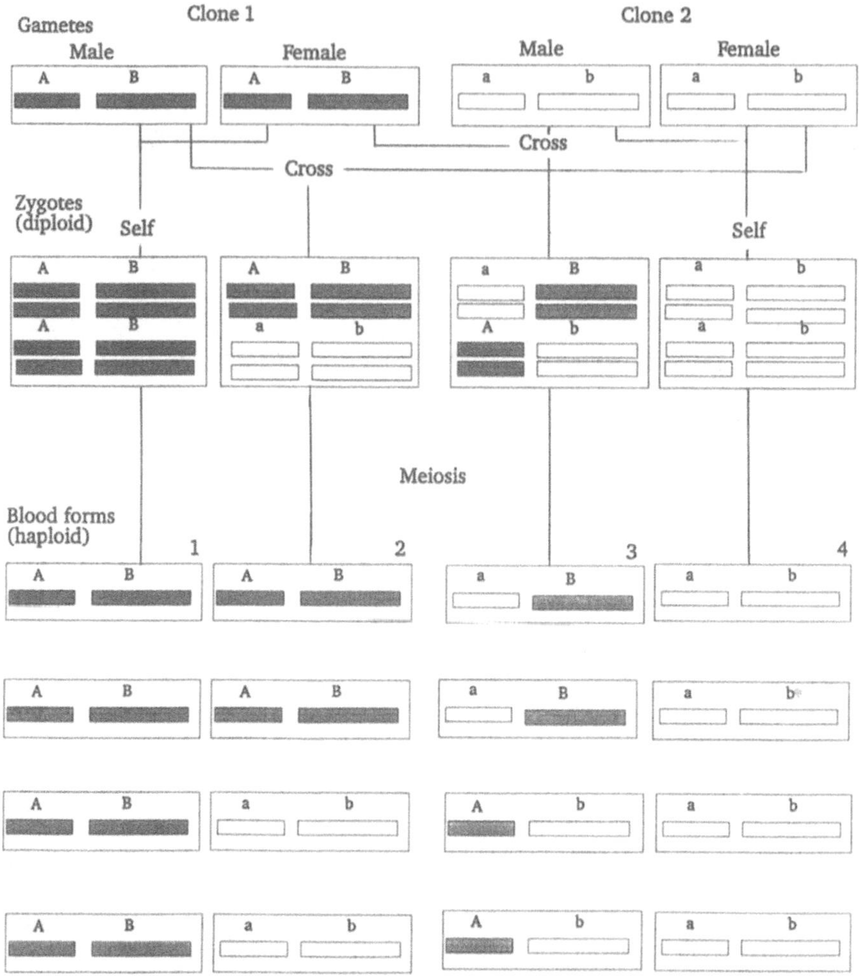

FIGURE 3. Inheritance patterns in *Plasmodium*. The diagrams illustrate the segregation of alleles of two genes, A/a and B/b, and recombination between them during meiosis in the mosquito phase of the life cycle. (A) The two genes are unlinked. Recombinant types Ab and aB in case 3 are produced by independent assortment of chromosomes. (B) The two genes are linked. In this instance, case 3 produces recombinant types Ab and aB by crossing-over between the linked genes.

there are many phenotypes and rapid selection systems for isolating mutants. Malaria parasites are not easily maintained in large quantities and it is therefore difficult to isolate and grow mutants. Until now, most genetic studies have exploited preexisting natural diversity. By analyzing a large number of geographically distinct isolates, it is possible to identify variable characters such as anti-

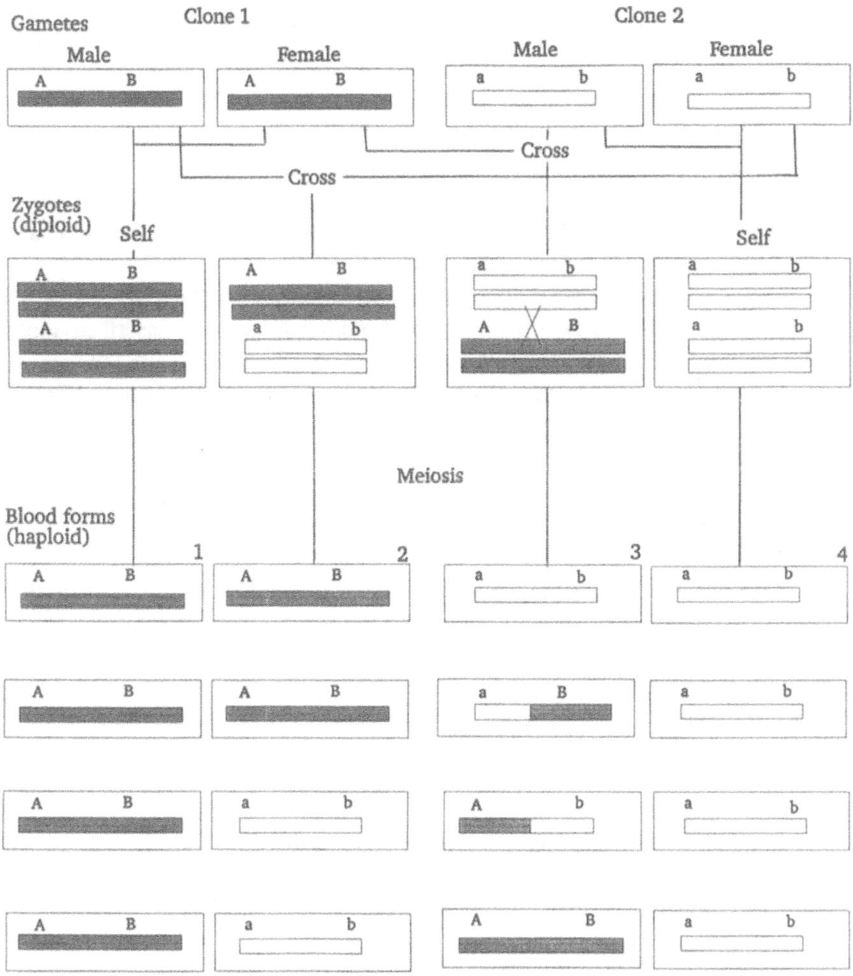

FIGURE 3 (*Continued*)

gens, enzymes, and other proteins. The inheritance patterns of such variant characters in crosses can be used to determine their genetic basis, for example, whether they are determined by alleles of the same gene or by two different genes. By crossing parasites that differ by more than one gene, linkage studies can be made to determine whether they are located on the same chromosome. However, this is a difficult procedure for malaria parasites as the numbers of cross progeny that it is possible to examine are usually small.

Figures 3A and B show typical patterns of inheritance for two genes in the malaria parasite. In Figure 3A the two genes are on different chromosomes. At meiosis, alleles of each gene reassort independently, giving four possible combinations at equal frequency. When the two genes are on the same chromosome, however, it would be usual to expect a predominance of the parental combinations, as shown in Figure 3B.

In the crossing technique used for malaria parasites, it is possible for both cross- and self-fertilization between male and female gametes of each parent. Consequently, the progeny of any cross can be expected to show a preponderance of parental types (see Figure 3A).

New techniques have now made it possible to examine the molecular genetics of this organism without using the conventional approach described above. For example, techniques such as pulsed field gradient gel electrophoresis, restriction analysis, and Southern blotting are allowing the karyotypes and linkage of genes to be examined directly. The principles of these techniques will be explained in the relevant sections.

3.2.2. Enzymes

One of the principle characters that has been used to study genetic variation in malaria parasites has been enzyme variation revealed by electrophoretic techniques. Enzymes are normally the direct products of nuclear genes. Alloenzymes, which represent allelic variants of a given enzyme distinguishable by charge differences, can be separated by techniques such as starch or cellulose acetate electrophoresis. The enzyme activity is subsequently visualized by supplying substrates and cofactors to give a color reaction, forming a band on the gel at the position to which the enzyme has migrated (Carter, 1978). Different electrophoretic forms are given different numbers; for example, ADA1 and ADA2 are two electrophoretic forms of adenosine deaminase. Six enzymes have proved useful in the study of malaria parasites, these are:

1. Adenosine deaminase (ADA-EC 3.5.4.4)
2. NADP-dependent glutamate dehydrogenase (GDH-EC 1.4.1.2)
3. Glucose phosphate isomerase (GPI-EC 5.3.1.9)
4. Lactate dehydrogenase (LDH-EC 1.1.1.27)
5. Peptidase (PEP-EC 3.4.11 or 13)
6. 6-phosphogluconate dehydrogenase (PGD-EC 1.1.1.43)

The inheritance of these alloenzymes has been followed in genetic crosses between parasite clones. Their patterns of segregation have provided important information on the ploidy of the parasites, as well as their relationship to other genes (Walliker, 1983a). For example, in the cross that was carried out between 3D7 and HB3, two *P. falciparum* clones that differ in the form of ADA they

Table I
Patterns of Alloenzyme Variation in Rodent Malaria Species[a]

Species	Alloenzyme[b] GDH	GPI	LDH	PGD
P. berghei	3	3	1	1
P. chabaudi	5	4,8	2,3,4,5,8,10	2,3,7
P. vinckei	6	5,6,7,9,11,12,13	6,7,9,11	5,6
P. yoelii	1,2,4	1,2,10	1	4

[a]From Beale *et al.* (1978); Lainson (1983). The numbers represent different electrophoretic forms of each enzyme; see text for details.
[b]GDH, NADP-dependent glutamate dehydrogenase; GPI, glucose phosphate isomerase; LDH, lactate dehydrogenase; PGD, 6-phosphogluconate dehydrogenase.

possess, the products of the cross were found to contain one or the other of these forms in approximately equal numbers (Walliker *et al.*, 1987). This result was in accordance with the expected ratio for a haploid organism possessing only single alleles of a given gene. The ADA locus also appeared not to be linked to any other character studied, as recombination was found between this and the other markers used.

Alloenzymes can also be used to distinguish different species and subspecies of malaria parasites. As an example, Table I shows the distribution of alloenzyme forms in four rodent malaria species: *P. berghei, P. yoelii, P. chabaudi,* and *P. vinckei.* Each species contains, for the most part, unique enzyme variants. From this information it can be concluded that no genetic exchange is taking place between these groups, confirming their status as different species (Beale *et al.*, 1978; Lainson, 1983).

Using the same technique, it is now clear that *P. falciparum* is a single species worldwide because shared enzyme variants are common and are found in geographically separated isolates (Sanderson *et al.*, 1981; Thaithong *et al.*, 1981; Creasey *et al.*, 1990). In this and other work, it has been found that a considerable proportion of *P. falciparum* isolates contain more than one form of a given enzyme marker. By cloning these isolates, it has been shown that they are composed of more than one parasite genotype. Clones contain either one or the other of enzyme variants, but never both (Rosario, 1981). Again this observation confirms the haploidy of asexual blood stages.

3.2.3. Antigens

The number of antigens that have been identified in malaria parasites is large and new antigens are continually being characterized. For a review of the malaria antigens identified so far, see Kemp *et al.* (1986, 1990).

Most of the genes for these antigens have been identified by using molecular techniques. This is done using specific reagents such as monoclonal antibodies or immune sera to screen bacteria expressing these genes as foreign proteins (Kemp *et al.*, 1983; Young and Davies, 1983). Once a gene is isolated, information about copy number, gene expression, and the extent of allelic diversity can be obtained. By using pulsed field gel electrophoresis, it is also possible to determine the locus of antigen genes on specific chromosomes (see Section 3.2.5).

Two *P. falciparum* antigens, merozoite surface antigens 1 (also known as PMMSA, P190, PSA, and MSA1) and 2 (MSA2), have been the center of much research. Their immunology, biochemistry, DNA, and amino acid sequences have been elucidated (Schofield *et al.*, 1982; Hall *et al.*, 1984; Schwartz *et al.*, 1986; Peterson, M. G., *et al.*, 1988; Fenton *et al.*, 1989, 1991; Saul *et al.*, 1989; Smythe *et al.*, 1990). By using monoclonal antibodies (mAbs) that recognize different epitopes on these molecules, it has been possible to classify them into different allelic groups or serotypes (McBride *et al.*, 1982, 1985; Fenton *et al.*, 1991). Both MSA1 and MSA2 exist as two major groups, within which further minor divisions can be made. Details of the amino acid and DNA sequences show that in both cases the mAb groupings are a direct consequence of sequence differences. DNA sequences of alleles from the two major groups show that there are many differences between major serotypes and only minor differences within a serotype. These changes involve insertions, deletions, point mutations, and increases or decreases in the numbers and types of tandemly repeated units (Tanabe *et al.*, 1987; Smythe *et al.*, 1990; Fenton *et al.*, 1991). Figure 4 shows a simplified scheme of the amino acid sequence variation for the major MSA2 alleles.

In the *P. falciparum* cross, the two parent parasites have different alleles of MSA1 and of MSA2. The MSA1 alleles of each could be distinguished by two mAbs, denoted 7.3 and 9.2, 3D7 being 9.2+/7.3− in immunofluorescence tests, and HB3 being 7.3+/9.2− (Walliker *et al.*, 1987). MSA2 from 3D7 was found to react with mAb 12.3, while HB3 was negative for this mAb (see Table II). The use of a technique that indirectly labels two mAbs different colors (Walliker *et al.*, 1987; Carter *et al.*, 1989) enabled the reactivities of individual parasites in the uncloned mixture derived from the cross to be determined (see Table II). The results of these experiments illustrated clearly that the two mAbs 7.3 and 9.2 recognized alternative forms of the MSA1 antigen, no parasite molecule being found that was positive for both mAbs; it could be concluded that the MSA1 variants were encoded by alleles of the same gene. MSA2 on the other hand was encoded by a different gene as new combinations of reactivities were found between mAb 12.3 recognizing the MSA2 antigen and mAbs recognizing MSA1. Results from the uncloned mixture clearly show that large numbers of parasites exhibiting recombination between these genes were produced. This result was consistent with the two genes being located on different chromosomes, as subsequently found in pulsed field gel electrophoretic studies.

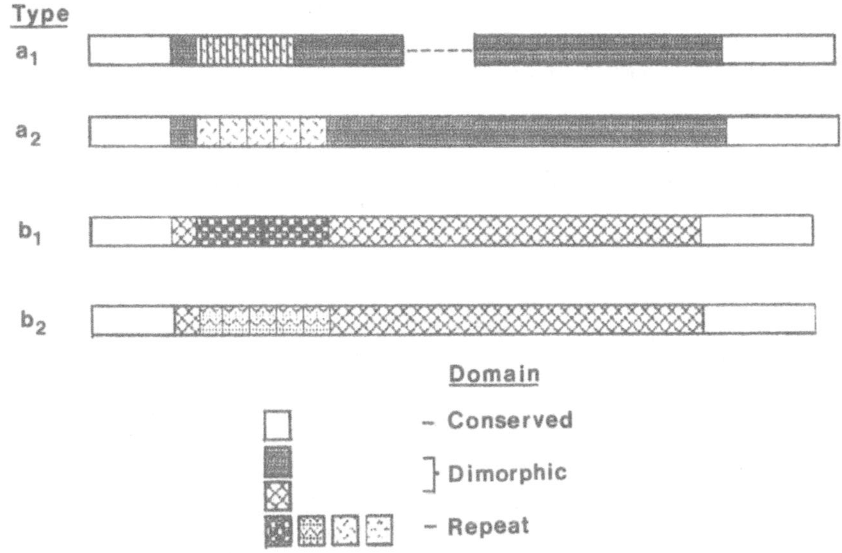

FIGURE 4. A simplified scheme of the variations in amino acid sequence of the *Plasmodium falciparum* merozoite surface antigen MSA2. A and B are the major dimorphic groupings. The variants of each group are formed by differences in the type of repeat, deletions, insertions, or amino acid substitutions.

Table II
Analysis of the Progeny of Cross between *P. falciparum* Clones 3D7 and HB3 for mAb Reactivities[a]

Products of cross	MSA1 mAbs 7.3	MSA1 mAbs 9.2	MSA2 mAb 12.3	Numbers of clones	Individual parasites in uncloned mixture
Parental (3D7)	−	+	+	5	91
Parental (HB3)	+	−	−	15	109
Recombinant (type 1)	+	−	+	3	105
Recombinant (type 2)	−	+	−	3	n/t
Recombinant (type 3)	+	+	+	0	0
Recombinant (type 4)	+	+	−	0	0
Recombinant (type 5)	−	−	−	n/t	n/t
Recombinant (type 6)	−	−	+	n/t	n/t

[a] After Walliker *et al.* (1987, unpublished data) and Fenton (1988). Results were obtained from (1) parasite clones grown in culture (Rosario, 1981) or (2) individual parasites analyzed in cultures of uncloned progeny (see Figure 2 and Walliker *et al.*, 1987).

Similar studies on antigens of *P. chabaudi* have shown that an analogue of MSA1 in these parasites exhibits antigenic and allelic variation of a similar nature to that described above (Newbold *et al.*, 1984; McLean *et al.*, 1991).

3.2.4. Proteins Detected by Two-Dimensional Electrophoresis

Two-dimensional polyacrylamide gel electrophoresis (PAGE) is a particularly useful technique for genetic analysis, as it can be used to screen numerous proteins for variation on a single gel. The proteins are separated in two dimensions, using isoelectric focusing to separate the proteins by charge and sodium dodecyl sulfate (SDS) PAGE to distinguish their relative mobility (O'Farrell, 1975). Two-dimensional PAGE has been very useful for examining the extent of variability of proteins in *P. falciparum* populations (Tait, 1981; Walker, 1984). Of 100 parasite proteins, 15 were found to exhibit variation in one or both dimensions in different parasite clones. Each of these protein groups was denoted by a letter A, B, C, and so on and each variant within a group was given a number 1, 2, 3, 4 and so on (for examples, see Table III). Some of these variants showed large variations in size and isoelectric point. One protein, designated B, was found in a different position in almost every isolate examined.

A large number of combinations exist for these markers and hence the technique has been particularly useful in discriminating isolates of *P. falciparum* from different countries (see Table III) (Pinswasdi *et al.*, 1987; Creasey *et al.*, 1990). Mixtures of two-dimensional variants are frequently found in an isolate. After cloning, each clone is found to possess only one variant of each protein, showing that each variant is encoded by an allele of the same gene (see Table III) (Thaithong *et al.*, 1984; Fenton *et al.*, 1985). Peptide digest studies have also shown that variant forms of each protein (e.g., C1, C2, C3, and so on) are structurally related to each other, a result that is also consistent with the variants in a group being encoded by alleles of the same gene (Fenton, 1987).

The identity of many of the proteins resolved by this technique is unknown. However, MSA1 and MSA2 have been identified (Anders *et al.*, 1983; Howard *et al.*, 1983; Brown *et al.*, 1983; Fenton *et al.*, 1989), as well as the parasite enzyme adenosine deaminase (Fenton and Walliker, 1990).

3.2.4.1. Inheritance of Two-Dimensional Proteins. The parent clones of the *P. falciparum* cross, 3D7 and HB3, differed in five two-dimensional marker proteins: C, D, G, K, and P. An analysis of the inheritance of these proteins is shown in Table IV. For each of the proteins D, G, K, and P, the alternative forms were inherited in a ratio of approximately 1 : 1, suggesting that they were encoded by alleles of nuclear genes. Protein C showed a predominance of the HB3 type in the progeny. Pairwise comparisons showed that recombination had occurred between the genes for all the proteins except P and ADA. From

Table III
Results of Analyzing Isolates and Clones of *P. falciparum* by Two-Dimensional PAGE

		Two-dimensional PAGE[a]														
Parasite	Country	A	B	C	D	E	F	G	H	I	J	K	L	M	N	P
IEC56/84	Brazil	4	—	8	1,8	3	1	2	—	1	—	8	1	1	2	2
IEC26/88	Brazil	2	—	8	3	4	1	2	—	1	—	1	3	1	—	2
IEC62/84	Brazil	2	—	7	8	2	1	1	—	1	—	3	3	2	2	2
WA26	Gambia	2	—	3	2	—	1	2	—	1	—	2	1	1	2	1
ZIM1	Zimbabwe	3	—	3	8	4	2	1	—	1	—	3	3	2	1	1,2
SL	Sierra Leone	—	—	3	4	4	1	2	—	1	—	3	1	1	—	2
MAD20	Papua New Guinea	2	1,12	2	4	2	1	2	3	1	2	7	1	1	1	1
W2	Indochina	—	—	2	2	2	1	1	—	1	—	2	2	1	1	1
PB1	Thailand	1	—	2	4,5	3	2	2	6	1	2	6	3	1	2	1,2
PB1 Clones																
PB1/1		1	—	2	5	3	2	2	6	1	2	6	3	1	2	1
PB1/4		1	—	2	4	3	2	2	6	1	2	6	3	1	2	2
PB1/5		1	—	2	4	3	2	2	6	1	2	6	3	1	2	2

[a] Each letter represents a protein exhibiting variant forms in different parasites and each number an allelic variant of each protein.

this it was concluded that the alternative forms of the two-dimensional proteins were encoded by alleles, and that none of the genes appeared to be in close proximity on a chromosome. It was subsequently demonstrated by electrophoretic and biochemical studies that the peptide P is adenosine deaminase, a result that explains the lack of recombination between these characters (Fenton and Walliker, 1990). This study on two-dimensional PAGE protein inheritance also highlighted the capacity of *P. falciparum* to form new genotypes after meiosis; the majority of progeny clones examined were recombinant for the parent clone markers.

3.2.5. Chromosomes Studied by Pulsed Field Gradient Gel Electrophoresis

Until 1985, little information was available about chromosomes of malaria parasites. This was partly because the parasite's chromosomes do not condense and are thus invisible by light microscopy and partly because genetic analysis could only be carried out using a small number of markers. However, ultrastructural studies revealed the presence of kinetochores, structures associated with chromosome division, to give estimates of chromosome numbers (Sinden, 1978; Prensier and Slomianny, 1986). In recent years, the technique of pulsed field gradient gel electrophoresis (PFGGE) which uses bidirectional pulsed currents to

Table IV
Proteins of Clones 3D7, HB3, and Their Progeny Analyzed Using Two-Dimensional PAGE

	Two-dimensional PAGE protein[a]					
	C	D	G	K	P	ADA
Parentals						
3D7	3	2	1	3	1	1
HB3	7	4	2	1	2	2
Progeny clones						
Parental types						
XP6,X10	7	4	2	1	2	2
X8,X14,X31,X36,X37	3	2	1	3	1	1
Recombinant types						
X2	3	4	2	1	1	1
X11,XP3	7	4	1	3	1	1
X6,X35	7	4	2	3	2	2
X12,XP8	7	4	2	3	1	1
X5,XP2	7	4	1	1	1	1
XP5,X34	7	4	2	1	1	1
XP1	7	2	2	3	2	2
X4	7	2	1	3	2	2
XP4	7	2	2	3	1	1
XP7,XP9	7	2	1	1	2	2
X13	7	4	2	3	2	2
X30	3	2	1	1	2	2
X33	3	2	1	3	2	2
Ratio	10:3	7:6	7:6	5:8	7:6	7:6

[a] Variable proteins C3 and 7, D2 and 4, G1 and 2, K3 and 5, and P1 and 2 differed in either Mr or IEP (from Fenton and Walliker, 1990).

separate DNA macromolecules, has been applied to malaria parasites (for a review, see Foote and Kemp, 1989). Fourteen DNA bands are found on PFGGE gels of an individual *P. falciparum* genome. This number agrees with predictions made using electron microscopy and it is therefore assumed that each band represents a chromosome (Kemp *et al.*, 1987; Wellems *et al.*, 1987). In *P. falciparum* the chromosomes vary in length from 630 kilobases to 3 megabases. Following chromosome separation, DNA can be transferred to nitrocellulose using Southern blotting (Southern, 1975). It is then possible to hybridize specific gene probes to chromosomes, thus localizing genes. When PFGGE is applied in combination with rare-cutting restriction enzymes, it is possible to construct linkage maps of plasmodial chromosomes (see Sinnis and Wellems, 1988; Foote and Kemp, 1989). For example, the genes for MSA1 and MSA2, and dihydrofolate reductase are located on chromosomes 9, 2, and 4, respectively.

An important finding has been the discovery that the sizes of homologous chromosomes of *P. falciparum* can vary in different isolates (Kemp *et al.*, 1985; Corcoran *et al.*, 1986). Certain changes in size can also occur during the continuous culture of a cloned line (Wellems *et al.*, 1988). Mechanisms such as nonhomologous recombination resulting in gene deletion, and changes in the structure of repeated segments of subtelomeric DNA are thought to be involved in such events (Corcoran *et al.*, 1988; Pologe and Ravetch, 1988).

Clones 3D7 and HB3 exhibited size differences in several chromosomes including number 4. In the progeny obtained after crossing these clones, it was shown that chromosomes that were intermediate in size between those of the parents were generated. Further investigation has shown that the intermediate-sized chromosome 4 found in one progeny clone resulted from a crossover between the two parental chromosomes (see Figure 3B) (Sinnis and Wellems, 1988).

In the rodent malaria parasite *P. chabaudi,* size heterogeneity of some homologous chromosomes has also been found (Sharkey *et al.*, 1988).

3.2.6. Restriction Fragment Length Polymorphism

Restriction enzymes cleave DNA at specific linear sequences. Using a large number of these restriction enzymes, variations in the structure of genes and intergenic regions can be detected, following electrophoresis of the cut DNA. Such variation in length of DNA fragments associated with specific genes are known as restriction fragment length polymorphisms (RFLPs). In the malaria parasite, RFLPs have been found for many genes. For example, Coppel *et al.* (1985) found that different isolates exhibited differences in the length of some of certain antigen gene-associated fragments. The technique was thus used to demonstrate genetic differences between isolates. Restriction fragment length polymorphisms have also been valuable in examining genes associated with drug resistance (see Section 3.2.7).

Repetitive DNA elements, analogous to the satellite DNA of higher eukaryotes (Jeffreys *et al.*, 1990), have also been found in *P. falciparum.* These sequences are generally nontranscribed DNA segments that are scattered throughout the genome, and in particular they are located near chromosome telomeres. Variations in the length of such repeated units can occur by increases or decreases in the numbers of such units. In *Plasmodium,* as with other eukaryotes, changes in repeat number appear to occur by recombination events at meiosis, as a result of unequal crossing-over (Oquendo *et al.*, 1986; Corcoran *et al.*, 1988; Vernick *et al.*, 1988). Changes in repetitive elements may also take place during mitosis (Jeffreys *et al.*, 1990). Evidence of such changes in the repetitive DNA of *P. falciparum* clones has been found (Bhasin *et al.*, 1985).

While meiosis contributes to changes in the patterns of RFLPs associated with the structure of repetitive DNA, this is not true of gene-associated RFLPs.

For example, genes for *P. falciparum* histidine-rich protein 2 (PfHRP2) and the circumsporozoite protein (CSP) were found to exhibit RFLPs in HB3 and 3D7. Among the recombinant progeny of the cross between these clones, both forms were found. No recombinant clone was found that exhibited new patterns or contained two forms (Wellems *et al.*, 1987). This suggested that the DNA regions encoding these molecules were being inherited as alternative forms in a stable manner.

3.2.7. Drug Resistance

3.2.7.1. Drug Resistance in Rodent Malaria Parasites. Rodent malaria parasites resistant to antimalarial drugs can be selected in the laboratory by exposing the parasites to either a single high dose or to increasing doses of drugs over periods of many weeks. The genetic basis of resistance to two drugs in particular, pyrimethamine and chloroquine, has been investigated most extensively.

Early work with the rodent malaria species *P. yoelii* and *P. chabaudi* showed that pyrimethamine-resistant parasites could be selected from sensitive cloned lines after a single high dose. When the genetics of this type of resistance was analyzed, it was found that the resistance and sensitivity characters were inherited in a manner consistent with their being determined by alleles of a single gene (Walliker *et al.*, 1973, 1975; Rosario, 1976).

The genetics of chloroquine resistance appeared less clear. Selection of chloroquine resistance in rodent malarias requires continuous treatment of sensitive parasites with gradually increasing doses of drug (Rosario, 1976). When resistant parasites selected in this way are crossed with drug-sensitive forms, some of their progeny have levels of resistance intermediate to both parental values (Padua, 1981). This suggests that chloroquine resistance may arise through mutations at more than one gene. The intermediate levels of resistance among progeny could be explained by recombination between the different genes.

3.2.7.2. Molecular Genetics of Drug Resistance in *P. falciparum*

Pyrimethamine Resistance. The biochemistry of pyrimethamine action suggested that this drug exerts its antiplasmodial effect by binding to parasite dihydrofolate reductase (DHFR) (Ferone, 1970; Sirawaraporn and Yuthavong, 1984; Chen *et al.*, 1987). In PFGGE and blotting experiments, the DHFR gene was located to chromosome 4 in *P. falciparum* (D. S. Peterson *et al.*, 1988). To examine this association in more detail, restriction fragment length polymorphisms (see Section 3.2.6) of the DHFR gene were obtained that distinguished the gene in the HB3 and 3D7 parent clones used in the *P. falciparum* cross (D. S. Peterson *et al.*, 1988). The DHFR gene in the resistant parasite (HB3) was carried on a DNA fragment of 4.2 kilobase (kb) after digestion with

Table V
The Association between Drug Response and Restriction Enzyme Sites Around the DHFR Gene in Recombinant Clones of *P. falciparum*[a]

DHFR HinfI fragment (kb)	Pyrimethamine response	Numbers	Type
4.2	Resistant	10	Parental
4.2	Sensitive	0	Recombinant
4.3	Resistant	0	Recombinant
4.3	Sensitive	2	Parental

[a]From D. S. Peterson *et al.* (1988).

the restriction enzyme Hinf1, while the gene from the sensitive 3D7 parasite was found on a 4.3-kb fragment. When recombinant progeny were analyzed, it was found that a resistant phenotype was invariably associated with the 4.2-kb fragment and a sensitive phenotype with the 4.3-kb fragment (see Table V). These results provided strong evidence that the DHFR gene was involved in pyrimethamine resistance. Alternatively, the result could be explained by close linkage between another gene determining pyrimethamine resistance lying close to the DHFR locus.

In further studies, DNA sequences of the DHFR gene were obtained in a number of resistant and sensitive parasites, including HB3, 3D7, and certain cross progeny. The results showed that an amino acid substitution at position 108 (serine or threonine altered to asparagine) appeared to be the cause of resistance, possibly by altering the ability of the DHFR molecule to bind to the drug (Cowman *et al.*, 1988; D. S. Peterson *et al.*, 1988; Snewin *et al.*, 1989). Other amino acid substitutions in the DHFR molecule have also been invoked to explain different levels of resistance to this drug in *P. falciparum*, although genetic crossing studies with such parasites have yet to be carried out.

Chloroquine Resistance. Recent studies have shown that a calcium channel blocking agent, verapamil, can apparently reverse chloroquine resistance in cultured forms of *P. falciparum* (Martin *et al.*, 1987). This has led to suggestions that ubiquitous protein molecules known as P-glycoproteins may be involved in this type of resistance. In resistant cells these molecules are thought to pump out toxic drugs, thus indirectly countering their action. Verapamil most probably blocks this process by binding to the P-glycoprotein, thus causing the antimalarial drug to remain inside the cell. The genes encoding P-glycoproteins are called *mdr* (multi-drug-resistant) genes.

Early reports using molecular techniques suggested that *P. falciparum* contained *mdr* genes, which were amplified in some chloroquine-resistant lines (Foote *et al.*, 1989; Wilson *et al.*, 1989). More recently it has been suggested

Table VI
The Lack of Association between Pfmdrl and Cloroquine Resistance[a]

Pfmdrl BamH1 fragment (kB)	Cloroquine response	Progeny numbers	Type
3.2	Sensitive	5	Parental
3.2	Resistant	4	Recombinant
3.3	Sensitive	3	Recombinant
3.3	Resistant	4	Parental

[a]From Wellems *et al.* (1990).

that there are several alleles of the *mdr* gene in the *P. falciparum* population distinguishable by base sequence variations. Some of these alleles have been claimed to be associated with drug resistance (Foote *et al.*, 1990). Wellems *et al.* (1990) have used genetic analysis to investigate the relationship between *mdr* genes and chloroquine resistance. This involved making a cross between a chloroquine-resistant clone denoted Dd2 (Oduola *et al.*, 1988; Wellems *et al.*, 1988) and the sensitive HB3 clone. The clones contained two *mdr* genes, denoted *pfmdr1* and *pfmdr2*. The alleles for *pfmdr1* in Dd2 and HB3 could be distinguished by an RFLP using the enzyme BamH1. The inheritance of these RFLPs was then analyzed among their recombinant progeny derived from the cross. In contrast to the results obtained with pyrimethamine resistance described above, no association was found between the chloroquine response of an individual progeny clone and its *pfmdr1* restriction pattern (see Table VI). A similar result was found for *pfmdr2*. These results thus suggest that these *mdr* genes are not by themselves responsible for the chloroquine resistance of the Dd2 clone. This work also showed that the levels of chloroquine resistance in progeny clones fell into the two parental classes, suggesting that in this instance chloroquine resistance may be governed by a single gene, or perhaps closely linked genes.

4. CONCLUSION

Genetic studies of malaria parasites have demonstrated that there is considerable variation in factors such as proteins, antigens, and drug sensitivity in all malaria species analyzed. The genes so far investigated undergo normal Mendelian inheritance, with different forms of enzymes, antigens, two-dimensional PAGE proteins, and some types of drug resistance being encoded by alleles of nuclear genes. Pulsed field gradient gel electrophoresis work has enabled many genes to be mapped to particular chromosomes (Kemp *et al.*, 1987; Wellems *et al.*, 1987). Studies on DNA that encodes certain variant proteins have led to a greater understanding of the genetic causes of the phenotypic variation. For example, this can occur as a result of a single base change in the gene that alters an amino acid, as in the altered DHFR molecule responsible for pyrimethamine resistance (Cowman *et al.*, 1988; D. S. Peterson *et al.*, 1988). Alternatively, many point mutations, deletions, insertions, and repetitiveness of the DNA can alter amino acid sequences; this may result in a multiplicity of alleles of certain genes as found, for example, in *P. falciparum* antigens MSA1, MSA2, and the "S" antigen (Kemp *et al.*, 1990). It is also evident that the subtelomeric regions of chromosomes are especially variable, being characterized by repetitive DNA sequences. Genes located at such subtelomeric sites may be deleted. The consequences of such variability have still to be fully explored.

The crossing experiments between genetically distinct *P. falciparum* clones have shown that new genotypes are produced at high frequency. The most likely occasion for cross hybridization to occur naturally is when a mosquito feeds from a person who has an infection of more than one parasite type. Such mixtures are common in field isolates (Trager *et al.*, 1981; Walliker, 1983a; Thaithong *et al.*, 1984; Creasey *et al.*, 1990).

Cross hybridization is important to the parasite as it allows new mutant genes such as novel antigen types or genes conferring drug resistance to become established in the parasite gene pool and to spread through the parasite population. Meiosis may in itself be important for amplification and/or aggregation of combinations of genes that confer drug resistance to the parasite.

The work carried out on genes and their products has provided useful markers for the study of parasite populations (McBride *et al.*, 1982; Forsyth *et al.*, 1989; Creasey *et al.*, 1990). This is likely to become an increasingly important area of research.

To analyze the implications behind the variation outlined in this chapter, let us consider the number of genetically different parasites that could be detected by the 30 different genes of *P. falciparum* that have been described. Allelic variation may be extensive, for example, in antigens such as MSA1, where at least 30 alleles have been identified (Conway and McBride, personal communication). Other genes may have as few as two alternative forms in the population, for example, the gene for the parasite enzyme adenosine deaminase. The number of possible genotypic combinations that could exist for the variants of these 30 markers can be determined by multiplying the numbers of alleles of each gene together (30 MSA1 types $\times$ 8 MSA2 types $\times$ 2 ADA types, etc.), and there are at least 1×10^{12}. It is estimated that there are 2×10^{8} people infected at any one time by malaria in Africa alone (Hyde, 1990). Assuming that each infection is caused by *P. falciparum*, consisting of a mixture of at least two parasite genotypes (Creasey *et al.*, 1990), then this figure corresponds to 4×10^{8} parasite clones. Worldwide, a figure of some 1×10^{9} parasite clones can potentially exist at any one time. If the alleles of the genes for the markers described here are undergoing random recombination, then any two parasites in the world population can be expected to be different for at least one marker. The available evidence indicates that this is probably the case (McBride *et al.*, 1982; Creasey *et al.*, 1990).

An alternative has been suggested, which is that populations of parasitic protozoa, including *Plasmodia,* may exhibit "clonality." Clonality is a restriction in the expected number of genotypes in an area (Tibayrence *et al.*, 1990). As the parasite can undergo self-fertilization and asexual multiplication, this is a possibility particularly where small numbers of parasites are expanding from a "founder" population in a previously uninfected area.

This subject could be examined further by studying the frequency of genetic exchange that occurs in nature. At present, the only information about the genetic structure of parasite populations is on the characters of asexual blood stages. These need to be isolated from patients and then cultured and/or cloned before characterization in the laboratory. This lengthy process could allow selection of parasite genotypes most suited to culture. With the advent of the polymerase chain reaction (Saiki *et al.*, 1985; Mullis and Faloona, 1987), it should now be possible to analyze parasite genes directly from patients' blood and from stages in the mosquito part of the life cycle which previously had not been available for study. Polymerase chain reaction "length alleles" have already been used to discriminate the type of *mdr* allele that *P. falciparum* isolates contain (Foote *et al.*, 1990). With the numbers of *P. falciparum* genes that have now been sequenced, it should soon be possible to make a detailed genetic analysis of malaria parasite populations.

ACKNOWLEDGMENT. We would like to thank Geoffrey Beale for reviewing this manuscript.

5. REFERENCES

Beale, G. H., and Walliker, D., 1988, Genetics of malaria parasites, in *Malaria* (W. H. Wernsdorfer and I. McGregor, eds.), pp. 379–393, Churchill Livingstone, Edinburgh.

Beale, G. H., Carter, R., and Walliker, D., 1978, Genetics, in *Rodent Malaria* (R. Killick-Kendrick and R. Peters, eds.), pp. 213–245, Academic Press, London.

Bhasin, V. K., Clayton, C., Trager, W., and Cross, G. A. M., 1985, Variations in the organisation of repetitive DNA sequences in the genomes of *Plasmodium falciparum* clones, *Mol. Biochem. Parasitol.* **15:**149–158.

Brown, G. V., Anders, R. F., and Knowles, G., 1983, Differential effect of immunoglobulin on the *in vitro* growth of several isolates of *Plasmodium falciparum, Infect. Immun.* **39:**1228–1235.

Bruce, M., Alano, P., Duthie, S., and Carter, R., 1990, Commitment of the malaria parasite *Plasmodium falciparum* to sexual and asexual development, *Parasitology* **100:**191–200.

Campbell, C. C., Collins, W. E., Nguyen-Dinh, P., Barber, A., and Broderson, J. R., 1982,

Plasmodium falciparum gametocytes from culture *in vitro* develop to sporozoites that are infectious to primates, *Science* **217:**:1048–1050.

Carter, R., 1978, Studies on enzyme variation in the murine malaria parasites *Plasmodium berghei, P. yoelii, P. vinckei,* and *P. chabaudi* by starch gel electrophoresis, *Parasitology* **76:**241–267.

Carter, R., Graves, P. M., Creasey, A., Byrne, K., Read, D., Alano, P., and Fenton, B., 1989, *Plasmodium falciparum:* An abundant stage specific protein expressed during early gametocyte development, *Exp. Parasitol.* **69:**140–149.

Chen, G., Mueller, C., Wendlinger, M., and Zolg, J. W., 1987, Kinetic and molecular properties of the dihydrofolate reductase from pyrimethamine sensitive and pyrimethamine resistant clones of the human malaria parasite *Plasmodium falciparum, Mol. Pharmacol.* **31:**430–437.

Coppel, R. L., Saint, R. B., Stahl, H. D., Langford, C. J., Brown, G. V., Anders, R. F., and Kemp, D. J., 1985, *Plasmodium falciparum:* Differentiation of isolates with DNA hybridisation using antigen gene probes, *Exp. Parasitol.* **60:**82–89.

Corcoran, L. M., Forsyth, K. P., Bianco, A. E., Brown, G. V., and Kemp, D. J., 1986, Chromosome size polymorphisms in *Plasmodium falciparum* can involve deletions and are frequent in natural parasite populations, *Cell* **44:**87–95.

Corcoran, L. M., Thompson, J. K., Walliker, D., and Kemp, D. J., 1988, Homologous recombination within subtelomeric repeat sequences generates chromosome size polymorphism in *Plasmodium falciparum, Cell* **53:**807–813.

Cowman, A. F., Morry, M. J., Biggs, B. A., Cross, G. A. M., and Foote, S. J., 1988, Amino acid changes linked to pyrimethamine resistance in the dihydrofolate reductase-thymidylate synthase gene of *Plasmodium falciparum, Proc. Natl. Acad. Sci. USA* **85:**9109–9113.

Creasey, A. C., Fenton, B., Walker, A., Thaithong, S., Oliveira, S., Mutambu, S., and Walliker, D., 1990, Genetic diversity of *Plasmodium falciparum* shows geographical variation, *Trans. Am. J. Trop. Med. Hyg.* **42:**403–413.

Downs, W. G., 1947, Infections of chicks with single parasites of *Plasmodium gallinaceum* Brumpt, *Am. J. Hyg.* **46:**41–44.

Fenton, B., 1987, Peptide digest studies of polymorphic proteins of *Plasmodium falciparum, Mol. Biochem. Parasitol.* **24:**13–22.

Fenton, B., 1988, Studies on polymorphic proteins of *Plasmodium falciparum,* Ph.D. Thesis, University of Edinburgh.

Fenton, B., Walker, A., and Walliker, D., 1985, Protein variation in clones of *Plasmodium falciparum* detected by two-dimensional electrophoresis, *Mol. Biochem. Parasitol.* **16:**173–183.

Fenton, B., and Walliker, D., 1990, Genetiic analysis of polymorphic proteins of the human malaria parasite *Plasmodium falciparum, Genet. Res. Cambridge* **55:**81–86.

Fenton, B., Clark, J. T., Wilson, C. F., McBride, J. S., and Walliker, D., 1989, Polymorphism of a 35–48 kDa *Plasmodium falciparum* merozoite surface antigen, *Mol. Biochem. Parasitol.* **34:**79–86.

Fenton, B., Clark, J. T., Anjam Khan, C. M., Robinson, J. V., Walliker, D., Ridley, R., Scaife, J. G., and McBride, J. S., 1991, Structural and antigenic polymorphism of the 35–48 kilodalton merozoite surface antigen (MSA-2) of the malaria parasite *Plasmodium falciparum, Mol. Cell. Biol.* **11:**963–971.

Ferone, R., 1970, Dihydrofolate reductase from pyrimethamine-resistant *Plasmodium berghei, J. Biol. Chem.* **245:**850–854.

Foote, S. J., and Kemp, D. J., 1989, Chromosomes of malaria parasites, *Trends Genet.* **5:**337–342.

Foote, S. J., Thompson, J. K., Cowman, A. F., and Kemp, D. J., 1989, Amplification of the multidrug resistance gene in some chloroquine-resistant isolates of *P. falciparum, Cell* **57:**921–930.

Foote, S. J., Kyle, D. E., Martin, R. K., Oduola, A. M. J., Forsyth, K., Kemp, D. J., and Cowman,

A. F., 1990, Several alleles of the multidrug resistance gene are closely linked to chloroquine resistance in *Plasmodium falciparum*, *Nature* **345**:255–258.

Forsyth, K. P., Anders, R. F., Kemp, D. J., and Alpers, M. P., 1988, New approaches to the serotypic analysis of the epidemiology of *Plasmodium falciparum*, *Phil. Trans. Roy. Soc. Ser. B.* **321**:485–493.

Garnham, P. C. C., 1966, *Malaria Parasites and Other Haemosporidia*, Blackwell Scientific Publications, Oxford, England.

Greenberg, J., and Trembley, H. L., 1954, The apparent transfer of pyrimethamine resistance from the BI strain of *Plasmodium galinaceum* to the M strain, *J. Parasitol.* **40**:667–672.

Hall, R., Hyde, J. E., Goman, M., Simmons, D. L., Hope, I. A., Mackay, M., Scaife, J., Merkli, B., Richle, R., and Stocker, J., 1984, Major surface antigen gene of a human malaria parasite cloned and expressed in bacteria, *Nature* **311**:379–382.

Haynes, J. D., Diggs, C. L., Hines, F. A., and Desjardins, R. E., 1976, Culture of human malaria parasites *Plasmodium falciparum*, *Nature* **263**:767–769.

Howard, R. J., Stephen, B. A., and Lemkin, P. F., 1983, High resolution comparison of *Plasmodium knowlesi* clones of different variant antigen phenotypes by two dimensional electrophoresis and computer analysis, *Electrophoresis* **4**:420–427.

Hyde, J. E., 1990, *Molecular Parasitology*, Open University Press, Buckingham, England.

Jeffreys, A., Neumann, R., and Wilson, V., 1990, Repeat unit sequence variation in minisatellites: A novel source of DNA polymorphism for studying variation and mutation by single molecule analysis, *Cell* **60**:473–485.

Kemp, D. J., Coppel, R. L., Cowman, A. F., Saint, R. B., Brown, G. V., and Anders, R. F., 1983, Expression of *Plasmodium falciparum* blood stage antigens in *Escherichia coli:* Detection with antibodies from immune humans, *Proc. Natl. Acad. Sci. USA* **80**:3787–3791.

Kemp, D. J., Corcoran, L. M., Coppel, R. L., Stahl, H. D., Bianco, A. E., Brown, G. V., and Anders, R. F., 1985, Size variation in chromosomes from independent cultured isolates of *Plasmodium falciparum*, *Nature* **315**:347–350.

Kemp, D. J., Coppel, R. L., Stahl, H. D., Bianco, A. E., Corcoran, L. M., McIntyre, P., Langford, C. J., Favaloro, J. M., Crewther, P. E., Brown, G. V., Mitchell, G. F., Culvenor, J. G., and Anders, R. F., 1986, Genes for antigens of *Plasmodium falciparum*, *Parasitology* **91**:S83–S108.

Kemp, D. J., Thompson, J. K., Walliker, D., and Corcoran, L. M., 1987, Molecular karyotype of *Plasmodium falciparum:* Conserved linkage groups and expendable histidine-rich protein genes, *Proc. Natl. Acad. Sci. USA* **84**:7672–7676.

Kemp, D., Cowman, A., and Walliker, D., 1990, Genetic diversity in *Plasmodium falciparum*, *Adv. Parasitol.* **29**:75–149.

Lainson, F. A., 1983, Observations on the morphology and electrophoretic variation of enzymes of the rodent malaria species of Cameroon, *Plasmodium yoeelii*, *P. chabaudi* and *P. vinckei*, *Parasitology* 86:221–229.

Martin, S. K., Oduola, A. M. J., and Milhous, W. K., 1987, Reversal of chloroquine resistance in *Plasmodium falciparum* by verapamil, *Science* **235**:899–901.

McBride, J. S., Walliker, D., and Morgan, G., 1982, Antigenic diversity in the human malaria parasite *Plasmodium falciparum*, *Science* **217**:254–257.

McBride, J. S., Newbold, C. I., and Anand, R., 1985, Polymorphism of a high molecular weight schizont antigen of the human malaria parasite *Plasmodium falciparum*, *J. Exp. Med.* **161**:160–180.

McLean, A., Lainson, F. A., Sharkey, A. M., and Walliker, D., 1991, Genetic studies on major merozoite surface antigen of rodent malaria *Plasmodium chabaudi*, *Parasite Immunol.* **13**:369–378.

Mullis, K. B., and Faloona, F., 1987, Specific synthesis of DNA *in vitro* via a polymerase catalysed chain reaction, *Methods Enzymol.* **155**:335–3500.

Newbold, C. I., Schryer, M., Boyle, D. B., McBride, J. S., McLean, A. P., Wilson, R. J. M., and Brown, K. N., 1984, A possible molecular basis for strain specific immunity to malaria, *Mol. Biochem. Parasitol.* **11**:337–347.

Oduola, A. M. J.., Milhouse, W. K., Weatherly, N. F., Bowdre, J. H., and Desjardins, R. E., 1988, *Plasmodium falciparum:* Induction of resistance to mefloquine in cloned strains by continuous drug exposure *in vitro, Exp. Parasitol.* **67**:354–360.

O'Farrell, P. H., 1975, High-resolution two-dimensional electrophoresis of proteins, *J. Biol. Chem.* **250**:4007–4021.

Oquendo, P., Goman, M., Mackay, M., Langsley, G., Walliker, D., and Scaife, J., 1986, Characterisation of a repetitive DNA sequence from the malaria parasite, *Plasmodium falciparum, Mol. Biochem. Parasitol.* **18**:89–101.

Padua, R. A., 1981, *Plasmodium chabaudi:* Genetics of resistance to cloroquine, *Exp. Parasitol.* **52**:419–426.

Peterson, D. S., Walliker, D., and Wellems, T. E., 1988, Evidence that a point mutation in dihydrofolate reductase-thymidylate synthase confers resistance to pyrimethamine in falciparum malaria, *Proc. Natl. Acad. Sci. USA* **85**:9114–9118.

Peterson, M. G., Coppel, R. L., Moloney, M. B., and Kemp, D. J., 1988, Third form of the precursor to the major merozoite surfaces of *Plasmodium falciparum, Mol. Cell. Biol.* **8**:2664–2667.

Pinswasdi, C., Thaithong, S., Beale, G. H., Fenton, B., Webster, H. K., and Pavanand, K., 1987, Polymorphism of proteins in malaria parasites following mefloquine treatment, *Mol. Biochem. Parasitol.* **23**:159–164.

Pologe, L. P., and Ravetch, J. V., 1988, Large deletions result from breakage and healing of *Plasmodium falciparum* chromosomes, *Cell* **55**:869–874.

Ponnudurai, T., Meuwissen, J. H. E. Th., Leeuwenberg, A. D. E. M., Verhave, J. P., and Lensen, A. H. W., 1982, The production of mature gametocytes of *Plasmodium falciparum* in continuous cultures of different isolates infective to mosquitoes, *Trans. Roy. Soc. Trop. Med Hyg.* **76**:242–250.

Prensier, G., and Slomianny, C., 1986, The karyotype of *Plasmodium falciparum* determined by ultrastructural serial sectioning and 3D reconstruction, *J. Parasitol* **72**:731–736.

Rosario, V. E., 1976, Genetics of cloroquine resistance in malaria parasites, *Nature* **261**:585–586.

Rosario, V., 1981, Cloning of naturally occurring mixed infections of malaria parasites, *Science* **212**:1037–1038.

Saiki, R., Scharf, S., Faloona, F., Mullis, K., Horn, G., Erlich, H., and Arnheim, N., 1985, Enzymatic amplification of β-globin genomic sequences and restriction site analysis for diagnosis of sickle cell anemia, *Science* **230**:1350–1354.

Sanderson, A., Walliker, D., and Molez, J-F., 1981, Enzyme typing of *Plasmodium falciparum* from some African and other old world countries, *Trans. Roy. Soc. Trop. Med. Hyg.* **75**:263–267.

Saul, A., Lord, R., Jones, G., Mario Geysen, H., Gale, J., and Mollard, R., 1989, Cross-reactivity of antibody against an epitope of the *Plasmodium falciparum* second merozoite surface antigens, *Parasite Immunol.* **11**:593–601.

Schofield, L., Saul, A., Myler, P., and Kidson, C., 1982, Antigenic differences among isolates of *Plasmodium falciparum* demonstrated by monoclonal antibodies, *Infect. Immun.* **38**:893–897.

Schwartz, R. T., Moreno, V. R., Lockyer, M. J., Nicholls, S. C., Davey, L. S., Hillman, Y., Sandhu, J. S., Freeman, R. R. and Holder, A. A., 1986, Structural diversity of the major surface antigen of *Plasmodium falciparum* merozoites, *Mol. Cell. Biol.* **6**:964–968.

Sharkey, A., Langsley, G., Patarapotikul, J., Mercereau-Puijalon, O., McLean, A., and Walliker, D., 1988, Chromosome size variation in the malaria parasite of rodents *Plasmodium chabaudi, Mol. Biochem. Parasitol.* **28**:47–54.

Sheppard, M., Thompson, J. K., Anders, R. F., Kemp, D. J., and Lew, A. M., 1989, Molecular karyotyping of the rodent malarias *Plasmodium chabaudi, Plasmodium berghei,* and *Plasmodium vinckei, Mol. Biochem. Parasitol.* **34**:45–52.

Sinden, R. E., 1978, Cell biology, in *Rodent Malaria* (R. Killick-Kendrick and W. Peters, eds.), pp. 85–168, Academic press, London.

Sinden, R. E., and Hartley, R. H., 1985, Identification of the meiotic division of malaria parasites, *J. Protozool.* **32**:742–744.

Sinnis, P., and Wellems, T. E., 1988, Long-range restriction maps of *Plasmodium falciparum* chromosomes: Crossing-over and size variation among geographically distinct isolates, *Genomics* **3**:287–295.

Sirawarapom, W., and Yuthavong, Y., 1984, Kinetic and molecular properties of dihydrofolate reductase from pyrimethamine-sensitive and pyrimethamine-resistant *Plasmodium chabaudi, Mol. Biochem. Parasitol.* **10**:355–367.

Smythe, J. A., Peterson, M. G., Coppel, R. L., Saul, A. J., Kemp, D. J., and Anders, R. F., 1990, Structural diversity in the 45-kilodalton merozoite surface antigen of *Plasmodium falciparum, Mol. Biochem. Parasitol.* **39**:227–234.

Snewin, V. A., England, S. M., Sims, P. F. G., and Hyde, J. E., 1989, Characterisation of the dihydrofolate reductase thymidylate synthetase gene from human malaria parasites highly resistant to pyrimethamine, *Gene* **76**:41–52.

Southern, E. M., 1975, Detection of specific sequences among DNA fragments separated by gel electrophoresis, *J. Mol. Biol.* **98**:503–517.

Tait, A., 1981, Analysis of protein variation in *Plasmodium falciparum* by two-dimensional gel electrophoresis, *Mol. Biochem. Parasitol.* **2**:205–218.

Tanabe, K., Mackay, M., Goman, M., and Scaife, J. G., 1987, Allelic dimorphism in a surface antigen gene of the malaria parasite *Plasmodium falciparum, J. Mol. Biol.* **195**:273–287.

Thaithong, S., Sueblinwong, T., and Beale, G. H., 1981, Enzyme typing of some isolates of *Plasmodium falciparum* from Thailand, *Trans. Roy. Soc. Trop. Med. Hyg.* **75**:268–270.

Thaithong, S., Beale, G. H., Fenton, B., McBride, J., Rosario, V., Walker, A., and Walliker, D., 1984, Clonal diversity in a single isolate of the malaria parasite *Plasmodium falciparum, Trans. Roy. Soc. Trop. Med. Hyg.* **78**:242–245.

Tibayrence, M., Kjellberg, F., and Ayala, F. J., 1990, A clonal theory of parasitic protozoa: The population structures of *Entamoeba, Giardia, Leishmania, Naegleria, Plasmodium, Trichomonas* and *Trypanosoma* and their medical and taxonomical consequences, *Proc. Natl. Acad. Sci. USA* **87**:2414–2418.

Trager, W., and Jensen, J. B., 1976, Human malaria parasites in continuous culture, *Science* **193**:673–675.

Trager, W., Tershakov, M., Lyandvert, L., Stanley, H., Lanners, N., and Gubert, E., 1981, Clones of the malaria parasite *Plasmodium falciparum* obtained by microscopic selection in their characterization with regard to knobs, chloroquine sensitivity, and formation of gametocytes, *Proc. Natl. Acad. Sci. USA* **78**:6527–6530.

Vernick, K. D., Walliker, D., and McCutchan, T. F., 1988, Genetic hypervariability of telomere related sequences is associated with meiosis in *Plasmodium falciparum, Nucleic Acids Res.* **16**:6973–6985.

Walker, A., 1984, "Protein variation in the malaria parasite *Plasmodium falciparum,*Thesis, University of Edinburgh.

Walliker, D., 1983a, *The Contribution of Genetics to the Study of Parasitic Protozoa,* Research Studies Press, Letchworth, Herts, England.

Walliker, D., 1983b, The genetic basis of diversity in malaria parasites, *Adv. Parasitol.* **22**:217–259.

Walliker, D., Carter, R., and Morgan, S., 1973, Genetic recombination in *Plasmodium berghei, Parasitology* **66**:309–320.

Walliker, D., Carter, R., and Sanderson, A., 1975, Genetic studies on *Plasmodium chabaudi:* Recombination between enzyme markers, *Parasitology* **70:**19–24.

Walliker, D., Quakyi, I. A., Wellems, T. E., McCutchan, T. F., Szarfman, A., London, W. T., Corcoran, L. M., Burkot, T. R., and Carter, R., 1987, Genetic analysis of the human malaria parasite *Plasmodium falciparum, Science* **236:**1661–1666.

Wellems, T. E., Walliker, D. W., Smith, C. L., Rosario, V., Maloy, W. L., Howard, R. J., Carter, R., and McCutchan, T. F., 1987, A histidine rich protein gene marks a linkage group favored strongly in a genetic cross of *Plasmodium falciparum, Cell* **49:**633–642.

Wellems. T. E., Oduola, A. M. J., Fenton, B., Desjardins, R., Panton, L. J., and Rosario, V., 1988, Chromosome size variation occurs in cloned *Plasmodium falciparum, Rev. Brasileira Genet.* **11:**813–825.

Wellems, T. E., Panton, L. J., Gluzman, I. Y., Rozario, V., Gwadz, R. W., Walker-Jonah, A., and Krogstad, D. J., 1990, Chloroquine resistance not linked to *mdr*-like genes in a *Plasmodium falciparum* cross, *Nature* **345:**253–255.

Wilson, C. M., Serrano, A. E., Wasley, A., Bogenschutz, M. P., Shankar, A. H., and Wirth, D. F., 1989, Amplification of a gene related to mammalian *mdr* genes in drug resistant *Plasmodium falciparum, Science* **224:**1184–1186.

Young, R. A., and Davies, R. W., 1983, Efficient isolation of genes by using antibody probes, *Proc. Natl. Acad. Sci. USA* **80:**1194–1198.

Chapter 10

How *Toxoplasma gondii* Gets In and Out of Host Cells

Joseph D. Schwartzman and Linda D. Saffer

1. *TOXOPLASMA GONDII:* THE UNIVERSAL COCCIDIAN

1.1. Importance of Toxoplasmosis

The obligate intracellular parasite *Toxoplasma gondii* is responsible for a wide spectrum of disease affecting humans and many animals. While the parasite is in general well adapted to its hosts, usually causing an asymptomatic lifelong infection, it occasionally produces severe disease, especially when the host is immunoincompetent. *Toxoplasma gondii* has been known for years as an important congenital infection in humans (Stagno, 1980), and it is now a major cause of opportunistic infection in people with AIDS. There have been several recent reviews of advances in knowledge of the cell biology (McLeod *et al.*, 1991), immunology, and molecular biology (Kasper and Boothroyd, 1991) of *T. gondii*. This chapter will review the progress of the last five years in understanding the mechanism of *T. gondii* invasion, since the review of Werk (1985).

The mechanism by which *T. gondii* infects host cells is of interest and

Joseph D. Schwartzman and Linda D. Saffer Department of Pathology, University of Virginia School of Medicine, Charlottesville, Virginia 22908.

Subcellular Biochemistry, Volume 18: Intracellular Parasites, edited by J. L. Avila and J. R. Harris. Plenum Press, New York, 1992.

importance in understanding the pathogenesis of toxoplasmosis. The parasite must invade host cells to survive and must exit the host cell to continue the infection. The latter point is not often appreciated, but active egress from host cells is likely to be important in both acute disease and in reactivation of dormant bradyzoites. Reactivation of encysted bradyzoites in the central nervous system of persons with AIDS causes most of the toxoplasmic encephalitis seen in this population (Luft *et al.*, 1983; Luft and Remington, 1988). In contrast to its nearest relatives, all of which have very limited host ranges, *T. gondii* infects essentially all vertebrates and will invade almost all host cell types (Frenkel, 1973). It has been shown to enter insect cells (Buckley, 1973), but not plant protoplasts (Werk and Fischer, 1982). All mammalian cell types, including red blood cells (Schupp *et al.*, 1978; Michel *et al.*, 1980; Tanabe *et al.*, 1980), can be invaded; this property suggests that *T. gondii* employs a generic mechanism for cell entry, perhaps utilizing a common molecule as a receptor for cell–parasite attachment. It is evident that host cell invasion is a key step in the pathogenicity of this widespread parasite, and an understanding of the mechanism by which *T. gondii* gets into and out of host cells may be useful in devising novel strategies to interfere with this infection.

1.2. The Process of Host Cell Invasion

Host cell penetration, as observed *in vitro* by light microscopy of cell monolayers, is a rapid, seamless progression that is not easily broken into discrete steps for analysis. Active parasite processes including gliding motility, shape changes mediated by the parasite cytoskeleton, the formation of a moving junction with the host plasma membrane, the secretion of factors from parasite organelles, and the formation of a specialized intracellular vacuole (in which the parasite resides and which does not fuse with host lysosomes) are all important to successful host cell penetration and intracellular survival. It is difficult to ascertain which of these steps is most important, if they are of equal importance, or even if they are sequential, but in this chapter parasite motility will be considered the primary effector of invasion, since nonmotile *T. gondii* do not get inside most host cells. It has been more difficult to interfere specifically with the function of other factors that are involved in host cell invasion, but none appear to influence penetration in an all-or-none fashion; therefore, for the purposes of this chapter these will be considered secondary determinants of successful invasion. The entire process evidently features considerable redundancy, which allows *T. gondii* to invade and parasitize a wide variety of hosts.

There are morphological similarities between the *T. gondii* tachyzoite and the malaria merozoite, and since malaria host cell invasion has been more thoroughly studied than that of *T. gondii*, some of the data relating to malaria will be reviewed. It is well to emphasize at the outset, however, that protists in general, and the Apicomplexa in specific, are not genetically a closely related grouping (Sogin *et al.*, 1986; Johnson *et al.*, 1988); therefore, mechanisms operating in malaria cannot be assumed to be identical to *T. gondii*. Even the more closely related Coccidia may not share all aspects of the *T. gondii* invasion mechanism, since their limited host range may have eliminated the necessity for flexibility that *T. gondii* must demonstrate.

1.3. Active Invasion versus Phagocytosis

There have been various interpretations of the role of phagocytosis in *T. gondii* host cell invasion. Extensive study of the interaction of macrophages with *T. gondii* has placed emphasis on phagocytosis, which certainly occurs when professional phagocytes meet this parasite, especially in the presence of antibody. Unambiguous identification of the route of entry of *T. gondii* that have successfully infected macrophages has been difficult, but study of other host cells has accumulated a body of data, which will be reviewed below, that is much in favor of active host cell penetration being the predominant route to successful host cell invasion. An argument for active penetration has also been made for the interaction of macrophages and malaria sporozoites (Vanderberg *et al.*, 1990).

Many studies have emphasized the interaction of *T. gondii* with macrophages, but the mechanism of infection of nonphagocytic cells is equally important, since it is the ability of *Toxoplasma* to infect a variety of cell types that leads to the serious sequelae of toxoplasmosis. The process of infection of nonphagocytic cells (and probably also of phagocytic cells) is an active function of the parasite. Cell penetration requires parasite metabolic energy (Werk and Bommer, 1980). Host metabolic energy may be required as well (Kimata and Tanabe, 1982; Kimata *et al.*, 1987), but studies using diffusable inhibitors make this difficult to determine. The organism is motile and possesses a group of organelles at its anterior pole, the "apical complex," (Chobotar and Scholtyseck, 1982), which is common to other members of the phylum Apicomplexa including *Plasmodium*. The complex consists of four to eight elongated, club-shaped electron-dense rhoptries, a variable number of electron-dense vacuoles called micronemes, a polar ring, and in the Coccidia, a truncated cone of filamentous structure called the conoid (Sheffield and Melton, 1968; Porchet-Hennere and Nicolas, 1983). Although understanding of the function of the apical structures is still incomplete, this chapter will review what has recently been learned about the role these organelles play in the process of cell entry.

1.4. The Parasitophorous Vacuole

In order to grow and replicate, *T. gondii* must live in a protected intracellular compartment and salvage host metabolites. This environment is provided by a parasitophorous vacuole that may arise at the time of invasion, at least in part from host membrane, but which is clearly modified by the parasite to allow efficient nutrient salvage and to prevent fusion with host cell hydrolytic compartments (Jones *et al.*, 1972; Joiner *et al.*, 1990). Understanding of this important compartment is still incomplete, but the parasitophorous vacuole is a complex, specialized organelle. All biochemical traffic must pass between host and parasite across this membrane; the tubulovesicular network that ramifies from its inner face may be a specialization for this task. Most ultrastructural evidence indicates that the parasitophorous vacuole membrane originates from host plasma membrane that is modified during and after parasite entry. The initial parasitophorous vacuole is poor in intramembranous particles (Cintra and De Souza, 1985) and lacks typical plasma membrane enzyme markers (De Carvalho and De Souza, 1989). Within an hour of parasite entry, intramembranous particles are evident in the parasitophorous vacuole membrane, and a specialized tubulovesicular network is forming within the vacuole.

The most remarkable attribute of the parasitophorous vacuole is that it fails to fuse with host hydrolytic compartments and it does not acidify (Jones *et al.*, 1972; Sibley *et al.*, 1985). This phenomenon has recently been shown to be a feature of parasites that have actively penetrated hosts cells, but not those that have been taken up through the intercession of Fc receptors (Joiner *et al.*, 1990). The parasitophorous vacuole lacks markers of lysosomal membranes and fails to fuse with host endosomes or lysosomes even when the enclosed parasites are killed (Joiner *et al.*, 1990). This last fact speaks against the necessity for *T. gondii* to continuously elaborate an inhibitory factor that would prevent host hydrolytic compartments from fusing with the parasitophorous vacuole. It seems more likely that *T. gondii* either provides the lipids for the parasitophorous vacuole membrane or excludes from it specific host molecules that would allow lysosome or endosome targeting and fusion.

Where does the parasitophorous vacuole membrane come from? It appears that in malaria both the parasite and the host contribute. Malaria merozoites appear to secrete membranous material from rhoptries (Stewart *et al.*, 1985, 1986; Bannister and Mitchell, 1986), and lipid transfer from host high density lipoproteins (Grellier *et al.*, 1991) and erythrocyte membrane (Haldar *et al.*, 1989) to the developing ring trophozoite can be traced by fluorescent lipid analogues either at the time of invasion or shortly thereafter. Parasite lipids are also contributed to the parasitophorous vacuole early in the intracellular life of malaria (Mikkelsen *et al.*, 1988). It seems likely that *T. gondii*, which demonstrates less pronounced membranous whorls in rhoptries than malaria after tannic acid–glutaraldehyde fixation, may not contribute the entire lipid requirement of the parasitophorous vacuole, but may insert some lipid moieties at or near the time of invasion. The lipids from isolated rhoptries of *T. gondii* have recently been analyzed (Foussard *et al.*, 1991). *Toxoplasma gondii* rhoptries are not rich in lipids, but the cholesterol to phospholipid ratio is high and lysophospholipids were detected.

Other *T. gondii* factors secreted at the time of invasion may function to prevent fusion with host organelles, especially since several of these molecules have been found associated with the parasitophorous vacuole membrane (Kimata and Tanabe, 1987). The tubulovesicular network that forms shortly after invasion contains parasite-specified components. It is first observed as a collection of vesicles near the parasite pellicle that evolve into a network of tubules with connections to the parasitophorous vacuole membrane (Porchet-Hennere and Torpier, 1983; Sibley *et al.*, 1986; Sibley, 1989). The stability of the tubulovesicular network depends on the calcium concentration (Sibley *et al.*, 1986); the calcium concentration of the parasitophorous vacuole is yet unknown, but the stability of the tubulovesicular network *in vivo* implies that the calcium concentration of the parasitophorous vacuole may be higher than that of the host cell cytoplasm.

All of the major *T. gondii* surface proteins can be found associated with the tubulovesicular network, as demonstrated by immunoblotting and immunoelectron microscopy (Sibley and Krahenbuhl, 1988; Sibley, 1989). A dense granule protein of 32 kDa is also found associated with the tubulovesicular network; the relationship of this protein with the major surface protein of 30 kDa remains unclear. Rhoptry proteins may be found associated with the inner parasitophorous vacuole membrane (J. F. Dubremetz, personal communication; J. D. Schwartzman, unpublished results). The functions of parasite-specified components of the parasitophorous vacuole are still unknown, as is the relationship between the tachyzoite parasitophorous vacuole and the cyst wall of the bradyzoites. The parasitophorous vacuole is a key element in the intracellular survival of *T. gondii,* the function of which is just beginning to be explored.

2. MORPHOLOGY OF INVASION AND EGRESS

2.1. Light Microscopy

2.1.1. General Observations

The adaptability of *T. gondii* to *in vitro* growth and its relatively large size have allowed a number of light microscopy studies that have defined the basic behavior of *T. gondii* in its interaction with cells in culture (Hirai *et al.*, 1966; Bommer *et al.*, 1969; Doran, 1973; Bogatchev *et al.*, 1980). Saltatory gliding motility of tachyzoites over a solid substrate is easily observed by phase microscopy, as is the rapid penetration of cultured cells of almost any type. The movement of the parasite is in a forward "corkscrew" fashion without obvious deformation of the parasite body, although torsion of the parasite external membrane has been noted by scanning electron microscopy. Anterior–posterior flexing of nongliding parasites and forward protrusion of the anterior tip of the organism can also be seen. These latter phenomena are not necessarily associated with forward movement. They are most often seen when the parasite is attached to a solid substrate by its posterior pole; it may also "spin" while in this configuration. It should be remembered that the virtually two-dimensional configuration of cultured cells limit the parasite to gliding in the plane of the monolayer, while *in vivo* the three-dimensional array of host cells allows the escaping parasite to proceed in almost any direction. Some of the activity noted *in vitro* may therefore be artifactual, such as spinning with only the posterior pole in contact with the substratum, which might in a three-dimensional setting be translated into forward movement with the body of the parasite in contact with extracellular matrix or cells.

The orientation of penetration of *T. gondii* appears to be associated with the forward motility of the parasite rather than being an active reorientation of the parasite, as has been described for the interaction of the malaria merozoite and the red blood cell (Aikawa *et al.*, 1978). The forward twisting movement of *T. gondii* through tissue has the effect of apposing the anterior tip of the parasite to the membrane of a host cell, while malaria interacts with host cellls in suspension and must first adhere and then orient for efficient invasion.

The shape of the host cell greatly influences the ease of observation of *T. gondii* host–parasite interaction. Although thinly spread host cell monolayers are probably artifactual for the reasons given above, the more nearly two-dimensional the cell monolayer is, the more easily the interaction of *T. gondii* with the cells is observed. Distinguishing intracellular parasites from extracellular parasites is more difficult in polygonal host cells. In fibroblast cell lines with a thinly spread phenotype, intracellular parasites clearly lack the refractive halo that extracellular parasites demonstrate and the transition is unambiguous.

2.1.2. Computer-Enhanced Video Microscopy

We have observed invasion and egress of *T. gondii* in cell cultures by means of computer-enhanced video microscopy (Schwartzman *et al.*, 1990). This technology allows a clear view of the process of penetration in which the deformation of the parasite body is obvious as it passes into the interior of the host cell, as is seen in Figure 1. The host cell structure responsible for impeding the passage of *T. gondii* is not visualized by this technique, but it is unlikely to be the host plasma membrane by itself. We suspect that the host cytoskeleton just beneath the plasma membrane is the structure most likely to possess the requisite strength to force the deformation of the *T. gondii* cytoskeleton. Once started, penetration is complete in a few seconds, but immotile *T. gondii* can be observed attached to

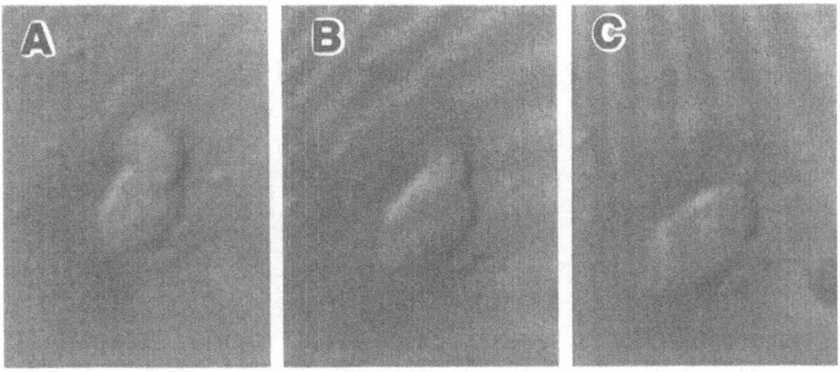

FIGURE 1. *Toxoplasma gondii* (RH strain) invading a human foreskin fibroblast, on glass, warmed during observation to 37°C with an infrared device. The fibroblast membrane is in the plane of the image; the parasite enters moving from the top of the frame to the bottom. Time elapsed between frames: approximately 1 sec. A Zeiss Axiomat microscope fitted with Nomarski differential interference contrast optics was coupled to a Dage model 65 video camera and the output transferred to a Quantex 9210 digital image processor. The output of the image processor was recorded on ¾-inch video tape, and the final images were produced by photography of processed images on a reference monitor. (A) Parasite halfway into host cell, note constriction of parasite body as it moves into cell. (B, C) Parasite completes crossing host cell membrane, the constricted zone moves to the posterior pole, but no parasitophorous vacuole is evident.

a host cell for over 10 min before the process of penetration commences. As penetration is observed in fibroblast culture, the process appears to lack the characteristics of phagocytosis, both in the speed of the process and in the lack of fibroblast cell membrane mobilized to engulf the parasite. Once inside the host, *T. gondii* continues to move for a short distance, then stops; within 30 min a lucent space is evident between the parasite and the parasitophorous vacuole membrane.

The process of egress, which is commonly thought to involve wholesale rupture of the host cell membrane with the release of motile tachyzoites, is in fact an active parasite function as well. This process was documented in infected host cells that were observed just prior to natural egress (the 64-128 parasite stage in human fibroblasts infected by the rapidly multiplying RH strain of *T. gondii*) or in earlier stages of infection when parasite egress can be induced by calcium ionophores (Endo *et al.*, 1982). Behaviors observed were identical in the spontaneous and ionophore-stimulated exiting parasites, but host cells with fewer parasites were much more clearly photographed and are shown in Figure 2. Egress began by individual parasites beginning to move; most of these immediately penetrated the parasitophorous vacuole membrane. Deformation of the parasite, similar to that seen on penetration, was noted. The parasitophorous

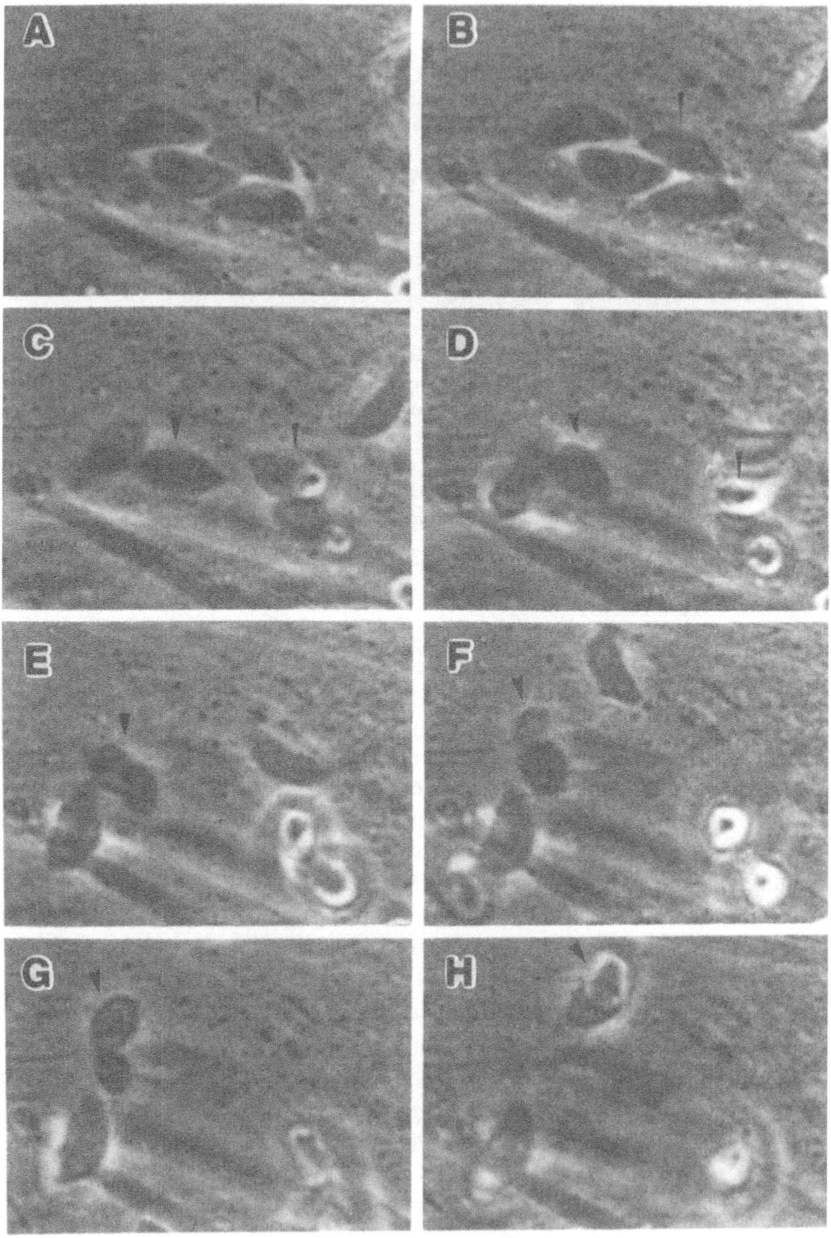
A
B
C
D
E
F
G
H

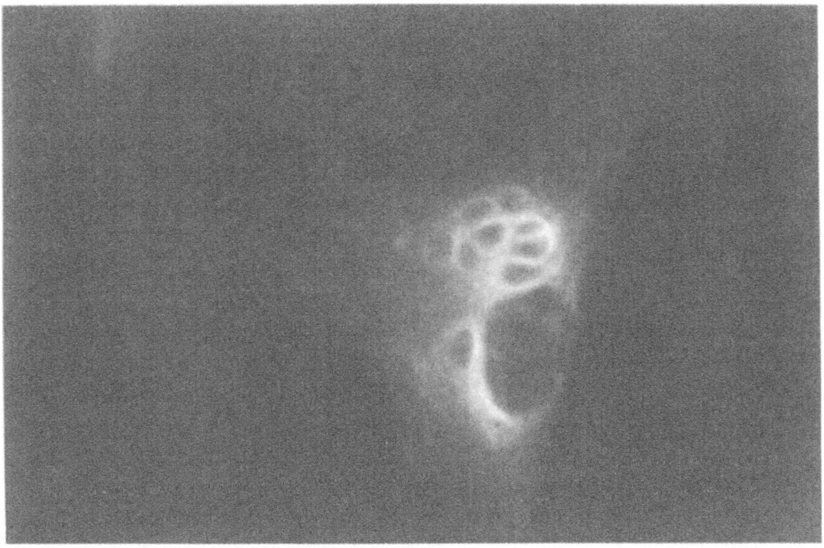

FIGURE 3. Immunofluorescence image of human fibroblast infected with multiple *T. gondii* and stained with rabbit antiserum to human intermediate filament proteins and FITC-conjugated antirabbit antisera. Intermediate filaments are seen surrounding the host nucleus and the parasitophorous vacuoles of the intracellular parasites.

vacuole is surrounded by host intermediate filaments (Figure 3), which may represent the barrier that is breached as *T. gondii* leaves the parasitophorous vacuole. Most parasites rapidly traversed the short space between the parasitophorous vacuole and the host plasma membrane and exited from the host cell, again with deformation. Some parasites moved through other intracellular compartments including the host nucleus, where they were seen deforming themselves as they passed through the nuclear envelope, displacing nucleoplasm as

FIGURE 2. Phase contrast image of four *T. gondii* within a parasitophorous vacuole of a fibroblast prepared as in Figure 1. The calcium ionophore A23187 (1 μM) had been introduced 15 sec prior to the first frame. Time elapsed between frames approximately 1 sec. (A) A single parasite within the vacuole begins to rotate and move to the right (thin arrow). (B) Moving parasite is against vacuole membrane, further rotated. (C) Parasite has penetrated the closely apposed vacuolar and host plasma membranes. Extracellular tip is refractile. A second parasite (thick arrow) moves toward the left. (D) First parasite is now completely extracellular, second parasite begins to penetrate vacuolar membrane (note constriction of parasite), emerging into host cytoplasm. (E–G) Parasite completes egress of vacuole, rotates, and traverses host cytoplasm. Parasites have been observed to enter and traverse other organelles, such as the host nucleus, in this process. (H) Parasite exits through host plasma membrane (becomes refractile).

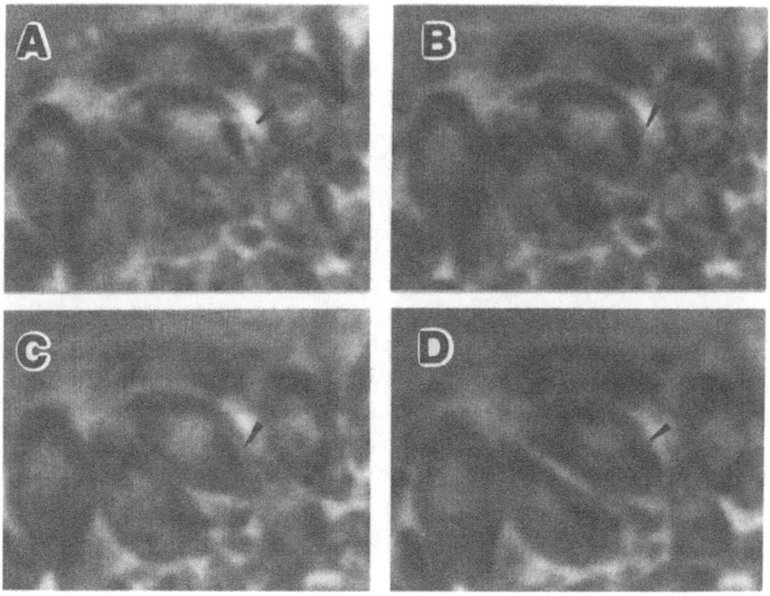

FIGURE 4. Phase contrast image of a portion of an end-stage parasitophorous vacuole containing approximately 64 parasites, 3 min prior to the start of unstimulated parasite egress. Phase-dense barlike bodies resembling rhoptries, the major secretory organelle, are seen at the anterior of many *T. gondii* (arrow). Approximately 5 sec elapse between frames. (A–D) Change in density of rhoptry over a period of 15 sec suggests that the contents may be discharged. Approximately 2–3 min later, parasites began to move within the vacuole and exited the host cell.

they crossed the nucleus, exiting the nuclear envelope, traversing the host cytoplasm and finally exiting the host plasma membrane. The host cell remained morphologically intact following the egress of several parasites, but when many parasites had exited, the host cell appeared to degenerate.

The process of egress shares many features of invasion, but the interaction of the parasite with various host membranes is likely to be somewhat different. No host membranes appear to enshroud escaping parasites, either by light or electron microscopic studies. The interaction with host cellular components need only be sufficient to provide forward propulsion, and no specialized junction need be formed with the host membrane.

Another phenomenon we noted in computer-enhanced video microscopy was the prominence of club-shaped anterior organelles that appeared to be rhoptries (shown in Figure 4). These were not noted until the later stages of intracellular parasite development, and these bodies lost their definition within minutes of the onset of motility in parasites spontaneously exiting host cells. This may indicate that rhoptry discharge occurs at the time of egress as well as invasion. Although rhoptrylike bodies were not noted in parasites in the early stages of intracellular development, parasites that had just entered host cells were competent to be stimulated to exit by calcium ionophore. The role of rhoptry discharge in parasite egress is therefore unclear.

2.2. Ultrastructure

There have been numerous studies of *T. gondii* morphology and parasite–host interaction by transmission, scanning, and freeze-fracture electron microscopy (reviewed by Chobotar and Scholtyseck, 1982). Recent studies by scanning microscopy have clearly shown the torsion of the invading *T. gondii,* which was suggested by the classic transmission and scanning electron microscopic study of Aikawa (Aikawa *et al.*, 1977). The interaction of the parasite and the erythrocyte membrane has been clearly shown (Michel *et al.*, 1980; Porchet-Hennere and Torpier, 1983) to form an electron-dense "moving junction" very similar to that observed in the malaria merozoite–red cell interaction (Aikawa *et al.*, 1978). More recently, immunoelectron microscopy has been used to establish the distribution of *T. gondii* antigens (Johnson *et al.*, 1983; Sibley and Sharma, 1987; Couvreur *et al.*, 1988; Sadak *et al.*, 1988; Sibley and Krahenbuhl, 1988; Schwartzman and Krug, 1989; Sibley, 1989; Charif *et al.*, 1990; Taylor *et al.*, 1990). These studies have been able to sort out the intracellular location and fate of a number of *T. gondii*-specified molecules, but much remains to be learned about the function of these molecules.

3. MOTILITY OF *T. GONDII*

Although motility is a key part of the process of active penetration in *Toxoplasma,* the mechanism has never been explained. Evidence for the central role of parasite motility in host cell penetration is provided by the observation that parasites transiently immobilized by low temperature, change of pH, change of extracellular ion milieu, or by reversible inhibitors of cytoskeletal function are unable to enter host cells, but regain their ability to invade when the inhibiting factor is removed (Doran, 1973; Russell and Sinden, 1981; Schwartzman and Pfefferkorn, 1983; Endo and Yagita, 1990). Gliding movement similar to that of *T. gondii* is seen in bacteria, fungi, algae, and other protozoa, but it is not well understood in any system (Bloodgood, 1989).

The transport of particles attached to the surface of algal flagellae along tracks defined by submembranous microtubules (Bloodgood, 1977, 1981) presents one promising model of gliding locomotion; another that has been suggested is the capping of ligands such as antibody molecules on cell surfaces.

Free-living gregarine sporozoans are the organisms most closely related to *Toxoplasma* in which locomotion has been studied (King, 1981, 1988). Like *Toxoplasma,* the gregarines only move when the organism is in contact with a substratum. In this system, inhibitors of actin and calmodulin block motility (King and lee, 1982).

The question of what component *Toxoplasma* interacts with as it glides across a cell surface or solid substrate or as it invades a host cell is unclear. It has been suggested that laminin may have an important role in this regard (Joiner *et al.*, 1989), but the issue is unsettled. The role of *Toxoplasma* surface antigens or secreted factors in this process is unknown. Malaria sporozoites appear to leave trails of circumsporozite protein on solid substrates over which they have moved (Stewart and Vandenberg, 1988), which could be related to their mode of locomotion and raises the question of whether *Toxoplasma* surface antigens could play a similar role. A protein has recently been described in malaria that may have a role in forming the moving junction seen in the invasion of the merozoite into the erythrocyte (Klotz *et al.*, 1989).

Complicating the application of any suggested model of gliding locomotion to *T. gondii* is the unusual arrangement of membranes of *T. gondii* tachyzoites. An apparently ordinary plasma membrane surrounds the organism. Two additional unit membranes have been noted internal to the plasma membrane, which have been shown in *Eimeria* and *Sarcocystis* (Porchet-Hennere, 1975) to be arranged as side-by-side envelopes appearing like "pavement blocks," with cross sections as illustrated schematically in Figure 5. Freeze-fracture electron micrographs (Figure 6) demonstrate similar patches of inner membrane complex in *T. gondii.* The somewhat irregular blocks are not continuous over the surface of the tachyzoite: the inner membrane complex is absent at the poles of the organism. Beneath the inner membrane complex are the cytoplasm and 22 pairs of longitudinal microtubules (Nichols and Chiappino, 1987). If the outer membrane must be coupled to a surface in order to transduce force and if the motor is located in the cytoplasm, the inner membrane complex is interposed. Transmembrane glycoproteins have been shown to couple microtubule-associated motors to external ligands in *Chlamydomonas* flagella (Bloodgood and Salomonsky, 1989). The multiple membrane layers of sporozoans would appear to complicate such a scheme of parasite motility. It is possible that transmembrane proteins in the inner membrane complex interact with a cytoplasmic motor and then transfer motive force to transmembrane bridges crossing the outer membrane to couple with the substratum. There are many organizational and topological problems in such a model, for instance, how the inner membrane complex, which is essentially made of collapsed envelopes of membrane, might rotate in a coordinated fashion and what happens to the force transduction apparatus at the seams between inner membrane blocks. It would be simpler if the motor proteins were

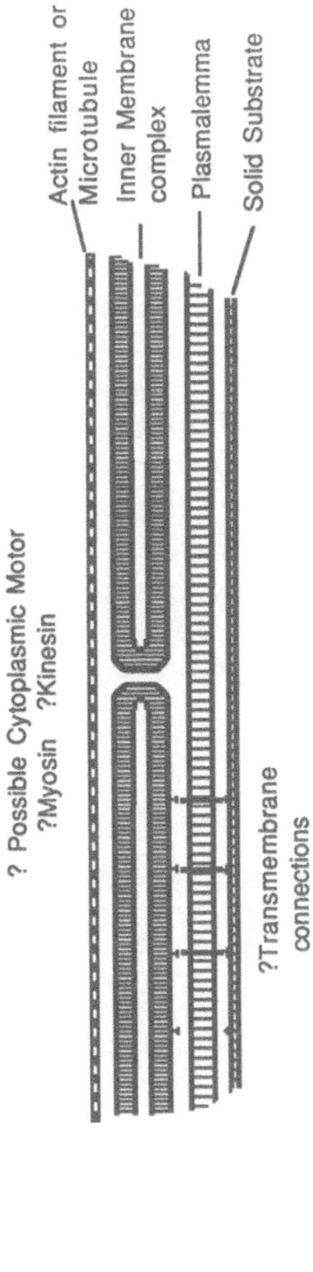

FIGURE 5. Schematic diagram of a cross section of *T. gondii* membranes.

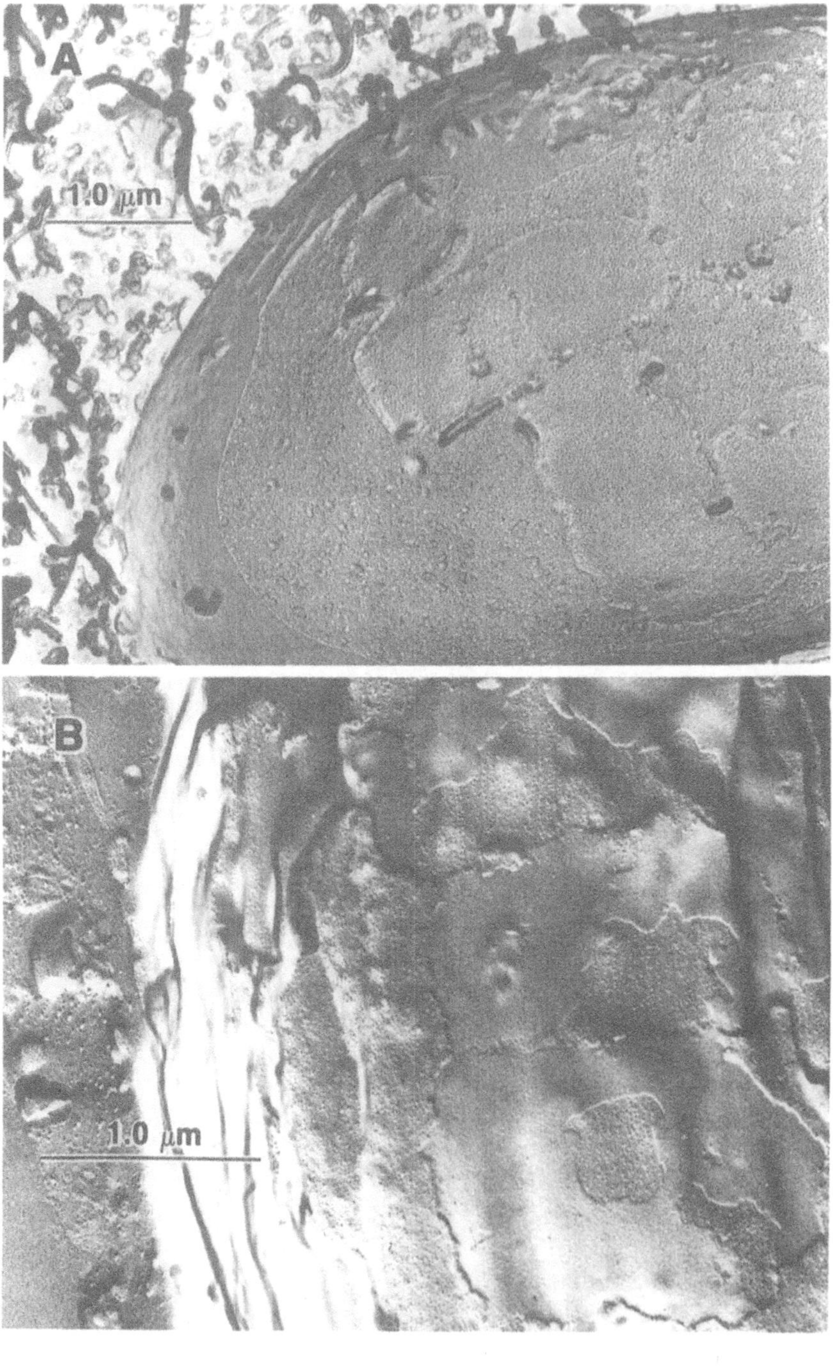
A
1.0 μm
B
1.0 μm

sited beneath the outer membrane of the parasite. There is, so far, no evidence to support any proposed model of locomotion in *T. gondii.*

T. gondii motility has been reported to be stimulated by changes in the internal pH and K^+ concentration, as studied by the use of the fluorescent weak base 9-amino acridine (Endo *et al.*, 1987). Acidification of the external medium can induce motility in extracellular parasites. Manipulation of the extracellular pH and ionic milieu can temporarily immobilize extracellular *Toxoplasma* (Endo and Yagita, 1990). The calcium ionophore A23187 induces motility in intracellular *T. gondii* (Endo *et al.*, 1982), and the calmodulin inhibitor trifluoperazine inhibits gliding motility in gregarine sporozoans (King and Lee, 1982). The interaction of ions in the initiation of motility of *Toxoplasma* remains to be elucidated.

3.1. Cytoskeleton of Coccidians

Cytoskeletal proteins have been implicated in the mechanism of motility of gregarine sporozoans. Actin is localized at the internal cytomembranes and in the cortical ectoplasm of gregarines (Baines and King, 1989b). By immunoblot using antiactin antibody, both the expected 43-kDa protein and a higher molecular weight species were detected in gregarines and *Eimeria* (Baines and King, 1989a,b). We have studied the distribution and organization of myosin, actin, and tubulin in *T. gondii* by immunofluoresence microscopy utilizing antibodies specific for cytoskeletal proteins of heterologous organisms (Schwartzman and Pfefferkorn, 1983; Schwartzman *et al.*, 1985). Myosin appears to be concentrated at the anterior pole of *T. gondii* (Figure 7A). Actin appeared uniformly distributed in our studies (unpublished results) either in the cytoplasm or perhaps associated with membranes, but not organized into filaments. Endo has reported that actin is localized to the anterior of *T. gondii* (Endo *et al.*, 1988; Yasuda *et al.*, 1988), and Taylor *et al.* (1990) have found an antigen in many sporozoans that has an anterior distribution and that they suspect may be actin. By immunoblot we detected actin as the expected 43-kDa protein, but myosin is recognized as an approximately 125-kDa species (Schwartzman, unpublished results) resembling the small myosins described in *Acanthamoeba* and *Dictyostelium* (Korn and Hammer, 1988). Such small myosins have been found to be membrane associated, and may be implicated in whole cell locomotion (Fukui *et al.*, 1989; Miyata *et al.*, 1989). The actin genes of malaria have recently been cloned and

FIGURE 6. Freeze-fracture preparations of *T. gondii* revealing details of the pellicle membranes. (A) Extracellular organism rapidly frozen, fractured, and etched for 1 min to reveal outside face of plasma membrane. Platinum carbon replica. (B) Organism within human fibroblast fractured to reveal multiple layers of inner membrane complex. Rows of intermembranous particles align with vertical folds of membrane corresponding to subpellicular microtubules that underlay inner membrane complex.

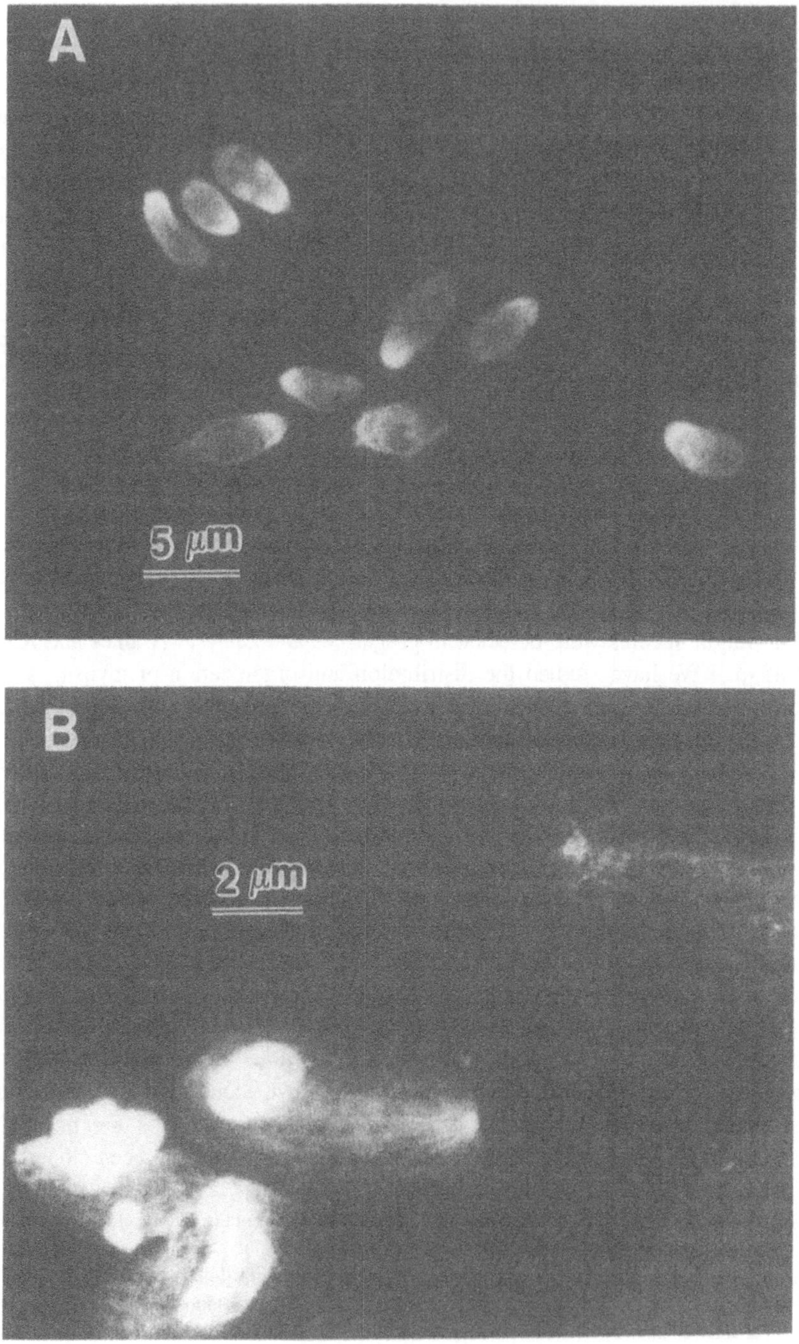
A
5 μm
B
2 μm

sequenced, and it is of interest that one of the two genes displays the most divergent sequence of any actin gene so far studied (Wesseling *et al.*, 1988a,b).

Although myosin was first described in 1864 in muscle extracts, it has now become evident that there are multiple myosinlike motors that interact with actin to power numerous intracellular tasks (Spudich, 1989). Many of these molecules have been described in simple eukaryotes such as *Acanthamoeba* and *Dictyostelium*, but homologues are now being found in mammalian cells. *Toxoplasma* may well have more than one myosin type.

The distribution of tubulin in *T. gondii* corresponds to the described subpellicular microtubules, but the conoid was not always detected by the monoclonal antibodies to β-tubulin that we employed (Schwartzman *et al.*, 1985). Figure 7B demonstrates the distribution of tubulin as detected by a polyclonal antiserum that also seems to react with conoid microtubules.

It is not yet understood how the parasite's cytoskeletal and contractile proteins interact to create and transfer force in an organized manner. It is likely that an actin-based motor is important in the mechanism of motility because parasites treated with cytochalasin D, which interferes with actin function in a number of organisms, are paralyzed (Jones *et al.*, 1972; Ryning and Remington, 1978; Schwartzman and Pfefferkorn, 1983). Cytochalasin paralysis has also been noted in other Coccidia (Russell and Sinden, 1981). Electron microscopy (Aikawa *et al.*, 1977) demonstrates protrusion of the conoid during penetration. The conoid appears to be composed of microtubules, which may therefore have a role in host cell penetration.

3.2. Possible Motors and Mechanisms

Although actin–myosin appears the most likely motive system for *T. gondii* motility, all of the movements of the parasite may not be based on a single system. Gliding, flexing, torsion, and protrusion of the conoid may be powered by different motors. Microtubule-based motors, such as kinesin, that have been shown to operate in a number of intracellular movements (Vale and Goldstein, 1990) could power flexing, torsion, and protrusion of the conoid. The relationship between the microtubule system and the linear motor proposed by King (1988), which is based on actin–myosin, is unclear. Further work to define the distribution and functions of motor proteins will give us more direction for the construction of a model for gliding motility. King (personal communication) has performed calculations of the bioenergetics of gliding motility to show that a

FIGURE 7. Immunofluorescence photomicrographs of *T. gondii* prepared with anticytoskeleton antisera. (A) Antiinsect myosin, demonstrating detection of cross-reacting *T. gondii* myosin localized in a polar distribution. (B) Antibovine brain tubulin, demonstrating subpellicular microtubules and conoid.

reasonable amount of ATP hydrolyzed by a relatively small number of myosin molecules could provide sufficient force to overcome the viscous forces that would be required to move the parasite.

4. FACTORS SECRETED AT THE TIME OF INVASION

4.1. Penetration-Enhancing Factor

The function of the secretory product of rhoptries is unknown, but it has been assumed since the work of Lycke and Norrby (1966) that the "penetration-enhancing factor" (PEF) that they described was stored in rhoptries and secreted at the time of cell invasion, altering the host cell membrane and promoting the entry of the parasite. The nature of PEF and its mode of action remain unclear. It seems likely that more than one factor may be involved in promoting penetration. Lycke and Norrby (Norrby and Lycke, 1967; Lycke *et al.*, 1968; Norrby, 1970) isolated an acidic protein from *T. gondii* that showed PEF activity in both *in vitro* and *in vivo* systems and Norrby (1971) identified two peaks with PEF activity within a 70- to 150-kDa size range by molecular sizing chromatography. In high concentration this fraction caused nonspecific ultrastructural lesions in the membranes of cultured cells. A rabbit antiserum to proteins from this fraction localized to the anterior half of the parasite (Norrby, 1971). Since lysozyme, hyaluronidase, and five lysosomal enzymes exhibited some penetration-enhancing activity (Lycke *et al.*, 1965, 1975), these authors felt that PEF was likely an enzyme. Others (De Souza and Souto-Padron, 1978) demonstrated that the contents of *T. gondii* rhoptries had the cytochemical properties of a basic protein, which was in accord with the findings of Kilejian (1976) of a histidine-rich basic protein associated with the anterior organelles of *Plasmodium lophurae*.

4.2. Apical Organelles

4.2.1. Rhoptry Proteins

Several rhoptry-associated proteins of *T. gondii* have been reported. Sadak *et al.* (1988) have characterized a family of rhoptry antigens by immunoblotting and immunoelectron microscopy that consists of at least two related members. A 55-kDa and a 60-kDa protein are processed from 66 and 68 kDa precursors. Other proteolytic species are also found that share epitopes with these two proteins. All stages of the life cycle share these rhoptry antigens, but the function of these molecules is not yet known. Leriche and Dubremetz (1991) have extended this study to identify a total of 11 rhoptry proteins in *T. gondii* by means of subcellular fractionation and monoclonal antibodies. Several rhoptry antigens of malaria have also been described (Holder *et al.*, 1985; Schofield *et al.*, 1986; Roger *et al.*, 1988; Sam-Yellowe *et al.*, 1988; Ridley *et al.*, 1990), but the relationship of these to the *T. gondii* antigens is unclear.

We produced a series of monoclonal antibodies that recognized rhoptries of *T. gondii* (Schwartzman, 1986) by immunofluorescence (Figure 8) and by immunoelectron microscopy. Although none of the antibodies could completely prevent parasite invassion of human fibroblast monolayers (either alone or in combination with one another), each of the antibodies interfered with the phenomenon of penetration enhancement, which had been described earlier by Lycke and Norrby (Lycke *et al.*, 1975), and all of the antibodies partially cross-reacted with each other in solid-phase competition assays using whole *T. gondii* antigen.

We further characterized the molecules recognized by the three most avid of the four original antibodies (Schwartzman and Krug, 1989). These antibodies immunoprecipitate a series of proteins that migrate in the range of 35–60 kDa on both reducing and nonreducing polyacrylamide gel electrophoresis. A transient precursor molecule of 75 kDa is seen in some preparations. The complex pattern recognized by one of the antibodies (Figure 9) can be simpliffied to a single 60-kDa band by inhibiting serine proteinases. This antigen, referred to as rhoptry protein 1 (ROP-1), can be cleaved by pronase and trypsin and contains little or no carbohydrate. It has a pI of approximately 5.85. Charge shift electrophoresis, utilizing various detergents to determine whether the antigen behaves like a molecule with a hydrophobic, membrane-anchored domain (Helenius and Simons, 1977), indicates that the antigen does not resemble a typical intrinsic membrane protein. Although both immunofluorescence microscopy and immunoelectron microscopy showed the presence of extracellular antigen in a pattern suggestive of secretion, and although rhoptry contents appear to diminish in the process of host cell invasion (Nichols *et al.*, 1983; Porchet-Hennere and Nicholas, 1983), the majority of ROP-1 remained with the insoluble fraction of the parasite when *T. gondii* was extracted in a number of physiological buffers, with nonionic detergents, or with all except the highest concentrations of chaotropic salts. Only 8 M guanidine hydrochloride or the ionic detergent sodium dodecyl sulfate could consistently put all of the antigen into the supernatant phase of 13,000 × G centrifuged extracts. Antigen so denatured is still recognized by monoclonal antibody T*g*49, but it lacks biological activity.

We purified ROP-1 by means of several chromatographic techniques and assayed the resultant fractions in a bioassay for penetration enhancement, which relies on the specific incorporation of tritiated uracil into parasites that have successfully invaded host cells. Penetration-enhancing activity from unfractionated *T. gondii* extracts is labile. The activity in fresh extracts or supernatant media from *T. gondii* freshly released from host cells disappears at room temperrature within 4–6 hr, is stable at 4 °C for 12 hr, but is lost after freezing. The

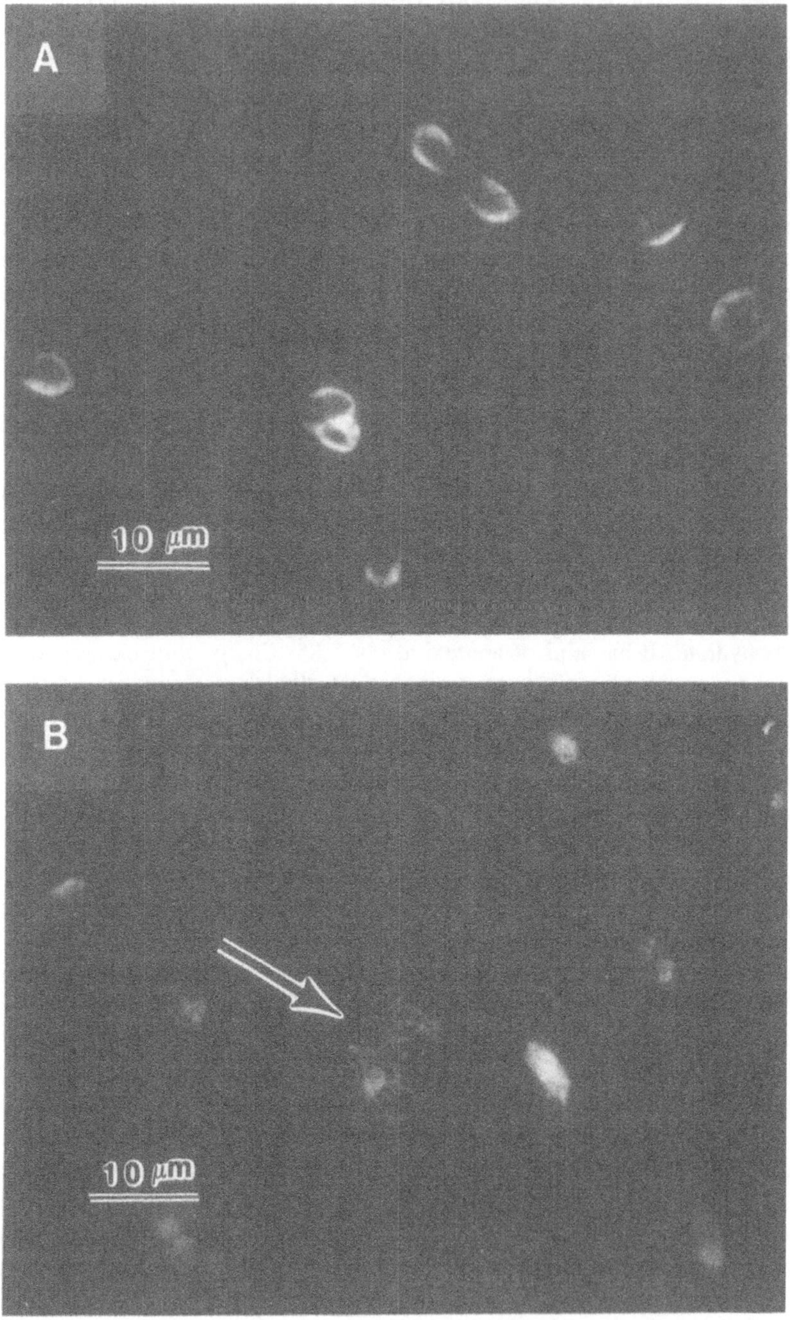
A
10 µm
B
10 µm

activity remains in the supernatant fraction after 30 min of centrifugation at 100,000 × G, but is lost upon filtration through 0.22 mM nitrocellulose membrane. Purification of ROP-1 by affinity chromatography produces large quantities of relatively pure antigen that has little biological activity. Most characterization of biological activity of ROP-1 containing extracts has therefore been performed with preparations which are partially purified by rapid size exclusion chromatography on a Superose-12™ column (Pharmacia).

The apparent contradictions between the biological and biochemical properties of ROP-1 have been explained in part by analysis of the amino acid sequence predicted by molecular cloning of ROP-1 cDNA (Ossorio *et al.*, 1992). The predicted protein demonstrates an unusual charge asymmetry, with 37 acidic to 4 basic amino acids in a domain of the amino terminal portion of the sequence, and 31 basic to 4 acidic amino acids in a domain of the carboxyterminal portion of the sequence. The aminoterminal acidic domain is also 20% proline. The highly charged nature of this polypeptide may explain its apparent low solubility in the absence of a substantial hydrophobic domain. The predicted size of the protein is 44 kDa much smaller than the major species identified on polyacrylamide gel electrophoresis (PAGE) of *T. gondii* lysates or ROP-1 gene product expressed in *Escherichia coli*. The charged nature and proline richness of the ROP-1 polypeptide may influence its migration in PAGE. Functional characterization of the cloned ROP-1 product is in progress.

One suggested mechanism for the action of PEF has been that it allows the parasite to create a hole in the host membrane. To assay the ability of *T. gondii* to produce holes in membranes, we tested the ability of the parasite or parasite extracts to damage a well-characterized membrane: that of the erythrocyte. *Toxoplasma gondii* extracts or live, infectious organisms were mixed with human, dog, or chicken erythrocytes suspended in Hank's balanced salts. Hemolysis was not detected. To increase sensitivity, erythrocytes were treated with sublytic concentrations of the detergent deoxycholate and exposed to *T. gondii* extracts or live organisms, but no hemolysis was detected. Since *T. gondii* infects mature erythrocytes inefficiently, human fibroblasts (a well-characterized target of *T. gondii* invasion) were loaded with 5,6,carboxyfluorescein diacetate and treated in a similar fashion. Neither extracts of *T. gondii* nor live organisms produced detectable egress of indicator dye above the level of controls. Whatever the damage to the host cell membrane produced by PEF or whole organisms, these treatments do not appear to open channels for the escape of uncharged molecules.

FIGURE 8. Immunofluorescence photomicrographs of *T. gondii* labeled with monoclonal antibody Tg49, demonstrating localization of antibodies to rhoptries. (A) Formaldehyde-fixed organisms demonstrating anterior pole localization of antibody. (B) Triton X-100 extracted organisms demonstrating rodlike rhoptries (arrow).

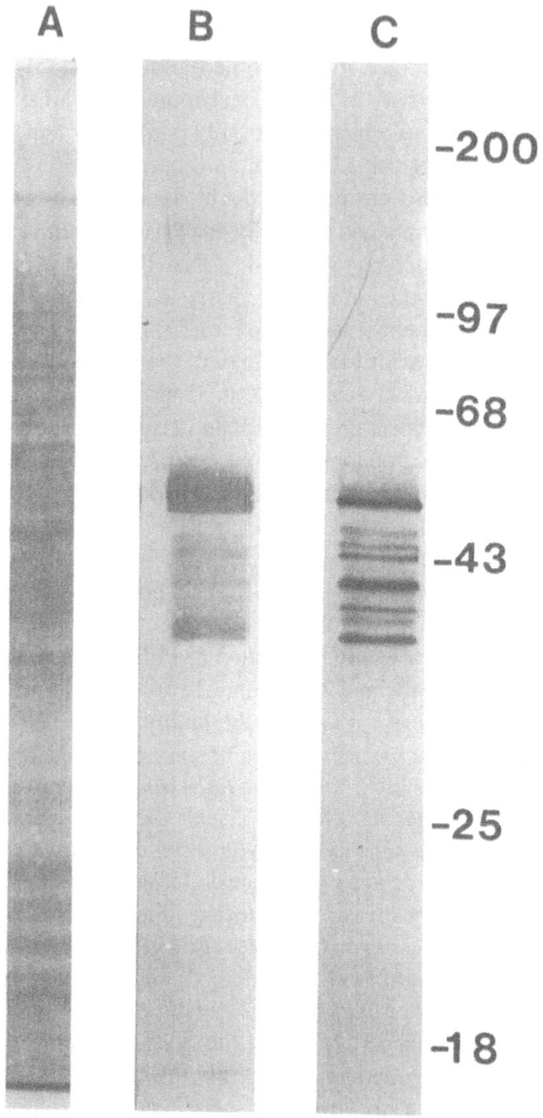

FIGURE 9. Western immunoblot of *T. gondii* proteins on nitrocellulose paper detected with monoclonal antibody Tg49. (A) Amido black stain of nitrocellulose demonstrating total transferred proteins. (B) Nonreduced *T. gondii* proteins. (C) Reduced *T. gondii* proteins. Molecular weight markers $\times\ 10^3$.

There is little information on the control of secretion of rhoptry contents or the contents of other exocytotic compartments of *Toxoplasma*. The control of secretion in protozoa has been extensively studied in the ciliates, which discharge visible hairlike projections called trichocysts upon various stimuli. This process appears to be under the control of calcium-linked mechanisms (Satir *et al.*, 1988). The calcium messenger system is quite complex and appears to operate in a wide variety of organisms (Rasmussen, 1986a,b). In *Paramecium*, external calcium acting through calcium channels interacts with calmodulin, and results in discharge of trichocysts (Gilligan and Satir, 1983; Garofalo and Satir, 1984). Specific dephosphorylation of a 63-kDa protein (Gilligan and Satir, 1982; Murtaugh *et al.*, 1987) is also associated with exocytosis of the trichocysts; the relationship of this process with the phosphoinositol transmembrane-signaling pathway (Low and Saltiel, 1988; Nishizuka, 1988) and protein kinase C is as yet unknown. Antagonists of calcium and calmodulin interfere with this process (Garofalo *et al.*, 1983) and also have been found to inhibit malarial growth (Scheibel *et al.*, 1987). *Paramecium* mutants with defects in exocytosis have been selected (Orias *et al.*, 1983; Garofalo and Satir, 1984). ATP has been determined to be inhibitory to exocytosis but required for recovery (Vilmart-Seuwen *et al.*, 1986). Little work has been reported in other protozoans, but the findings in the ciliata may provide a starting point for considering the mechanism of secretion of *T. gondii*. It has been noted that calmodulin appears to be localized by immunoelectron microscopy at the anterior tip of *Toxoplasma* (J. F. Dubremetz, personal communication) which suggests that a calcium-based mechanism may be operative.

4.2.2. Other Exocytotic Organelles: Dense Granules and Micronemes

Other *Toxoplasma* organelles have been shown to discharge contents, but usually not in conjunction with invasion. Dense granules, of which there are two morphological types, appear to discharge their contents into the parasitophorous vacuole (Sibley and Krahenbuhl, 1988; Sibley, 1989). A calcium-binding protein has been identified as a member of the dense granule family (Sibley, 1989; Charif *et al.*, 1990; Leriche and Dubremetz, 1990). Other dense granule proteins without defined functions have also been identified. (Dubremetz and Dissous, 1980; Leriche and Dubremetz, 1991). Micronemes are even more mysterious in function, but several antigens have been localized to this compartment (Dubremetz and Dissous, 1980; Dubremetz *et al.*, 1989; Pohl *et al.*, 1989). An erythrocyte-binding protein of malaria (the Duffy receptor) has been localized to micronemes (Adams *et al.*, 1990) so it is possible that this organelle functions early in the invasion process.

4.3. Unlocalized Factors

4.3.1. Phospholipases

Phospholipase A2 (PLA2) plays an important pathogenic role in infections caused by several microorganisms (Turco and Winkler, 1986) and has been implicated in host cell invasion by intracellular bacteria of the genus *Rickettsia*. We have investigated the role of PLA2 in cellular invasion by the tachyzoite stage of *T. gondii* (Saffer *et al.*, 1989). Phospholipase A2 from *Naja naja* venom increased the penetration of fibroblasts by *T. gondii*, while horse antiserum to *Naja hannah* snake venom (which contains anti-PLA2) inhibited penetration in a dose-dependent manner. An irreversible PLA2 inhibitor, *p*-bromophenacyl bromide (BPB), blocked penetration without metabolically disabling the parasite. When host fibroblasts were preincubated with BPB, penetration was not prevented, supporting a role for parasite rather than host cell PLA2 in the penetration process. Another PLA2 inhibitor, nordihydroguaiaretic acid, also inhibited penetration. We assayed extracellular *T. gondii* tachyzoites, purified from host cell debris, for PLA2 activity by radiometric detection of fatty acid release from labeled *E. coli* membranes. Sonically disrupted parasites contained a calcium-dependent PLA2 with maximum activity at pH 8.5–9.0.

These experiments suggested that a phospholipase was involved in the penetration of *T. gondii* into human fibroblasts grown in culture. We also assayed the effect of lysolecithin in this system, which might have been expected to have promoted increased penetration: it did not, but an effect may have been obscured by the potent lytic effect that this compound had on the host fibroblasts. Similarly, phospholipase C from *Clostridium perfringens* did not enhance penetration in concentrations that were not lytic to the host cell. It seems unlikely that phospholipase C is the important phospholipase in this situation, although it may be of importance in the release of phosphoinositol-anchored surface proteins from the parasite surface.

The relatively low levels of PLA2 activity found in sonically disrupted *T. gondii* are similar to those found in similarly treated *Entamoeba histolytica* (Ravdin *et al.*, 1985). Purified fractions of *E. histolytica* show higher PLA2 activities. The calcium dependence and alkaline activity maximum are consistent with a secreted PLA.

Phospholipase A2 may be involved in host cell penetration in one of several ways. The products of the enzyme could cause membrane lysis or membrane fusion or could effect membrane fluidity. Phospholipase A2 may be secreted by the rhoptries or it may be membrane bound. It could act independently or in conjunction with rhoptry secretions. The relationship of phospholipase and previously described PEFs is at present uncertain, although the evidence that we have so far favors the two being separate entities.

4.3.2. Proteases

Another possible mode of PEF action might be as a proteinase, or PEF may require the activity of a proteinase. We therefore assayed the effect of a series of proteinase inhibitors on penetration of *T. gondii* into host cells. We tested representatives of four different classes of proteinase inhibitors: (1) pepstatin, an inhibitor of acid proteases; (2) leupeptin, an inhibitor of serine and thiol proteases; (3) aprotinin, an inhibitor of a wide range of serine proteases; and (4) phenylmethylsulfonyl fluoride (PMSF), an inhibitor of serine proteases, and 1,10,phenanthroline, an inhibitor of metalloproteases, were studied. Each inhibitor, at the indicated concentration (Table I) in Eagle's minimum essential medium without serum, was mixed with an equal inoculum of *T. gondii* and allowed to infect fibroblast monolayers. Growth of intracellular *T. gondii* was measured by the incorporation of [^{3}H]uracil, as in our usual PEF assay. Other agents tested included iodoacetamide, a thiol protease inhibitor, and EDTA, a metalloprotease inhibitor, but these agents disrupted the host cell monolayers and made the determination of host cell penetration impossible.

The only agent that decreased penetration and subsequent growth was 1,10,phenanthroline. Several concentrations of 1,10,phenanthroline, from 1–50 μM, were tested with results similar to those shown in Table I. Inhibition was seen at concentrations above 1 μM. Attempts to reverse this inhibition by the addition of 25–75 μM zinc chloride did not produce statistically significant results, but there was a trend toward cancellation of the inhibitory effect of 1,10,phenanthroline. The amount of zinc in solution was not determined; in the presence of carbon dioxide, zinc precipitates as a carbonate and this may have influenced the experiment. Although serine protease inhibitors can prevent the hydrolysis of ROP-1, they had little effect on *T. gondii* penetrration.

We also assayed *T. gondii* lysates, which had been separated by PAGE, for proteolysis of substrates immobilized in the acrylamide matrix. We detected no proteolysis of gelatin or casein by this technique, although positive bands could be detected from control host cell proteins. The role of proteolysis in the process of host cell penetration remains unclear.

Table I
Effect of Proteinase Inhibitors on Invasion

Treatment	Concentration (μM)	Percent of control
Control	NA	100
Pepstatin	5	111
Leupeptin	25	100
Aprotinin	20	101
PMSF	100	113
1,10,Phenanthroline	50	44

5. CONCLUSIONS

The mechanism of motility of *T. gondii* and the identification of the function of the molecules identified in its secretory organelles present important targets for future research. Understanding how *Toxoplasma* glides may provide a novel target for pharmacological interruption of toxoplasmosis. The mechanism of gliding motility, or the motors that power it, may be sufficiently different from mechanisms utilized by mammalian cells that they could be interrupted without interference with host cell function.

The *Toxoplasma*-specified products that facilitate host cell invasion and that establish the function of the parasitophorous vacuole might be targets for chemotherapy or vaccine development. As functions are discovered for some of the molecules that have been defined by morphological and immunological means, and molecules are assigned to functional activities that are defined biologically, we will uncover opportunities to attack both the acute phase of infection and the maintenance or reactivation of chronic infection.

ACKNOWLEDGMENTS. I would like to thank J. Boothroyd, J. F. Dubremetz, K. Haldar, L. Kasper, C. King, and D. Sibley for useful discussions during the writing of this manuscript. Some of the experiments reported were supported by PHS/NIH grant AI 23074 to J. D. S.

6. REFERENCES

Adams, J. H., Hudson, D. E., Torii, M., Ward, G. E., Wellems, T. E., Aikawa, M., and Miller, L. H., 1990, The Duffy receptor family of *Plasmodium knowlesi* is located within the micronemes of invasive malaria merozoites, *Cell* **63**;141–153.

Aikawa, M., Komata, Y., Asai, T., and Midorikawa, O., 1977, Transmission and scanning electron microscopy of host cell entry by *Toxoplasma gondii, Am. J. Pathol.* **87**:285–296.

Aikawa, M., Miller, L. H., Johnson, J., and Rabbage, J., 1978, Erythrocyte entry by malarial parasites. A moving junction between erythrocyte and parasite, *J. Cell. Biol.* **77**:72–82.

Baines, I., and King, C. A., 1989a, Demonstration of actin in sporozoites of the protozoon *Eimeria, Cell Biol. Int. Rep.* **13**:639–641.

Baines, I., and King, C. A., 1989b, Demonstration of actin in the protozoon *Gregarina, Cell Biol. Int. Rep.* **13**:679–686.

Bannister, L. H., and Mitchell, G. H., 1986, Lipidic vacuoles in *Plasmodium knowlesi* erythrocytic schizonts, *J. Protozool.* **32**:271–275.

Bloodgood, R. A., 1977, Motility occurring in association with the surface of the *Chlamydomonas* flagellum, *J. Cell Biol.* **75**:983–989.

Bloodgood, R. A., 1981, Flagellum as a model system for studying dynamic cell-surface events, *Cold Spring Harbor Symp. Quant. Biol.* **46**:683–693.

Bloodgood, R. A., 1989, Gliding motility: Can regulated protein movements in the plasma membrane drive whole cell locomotion? *Cell Motil. Cytoskel.* **14**:340–344.

Bloodgood, R. A., and Salomonsky, N. L., 1989, Use of a novel *Chlamydomonas* mutant to demonstrate that flagellar glycoprotein movements are necessary for the expression of gliding motility, *Cell Motil. Cytoskel.* **13**:1–8.

Bogatchev, Y. V., Khavkin, T. N., Shustrov, A. K., and Freidlin, I. S., 1980, Motion picture study of the response of cultured peritoneal macrophages to the invasion of endozoits of *Toxoplasma gondii*, RH strain, *Acta Microbiol. Acad. Sci. Hung.* **27**:1–8.

Bommer, W., Heunert, H. H., and Milthaler, B., 1969, Cinematographic studies on the movement of *Toxoplasma gondii, Z. Tropenmed. Parasitol.* **20**:450–458.

Buckley, S. M., 1973, Survival of *Toxoplasma gondii* in mosquito cell lines and establishment of continuous infection in vero cell cultures, *Exp. Parasitol.* **33**:23–26.

Charif, H., Darcy, F., Torpier, G., Cesbron-Delauw, M.-F., and Capron, A., 1990, *Toxoplasma gondii:* Characterization and localization of antigens secreted from tachyzoites, *Exp. Parasitol.* **71**:114–124.

Chobotar, B., and Scholtyseck, E., 1982, Ultrastructure, in *The Biology of the Coccidia* (P. L. Long, ed.), pp. 101–156, University Park Press, Baltimore.

Cintra, W. M., and De Souza, W. M., 1985, Immunocytochemical localization of cytoskeletal proteins and electron microscopy of detergent extracted tachyzoites of *Toxoplasma gondii, J. Submicrosc. Cytol.* **17**:503–508.

Couvreur, G., Sadak, A., Fortier, B., and Dubremetz, J. F., 1988, Surface antigens of *Toxoplasma gondii, Parasitology* **97**:1–10.

De Carvalho, L., and De Souza, W., 1989, Cytochemical localization of plasma membrane enzyme markers during interiorization of tachyzoites of *Toxoplasma gondii* by macrophages, *J. Protozool.* **36**:164–170.

De Souza, W., and Souto-Padron, T., 1978, Ultrastructural localization of basic proteins on the conoid, rhoptries and micronemes of *Toxoplasma gondii, Z. Parasitenkd.* **56**:123–129.

Doran, D. J., 1973, Cultivation of Coccidia in avian embryos and cell culture, in *The Coccidia: Eimeria, Isopora, Toxoplasma, and Related Genera* (D. M. Hammond and P. L. Long, eds.), pp. 185–252, University Park Press, Baltimore.

Dubremetz, J. F., and Dissous, C., 1980, Characteristic proteins of micronemes and dense granules from *Sarcocystis tenella* zoites (Protozoa, Coccidia), *Mol. Biochem. Parasitol.* **1**:279–289.

Dubremetz, J. F., Ferreira, E., and Dissous, C., 1989, Isolation and partial characterization of rhoptries and micronemes from *Eimeria nieschulzi* zoites (Sporozoa, Coccidia), *Parasitol. Res.* **75**:449–454.

Endo, T., and Yagita, K., 1990, Effect of extracellular ions on motility and cell entry in *Toxoplasma gondii, J. Protozool.* **37**:133–138.

Endo, T., Sethi, K. K., and Piekarski, G., 1982, *Toxoplasma gondii:* Calcium ionophore A23187-mediated exit of trophozoites from infected murine macrophages, *Exp. Parasitol.* **53**:179–188.

Endo, T., Tokuda, H., Yagita, K., and Koyama, T., 1987, Effects of extracellular potassium on acid release and motility initiation in *Toxoplasma gondii, J. Protozool.* **34**:291–295.

Endo, T., Yagita, K., Yasuda, T., and Nakamura, T., 1988, Detection and localization of actin in *Toxoplasma gondii, Parasitol. Res.* **75**:102–106.

Foussard, F., Leriche, M. A., and Dubremetz, J. F., 1991, Characterization of the lipid content of *Toxoplasma gondii* rhoptries, *Parasitology* **102**:367–370.

Frenkel, J. K., 1973, Toxoplasmosis: Parasite life cycle, pathology, and immunology, in *The Coccidia: Eimeria, Isopora, Toxoplasma, and Related Genera* (D. M. Hammond and P. L. Long, eds.), pp. 344–310, University Park Press, Baltimore.

Fukui, Y., Lynch, T. J., Brzeska, H., and Korn, E. D., 1989, Myosin 1 is located at the leading edges of locomoting *Dictyostelium* amoebae, *Nature* **341**:328–331.

Garofalo, R. S., and Satir, B. H., 1984, *Paramecium* secretory granule content: Quantitative studies on *in vitro* expansion and its regulation by calcium and pH, *J. Cell Biol.* **99**:2193–2199.

Garofalo, R. S., Gilligan, D. M., and Satir, B. H., 1983, Calmodulin antagonists inhibit secretion in *Paramecium, J. Cell Biol.* **96:**1072–1081.

Gilligan, D. M., and Satir, B. H., 1982, Protein phosphorylation/dephosphorylation and stimulus-secretion coupling in wild type and mutant *Paramecium, J. Biol. Chem.* **257:**13903–13906.

Gilligan, D. M., and Satir, B. H., 1983, Stimulation and inhibition of secretion in *Paramecium:* Role of divalent cations, *J. Cell Biol.* **97:**224–234.

Grellier, P., Rigomier, D., Clavey, V., Fruchart, J.-F., and Schrevel, J., 1991, Lipid traffic between high density lipoproteins and *Plasmodium falciparum*-infected red blood cells, *J. Cell Biol.* **112:**267–277.

Haldar, K., De Amorim, A. F., and Cross, G. A. F., 1989, Transport of fluorescent phospholipid analogues from the erythrocyte membrane to the parasite in *Plasmodium falciparum*-infected cells, *J. Cell Biol.* **108:**2183–2192.

Helenius, A., and Simons, K., 1977, Charge shift electrophoresis: Simple method for distinguishing between amphiphilic and hydrophilic proteins in detergent solution, *Proc. Natl. Acad. Sci. USA* **74:**529–532.

Hirai, K., Hirato, K., and Yanagwa, R., 1966, A cinematographic study of the penetration of cultured cells by *Toxoplasma gondii, Jpn. J. Vet. Res.* **14:**81–90.

Holder, A. A., Freeman, R. R., Uni, S., and Aikawa, M., 1985, Isolation of a *Plasmodium falciparum* rhoptry protein, *Mol. Biochem. Parasitol.* **14:**293–303.

Johnson, A. M., Haynes, W. D., Leppard, P. J., McDonald, P. J., and Neoh, S. H., 1983, Ultrastructural and biochemical studies on the immunohistochemistry of *Toxoplasma gondii* antigens using monoclonal antibodies, *Histochemistry* **77:**209–215.

Johnson, A. M., Illana, S., Hakendorf, P., and Baverstock, P. R., 1988, Phylogenetic relationships of the apicomplexan protist *Sarcocystis* as determined by the small subunit ribosomal RNA comparison, *J. Parasitol.* **74:**847–860.

Joiner, K. A., Furtado, G., Mellman, I., Kleinman, H., Miettinen, H., Kasper, L., Hall, L., and Fuhrman, S., 1989, Cell attachment and invasion by tachyzoites of *Toxoplasma gondii, J. Cell. Biochem. Suppl.* **13:**64.

Joiner, K. A., Furhman, S. A., Miettinen, H. M., Kasper, L. H., and Mellman, I., 1990, *Toxoplasma gondii:* Fusion competence of parasitophorous vacuoles in Fc receptor-transfected fibroblasts, *Science* **249:**641–646.

Jones, T. C., Yeh, S., and Hirsch, J. G., 1972, The interaction between *Toxoplasma gondii* and mammalian cells, *J. Exp. Med.* **136:**1157–1172.

Kasper, L. H., and Boothroyd, J. C., 1991, *Immunology and Molecular Biology of Toxoplasma gondii* (K. Warren and N. Agabian, eds.), Blackwell Scientific, London (in press).

Kilejian, A., 1976, Does a histidine-rich protein from *Plasmodium lophurae* have a function in merozoite penetration, *J. Protozool.* **23:**272–277.

Kimata, I., and Tanabe, K., 1982, Invasion by *Toxoplasma gondii* of ATP-depleted and ATP-restored chick embryo erythrocytes, *J. Gen. Microbiol.* **128:**2599–2501.

Kimata, I., and Tanabe, K., 1987, Secretion by *Toxoplasma gondii* of an antigen that appears to become associated with the parasitophorous vacuole membrane upon invasion of the host cell, *J. Cell Sci.* **88:**231–239.

Kimata, I., Tanabe, K., Izumo, A., and Takada, S., 1987, Host cell ATP level and invasion of *Toxoplasma gondii, Trans. Royal Soc. Trop. Med. Hyg.* **81:**377.

King, C. A., 1981, Cell surface interaction of the protozoan *Gregarina* with Concanavalin A beads—implications for models of gregarine gliding, *Cell Biol. Int. Rep.* **5:**297.

King, C. A., 1988, Cell motility of sporozoan protozoa, *Parasitol. Today* **4:**315–319.

King, C. A., and Lee, K., 1982, Effect of trifluoperazine and calcium ions on gregarine gliding, *Experientia* **38:**1051–1052.

Klotz, F. W., Hadley, T. J., Aikawa, M., Leech, J., Howard, R. J., and Miller, L. H., 1989, A 60-kDa *Plasmodium falciparum* protein at the moving junction formed between merozoite and erythrocyte during invasion, *Mol. Biochem. Parasitol.* **36:**177–185.

Korn, E. D., and Hammer, J. A., 1988, Myosins of nonmuscle cells, *Annu. Rev. Biophys. Chem.* **17:**23–45.

Leriche, M. A., and Dubremetz, J. F., 1990, Exocytosis of *Toxoplasma gondii* dense granules into the parasitophorous vacuole after host cell invasion, *Parasitol. Res.* **76:**559–562.

Leriche, M. A., and Dubremetz, J. F., 1991, Characterization of the protein contentts of rhoptries and dense granules of *Toxoplasma gondii* tachyzoites by subcellular fractionation and monoclonal antibodies, *Mol. Biochem. Parasitol.* **45:**249–260.

Low, M. G., and Saltiel, A. R., 1988, Structural and functional roles of glycosylphosphatidylinositol in membranes, *Science* **239:**268–275.

Luft, B. J., and Remington, J. S., 1988, Toxoplasmic encephalitis, *J. Infect. Dis.* **157:**1–6.

Luft, B. J., Conley, F., and Remington, J. S., 1983, Outbreak of central-nervous-system toxoplasmosis in Western Europe and North America, *Lancet* **1:**781–783.

Lycke, E., and Norrby, R., 1966, Demonstration of a factor of *Toxoplasma gondii* enhancing the penetration of *Toxoplasma* parasites into cultured host cells, *Br. J. Exp. Pathol.* **47:**248–256.

Lycke, E., Lund, E., and Strannegard, O., 1965, Enhancement by lysozyme and hyaluronidase of the penetration by *Toxoplasma gondii* into cultured host cells, *Br. J. Exp. Pathol.* **46:**189–199.

Lycke, E., Norrby, R., and Remington, J., 1968, Penetration-enhancing factor extracted from *Toxoplasma gondii* which increases its virulence for mice, *J. Bacteriol.* **96:**785–788.

Lycke, E., Carlberg, K., and Norrby, R., 1975, Interactions between *Toxoplasma gondii* and its host cells: Function of the penetration-enhancing factor of *Toxoplasma, Infect. Immun.* **11:**853–861.

McLeod, R., Mack, D., and Brown, C., 1991, *Toxoplasma gondii*—New advances in cellular and molecular biology, *Exp. Parasitol.* **72:**109–121.

Michel, R., Schupp, K., Raether, W., and Bierther, F. W., 1980, Formation of a close junction during invasion of erythrocytes by *Toxoplasma gondii in vitro, Int. J. Parasitol.* **10:**309–313.

Mikkelsen, R. B., Kamber, M., Wadwa, K. S., Lin, P. S., and Schmidt-Ullrich, R., 1988, The role of lipids in *Plasmodium falciparum* invasion of erythrocytes: A coordinated biochemical and microscopic analysis, *Proc. Natl. Acad. Sci. USA* **85:**5956–5960.

Miyata, H., Bowers, B., and Korn, E. D., 1989, Plasma membrane association of Acanthamoeba myosin 1, *J. Cell Biol.* **109:**1519–1528.

Murtaugh, T. J., Gilligan, D. M., and Satir, B. H., 1987, Purification of and production of an antibody against a 63,000 Mr stimulus-sensitive phosphoprotein in *Paramecium, J. Biol. Chem.* **262:**15734–15739.

Nichols, B. A., and Chiappino, M. L., 1987, Cytoskeleton of *Toxoplasma gondii, J. Protozool.* **34:**217–226.

Nichols, B. A., Chiappino, M. L., and O'Connor, G. R., 1983, Secretion from the rhoptries of *Toxoplasma gondii* during host-cell invasion, *J. Ultrastruct. Res.* **83:**85–98.

Nishizuka, Y., 1988, The molecular heterogeneity of protein kinase C and its implications for cellular regulation, *Nature* **334:**661–665.

Norrby, R., 1970, Host cell penetration of *Toxoplasma gondii, Infect. Immun.* **2:**250–255.

Norrby, R., 1971, Immunological study on the host cell penetration factor of *Toxoplasma gondii, Infect. Immun.* **3:**278–286.

Norrby, R., and Lycke, E., 1967, Factors enhancing the host-cell penetration of *Toxoplasma gondii, J. Bacteriol.* **93:**53–58.

Orias, E., Flacks, M., and Satir, B. H., 1983, Isolation and ultrastructural characterization of secretory mutants of *Tetrahymena thermophila, J. Cell Sci.* **64:**49–67.

Ossorio, P. N., Schwartzman, J. D., and Boothroyd, J. C., 1992, A *Toxoplasma gondii* rhoptry protein associated with host cell penetration has unusual charge assymetry, *Mol. Biochem. Parasitol.* **50:**1–16.

Pohl, U., Dubremetz, J. F., and Entzeroth, R., 1989, Characterization and immunolocalization of the protein contents of micronemes of *Sarcocystis muris* cytstozoites (Protozoa, Apicomplexa), *Parasitol. Res.* **75:**199–205.

Porchet-Hennere, E., 1975, Quelques precisions sur l'ultrastructure de *Sarcocystis tenella.* L'endozoite (apres coloration negative), *J. Protozool.* **22:**214–220.

Porchet-Hennere, E., and Nicolas, G., 1983, Are rhoptries of Coccidia really extrusomes, *J. Ultrastruct. Res.* **84:**194–203.

Porchet-Hennere, E., and Torpier, G., 1983, Relations entre *Toxoplasma* et sa cellule-hote, *Protistologica* **19:**357–370.

Rasmussen, H., 1986a, The calcium messenger system, *N. Engl. J. Med.* **314:**1164–1170.

Rasmussen, H., 1986b, The calcium messenger system, *N. Engl. J. Med.* **314:**1094–1101.

Ravdin, J. I., Murphy, C. F., Guerrant, R. L., and Long-Krug, S. A., 1985, Effects of antagonists of calcium and phospholipase A on the cytopathogenicity of *Entamoeba histolytica, J. Infect. Dis.* **152:**542–549.

Ridley, R. G., Takacs, B., Lahm, H.-W., Delves, C. J., Goman, M., Certa, U., Matile, H., Woollett, G. R., and Scaife, J. G., 1990, Characterization and sequence of a protective rhoptry antigen from *Plasmodium falciparum, Mol. Biochem. Parasitol.* **41:**125–134.

Roger, N., Dubremetz, J. F., Delplace, P., Fortier, B., Tronchin, G., and Vernes, A., 1988, Characterization of a 225 kilodalton rhoptry protein of *Plasmodium falciparum, Mol. Biochem. Parasitol.* **27:**135–141.

Russell, D. G., and Sinden, R. E., 1981, The role of the cytoskeleton in the motility of Coccidian sporozoites, *J. Cell Sci.* **50:**345–359.

Ryning, F. W., and Remington, J. S., 1978, Effect of cytochalasin D on *Toxoplasma gondii* cell entry, *Infect. Immun.* **20:**739–743.

Sadak, A., Taghy, Z., Fortier, B., and Dubremetz, J. F., 1988, Characterization of a family of rhoptry proteins of *Toxoplasma gondii, Mol. Biochem. Parasitol.* **29:**203–211.

Saffer, L. D., Long Krug, S. A., and Schwartzman, J. D., 1989, The role of phospholipase in host cell penetration by *Toxoplasma gondii, Am. J. Trop. Med. Hyg.* **40:**145–149.

Sam-Yellowe, T. Y., Shio, H., and Perkins, M. E., 1988, Secretion of *Plasmodium falciparum* rhoptry protein into the plasma membrane of host erythrocytes, *J. Cell Biol.* **106:**1505–1513.

Satir, B. H., Busch, G., Vuoso, A., and Murtaugh, T. J., 1988, Aspects of signal transduction in stimulus exocytosis-coupling in *Paramecium, J. Cell Biochem.* **36:**429–443.

Scheibel, L. W., Colombani, P. M., Hess, A. D., Aikawa, M., Atkinson, C. T., and Milhous, W. K., 1987, Calcium and calmodulin antagonists inhibit human malaria parasites (*Plasmodium falciparum*): Implications for drug design, *Proc. Natl. Acad. Sci. USA* **84:**7310–7314.

Schofield, L., Bushell, G. R., Cooper, J. A., Saul, A. J., Upcroft, J. A., and Kidson, C., 1986, A rhoptry antigen of *Plasmodium falciparum* contains conserved and variable epitopes recognized by inhibitory monoclonal antibodies, *Mol. Biochem. Parasitol.* **18:**183–195.

Schupp, E., Michel, R., Raether, W., Niemeitz, H., and Uphoff, M., 1978, Invasion von erythrozyten durch *Toxoplasma gondii, Z. Parasitenkd.* **55:**189–193.

Schwartzman, J. D., 1986, Inhibition of a penetration-enhancing factor of *Toxoplasma gondii* by monoclonal antibodies specific for rhoptries, *Infect. Immun.* **51:**760–764.

Schwartzman, J. D., and Krug, E. C., 1989, *Toxoplasma gondii:* Characterization of monoclonal antibodies that recognize rhoptries, *Exp. Parasitol.* **68:**74–82.

Schwartzman, J. D., and Pfefferkorn, E. R., 1983, Immunofluorescent localization of myosin at the anterior pole of the Coccidian, *Toxoplasma gondii, J. Protozool.* **30:**657–661.

Schwartzman, J. D., Krug, E. C., Payne, M. R., and Binder, L. I., 1985, Detection of the microtubule cytoskeleton of the coccidian *Toxoplasma gondii* and the hemoflagellate *Leishmania donovani* by monoclonal antibodies specific for beta-tubulin, *J. Protozool.* **32:**747–749.

Schwartzman, J. D., Sullivan, J. D., and Mandell, G. D., 1990, Computer-enhanced microscopic observation of cell penetration and egress by *Toxoplasma gondii, Clin. Res.* **38:**353a.

Sheffield, H. G., and Melton, M. L., 1968, The fine structure and reproduction of the *Toxoplasma gondii, J. Parasitol.* **54:**209–226.

Sibley, L. D., 1989, Active modification of host cell phagosomes by *Toxoplasma gondii* in *Intracellular Parasitism* (J. W. Moulder, ed.), pp. 245–255, CRC Press, Boca Raton.

Sibley, L. D., and Krahenbuhl, J. L., 1988, Modification of host cell phagosomes by *Toxoplasma gondii* involves redistribution of surface proteins and secretion of a 32 kDa protein, *Eur. J. Cell Biol.* **47:**81–87.

Sibley, L. D., and Sharma, S. D., 1987, Ultrastructural localization of an intracellular *Toxoplasma* protein that induces protection in mice, *Infect. Immun.* **55:**2137–2141.

Sibley, L. D., Weidner, E., and Krahenbuhl, J. L., 1985, Phagosome acidification blocked by intracellular *Toxoplasma gondii, Nature* **315:**416–419.

Sibley, L. D., Krahenbuhl, J. L., Adams, G. M. W., and Weidner, E., 1986, *Toxoplasma* modifies macrophage phagosomes by secretion of a vesicular network rich in surface proteins, *J. Cell Biol.* **103:**867–874.

Sogin, M. L., Elwood, H. J., and Gunderson, J. H., 1986, Evolutionary diversity of eukaryotic small-subunit rRNA genes, *Proc. Natl. Acad. Sci. USA* **83:**1383–1387.

Spudich, J. A., 1989, In pursuit of myosin function, *Cell Regulation* **1:**1–11.

Stagno, S., 1980, Congenital toxoplasmosis, *Am. J. Dis. Child.* **134:**635–637.

Stewart, M. J., and Vandenberg, J. P., 1988, Malaria sporozoites leave behind trails of circumsporozoite protein during gliding motility, *J. Protozool.* **35:**389–393.

Stewart, M. J., Schulman, S., and Vanderberg, J. P., 1985, Rhoptry secretion of membranous whorls by *Plasmodium berghei* sporozoites, *J. Protozool.* **32:**280–283.

Stewart, M. J., Schulman, S., and Vanderberg, J. P., 1986, Rhoptry secretion of membranous whorls by *Plasmodium falciparum* merozoites, *Am. J. Trop. Med. Hyg.* **35:**37–44.

Tanabe, K., Kimata, I., and Takada, S., 1980, Multiplication of *Toxoplasma gondii* in maturing erythroid cells, *J. Gen. Microbiol.* **118:**45–49.

Taylor, D. W., Evans, C. B., Aley, S. B., Barta, J. R., and Danforth, H. D., 1990, Identification of an apically located antigen that is conserved in sporozoan parasites, *J. Protozool.* **37:**540–545.

Turco, J., and Winkler, H. H., 1986, Gamma-interferon-induced inhibition of the growth of *Rickettsia prowazekii* in fibroblasts cannot be explained by the degradation of tryptophan or other amino acids, *Infect. Immun.* **53:**38–46.

Vale, R. D., and Goldstein, L. S. B., 1990, One motor, many tails: An expanding repertoire of force-generating enzymes, *Cell* **60:**883–885.

Vanderberg, J. P., Chew, S., and Stewart, M. J., 1990, *Plasmodium* sporozoite interaction with macrophages *in vitro:* A videomicroscopic analysis, *J. Protozool.* **37:**528–536.

Vilmart-Seuwen, J., Kersken, H., Sturzl, R., and Plattner, H., 1986, ATP keeps exocytosis sites in a primed state but is not required for membrane fusion: An analysis with *Paramecium* cells *in vivo* and *in vitro, J. Cell Biol.* **103:**1279–1288.

Werk, R., 1985, How does *Toxoplasma gondii* enter host cells, *Rev. Infect. Dis.* **7:**449–457.

Werk, R., and Bommer, W., 1980, *Toxoplasma gondii:* Membrane properties of active energy-dependent invasion of host cells, *Tropenmed. Parasitol.* **31:**417–420.

Werk, R., and Fischer, S., 1982, Attempts to infect plant protoplasts with *Toxoplasma gondii, J. Gen. Microbiol.* **128:**211–213.

Wesseling, J. G., de Ree, J. M., Ponnudurai, T., Smits, M. A., and Schoenmakers, J. G. G., 1988a,

Nucleotide sequence and deduced amino acid sequence of a *Plasmodium falciparum* actin gene, *Mol. Biochem. Parasitol.* **27**:313–320.

Wesseling, J. G., Smits, M. A., and Schoenmakers, J. G. G., 1988b, Extremely diverged actin proteins in *Plasmodium falciparum, Mol. Biochem. Parasitol.* **30**:143–154.

Yasuda, T., Yagita, K., Nakamura, T., and Endo, T., 1988, Immunocytochemical localization of actin in *Toxoplasma gondii, Parasitol. Res.* **75**:107–113.

Chapter 11

Biochemical and Ultrastructural Observations of Coccidian Parasite and Host Cell Interactions

Michael W. White, Alwi M. Shatry, Mark A. Jutila, and Clarence A. Speer

1. INTRODUCTION

In general, the interactions of protozoan or metazoan parasites and their animal hosts are less understood than animal–virus interactions. Consequently, there is little biochemical information about the effects of these parasites on their animal hosts. One critical aspect of this field that is only beginning to receive attention is the study of the biochemical mechanisms involved in the communication between intracellular protozoa and their host cells.

An interesting example of biochemical communication has been provided by studies on *Theileria parva*, an obligate intracellular protozoan that induces the clonal expansion of infected blood leukocytes, which may result in lymphoproliferative disease and death of the bovine host. Although the mechanisms of this induction of leukocyte transformation are not well understood, the killing of the parasite by drug treatment reverses the lymphocyte proliferative response

---

Michael W. White, Alwi M. Shatry, Mark A. Jutila, and Clarence A. Speer Veterinary Molecular Biology, Marsh Laboratory, Montana State University, Bozeman, Montana 59717.
Subcellular Biochemistry, Volume 18: Intracellular Parasites, edited by J. L. Avila and J. R. Harris. Plenum Press, New York, 1992.

(Hudson *et al.*, 1985). Recent studies have also shown that *T. parva* induces constituitive expression of the interleukin-2 receptor (Coquerelle *et al.*, 1989) and a high level of the lymphocyte-specific transcription factor, NF-kB (Ivanov *et al.*, 1989). NF-kB activity in infected cells cannot be further stimulated with phorbol esters, indicating that this transcription system is fully expressed in these cells. Killing of the intracellular parasite, however, causes a reduction in the NF-kB levels and the restoration of the phorbol ester response. The mechanisms by which this parasite influences the expression of lymphocyte genes remain to be elucidated.

In another example, Jasmer (1990) recently reported alterations in muscle-specific gene expression in bovine cells infected with the metazoan, *Trichinella spiralis*. Larvae of *T. spiralis* infect mammalian muscle cells causing them to become nurse cells for the parasite. Infected nurse cells lose some of their muscle cell characteristics, including the expression of muscle-specific genes indicated by decreases in the heavy chain of muscle-specific myosin and α-actin.

An intriguing phenomenon is the survival and replication of *Leishmania donovani* in the phagolysosomes of mammalian macrophages. This protozoan parasite circumvents the normal monocytic defense mechanisms by producing a unique lipophosphoglycan that inhibits the activation of protein kinase C, which is required for the macrophage oxidative burst (McNeely *et al.*, 1989) and may also serve as a scavenger of oxygen radicals (Chan *et al.*, 1989).

2. COCCIDIAN BIOLOGY

The Coccidia belong to the phylum Apicomplexa and comprise a diverse group of intracellular protozoan parasites of vertebrates and invertebrates. Certain stages in the life cycle of the Apicomplexa are motile, invasive, and contain an apical complex (polar rings, conoid, rhoptries, micronemes, and subpellicular microtubules) that is used in an unknown way to actively penetrate host cells. Numerous species of Coccidia are of medical and veterinary importance, especially those belonging to the genera *Besnoitia, Cryptosporidium, Eimeria, Hammondia, Isospora, Neospora, Sarcocystis,* and *Toxoplasma.*

The Coccidia have complex life cycles with certain species being homoxenous in which all endogenous stages of the parasite occur within a single host. Others are facultatively or obligatorily heteroxenous in that certain stages occur within an intermediate host and other stages occur in a definitive host. Regardless of whether a species is homoxenous or heteroxenous, the Coccidia are known to progress sequentially through certain stages of their life cycle. The exogenous stages are oocysts that are shed in the feces by the definitive host. Oocysts are environmentally resistant forms that transmit the infection to other susceptible hosts. When oocysts (or sporocysts as in the case of *Sarcocystis* spp.)

are ingested by an appropriate host, the sporozoites excyst from oocysts and actively penetrate cells of the intestinal tract and move to those host cells (enterocytes, mesenteric artery endothelial cells, endothelial cells of the central lacteal, etc.) in which they will undergo asexual reproduction by schizogony, endodyogeny, or endopolygeny to form merozoites. After one or more asexual generations, merozoites develop into male and female gametocytes, called microgametocytes and macrogametocytes, respectively, or into tissue cysts containing bradyzoites that when ingested by the appropriate definitive host will develop into gametocytes in that host's enterocytes. After fertilization of a macrogametocyte by a microgamete, the zygote becomes an oocyst that causes the host cell to rupture and is shed from the host in the feces. Except for sporogony, which occurs within the oocyst, all Coccidia are obligatorily intracellular parasites that must be within cells of the host in order to undergo asexual or sexual development.

Coccidians vary considerably in their ability to infect a range of hosts. Most *Eimeria* species have rigid host specificities, whereas *T. gondii* will infect a wide variety of intermediate hosts. Only felines are used as definitive hosts by *T. gondii*. In general, *Sarcocystis* species demonstrate a relatively high degree of specificity for their intermediate hosts, but may use any canine or feline as a definitive host.

Most coccidian parasites demonstrate a high degree of tissue site and host cell specificity. For example, after sporozoites of *E. bovis* excyst from oocysts in the gut of the bovine host, they migrate into the lamina propria of the small intestine, especially that region just anterior to the ileocecal valve, where they penetrate endothelial cells of the central lacteal and develop into schizonts. Following schizogony, merozoites travel by an unknown route to the large intestine and cecum, where they undergo a second generation of schizogony and then gametogony in glandular enterocytes. After fertilization, zygotes develop into oocysts, which are shed in the feces. The cells that serve as hosts for parasite development are not simply a consequence of the oocysts being ingested and sporozoites excysting in the small intestine. If oocysts or sporozoites of a particular *Eimeria* species are injected parenterally, the protozoa will travel to their usual site of development. Haberkorn (1970) parenterally injected sporulated oocysts of *E. falciformis* into mice and observed developing parasites in the cecum and large intestine, which are the usual sites for this species. Similarly, Pellérdy and Durr (1969) found that intravenous inoculation of rabbits with oocysts, sporozoites, or merozoites of *E. stiedai* resulted in development of the protozoan in the liver, the normal site. These studies clearly indicate that there are strong lines of biochemical communication between the parasite and its host and host cells.

The Coccidia provide excellent models to study biochemical communication between host cells and parasites because, in general, their asexual stages and in some cases their sexual stages can be grown relatively easily in a wide variety of cell lines (for review, see Speer, 1983). Some species will multiply exclusively in cultured cells by endodyogeny (*Toxoplasma gondii, Hammondia heydorni*), whereas others will undergo schizogony (*Eimeria* spp., *Sarcocystis* spp.) and a few species will form gametocytes (*E. tenella, Sarcocystis* spp.).

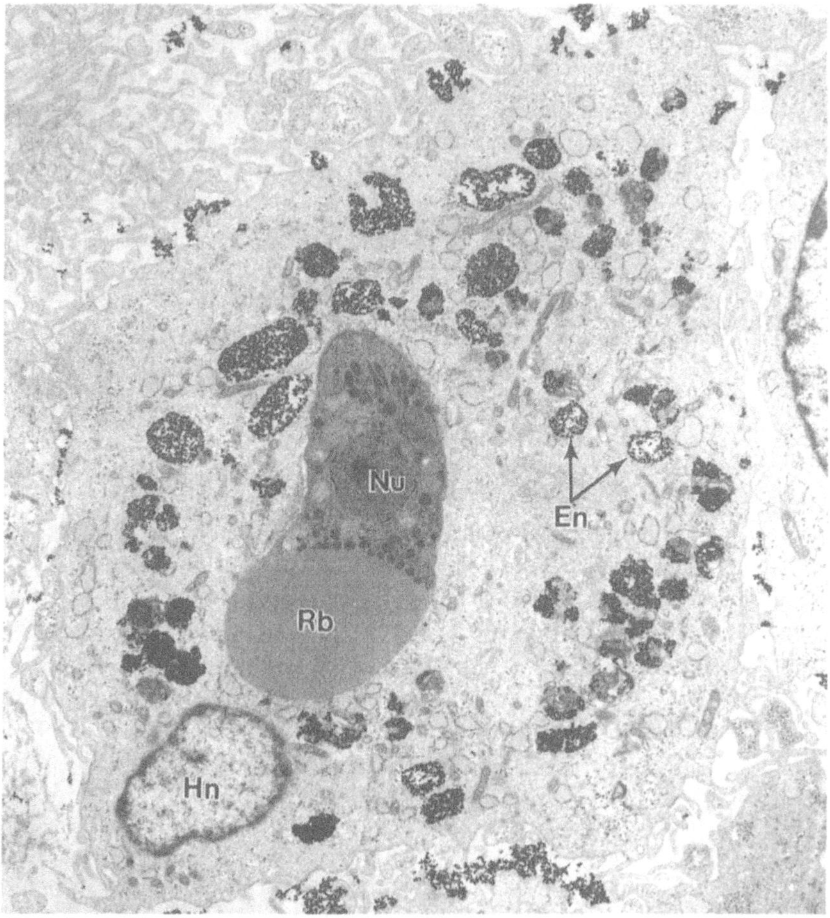

FIGURE 1. Transmission electron micrograph (TEM) of a killed sporozoite of *Eimeria bovis* that was phagocytosed by a cultured bovine M617 macrophage. The sporozoite was killed by treatment with 2.5% glutaraldehyde and then inoculated into the macrophage culture. Thorium dioxide was added to the culture medium at seven days after sporozoite inoculation and at eight days the cultured cells and parasites were fixed and prepared for TEM. Abbreviations: En, endosomes containing thorium dioxide; Hn, host cell nucleus; Nu, nucleus of sporozoite; Rb, refractile body of sporozoite. Magnification × 6000.

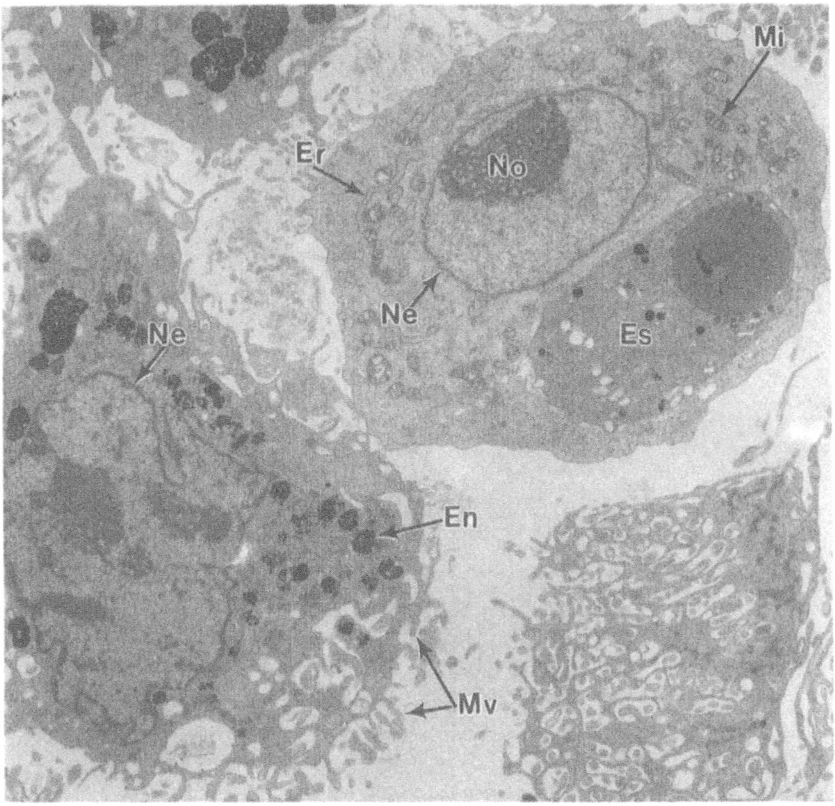

FIGURE 2. TEM showing the effects of intracellular parasitism on a cultured bovine M617 macrophage at eight days after inoculation of sporozoites. The uninfected cell in the left half of the micrograph contains numerous microvilli (Mv), a nucleus with an irregular nuclear envelope (Ne), and numerous endosomes (En) containing thorium dioxide. The macrophage containing an early schizont (Es) of *Eimeria bovis* has few microvilli, no thorium dioxide-labeled endosomes, numerous mitochondria (Mi), endoplasmic reticulum (Er) with distended cisternae, an oval-shaped nucleus with a smooth nuclear envelope (Ne), and an enlarged nucleolus (No). Thorium dioxide was added to the culture medium at seven days after sporozoite inoculation and the macrophage cultures were fixed for TEM at 8 days. Magnification × 5000.

3. ULTRASTRUCTURAL ALTERATIONS IN INFECTED HOST CELLS

Sporozoites and merozoites are motile, fusiform stages capable of selecting, penetrating, and initiating intracellular development within cells of the host. The process of host cell penetration is preceded by flexing and gliding movements of

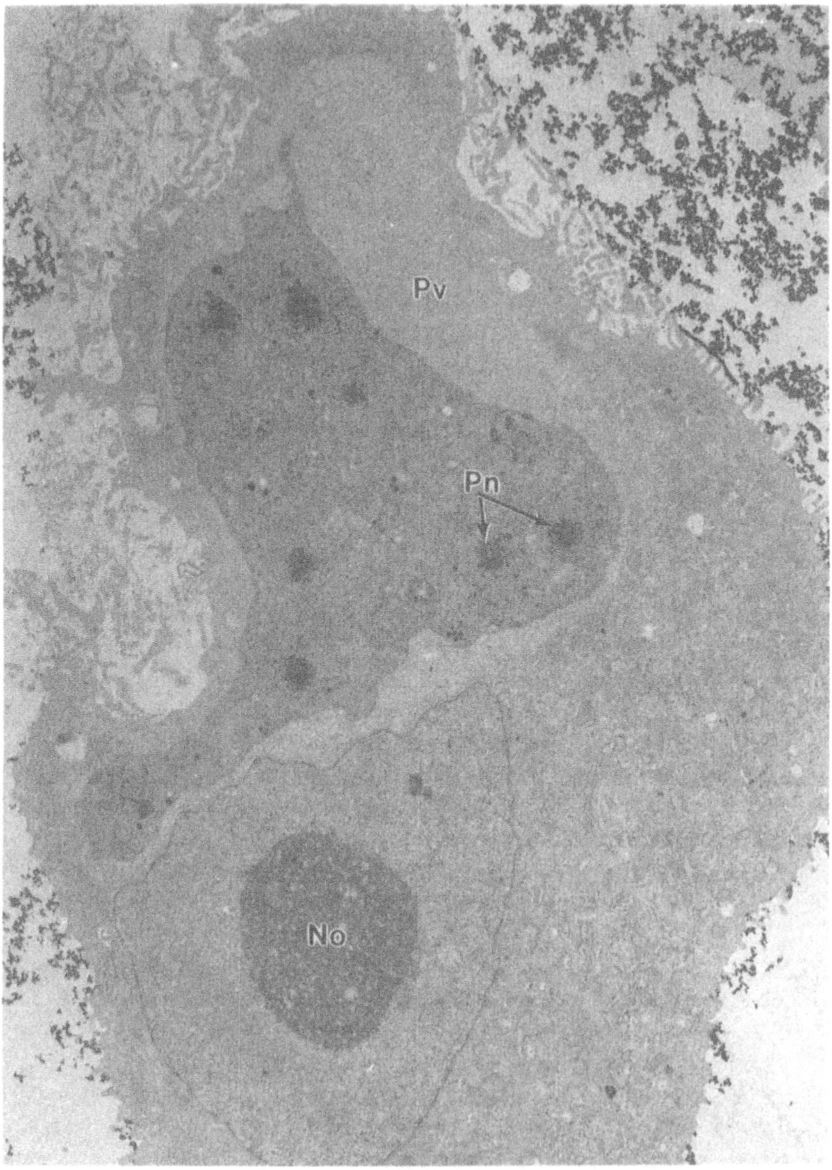

FIGURE 3. TEM of a bovine M617 macrophage infected with an intermediate schizont of *Eimeria bovis* at eight days after sporozoite inoculation. The macrophage contains a nucleus with a prominent nucleolus (No) and lacks thorium dioxide-labeled endosomes. Thorium dioxide was added to the culture medium at seven days and the culture was fixed and prepared for TEM at eight days. Abbreviations: Pn, nucleolus of parasite nuclei; Pv, parasitophorous vacuole. Magnification × 4800.

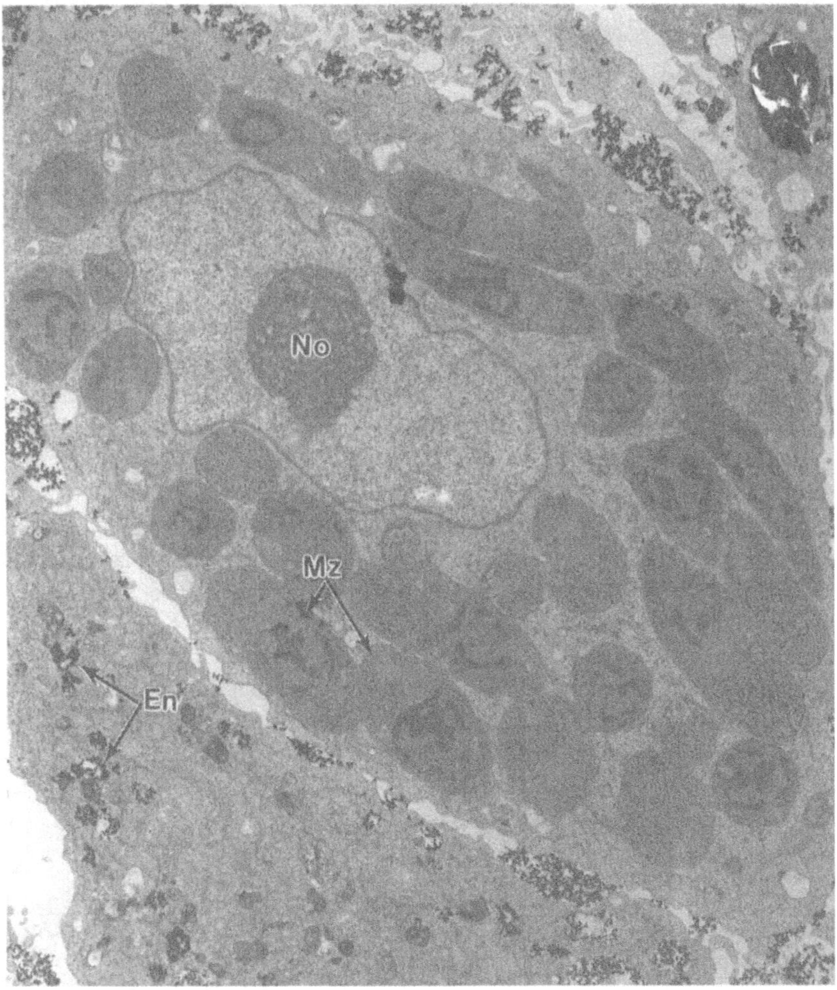

FIGURE 4. TEM of a bovine M617 macrophage containing a mature schizont of *Sarcocystis cruzi* with fully formed merozoites (Mz). The host cell lacks thorium dioxide-labeled endosomes (En) and contains a nucleus with a large nucleolus (No). Thorium dioxide was added to the culture medium at 35 days after sporozoite inoculation and the cultures were fixed for TEM at 36 days. Mgnification × 6000.

the parasites and the formation of an anterior styletlike protuberance, the extended conoid, that can be thrust forward or retracted (Speer *et al.*, 1971). Penetration is usually active and rapid, occurring within a few seconds, and begins when the anterior protuberance comes in contact with the host cell plasmalemma. In response to the advancing parasite, the area of the host cell plasmalemma in contact with the conoid region invaginates to form a cavity. On completion of penetration, the parasite is enclosed within a membrane-bound vacuole called the parasitophorous vacuole (Roberts *et al.*, 1971). In *E. magna* and *Isospora canis,* the original parasitophorous vacuolar membrane (PVM), which was composed of the original invaginated host cell plasmalemma, breaks down quickly and is replaced by another membrane of host cell origin that then becomes the PVM (Jensen and Hammond, 1975; Jensen and Edgar, 1976, 1978). Although most species of Coccidia reside within a parasitophorous vacuole, there are exceptions. The schizont stages that occur before cyst formation in *Sarcocystis* species occur free in the host cell cytoplasm, not surrounded by a parasitophorous vacuole (Dubey *et al.*, 1989).

After penetration by coccidian parasites, host cells exhibit a number of dramatic ultrastructural changes (Figures 1–4). For example, in cells infected with sporozoites or schizonts of *E. bovis* or *S. cruzi,* the host cell nucleus and nucleolus become enlarged and the nuclear envelope is smooth in outline compared to the irregular nuclear envelope of noninfected cells (Figures 2–4). The nucleus of infected cells contains little or no heterochromatin and that which is normally associated with the inner surface of the nuclear envelope is absent (Figures 2–3). The cytoplasm of infected cells contains fewer vacuoles and usually more numerous and active mitochondria, as indicated by an electron-dense matrix and endoplasmic reticulum with distended cisternae (Figure 2). The plasmalemma of noninfected cells has numerous microvilli and coated pits, whereas that of infected cells is smooth (Figure 2) and essentially devoid of coated pits. Alterations in the plasmalemma are also reflected in the inability of cultured cells to endocytose (see Section 4).

4. INITIAL PARASITE AND HOST CELL INTERACTIONS

The mechanisms that govern host and host cell specificity are not well understood but probably involve the host immune response, host genetic background, parasite biochemistry and nutrition, and the ability of parasites to attach to surface receptors and to penetrate host cells. Our research group has recently initiated studies designed to obtain new biochemical information on host cell–parasite relationships with emphasis on identifying parasite and host cell molecules that are involved in these interactions.

4.1. Adherence and Penetration

Recent evidence indicates that plasma membrane-associated molecules play an important role in the adhesion and invasion of host cells by certain infectious agents. It is well known that the pathogenicity of *Escherichia coli* is due largely to the action of enterotoxins (Smith and Halls, 1967), but bacterial virulence is also dependent on the ability of the bacteria to adhere to cells of their host (Isberg, 1991). In a recent report, Mouricout *et al.* (1990) obtained a significant reduction (about two orders of magnitude) in the adhesion of *E. coli* to the intestines of calves treated with glycoprotein glycans derived from bovine plasma, which correlated with prevention of enteric disease. Furthermore, the attachment and subsequent invasion of cultured cells by *Trypanosoma cruzi* can be affected by the treatment of cells with proteases, lectins, and specific sugars as well as the phase of the host cell cycle. In a recent report concerned with binding interactions between *T. cruzi* and host cell membranes, two polypeptides with molecular weights of 32 and 34 kDa were found to bind parasite antigens and live parasites in a highly specific manner (Davis and Kuhn, 1990).

Experiments involving electron microscopy, cinemicrography, and basic biochemistry have been performed to study the penetration of coccidian parasites into their host cells (Speer *et al.*, 1971; Long and Speer, 1977; Long, 1982; Whitmire *et al.*, 1988), but essentially nothing is known regarding the mechanisms that determine host tissue and cell specificity or the nature of the biochemical interaction between these parasites and their host cells. Recently, we used an *ex vivo* binding assay to study *E. bovis*–host cell interactions in unfixed tissues (Shatry, Jutila, and Speer, unpublished data). The *ex vivo* assay was originally used to elucidate lymphocyte homing receptors and phagocyte–endothelial cell interactions (reviewed in Berg *et al.*, 1989). The assay involved flooding 10 μm frozen cryostat sections with 5×10^5 live sporozoites of *E. bovis* and incubating these sections at 4 °C with constant agitation. After incubation, the preparations were rinsed with phosphate buffered saline (PBS), fixed with 1% glutaraldehyde in PBS, rinsed with cold tap water, and stained with hematoxylin and eosin.

Results obtained with this assay (Table I) showed that sporozoites and merozoites of *E. bovis* bound preferentially in high numbers to cells (especially enterocytes) of the small and large intestines and cecum, whereas few parasites bound to cells in the thymus, mesenteric and prescapular lymph nodes, Peyer's patch, liver, spleen, and kidney. The binding pattern exhibited by the sporozoites and merozoites of *E. bovis* corresponded to their normal sites of penetration and asexual development.

In order to identify the parasite ligands involved in attachment, *E. bovis* sporozoites and merozoites were solubilized in sodium dodecyl sulfate (SDS), their proteins were separated on a 12.5% SDS–polyacrylamide gel (Reduker and Speer, 1986a,b), and transferred to nitrocellulose (Whitmire *et al.*, 1988). The resulting blot was blocked with 10% gamma globulin-free horse serum, probed with biotinylated-lysates of bovine small intestine or large intestine enterocytes (Nagi and Babiuk, 1988), or thymus cells and protein binding detected with avidin-conjugated horseradish peroxidase. Figure 5 shows the results from this

Table I
Binding of *Eimeria bovis* Sporozoites to Various Bovine Tissues

Tissue type	Number of sporozoites: Nylon-purified[a]	Number of sporozoites: Nonpurified[b]
Duodenal epithelium	2540	497.5 (39.1)
Duodenal lamina propria	23	53.5 (7.7)
Jejunal epithelium	130	174.7 (106.7)
Jejunal lamina propria	116	64.7 (35.1)
Ileal epithelium	3276	1,177 (75.6)
Ileal lamina propria	340	140 (119)
Colonal epithelium	402	536 (186)
Colonal lamina propria	9	54.5 (64.3)
Cecal epithelium	2330	515.5 (82.3)
Cecal lamina propria	225	70 (51.9)
Kidney	50	64.3 (18.7)
Liver	5	21.3 (21)
Mesenteric lymph node	9	6.5 (3.5)
Peyer's patch	11	22.7 (9.8)
Spleen	102	68.3 (35.3)
Thymus	7	6.5 (7.8)

[a]Total number of sporozoites in 10 miroscopic fields at 400X magnification.
[b]Mean value obtained from 2 or 3 independent experiments with each experiment representing all sporozoites in 10 microscopic fields at 400X magnification; standard deviation given in parentheses.

experiment. Lysates from thymus cells showed very little binding activity (Figure 5, lanes 4 and 8), which was consistent with the absence of binding in the *ex vivo* assay (Table I). The protein-binding patterns of lysates from small and large intestine enterocytes, although complex, revealed clear differences. The lysate from the large intestine bound a > 110-kDa protein in sporozoites (Figure 5, lane 2), which was low or absent in merozoites (Figure 5, lane 5). Conversely, this lysate bound an approximately 70-kDa merozoite protein (Figure 5, lane 6), which was absent in sporozoites and merozoites probed with the small intestinal lysate (Figure 5, lane 5). The lysate from the large intestine bound a complex set of merozoite proteins above the 110 kDa standard (Figure 5, lane 6), which was not seen in sporozoites (Figure 5, lane 2). Based on these preliminary studies, several unique sporozoite and merozoite proteins have been identified that bind in a tissue-specific manner to enterocytes of the bovine large and small intestines.

Our findings with the *ex vivo* binding and protein-binding assays are consistent with *in vivo* development of *E. bovis*. Thus, these assays provide a means of identifying the molecular components involved in parasite attachment and pen-

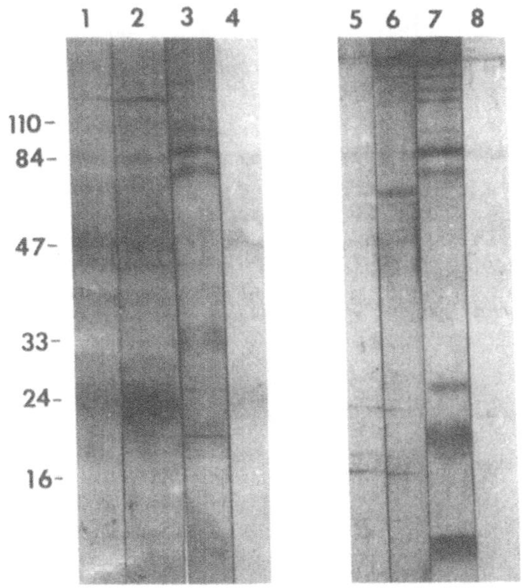

FIGURE 5. Protein binding assay. Protein extracts from *E. bovis* sporozoites (lanes 1–4) and merozoites (lanes 5–8) were prepared and fractionated on a 12.5% nonreducing SDS–PAGE gel and then transferred to nitrocellulose, blocked with 10% gamma globulin-free horse serum, and incubated with biotinylated lysates from bovine small intestine (lanes 1, 5), large intestine (lanes 2, 6), hyperimmune serum (lanes 3, 7), and thymus (lanes 4, 8). Protein binding was detected by developing the filters with avidin-conjugated horseradish peroxidase. The indicated molecular weight standards were fractionated and stained with amido black.

etration. These assays are important advances because they provide new ways of identifying potential targets for vaccine or chemoprophylactic therapies. Their major advantage is their ability to identify ways of intervening with parasite function rather than relying on the host immune response.

In previous work, we found that treatment of *E. bovis* sporozoites with either of two monoclonal antibodies (MAbs EbS9 and EbS11) resulted in an approximately 75% decrease in sporozoite penetration of cultured bovine cells (Figure 6) (Whitmire *et al.*, 1988). Both EbS9 and EbS11 reacted in Western blots of sporozoites with P20, an immunodominant surface protein as determined by immune serum (Figure 7). We also found that MAbs EbS9 and EbS11 have a deleterious effect on intracellular parasites by inhibiting developments of sporozoites to schizonts (Figure 6). Interestingly, when MAb EbS9 was tested in the *ex vivo* binding assay we found that it did not block the binding of *E. bovis* sporozoites to enterocytes, indicating that P20 may not be critical for host cell–parasite adhesion.

4.2. Influence of Parasites on Endocytosis

The coccidians studied thus far have been found to interfere in an unknown way with the fusion of the host cell endosomes with the PVM surrounding the parasite (Jones and Hirsch, 1972). Recent work on *T. gondii* by Joiner *et al.*

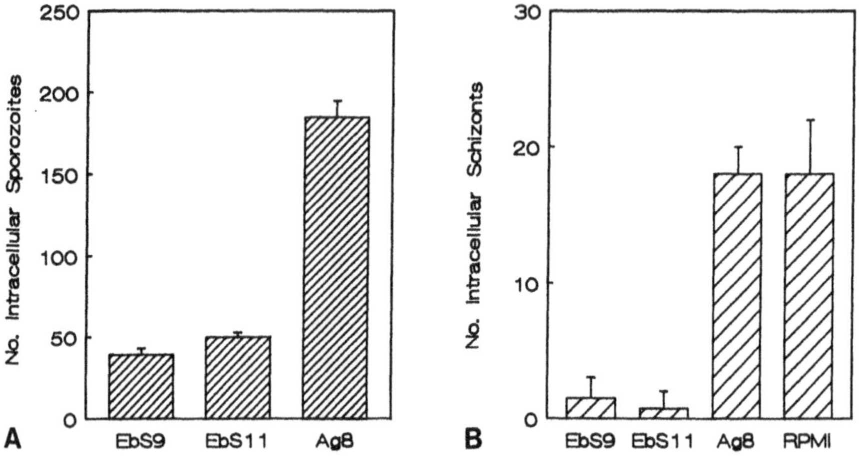

FIGURE 6. Effects of surface-reactive MAbs EbS9 and EbS11 on the number of intracellular schizonts of *E. bovis* at ten days after sporozoite inoculation of bovine M617 cells. Ag8 is a monoclonal antibody that does not react with *E. bovis* sporozoites. RPMI is normal culture medium.

(1990) indicates that the prevention of phagolysosome fusion occurs at the time of the formation of the parasitophorous vacuole and may be mediated by the initial parasite–host cell interaction. Parasites coated with antibodies followed a different route of entry in that the PVM fused with the host cell lysosomes to form a phagolysosome. The inhibition of PVM–lysosome fusion did not require the continued synthesis of a parasite factor since the parasitophorous vacuole surrounding parasites that were killed intracellularly remained intact. Their studies have suggested that the initial parasite interaction may involve one or more host cell membrane receptor molecules that generate a cascade of events leading to the penetration by the parasite and the formation of the parasitophorous vacuole. As mentioned above, sporozoites of *E. bovis* bind *in vivo* in specific tissue patterns that are consistent with their normal sites of penetration and development in the bovine host. These results strongly suggest that there are specific adhesins on the surfaces of *E. bovis* parasites and bovine epithelial cells that control cell-parasite surface interactions.

Animal cell endocytosis involves either clathrin-mediated or clathrin-independent mechanisms (see Hubbard, 1989, for a recent review) and is either a constituitive or ligand–receptor-driven process. Once inside the cell these vesicles travel through one or more subcompartments and may result in the transport or recycling of the vesicle contents or the degradation of the contents in lysosomes. The conclusion that endocytosis may take place in the absence of clathrin stems from experiments involving the inhibition of normal coated pit formation by cytoplasmic acidification (Sandvig *et al.*, 1987; Cosson *et al.*,

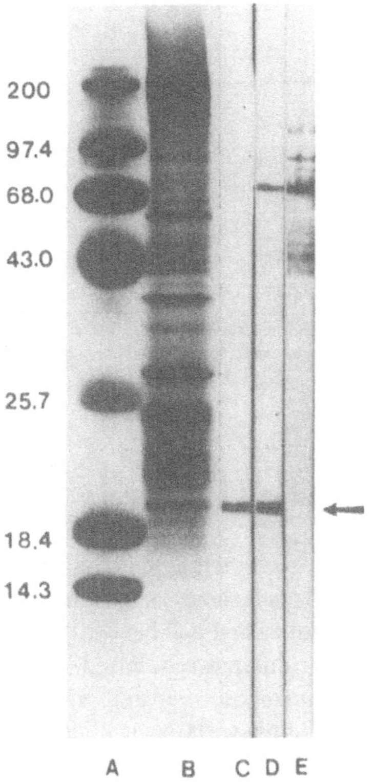

FIGURE 7. Nonreduced profile of *E. bovis* sporozoites from a 12.5% SDS–polyacrylamide gel stained with Coomassie brilliant blue (lane B), and Western blots of sporozoite proteins probed with the monoclonal antibody EbS9 (lane C), EbS11 (lane D), and Ag8 (lane E). The arrow indicates the sporozoite protein 20 band (M_r 20 kDa) against which EbS9 and EbS11 react. The M_r protein bands between 40 and 100 kDa in lanes C–E are due to nonspecific cross-reactivity of serum proteins in the concentrated culture medium. Lane A consists of prestained BRL molecular weight standards (× 10^3).

1989) or conditions of hypertonicity (Daukas and Zigmond, 1985). We have begun to examine the effects of *E. bovis* infection on host cell endocytosis by using an *in vitro* cultivation system (Speer *et al.*, 1985).

Bovine monocytes (M617 cells) were inoculated with culture medium (RPMI 1640 plus 3% fetal bovine serum) containing freshly excysted sporozoites of *E. bovis*, incubated (38.5 °C, 95% air, 5% CO_2) for seven days and then exposed to culture medium containing 0.05% thorium dioxide to label the endosomes. At 24 hr after the addition of thorium dioxide, the M617 cells were fixed in 2.5% glutaraldehyde in Millonig's phosphate buffer and prepared for transmission electron microscopy (TEM) as described previously (Speer and Dubey, 1989). On examination by electron microscopy, the endosomes of the M617 cells appeared electron-dense due to the presence of thorium dioxide (Figures 1–4).

Transmission electron microscopies with final magnifications of 5,000–10,000× of noninfected M617 and parasite-infected M617 cells containing sporozoites or first-generation schizonts of *E. bovis* were then evaluated for cell size, number, and size of endosomes and total endosome area by using a Zeiss

Table II
Effect of Intracellular Stages of *Eimeria bovis* on Phagocytosis by Bovine Monocytes

Parasite stage	Mean cell area ($\times 10^{-4}$ mm^2)	Mean no. endosomes	Mean[a] endosome size ($\times 10^{-8}$ mm^2)	Mean total endosome area ($\times 10^{-6}$ mm^2)
Noninfected	1.9	63.9	15.4	9.9
Sporozoite	1.7	28.6	7	1.8
Schizont	5	18.4	1.4	0.3

[a]Mean endosome size was first calculated for each cell examined. Next, mean values for individual in each category were averaged and are presented as grand means.

Interactive Digital Analysis System (ZIDAS) (Table II). Only those TEMs that showed the host cell nucleus cut in cross-section were used for ZIDAS analysis. The ZIDAS data showed that M617 cells infected with *E. bovis* sporozoites or schizonts had a highly significant decrease in both the number and size of host cell endosomes when compared to noninfected cells or cells harboring sporozoites that had been killed before inoculation with glutaraldehyde. No significant difference occurred in endosome number nor size in cells harboring killed sporozoites compared with noninfected cells (data not shown). In M617 cells containing first-generation schizonts, the change was dramatic with a 3.5-fold decrease in the number of endosomes and an 11-fold decrease in mean endosome size, resulting in a 33-fold decrease in total endosomal area. Mean cell area did not differ between noninfected and sporozoite-infected M617 cells, but schizont-infected cells had significantly larger cell areas that did not include the parasite nor the area within the parasitophorous vacuole. The host cell plasmalemma showed ultrastructural changes such as loss of clathrin-coated pits, loss of microvilli, and a general smoothing of the membrane.

Based on these ultrastructural observations we initiated studies on the effects of intracellular parasitism by *E. bovis* on cellular clathrin. Clathrin forms the basic structure of coated pits and is composed of a triskeleton of three heavy chains (M_r 192 kDa) and three light chains (M_r 30–40 kDa) which assemble into a hexagonal lattice (Kirchhausen and Harrison, 1981; Ungewickell and Branton, 1981) The function of the light chains is unclear since under appropriate conditions *in vitro* they are not required for triskeleton assembly. Some evidence indicates that the light chains may play a role in the assembly of clathrin structures since monoclonal antibodies that inhibit clathrin assembly have been shown to bind epitopes on the heavy chain that corresponds to light chain binding sites (Blank and Brodsky, 1987).

Because of the potential role of the light chains in the assembly of clathrin, we examined the levels of mRNAs encoding the light chains LC_a and LC_b in M617 cells infected with *E. bovis* sporozoites and schizonts using cDNA probes kindly provided by Peter Parham (Stanford University). At one, four, and ten days after sporozoite inoculation, total RNA was prepared from *E. bovis*-infected and noninfected M617 cells, and separated on a formaldehyde-agarose gel (Man-

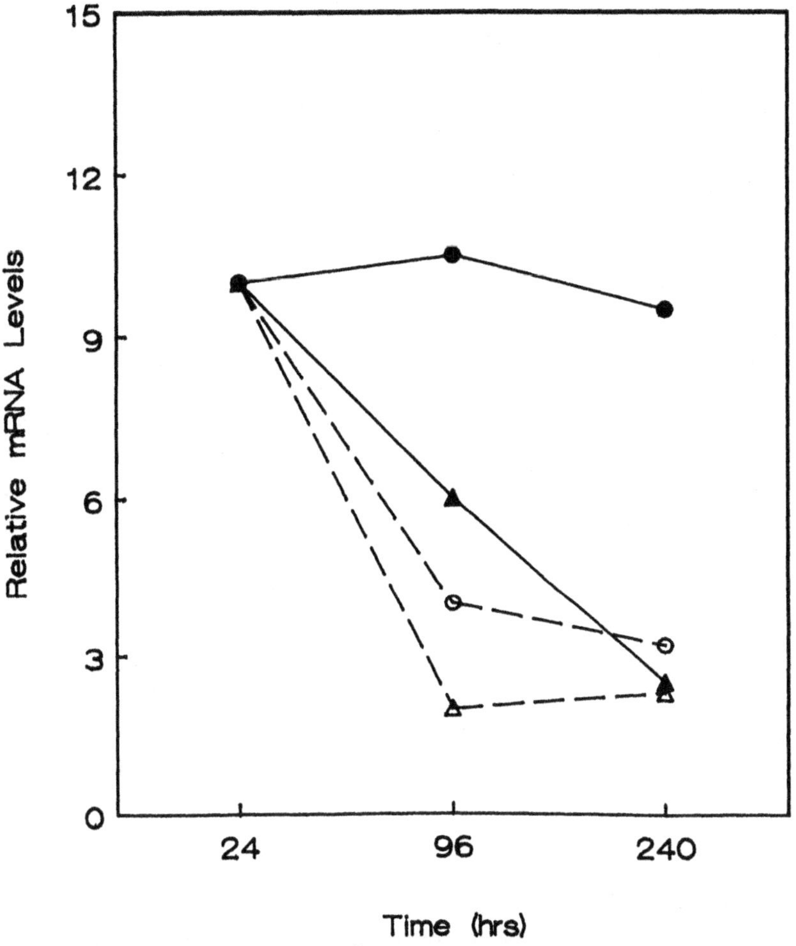

Figure 8. Northern analysis of mRNAs encoding clathrin light chains LC_a and LC_b. Total RNA was prepared from infected or noninfected M617 cells and separated on 1.2% formaldehyde gels. The RNA was then transferred to nitrocellulose filters and the filters with ^{32}P-labeled DNA fragments. The autoradiographs from this experiment were quantitated on a laser densitometric scanner (Pharmacia) and the results plotted as relative change in RNA levels. (△) Infected cells, LC_a; (▲) noninfected cells, LC_a; (○) infected cells, LC_b; (●) noninfected, LC_b.

iatis *et al.*, 1982). The RNA was then transferred to nitrocellulose (Thomas, 1983) and probed with [^{32}P]-cDNA fragments encoding LC_a and LC_b. The autoradiograph from the above Northern blot was quantitated by densitometric scanning using a Pharmacia Laser scanner and the results plotted in Figure 8. At one to ten days, the level of LC_b mRNA did not change in the control cultures of M617 cells, whereas at four days LC_a mRNA decreased over threefold. This selective decrease in one light chain mRNA type did not correlate with a loss in the endosomal area since freshly plated and eight-day cultured cells had nearly identical endosomal areas (data not shown). In contrast, in *E. bovis*-infected M617 cells LC_a and LC_b mRNA levels were fivefold and threefold lower, respectively, by four days, which are consistent with the decrease in the endosomal area seen in the thorium dioxide experiment. While these results are consistent with the endosome staining experiments, further study is needed to confirm these observations at the protein level.

Coccidian parasites are known to depend heavily on carbohydrate metabolism for their energy requirements, which results in the depletion of oxygen content of the host cell cytoplasm (Beyer, 1970). Once oxygen levels are depleted the parasite must rely increasingly on glycolysis, resulting in massive production of lactic acid. Animals infected with *Eimeria* species show dramatic acidification of intestinal tissues (Stephens, 1965; Van der Horst and Kouwenhoven, 1973), which is evidently due to parasite-produced lactic acid. Since acidification of animal cells is known to inhibit clathrin function, the block of endocytosis in *E. bovis*-infected cells may be simply the indirect consequence of parasite metabolism. However, cells that have been acidified experimentally by various methods have clathrin-coated pits attached to the inner surface of the plasma membrane (Heuser, 1989). These coated pits, which have been termed microcages, are apparently frozen on the membrane surface unable to release. We did not observe these types of structures in parasite-infected cells, suggesting that the type of pH change induced by the parasite may be qualitatively different or there are other explanations for the inhibition or endocytosis in cells infected with *E. bovis*.

5. CONCLUSIONS

There is little information concerning the biochemical events involved in the interaction between heterospecific organisms in a symbiotic relationship. Even though not all symbiotic relationships are parasitic, parasitism has received the most attention because of its importance in human medicine and economics. Much of the scientific focus has been directed toward developing methods of neutralizing host–parasite relationships. Progress has been slow because the biological principles involved in host–parasite relationships are still not well understood.

After nearly a century of study of parasitic protozoan diseases, malaria, trypanosomiasis, and leishmaniasis remain leading causes of death for millions of humans. Coccidian parasites lead the list of opportunistic human infections (i.e., *Cryptosporidium* spp. and *Toxoplasma gondii*) and are leading causes of livestock losses (i.e., *Eimeria* spp.). The recent resurgence of research in this field promises new hope for combating these diseases that can be attributed to the new emphasis on understanding the biochemistry of protozoan parasites and the mechanisms of host–parasite interaction.

Over the last five years the field of parasitic protozoan–metazoan biochemistry has expanded rapidly and provided interesting information concerning the diversity of biochemical mechanisms utilized by parasites, including RNA editing and variant somatic gene expression in *Trypanosoma* species (Englund *et al.*, 1982; Simpson and Shaw, 1989), and the presence of ribosomal RNA genes on extrachromosomal DNA in *Naegleria* (Clark and Cross, 1982) and *Entamoeba histolytica* (Bhattacharya *et al.*, 1989). Many of these findings are so novel and extraordinary that profound changes are being made in our understanding of cellular biochemical mechanisms.

Our studies as well as those of others are beginning to elucidate the mechanisms involved in host–parasite interactions. With new assays such as the *ex vivo* binding assay we have begun to identify the molecules involved in host cell–parasite adhesions. Further characterization of these molecules may shed light on the remarkable host and tissue specificity of these protozoan parasites. Our partial characterization of these adhesins have already given us a tool to discriminate among the unique properties of the different developmental stages of coccidian parasites. The clear differences in protein binding we observed between merozoites and sporozoites and the large and small intestine enterocytes indicate that there are specific molecules that may control host cell–parasite interactions with different developmental parasite stages. The biochemical mechanisms involved in development of the parasitic protozoa are still not well understood. The complex life cycles of these parasites, which may involve two or more hosts, have been major deterrents to this area of study. The challenge to researchers will be to apply the techniques now available in molecular biology to develop tools that may be used to solve the intricate processes involved in parasitic protozoan development.

6. REFERENCES

Berg, E. L., Goldstein, L. A., Jutila, M. A., Picker, L. J., Streeter, P. R., Wu, N. W., Zhou, D., and Butcher, E. C., 1989, Homing receptors and vascular addressing: Cell adhesion molecules that direct lymphocyte traffic, *Immunol. Rev.* **108**:5–18.

Beyer, T. V., 1970, Coccidia of domestic animals. Some metabolic peculiarities of particular stages of the life cycle, *J. Parasitol.* **56**:28–29.

Bhattacharya, S., Bhattacharya, A., Diamond, L. S., and Solido, A. T., 1989, Circular DNA of *Entamoeba histolytica* encodes ribosomal RNA, *J. Protozool.* **36**:455–458.

Blank, G. S., and Brodsky, F. M., 1987, Clathrin assembly involves a light chain-binding region, *J. Cell Biol.* **105**:2011–2019.

Chan, J., Fujiwara, T., Brennan, P., McNeil, M., Turco, S. J., Sibille, J. C., and Snapper, M., 1989, Microbial glycolipids: Possible virulence factors that scavenge oxygen radicals, *Proc. Natl. Acad. Sci. USA* **86**:2453–2457.

Clark, C. G., and Cross, G. A. M., 1987, rRNA genes of *Naegleria gruberi* are carried exclusively on a 14-kilobase-pair plasmid, *Mol. Cell. Biol.* **7**:3027–3031.

Coquerelle, T. M., Eichhorn, M., Magnuson, N. S., Reeves, R., Williams, R. O., and Dobbelaere, D. A., 1989, Expression and characterization of the interleukin-2 receptor in *Theileria parva* infected bovine lymphocytes, *Eur. J. Immunol.* **9**:655–659.

Cosson, P., Curtis, I., Pouyssegur, J., Griffiths, G., and Davoust, J., 1989, Low cytoplasmic pH inhibits endocytosis and transport from the trans-golgi network to the cell surface, *J. Cell Biol.* **108**:377–387.

Daukas, G., and Zigmond, S. H., 1985, Inhibition of receptor-mediated but not fluid-phase endocytosis in polymorphonuclear leucocytes, *J. Cell Biol.* **101**:1673–1679.

Davis, C. D., and Kuhn, R. E., 1990, Selective binding of *Trypanosoma cruzi* to host cell membrane polypeptides, *Infect. Immun.* **58**:1–6.

Dubey, J. P., Speer, C. A., and Fayer, R., 1989, *Sarcocystosis of Animals and Man*, CRC Press, Boca Raton, Florida.

Englund, P. T., Hajduk, S. L., and Marini, J. C., 1982, The molecular biology of trypanosomes, *Annu. Rev. Biochem.* **51**:695–726.

Haberkorn, A., 1970, Zur Empfanglichkeit nicht spizefischer Wirte fur Schizogonie-Stadien verschiedener *Eimeria*-Arten, *Z. Parasitenkd.* **35**:156–161.

Heuser, J., 1989, Effects of cytoplasmic acidification on clathrin lattice morphology, *J. Cell Biol.* **108**:401–411.

Hubbard, A. L., 1989, Endocytosis, *Curr. Opin. Cell Biol.* **1**:675–683.

Hudson, A. T., Randall, A. W., Fry, M., Ginger, C. D., Hill, B., Latter, V. S., McHardy, N., and Williams, R. B., 1985, Novel anti-malarial hydronaphthoquinones with potent broad spectrum anti-protozoal activity, *Parasitology* **90**:45–55.

Isberg, R., 1991, Discrimination between intracellular uptake and surface adhesion of bacterial pathogens, *Science* **252**:934–938.

Ivanov, V., Stein, B., Dirk, I. B., Bobbelaere, A. E., Herrlich, P., and Williams, R., 1989, Infection with the intracellular protozoan parasite, *Theileria parva,* induces constituively high levels of NF-kB in bovine T lymphocytes, *Mol. Cell Biol.* **9**:4677–4686.

Jasmer, D. P., 1990, *Trichinella spiralis:* Altered expression of muscle proteins in trichinosis, *Exp. Parasitol.* **70**:452–465.

Jensen, J. B., and Edgar, S. A., 1976, Possible secretory function of the rhoptries of *Eimeria magna* during penetration of cultured cells, *J. Parasitol.* **62**:988–992.

Jensen, J. B., and Edgar, S. A., 1978, Fine structure of penetration of cultured cells by *Isospora canis* sporozoites, *J. Protozool.* **25**:169–173.

Jensen, J. B., and Hammond, D. M., 1975, Ultrastructure of the invasion of *Eimeria magna* sporozoites into cultured cells, *J. Protozool.* **22**:411–415.

Joiner, K. A., Fuhrman, S. A., Miettinen, H. M., Kaspar, L. H., and Mellman, I., 1990, *Toxoplasma gondii:* Fusion competence of parasitophorous vacuoles in Fc receptor-transfected fibroblasts, *Science* **249**:641–646.

Jones, T. C., and Hirsch, J. G., 1972, The interaction between *Toxoplasma gondii* and mammalian cells. II. The absence of lysosomal fusion with phagocytic vacuoles containing living parasites, *J. Exp. Med.* **136**:1173–1194.

Kirchhausen, T., and Harrison, S. C., 1981, Protein organization in clathrin trimers, *Cell* **23**:755–761.

Long, P. L., 1982, *The Biology of the Coccidia*, University Park Press, Baltimore.

Long, P. L., and Speer, C. A., 1977, Invasion of host cells by coccidia, in *Fifteenth Symposium of the British Society of Parasitology*, pp. 1–26, London, Blackwell Publishers.

Maniatis, T., Fritsch, E. F., and Sambrook, J., 1982, *Molecular Cloning; A Laboratory Manual*, Cold Spring Harbor Laboratory, Cold Spring Harbor, New York.

McNeely, T. B., Rosen, G., Londner, M. V., and Turco, S. J., 1989, Inibitory effects on protein kinase C activity by lipophosphoglycan fragments and glyocosylphosphadylinositol antigens of the protozoan parasite *Leishmania*, *Biochem. J.* **259**:601–604.

Mouricout, M., Petit, J. M., Carias, J. R., and Julien, R., 1990, Glycoprotein glycans that inhibit adhesion of *Escherichia coli* mediated by K99 fimbriae: Treatment of experimental colibacillosis, *Infect. Immun.* **58**:98–106.

Nagi, M. N., and Babiuk, L. A., 1988, Preparation, purification and characterization of bovine Peyer's patch leukocytes, *Can. J. Vet. Res.* **52**:249–257.

Pellérdy, L. P., and Durr, U., 1969, Orale and perenterale Ubertragungsversuche von Kokzidien auf nicht spezifische Wirte, *Acta Vet. Acad. Sci. Hung.* **19**:171–182.

Reduker, D. W., and Speer, C. A., 1986a, Antigens of *in vitro*-produced first-generation merozoites of *Eimeria bovis* (Apicomplexa), *J. Parasitol.* **72**:782–785.

Reduker, D. W., and Speer, C. A., 1986b, Proteins and antigens of merozoites and sporozoites of *Eimeria bovis* (Apicomplexa), *J. Parasitol.* **72**:901–907.

Roberts, W. L., Speer, C. A., and Hammond, D. M., 1971, Penetration of *Eimeria larimerensis* sporozoites into cultured cells as observed with the light and electron microscopes, *J. Parasitol.* **57**:615–625.

Sandvig, K., Olsnes, S., Petersen, O. W., and van Deurs, B., 1987, Acidification of the cytosol inhibits endocytosis from coated pits, *J. Cell Biol.* **105**:679–689.

Simpson, L., and Shaw, J., 1989, RNA editing and the mitochondria cryptogenes of kinetoplastid protozoa, *Cell* **57**:355–366.

Smith, H. W., and Halls, S., 1967, Observations by the ligated intestinal segment and oral inoculation methods on *Escherichia coli* infections in pigs, calves, lambs and rabbits, *J. Pathol. Bacteriol.* **93**:531–543.

Speer, C. A., 1983, The Coccidia, in *In Vitro Cultivation of Protozoan Parasites* (J. B. Jensen, ed.), pp. 1–65, CRC Press, Boca Raton, Florida.

Speer, C. A., and Dubey, J. P., 1989, Ultrastructure of sporozoites and zoites of *Hammondia heydorni*, *J. Protozool.* **36**:488–493.

Speer, C. A., Reduker, D. W., Burgess, D. E., Whitmire, W. M., and Splitter, G. A., 1985, Lymphokine-induced inhibition of growth of *Eimeria bovis* and *Eimeria papillata* (Apicomplexa) in cultured bovine monocytes, *Infect. Immun.* **50**:566–571.

Speer, C. A., Davis, L. R., and Hammond, D. M., 1971, Cinemicrographic observations on the development of *Eimeria larimerensis* in cultured bovine cells, *J. Protozool.* **18** (Suppl.):19.

Stephens, J. P., 1965, Some physiological effects of coccidiosis caused by *Eimeria necatrix* in the chicken, *J. Parasitol.* **51**:331–335.

Thomas, P. S., 1983, Hybridization of denatured RNA transferred or dotted to nitrocellulose paper, *Meth. Enzymol.* **100**:255–266.

Ungewickell, E., and Branton, D., 1981, Assembly units of clathrin coats, *Nature* **239**:420–422.

Van der Horst, C. T. G., and Kouwenhoven, B., 1973, Biochemical investigation with regard to infection and immunity of *Eimeria acervulina* in the fowl, *Z. Parasitenkd.* **42**:23–38.

Whitmire, W. M., Kyle, J. E., Speer, C. A., and Burgess, D. E., 1988, Inhibition of penetration of cultured cells by *Eimeria bovis* sporozoites by monoclonal immunoglobulin G antibodies against the parasite surface protein P20, *Infect. Immun.* **56**:2538–2543.

Chapter 12

Cytoskeletal Proteins Expressed by Microsporidian Parasites

Earl Weidner

1. INTRODUCTION

1.1. Microsporidians as Intracellular Parasites

Microsporidians are intracellular parasites characterized by a small missile cell spore that when activated, fires a discharge tube through which infective sporoplasm is injected into a target cell (Weidner, 1972). Microsporidians develop directly in the cytoplasm of host cells and therefore are not confined to a parasitophorous vacuole.

1.2. Microsporidians in Epithelial Cells

Microsporidians undergo initial development in epithelial cells of most metazoan animals. However, many of the approximately 1000 species undergo sporogenesis in additional host tissues such as Malpighian tubules, muscle, and adipose cells (see Thomson, 1960).

Earl Weidner Department of Zoology and Physiology, Louisiana State University, Baton Rouge, Louisiana 70803, and The Marine Biological Laboratory, Woods Hole, Massachusetts 02543.
Subcellular Biochemistry, Volume 18: Intracellular Parasites, edited by J. L. Avila and J. R. Harris. Plenum Press, New York, 1992.

1.3. Microsporidian Range in Phyletic Host Groups

Microsporidian parasites are found in nearly all phyletic animal categories but are particularly common in arthropods and fish species (Sprague, 1977; Canning and Lom, 1986). The majority of these parasites are in arthropods, infecting the midgut and adipose cells (Weiser, 1976).

1.4. Microsporidians with Sporophorous Vesicles

A majority of microsporidians develop a sporophorous vesicle (spore sac, not to be confused with parasitophorous vacuole) (see Canning *et al.*, 1982). During spore sac development, a spore-forming cell (sporont) buds internally within a progenitor cell and the sporont segregates from the external cytoplasmic domain of the progenitor by an additional membrane. The cytoplasmic domain of the progenitor external to the sporont or spores develops into a spore sac. The extrasporont or extrasporular space within the spore sac characteristically express envelopes that surround the spores. Sometimes the extrasporular envelopes within the spore sacs are remarkably stable, as a result of the addition of some structural binding proteins.

1.5. Cytoskeletal Proteins

A few of the many species of spore sac-forming microsporidians in arthropods have been examined and appear to be equipped with vertebrate epithelial cell-like cytoskeletal elements found in the spore sac compartment. We have looked at *Thelohania butleri* in *Pandalus borealis, Thelohania* species in *Pandalus jordani,* and *Thelohania* species from *Callinectes sapidus*. A report by Larsson (1981) shows that *Amblyospora undulata* may also be equipped with a cytoskeletal apparatus. Our study has been primarily on *Thelohania* species from *C. sapidus.* We have selected the spore sac cytoskeletal systems of microsporidians from arthropods since this host group does not express cytoplasmic intermediate filaments (CIFs) or demosomes (Bartnik and Weber, 1989).

2. CYTOSKELETAL PROTEINS OF VERTEBRATE KERATINOCYTES AND *THELOHANIA* SPORE SACS

2.1. Cytokeratins, Desmosomal Proteins, and Involucrin

Cytoskeletal proteins of vertebrate keratinocytes and the analogues found in arthropod-infecting microsporidian spore sacs are the cytokeratins, desmosomal proteins, and involucrin.

2.1.1. Cytokeratin Intermediate Filaments

Histologically, epithelial cells are in layers and are characterized by the presence of intermediate filaments (IFs) (7–11 nm) of the cytokeratin type and the presence of desmosomes, that is, junctions that consist of proteins that are distinct from other junctions (see Romano *et al.*, 1986). Both cytokeratins and desmosomal proteins are referred to as specific markers for epithelial cells, although they have been found in some nonepithelial cells. Cytokeratins are products of a multigene family and expressed as acid or base subfamilies. Simple epithelial cells frequently have human cytokeratins 7, 8, 18, or 19. Comparison of CIF homologies of vertebrate cytokeratins with possible invertebrate analogues indicate that little homology exists; moreover, antibodies directed to specific mammalian cytokeratins do not appear to react with CIFs in epithelial cells of invertebrates of the various metazoan phyletic categories (Bartnik and Weber, 1989). We will address the curious phenomenon of specific mammalian cytokeratin and desmosomal antibody responses to cytoskeletal counterparts found in microsporidian spore sacs recovered from arthropods, a large phyletic group that does not express cytokeratins or desmosomes.

2.1.2. Desmosomes

These protein assemblages are generally associated with simple epithelial cells. These structures consist of components resistant to solubilization by nonionic detergents and buffers of low or high ionic strength (Pasdar and Nelson, 1988). Desmosomes consist of 5–11 proteins segregated into two elements, a plaque and membrane foci. The plaque consists of desmoplakin (240–250 kDa), 83- and 75-kDa proteins that bind to the cytokeratins. The membrane-associated foci proteins of desmosomes are glycoproteins and will not be addressed here (see Pasdar and Nelson, 1988). Desmosomal proteins, particularly desmoplakins, are conserved and show antibody cross-reactivity among desmosomal elements present in epithelia of the various metazoan phyletic categories (see Bartnik and Weber, 1989); however, arthropods do not have desmosomes.

2.1.3. Involucrin Analogues

Both vertebrate keratinocytes and microsporidian spore sac domains of the genus *Thelohania* (from *Callinectes sapidus*) express lipid-filled lamellar centers (see Eckert, 1989), which appear to release membrane-coating material that confers a remarkable stability to cells. In keratinocytes, this tough material is termed *involucrin*, with a molecular weight of 80–85 kDa. Involucrin appears to cause a cornification of the envelopes in the cell; these become highly stable and resist treatments to alkali (Matoltsy and Matoltsy, 1966), detergents, and reducing agents (Sun and Green, 1976). The involucrin analogue is hydrophilic but soluble in 67% ethanol, indicating it is also affected by hydrophobic elements (Parenteau *et al.*, 1987).

2.2. *Thelohania* Species Cytokeratin and Desmosomal Protein Expression

The expression of cytokeratins and desmosomal elements in *Thelohania* species from *Callinectes sapidus* have been studied by various techniques. Immunolabeling data, supported by gel electrophoretic analyses and immunoblotting, indicate cytokeratin and desmosomal protein analogues of *Thelohania* species respond to specific antibodies directed to bovine cytokeratin protein counterparts. Between 60–80% of the total protein in the spore sacs of *Thelohania* species from *Callinectes sapidus* is expressed as cytokeratin and desmosomal analogues; this is nearly equivalent to the amount of cytoskeletal protein expressed by vertebrate keratinocytes (Eckert, 1989). Immunofluorescence localization and immunoblotting was done on proteins recovered from purified *Thelohania* species spore sacs using monoclonal (see Figure 1) antibodies prepared against mammalian counterparts (monoclonal antibodies AE1 and AE3, K8.13 and Lu5). The immunoblots indicate that the cytokeratins are in the 50–58 kDa range. To identify desmosomal proteins, monoclonal antibody DP2.15 was used to probe desmoplakin I. Immunoblots showed a desmoplakin I analogue in the 240–250 kDa range; by immunofluorescence microscopy, this protein is associated with the plaques surrounding the eight spores within each spore sac of *Thelohania*. Immunofluorescence microscopy also indicates that the cytokeratin analogues are confined to the fine filaments extending from the envelopes surrounding the spores within the spore sacs of *Thelohania* species. The desmoplakin analogues are confined to the plaques expressed as continuous

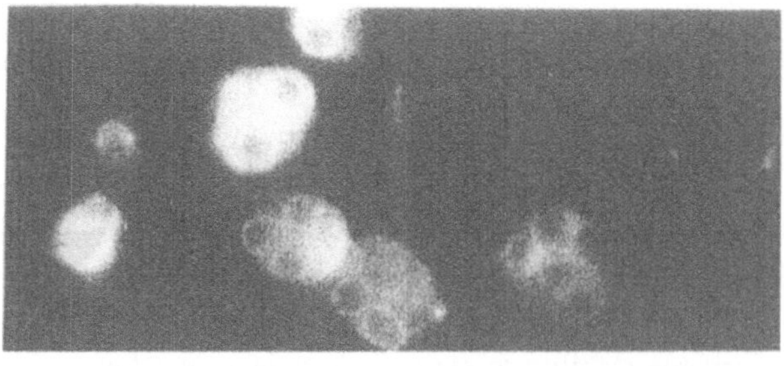

FIGURE 1. Indirect immunofluorescence to cytokeratins in *Thelohania* species spore sacs recovered from blue crab muscle. Antibody K8.13 (Sigma Chemical Co., St. Louis) was used. Figure shows fluorescence dominant around spores. *Thelohania* spores are 1.5 μm in diameter; spore sacs are 12–14 μm across.

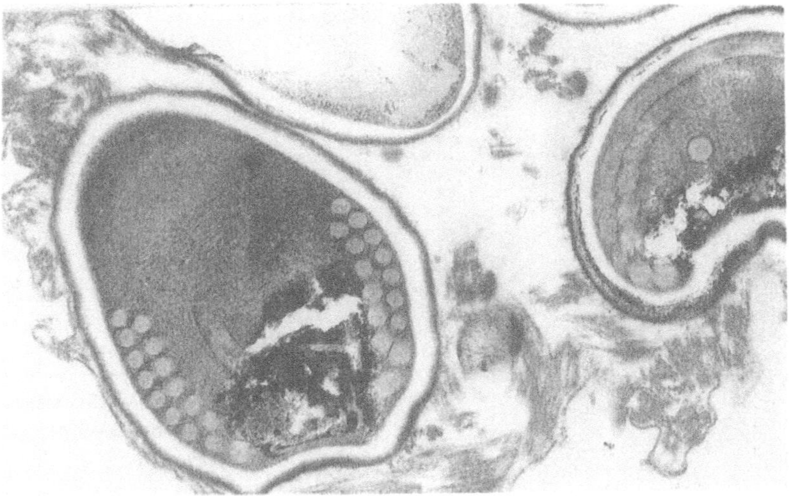

FIGURE 2. Electron micrograph of a cross-section of a *Thelohania* species spore sac depicting bundles of cytokeratin analogues attached to envelopes surrounding spores.

half-desmosomes enveloping each of the eight *Thelohania* species spores within a spore sac. Since the *Thelohania* species cytokeratin and desmoplakin analogues are immunologically responsive to antibodies prepared against mammalian counterparts, it indicates some significant degree of homology in these proteins. The absence of cytokeratin or desmoplakin antibody activity in the arthropod host is not surprising, since cytokeratins and desmosomes have not been found in arthropods (Bartnik and Weber, 1989).

2.3. The Cytoskeletal Analogue Structure in *Thelohania* Species

The structure of the cytoskeletal analogues within *Thelohania* species spore sacs resembles the expression in the vertebrate keratinocyte (see Eckert, 1989) (Figure 2). In both instances, the newly synthesized IFs are unbound to membrane but subsequently attach to the membrane-associated plaques. Pasdar and Nelson (1988) believe that desmoplakins can bind to cytokeratins before the assemblage actually attaches to the membrane–membrane contact points between epithelial cells. The situation for the binding of cytokeratin analogues to plaque assemblages in *Thelohania* species spore sacs is different because the arrangement is a membrane–spore surface. Ultrastructural images depicting *Thelohania* species spore sacs at different stages in development would seem to indicate that cytokeratin analogues bind only to membrane surrounding *Thelohania* species sporoblasts, but only after the apparent attachment of the

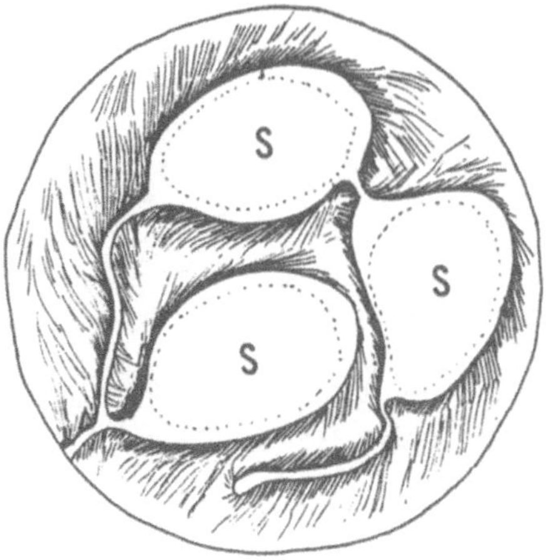

FIGURE 3. A conceptual view of cytokeratin-plaque envelopes surrounding *Thelohania* species spores (S) within spore sac. Note plaque-bearing envelope is continuous within spore sac.

membrane to the spore; there is apparently a stabilization of the membrane surface onto an adjoining surface requirement before the IFs and desmoplakin can attach. The plaque arrangements within the *Thelohania* species spore sac are primarily half-desmosomes, resembling somewhat the desmosomes extending the length of cells in bovine mammary gland epithelial cells (Cowin *et al.*, 1985); however, in *Thelohania* species spore sacs, the half-desmosomes extend over the perimeter of each spore (Figure 3).

2.4. Origins of Desmoplakin–Cytokeratin Assemblages in *Thelohania* Species

Do the desmoplakin–cytokeratin assemblages of the microsporidian *Thelohania* species originate from the parasite? Of the approximately 1000 species of microsporidians reported, only a fraction have been examined for the presence of cytoskeletal elements. Nevertheless, cytoskeletal assemblages are relatively obvious structures; hence, it is already apparent that these structures are not dominant in the majority of microsporidians. More likely, the cytoskeletal analogues have been acquired from host epithelial cells. However, most of the microsporidian species so far examined that have the cytokeratins and desmosomes are parasitic in arthropods, a host group that does not appear to have the capabilities for cytokeratin or desmosomal expression (Fuchs and Marchuk, 1983; Bartnik and Weber, 1989). However, approximately 100 species of microsporidians are parasitic in aquatic vertebrates and most of these likely go through initial growth in the epithelial areas of the gut. Hence, the possibility exists that a few of the *Thelohania* species initially parasitic in vertebrate hosts are acquired and incorporated vertebrate cytoskeletal genes into their own genome. The cross-reactivity data would indicate at least a possibility that vertebrate host genes may have been acquired sometime in the past (by *Thelohania* species) since there is cross-reactivity of the *Thelohania* cytoskeletal analogues to monoclonal antibodies directed to bovine counterparts.

3. INTERMEDIATE FILAMENT CAGE ASSEMBLY IN *SPRAGUEA LOPHII*

The intermediate filament cage assembly (IFCA) forms a remarkable envelope on spores of the microsporidian, *Spraguea lophii*. The IFCA has cytokeratins, clathrin, calmodulin, and an apparent ATPase. Below is an analysis of this microsporidian spore surface cage.

3.1. Intermediate Filament Cage Assemblages

Vimentin, cytokeratins, and lamins mobilize from an IF bundle order into a lattice, cagelike arrangement by directed phosphorylation–dephosphorylation events (Franke *et al.*, 1987; Gerace and Burke, 1988; Mandeville and Rieder, 1990). This assembly can take place in part due to intrinsic molecular or ionic signals such as clock events including mitosis (Franke *et al.*, 1982; Tolle *et al.*, 1987); but CIF assemblies can additionally be affected by extrinsic factors, such as sudden temperature shifts (Schliwa and Euteneuer, 1979). The IF monolayer cage conformation, compared to IF bundle assemblages, has a pronounced binding capacity for a number of proteins. Thus, IFs have associated proteins that become attached into the cage and potentially can become important functional entities. A number of proteins are known to bind IFs, including desmoplakin (Jones and Goldman, 1985), ankyrin (Georgatos *et al.*, 1985), and heat shock proteins (70 kDa hsp) (Georgatos *et al.*, 1989). Recent evidence indicates clathrin light chains may have a binding capacity for cytokeratins, as indicated by the close homologies in the functional groups in these molecules (Brodsky *et al.*, 1987; Jackson *et al,* 1987). We have observed a cytokeratin IF cage with clathrin and calmodulin associations within the IF envelope on the spores of the microsporidian, *Spragues lophii* (Figures 4 and 5). *Spraguea lophii* grows as an intracellular parasite with a plasmodium stage within ganglionic cells of the peripheral and central nervous system of the anglerfish, *Lophius americanus, L. budegassa,* and *L. piscatorius*. The infection is devastating and has an extremely high incidence (above 90% of the *L. americanus* off the new England coast) in these fish. During plasmodial growth the IFs are expressed in the extrasporular domain with IF cage assemblies as bundles or monolayer cages; the cage assemblages are confined to the surface of the developing spore. After spore maturation and subsequent liberation from the host, the IF monolayer cage is

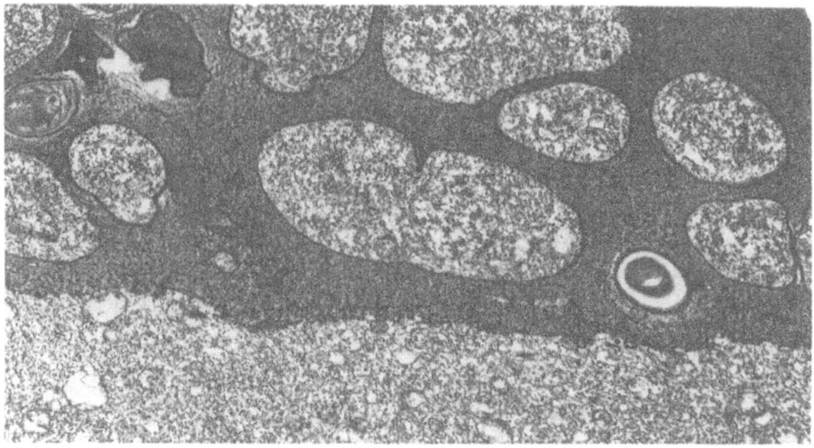

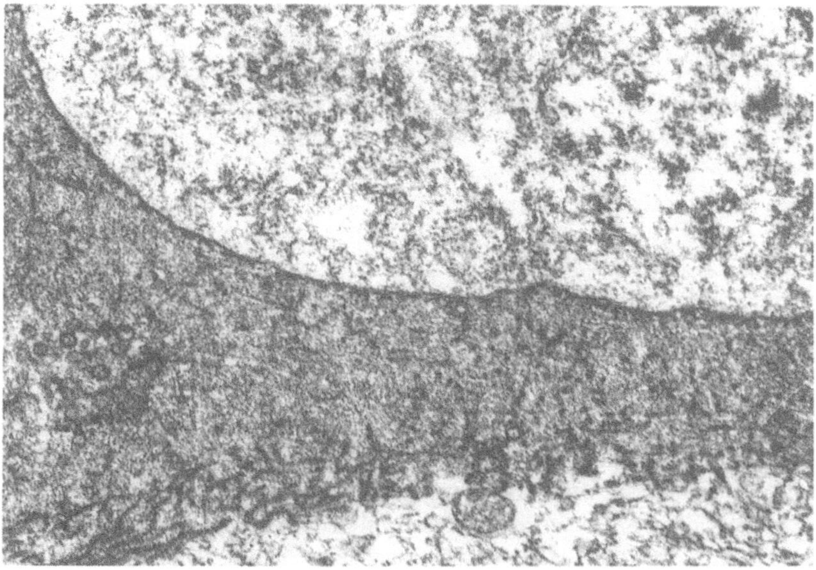

FIGURES 4 AND 5. Electron micrographs of peripheral region of *S. lophii* plasmodium with internal sporonts. Sporonts develop into the spore stage. The extrasporular domain consists of intermediate filaments that respond to cytokeratin antibody K8.13. Figure 5 is a blowup of part of Figure 4, showing cytokeratin bundles at the edge of plasmodium. Diameter of the *S. lophii* sporont in Fig. 5 is 3.5 μm.

retained as part of the spore wall. Additional information on *S. lophii* development is presented by Weissenberg (1976).

3.2. The Cytoskeletal Envelope on *S. lophii* Spores

The cytokeratin monolayer on *S. lophii* spores is removed from the wall by incubation in 0.5 mM EGTA with intermittent washes in distilled water followed by a dehydration cycle. Recovered spore cages, negatively stained with uranyl acetate (Figure 6) show little IF substructure of protofilaments, but IF cage protofilaments can be resolved by sodium phosphotungstate, although this negative stain caused an unraveling and apparent disassembly of the IFs. Staining the *S. lophii* sphore IF cages first with uranyl acetate and poststaining with phosphotungstate visualizes a row of clathrinlike triskelions. These structures responded well to monoclonal antibody directed against clathrin isolated from bovine brain coated vesicles (Figure 7). Cationized ferritin has a strong binding affinity for clathrin and this made it possible to use it as a probe; the cationized ferritin effectively labeled only the areas that were lined with triskelion proteins on the *S. lophii* IF cage (Figure 8). Pretreatment of the IF cage envelopes with anticlathrin significantly impaired the binding of cationized ferritin to the triskelions on the spore surface. Native (anionic) ferritin did not bind to the clathrinlike proteins.

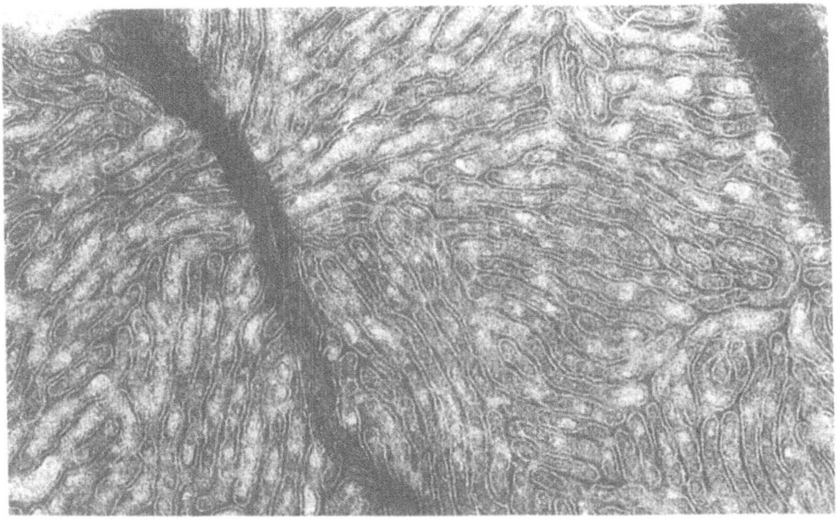

FIGURE 6. Electron micrograph of discarded cage envelope from *S. lophii* spore. Cage is on a formvar-coated grid and negatively stained with uranyl acetate. Note protofilaments do not take up stain.

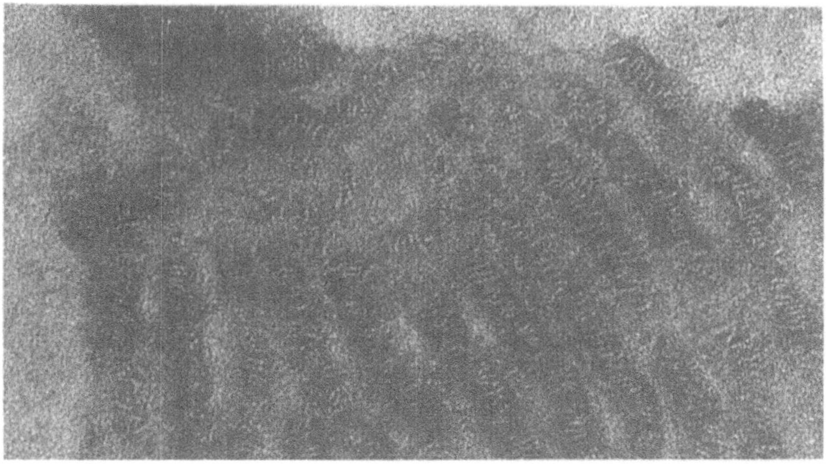

FIGURE 7. Electron micrograph of shed IF cage envelope of *S. lophii* spore. Cage prestained with uranyl acetate and poststained with phosphotungstic acid (PTA). The PTA causes significant unraveling of filaments and they are nearly invisible; however, between the filaments rows of triskelionlike assemblages are now apparent. The triskelions respond to clathrin antibody.

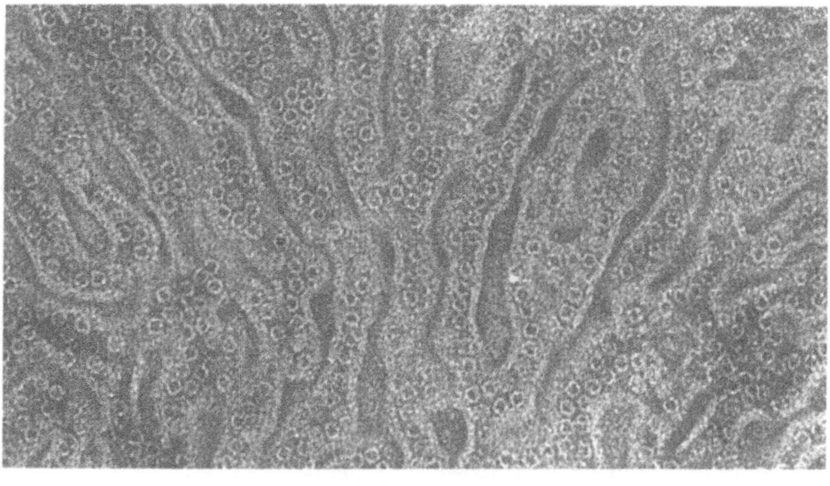

FIGURE 8. Electron micrograph of IF cage envelope showing IF filaments and cationized ferritin particles attached to positions where triskelions normally locate. Cationized ferritin has a binding affinity for clathrin and hence it is a useful probe here for monitoring the alignment patterns of the triskelions within the *S. lophii* spore envelope.

3.3. Cytokeratins, Clathrin, and Calmodulin in the *S. lophii* Cage

Identification of *S. lophii* cage cytokeratins, clathrin, and calmodulin has been achieved by immunoblotting. For further analyses of cytokeratin IFs reactive to specific antibodies, we applied electrophoresis and immunoblotting against proteins recovered from *S. lophii* spore cage envelopes. Monoclonal antibody K8.136 reacted to epitopes to cytokeratin IFs with approximate molecular weights of 50, 52, 56, and 58kDa. Antibodies to neurofilament or vimentin IFs were unreactive. To identify clathrin, monoclonal antibody clone C5.9 labeled the *S. lophii* spore surface. Blots of spore cage proteins showed reactions at 180, 33, and 38 kDa, similar to the control blots against bovine brain clathrin. For the identification of *S. lophii* spore surface calmodulin, goat antibody directed against bovine brain calmodulin was used (Figure 9). Clathrin antibody directed to proteins recovered from *S. lophii* spore envelopes indicates a strong reaction at 15–18 kDa.

3.4. Discussion of *S. lophii* Spore IF Cage Assemblage

Cytokeratin, vimentin, and lamin IFs can organize into a monolayer cage (Klymkowsky *et al.*, 1989). The IF cage can be a dynamic assemblage as shown by: nuclear lamins during mitosis (Gerace and Burke, 1988); cytokeratin remobilization from a bundle to a cage order at the nuclear–cytoplasmic interface after nuclear envelope disassembly (Mandeville and Rieder, 1990); and vimentin

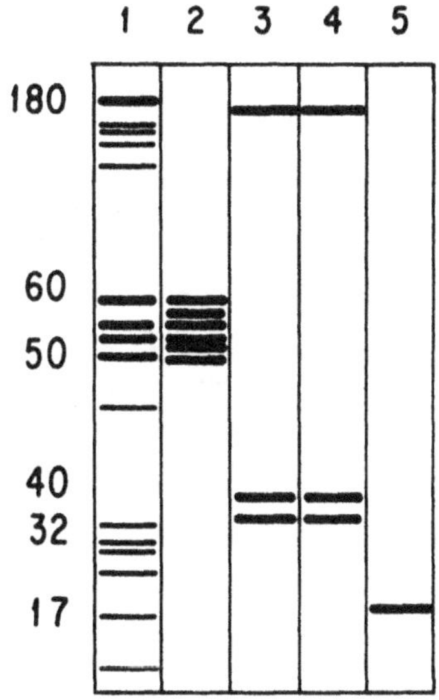

FIGURE 9. Diagram of SDS-PAGE gels and immunoblots of *S. lophii* spore wall proteins. Lane 1 shows the major proteins from the spore envelope. Lane 2 shows the results of an immunoblot response to cytokeratin antibody; note that the 50–58 kDa bands are reactive. Lane 3 shows blot with clathrin antibody directed to purified clathrin from bovine brain coated vesicles. Lane 4 shows blot response of clathrin antibody to *S. lophii* cage proteins. Lane 5 denotes a blot analysis with anticalmodulin to *S. lophii* spore cage proteins.

at the boundaries of lipid globules (Franke *et al.*, 1987). Our results indicate cytokeratins, with associated clathrin and calmodulin, comprise the cage envelope of *S. lophii* spores. The cage envelope appears essential for the activation of microsporidian spore discharge since removal of the cage irreversibly inactivates the spore and it cannot fire. Monoclonal antibody labeling, applied to immunofluorescence and immunoelectron microscopy, indicates that the IF cage envelope on *S. lophii* spores is composed of cytokeratin with accompanying clathrin and calmodulin analogues (E. Weidner, unpublished data). *Spraguea lophii* IF cage proteins respond to monoclonal antibodies K8.13, AE1, and AE3, known to have specificity for cytokeratin epitopes. The IF cage-associated clathrin appears to have a triskelion arrangement in rows adjoining the IFs, as indicated by negative-staining and immunoperoxidase labeling. The clathrin alignment pattern within the cage is stable as indicated by its resistance to dispersion at pH shifts from 4–10; nevertheless, the clathrin is destabilized and displaced from the spore IF cage with exposure to EGTA. Immunolabeling indicates calmodulin analogues are also present in the *S. lophii* IF cage; the calmodulin is most likely clathrin-associated, since it has specific binding sites for clathrin light chains (Lisanti *et al.*, 1982; Brodsky *et al.*, 1987; Kohtz *et al.*, 1987). Spore cage immunoblot analyses, using cytokeratin antibody K8.136, indicate the presence of 50, 48–52, and 56–58 kDa analogues. Immunoblots of spore cage proteins, using goat antibody directed to bovine brain clathrin and calmodulin, indicate these analogues are present in the spore cage and antigenically similar to their bovine counterparts.

The IF cages of *S. lophii* spores consist of laterally attached filament loops, resembling the vimentin cages of IFs in stacking order (Franke *et al.*, 1987); also, the spore cage IFs appear to be supported by a stable but uncharacterized substrate. At the onset of spore activation, the monolayer cytokeratin cage instantaneously disassembles by a presumptive phosphorylation of the IFs, resembling the disassemblies reported for other kinds of IF cages (Franke *et al.*, 1987; Gerace and Burke, 1988; Mandeville and Riedle, 1990). The available evidence supports a phosphorylation-induced IF cage disassembly event, as indicated by the *in vitro* collapse in the IF cage order in phosphate media, known to unravel IF bundles (Aebi *et al.*, 1983). The *S. lophii* spore cage clathrin, as evidenced by antibody probes and negative staining, is expressed as triskelions that have a stable positioning within the cage. This stable positioning of the apparent triskelion clathrin molecules would seem to indicate that these proteins are likely to be attached to substrate. Whether clathrin is directly attached to the cytokeratins is not clear, although clathrin light chains appear to have site specificity for the so-called active sites of cytokeratin polypeptides (Brodsky *et al.*, 1987; Jackson *et al.*, 1987). The presence of clathrin in the IF cage probably increases the functional binding capabilities in the assemblage for such molecules as calmodulin and the 70-kDa heat shock protein (an ATPase) (Brodsky *et al.*, 1987; Kohtz *et al.*, 1987).

3.5. Inactivation of the *S. lophii* Spore Extrusion Apparatus

The microsporidian spore is an infective missile cell equipped with an extrusion apparatus (Weidner, 1989). When activated, the spore discharges an invasion tube with millisecond velocity through which the interior of the spore cell extrudes. The *S. lophii* spore IF cage-associated calmodulin is believed to be the primary messenger in the Ca^{2+}-mediated spore activation since the calmodulin antagonists trifluoperazine and chlorpromazine effectively block *S. lophii* spore activation and impair the Ca^{2+} influx associated with spore discharge (Pleshinger and Weidner, 1985). The removal of clathrin and calmodulin, or the removal of the entire IF cage, causes irreversible inactivation of the spore. Since phosphate-induced disassembly of the *S. lophii* IF cage in preactivated spores does not initiate spore discharge, it is believed that cage disassembly does not play a primary role in spore activation. The spore IF cage disassembly may be a secondary response to a sequence of reactions leading to a Ca^{2+} influx into the spore at the onset of spore activation. Our preliminary evidence indicates a Ca^{2+}-dependent ATPase is associated with the calcium import into the spore; intracellular phosphate release might account for the subsequent phosphorylation of the IFs in the cage.

4. REFERENCES

Aebi, U., Fowler, W. E., Rew, P., and Sun, T., 1983, The fibrillar substructure of keratin filaments unraveled, *J. Cell Biol.* **97:**1131–1143.

Bartnik, E., and Weber, K., 1989, Widespread occurrence of intermediate filaments in invertebrates; common principles and aspects of diversion, *Eur. J. Cell Biol.* **50:**17–33.

Brodsky, F. M., Galloway, C. J., Blank, G. S., Jackson, A. P., Seow, H., Drickawer, K., and Parham, P., 1987, Localization of clathrin light chain sequences mediating heavy chain binding and coated vesicle diversity, *Nature* **326:**203–205.

Canning, E. U., and Lom, J., 1986, *The Microsporidae of Vertebrates*, Academic Press, New York.

Canning, E. U., Lom, J., and Nicholas, J. P., 1982, Genus *Glugea* Thelohan 1891. (phylum Microspora): Redescription of the type species *Glugea anomala* (Moniez 1887) and recognition of its sporogonic development within sporophorous vesicles (pansporoblastic membranes), *Protistologica* **18:**192–210.

Cowin, P., Kapprell, H., and Franke, W. W., 1985, The complement of desmosomal plaque proteins in different cell types, *J. Cell. Biol.* **101:**1442–1454.

Eckert, R. L., 1989, Structure, function, and differentiation of the keratinocyte, *Physiol. Rev.* **69:**1316–1345.

Franke, W. W., Schmid, E., Grund, C., and Geiger, B., 1982, Intermediate filamentous structures: Transient disintegration and inclusion of subunit proteins in granular aggregates, *Cell* **30:**103–113.

Franke, W. W., Hergt, M., and Grund, C., 1987, Rearrangement of the vimentin cytoskeleton during adipose conversion: Formation of an intermediate filament cage around lipid globules, *Cell* **39:**131–141.

Fuchs, E., and Marchuk, D., 1983, Type 1 and type 11 keratins have evolved from lower eukaryotes to form the epidermal intermediate filaments in mammalian skin, *Proc. Natl. Acad. Sci. USA* **80:**5857–5861.

Georgatos, S. D., Weaver, D. C., and Marchesi, W. T., 1985, Site-specificity in vimentin–membrane interactions: Intermediate filament subunits associate with the plasma membrane via their head domains, *J. Cell Biol.* **100:**1962–1967.

Georgatos, S. D., Blobel, G., and Chirico, M., 1989, Molecular interactions between intermediate filament proteins and heat-shock proteins, *J. Cell Biol.* **109:**1417A.

Gerace, L., and Burke, B., 1988, Functional organization of the nuclear envelope, in *Annual Review of Cell Biology,* Vol. 4 (G. E. Palade, B. M. Alberts, and J. A. Spudich, eds.), pp. 335–374, Annual Reviews Inc., Palo Alto, California.

Jackson, A. P., Seow, H., Holmes, N. J., Drickamar, K., and Parham, P., 1987, Clathrin light chains contain brain-specific insertion sequences and a region of homology with intermediate filaments, *Nature* **326:**154–159.

Jones, J. C. R., and Goldman, R. D., 1985, Intermediate filaments and the initiation of desmosome assembly, *J. Cell Biol.* **101:**506–517.

Klymkowsky, M. W., Bachant, J. B., and Domingo, A., 1989, Functions of intermediate filaments, *Cell Motil. Cytoskel.* **14:**309–331.

Kohtz, D. S., Georgieva-Hanson, V., Kohtz, J. D., Schook, W. J., and Paszkin, S., 1987, Mapping two functional domains of clathrin light chains with monoclonal antibodies, *J. Cell Biol.* **104:**897–903.

Larsson, R., 1981, An ultrastructural study of *Amblyospora undulata* N. sp., a microsporidian parasite of the caddis fly *Cyrnus trimaculatus* (Trichoptera, Polycentropidae), *Protistologica* **17:**511–523.

Mandeville, E. C., and Rieder, C. L., 1990, Keratin filaments restrict organelle migration into the forming spindle of newt pneumocytes, *Cell Motil. Cytoskel.* **15:**111–120.

Matoltsy, A. G., and Matoltsy, M. N., 1966, The membrane proteins of horny cells, *J. Invest. Dermatol.* **46:**127–129.

Parenteau, N. L., Eckert, R. L., and Rice, R. H., 1987, Primate involucrins: Antigenic relatedness and detection of multiple forms, *Proc. Natl. Acad. Sci. USA* **84:**7571–7575.

Pasdar, M., and Nelson, N. J., 1988, Kinetics of desmosome assembly in Madin-Darby canine kidney epithelial cells: Temporal and spatial regulation of desmoplakin organization and stabilization upon cell–cell contact, 1. Biochemical analysis, *J. Cell Biol.* **106:**677–685.

Pleshinger, J., and Weidner, E., 1985, The microsporidian spore invasion tube. IV. Discharge activation begins with pH-triggered Ca^{2+} influx, *J. Cell Biol.* **100:**1834–1838.

Romano, V., Hatzfeld, M., Magin, T. M., Zimbelmann, R., Franke, W. W., Maier, G., and Ponstingl, H., 1986, Cytokeratin expression in simple epithelia I. Identification of mRNA coding for human cytokeratin #18 by a cDNA clone, *Differentiation* **30:**244–253.

Schliwa, M., and Euteneuer, U., 1979, Structural transformation of cold epidermal tonofilaments upon cold treatment, *Exp. Cell Res.* **122:**93–101.

Sprague, V., 1977, Systematics of the Microsporidia, in *Comparative Pathobiology,* Vol. 2 (L. A. Bulla, Jr., and T. C. Cheng, eds.), pp. 2–14, Plenum Press, New York.

Sun, T. T., and Green, H., 1976, Differentiation of the epidermal keratinocyte in cell culture: Formation of the cornified envelope, *Cell* **9:**511–521.

Thomson, H. M., 1960, A list and brief description of the microsporidia infecting insects, *J. Insect Pathol.* **2:**346–385.

Tolle, H. G., Weber, K., and Osborn, M., 1987, Keratin filament disruption in interphase and mitotic cells—how is it induced? *Eur. J. Cell Biol.* **43:**35–47.

Weidner, E., 1989, Microsporidian spores as missile cells, in *Intracellular Parasitism* (J. Moulder, ed.), pp. 183–191, CRC Press, Boca Raton, Florida.

Weiser, J., 1976, Contribution to the classification of microsporidia, *Vestn. Cesk. Spol. Zool.* **61**:308–320.

Weidner, E., 1972, Ultrastructural study of microsporidian invasion into cells, *Z. Parasitenk.* **40**:227–242.

Weissenberg, R., 1976, Microsporidian interactions with host cells, in *Comparative Pathobiology* Vol. 1 (L. A. Bulla, Jr., and T. C. Cheng, eds.), pp. 203–207, Plenum Press, New York.

Chapter 13

Intracellular Parasites

New Developments in Chemotherapy

Steven R. Meshnick and J. Joseph Marr

1. INTRODUCTION

Intracellular protozoan parasites cause many of the most severe and widespread diseases of man, particularly malaria, leishmaniasis, Chagas' disease, and toxoplasmosis. This chapter will summarize some recent advances in the chemotherapy of these diseases, which have received considerable attention in earlier chapters of this volume.

2. MALARIA

2.1. Background

Prior to the twentieth century, malaria was endemic throughout most of the world including North America and Europe. In this century, particularly since the 1940s, malaria was eliminated from many regions through the use of a combina-

Steven R. Meshnick Department of Epidemiology, University of Michigan School of Public Health, Ann Arbor, Michigan 48109. **J. Joseph Marr** Discovery Research, Searle Pharmaceuticals, Skokie, Illinois 60077, and Department of Internal Medicine, Rush Medical College, Chicago, Illinois 60612.

Subcellular Biochemistry, Volume 18: Intracellular Parasites, edited by J. L. Avila and J. R. Harris. Plenum Press, New York, 1992.

tion of mosquito control and antimalarial drugs. However, malaria has made a comeback in the past 20 to 30 years due to the development of insecticide-resistant vectors and drug-resistant parasites. Currently, it is estimated that there are 500 million cases of malaria per year (Sturchler, 1989).

The chemotherapy and chemoprophylaxis of malaria have been reviewed in great detail in both monographs (Steck, 1971; Peters and Richards, 1984; Peters, 1987) and review articles (Sweeney and Strube, 1980; Warhurst, 1981; White, 1985; Warhurst, 1987; Keystone, 1990; Panisko and Keystone, 1990). Over the past several years, much as been learned about the biochemical mechanisms of current and potential antimalarial agents and it is the goal of this section to review this new biochemical information.

2.1.1. Microbiology

There are four species of malaria that affect man: *Plasmodium falciparum*, *P. vivax*, *P. malariae*, and *P. ovale*. *Plasmodium falciparum* and *P. vivax* are the most common. The parasite undergoes a sexual phase of its life cycle in its vector, the Anopheles mosquito, where it develops into infective forms known as sporozoites. Sporozoites are injected into the vertebrate host by the mosquito and quickly invade hepatocytes. Asexual reproduction then occurs in the liver (pre-erythrocytic schizogony). The infected hepatocytes then rupture and release merozoites, which invade erythrocytes. After infection by *P. vivax* or *P. ovale*, some parasites, hypnozoites, remain dormant in the liver and can later reactivate to cause relapse. The intraerythrocytic cycle (erythrocytic schizogony) can continue indefinitely, with parasites multiplying intracellularly, rupturing the host cell, and then releasing merozoites, which then reinvade new red cells. Some of the intraerythrocytic parasites develop into gametocytes, which are infective to mosquitoes.

The symptomatology of malaria is due to the intraerythrocytic parasites. Within an infected patient, these parasites tend to be synchronized and to rupture and release merozoites every 48 or 72 hr. It is this rupture of infected erythrocytes that is responsible for the periodic fevers and chills.

2.1.2. Classes of Antimalarial Agents

There are antimalarial agents that are active against each of the different stages of the parasite's life cycle. Some drugs, when administered to the vertebrate host, interrupt the development of the parasite in the mosquito feeding on that host. These are called sporontocides. However, specifically sporontocidal agents are not currently in clinical use. Other agents, such as proguanil, act on the preerythrocytic liver stages, and these are called causal prophylactics. Radical curative drugs, such as primaquine, are those that act on the persistent exoerythrocytic stage, or hypnozoite. Primaquine also has activity against gametocytes and aids in interrupting the transmission of the disease. The most commonly used drugs, however, are the schizonticides, such as chloroquine, quinine, and mefloquine, which act primarily on intraerythrocytic asexual stages.

2.1.3. Drug Resistance

Drug resistance is a particularly serious problem in the treatment and prophylaxis of *P. falciparum* infections. The sensitivity of a parasite strain to a particular drug is usually graded S (sensitive), RI (slightly resistant), RII (moderately resistant), or RIII (highly resistant). Resistance can be graded based on the clinical response to treatment or based on standard *in vitro* tests (Peters, 1987).

2.2. Biochemistry of Plasmodia

The biochemistry of malarial parasites has been reviewed elsewhere (Sherman, 1984). Some aspects of malarial biochemistry will be presented in order to help understand the mechanisms of action of antimalarial drugs. Specific areas that are relevant to understanding the mechanisms of action of antimalarial drugs are reviewed.

2.2.1. Hemoglobin Digestion

Intraerythrocytic malarial parasites are continually ingesting host erythrocyte cytoplasm by endocytosis. The endocytic vacuoles fuse with lysosomes, where hemoglobin-specific proteases break down hemoglobin and release amino acids. The hemoglobin-bound hemin, however, does not appear to be broken down. Instead, it becomes a part of large particles within the food vacuoles and is called hemozoin. The exact composition and structure of hemozoin is still unknown, but appears to consist largely of hemin (Sherman, 1984; Goldberg *et al.*, 1990). Intraparasitic hemin may play an important role in the mechanisms of action of chloroquine and artemisinin.

Although hemin contains iron, the parasites have not been shown to liberate iron from hemin. Some investigators have suggested that the parasites obtain their iron from serum transferrin, but this remains controversial (Rodriguez and Jungery, 1986; Pollack, 1989).

2.2.2. Oxidant Defenses

All aerobic organisms are exposed to toxic forms of oxygen known as activated oxygen, including superoxide (O_2-), hydrogen peroxide (H_2O_2), and hydroxyl radicals (OH·). Activated oxygen is toxic because it is capable of damaging nucleic acids, proteins, and lipids (Halliwell and Gutteridge, 1989). Accordingly, several enzymatic defense systems have been developed and are nearly ubiquitous in aerobic organisms.

Malarial parasites are particularly sensitive to drugs that generate activated oxygen (Clark *et al.*, 1989). In addition, their growth seems to be inhibited in erythrocytes that are particularly oxidant sensitive, such as glucose-6-phosphate dehydrogenase-deficient cells (Roth *et al.*, 1983), suggesting that the parasites may have impaired oxidant defenses. However, relatively normal levels of catalase, superoxide dismutase, and peroxidases have been found associated with parasites (Fairfield *et al.*, 1986a), although most of the parasites superoxide dismutase appears to be lysosomal (Fairfield *et al.*, 1986b; Ranz and Meshnick, 1989).

2.2.3. Membrane Transport

Parasites appear to induce alterations in the membrane transport properties of their host erythrocytes. Evidence for a parasite-induced permeation pathway has been presented (Ginsburg *et al.*, 1985).

2.3. Quinoline Methanols

2.3.1. History and Clinical Use

In the seventeenth century, Spanish settlers noticed that Peruvian Indians chewed the bark of wild chinchona trees and appeared to be protected from malaria. The active ingredient of this bark, quinine (Figure 1), was isolated in 1820. Quinine, as well as quinidine, its D-epimer, continue to be important in the therapy of malaria (Hoffheinz and Merkli, 1984). Another quinoline methanol, mefloquine (Figure 1), has recently been approved by the U.S. Food and Drug Administration for both therapy and prophylaxis. Mefloquine was developed by the U.S. Army in the 1970s (Sweeney, 1984).

2.3.2. Mechanism of Action

Quinine is taken up and concentrated by malarial parasites 200-fold (Polet and Barr, 1968). It has been suggested that this uptake, as well as the uptake of mefloquine, is due to the formation of a complex between the drug and parasite hemin (Warhurst, 1981; Fitch, 1986). Hemin binding may not mediate the antimalarial mechanism of quinine because the hemin–quinine complex has not been shown to be toxic. The antimalarial activity of quinine may be mediated by an

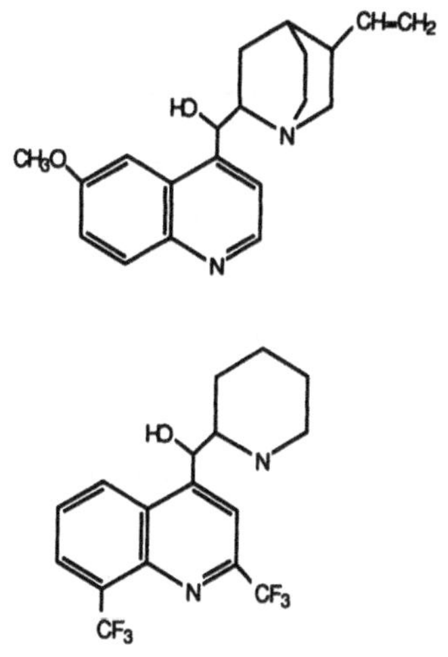

FIGURE 1. Quinoline methanols.

effect on DNA, because it has been found to be a strong DNA intercalator (Estensen *et al.*, 1969). Mefloquine, in contrast, does not intercalate into DNA (Davidson *et al.*, 1977). Mefloquine has also been found to bind to phospholipids (Chevli and Fitch, 1982).

2.3.3. Resistance

Quinine resistance has recently been reported (Malin and Hall, 1990). Mefloquine-resistance, on the other hand, appeared in 1982 and is spreading (Boudreau *et al.*, 1982).

2.3.4. Pharmacology

Quinine has a half-life of around 10 hr, but reported values vary widely (White, 1985). It is a fast-acting drug, but frequently associated with side effects, such as fever, confusion, amblyopia, delirium, respiratory arrest, and arrhythmias (Hoffheinz and Markli, 1984). Quinine is commonly given along with tetracycline for the therapy of *P. falciparum* malaria.

Mefloquine is most commonly used prophylactically. It has a very long half-life (15–33 days) (White, 1985). Adverse effects of mefloquine are generally mild and include dizziness, nausea, and diarrhea. Sinus bradycardia and neuropsychiatric changes have also been reported (Keystone, 1990; Sturchler *et al.*, 1990).

2.4. 4-Aminoquinolines

2.4.1. History and Clinical Use

Chloroquine (Figure 2) was first discovered in the late 1930s in Germany but developed by Americans in the 1940s. Soon after the war, the drug was in widespread use (McChesney and Fitch, 1984). It contributed greatly to the advances made in the control of malaria during the 1950s and 1960s, by virtue of the fact that it was an orally effective prophylactic and therapeutic agent with minimal side effects, a rapid onset of action, and a long serum half-life. Amodiaquine is another 4-aminoquinoline that has been widely used, but when used prophylactically has been shown to be associated with neutropenia (Hatton *et al.*, 1986).

Chloroquine-resistant strains of *P. falciparum* began appearing in the late 1950s and steadily increased in incidence until they became more common than sensitive strains in large parts of South America, Asia, and Africa (Peters, 1987). Recently, chloroquine-resistant *P. vivax* has been reported (Whitby *et al.*, 1989).

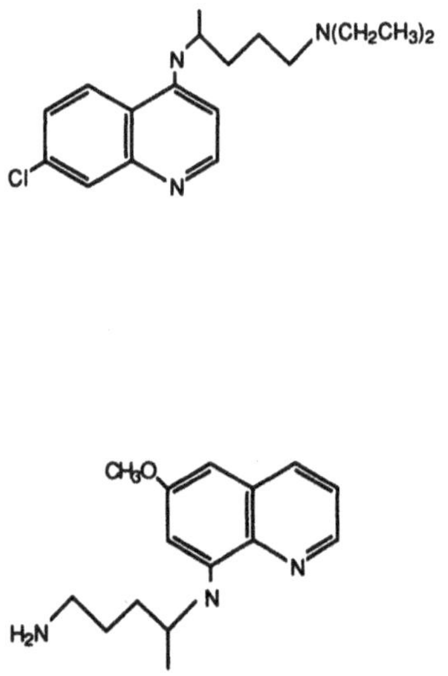

Chloroquine

Primaquine

FIGURE 2. Aminoquinolines.

2.4.2. Mechanism of Action

There is a great deal of controversy about the mechanism of action of chloroquine (Meshnick, 1990). The earliest proposed mechanism of action was binding to DNA. Indeed, the proposal that 8-aminoquinolines act in this manner was made in 1949 (Irvin and Irvin, 1949), predating Watson and Crick's discovery of the double helix. Homewood and collaborators (1972) suggested that chloroquine, acting as a weak base, accumulated in parasite lysosomes and caused their alkalinization. Further support for this hypothesis was supplied by Krogstad and collaborators (Krogstad and Schlesinger, 1987; Krogstad *et al.*, 1987). Another theory, proposed by Fitch (1986), suggested that chloroquine acts by binding to hemin, forming a complex that is toxic to membranes.

Although it was clear that chloroquine intercalates into DNA, the reported binding affinities varied by several orders of magnitude. These apparent discrepancies were probably due to the fact that the binding of chloroquine to DNA is highly salt-dependent and sequence-dependent (Kwakye-Berko and Meshnick, 1989, 1990). Nevertheless, dissociation constants (Kd's) under physiological salt conditions are in the millimolar range, suggesting that the binding is too weak to be important. However, Kds for DNA are expressed per binding site, and a single nucleus may contain millions of important binding sites. Thus, in the presence of therapeutic concentrations of drug, 0.03–1.0% of potential intercalation sites might be occupied and this could be toxic to the parasite. Once in the DNA, chloroquine might act by preventing the transition of DNA to the Z form (Kwakye-Berko and Meshnick, 1990).

The lysosome accumulation hypothesis suggests that the drug, in its unprotonated form, penetrates the cell membranes and reaches the lysosome where it is protonated by the low pH environment. Here, in its charged form, it accumulates and raises intralysosomal pH in chloroquine-sensitive but not -resistant parasites (Krogstad and Schlesinger, 1987). Aikawa (1972) showed by electron microscopic autoradiography that the drug is indeed concentrated in parasite lysosomes.

The hemin-binding hypothesis proposes that hemin, present in the parasite lysosome, binds to chloroquine and forms a toxic complex. The connection between hemin and chloroquine resistance was first based on the observation that chloroquine-resistant *P. berghei* lack hemozoin and that chloroquine treatment causes clumping of hemozoin (McChesney and Fitch, 1984). The binding of chloroquine to hemin is quite tight. Chou *et al.* (1980) found that chloroquine bound to hemin in solution with an affinity constant of 3×10^8. The binding of chloroquine to hemin may explain why the drug is capable of inhibiting hemozoin biosynthesis with an IC_{50} of 10^{-4}M (Slater and Cerami, 1992).

The chloroquine–hemin complex is indeed lytic to both red cells and malarial parasites *in vitro* (Fitch, 1986). However, the complex is not lytic if serum proteins are present (Zhang and Hempelmann, 1987).

A large series of chloroquine congeners have been tested for antimalarial activity. When structural properties and activities are compared, antimalarial activity appears to correlate with ability to intercalate into DNA (O'Brien and Hahn, 1965; Bass *et al.*, 1971). For example, when chloroquine intercalates with DNA, the two amines on its side chain are the correct distance apart to interact with phosphates on opposite strands. When congeners, which have shorter or longer distances between amine groups were tested, they had less antimalarial effect. This latter observation is not consistent with the lysosomal accumulation hypothesis because varying the length of the side chain does not affect the acid dissociation constant (pKa) of the ring nitrogen (pKa = 8.1) (Irvin and Irvin, 1947), the protonation and deprotonation of which is supposed to be involved in its concentration into lysosomes.

The reason for the drug's selective toxocity is probably that it is concentrated by parasites to the extent of 600-fold or more (Fitch, 1972; Fitch *et al.*, 1974). Trapping in lysosomes cannot be responsible for this gradient, since the same pH difference exists with respect to mammalian lysosomes. Uptake, therefore, is probably due to a membrane carrier, or permease, which is responsible for transporting the drug across the cell membrane of the parasite (Warhurst, 1986).

2.4.3. Resistance

Drug-resistant *P. falciparum* strains appear to be resistant because they do not concentrate the drug. Although resistant parasites take up drug at the same rate as sensitive strains, they export the drug at 40 times the rate of sensitive strains (Krogstad *et al.*, 1987). Inhibitors of this export pump, such as calcium channel blockers and phenothiazines, promote the activity of chloroquine against drug-resistant parasites (Martin *et al.*, 1987; Bitonti *et al.*, 1988). The malarial export pump may be homologous to *P*-glycoprotein, a mammalian protein responsible for a multiple-drug resistance phenotype in cancer cells (Foote *et al.*, 1990). However, this is still controversial (Wellems *et al.*, 1990).

2.4.4. Pharmacology

Chloroquine is a fast-acting orally absorbable drug with a half-life of approximately 50 hours. When used prophylactically or therapeutically, there are few serious side effects (McChesney and Fitch, 1984).

2.5. 8-Aminoquinolines

2.5.1. History and Clinical Use

8-Aminoquinolines were first studied by German scientists, who introduced pamaquine in 1925. Unlike quinine, this agent was found to be effective in preventing relapses. An analogue of pamaquine, primaquine (Figure 2), became the antirelapse drug of choice soon after World War II (Carson, 1984). Primaquine is also active against gametocytes and preerythrocytic liver stages (Carson, 1984).

2.5.2. Mechanism of Action

Primaquine is effective against exoerythrocytic stages of malaria *in vitro*, but not as active as some of its metabolites. Primaquine is demethylated and hydroxylated by the liver to produce highly redox-active compounds (Strother *et al.*, 1984). One such metabolite, 5-hydroxy-6-demethyl primaquine, is 20 times more active than primaquine *in vitro*. The mechanism of action of this metabolite may be due to its ability to undergo oxidation and reduction to form activated oxygen, since the metabolites with the greatest activity spontaneously generate superoxide to the greatest degree (Bates *et al.*, 1990).

2.5.3. Resistance

No clinically important resistance has been reported (Carsons, 1984).

2.5.4. Pharmacology

The half-life of the parent drug is about 7 hr in humans (Breckenridge *et al.*, 1984). Primaquine can cause hemolytic anemia in patients with G6PDH-deficiency (Carsons, 1984). The hemolysis appears to be mediated by activated oxygen produced by primaquine metabolites (Allahyari *et al.*, 1984; Strother *et al.*, 1984).

2.6. Antifolates: Sulfa Drugs

2.6.1. History and Clinical Use

Sulfa drugs were first developed as antibacterial agents in the 1930s and include sulfonamides, such as sulfadoxine, and sulfones, such as dapsone (Figure 3). The efficacy of these agents against malaria parasites were demonstrated in 1937. Sulfadoxine, a component of fansidar, is the most commonly used of the sulfa drugs in malaria chemotherapy. Dapsone, a component of maloprim, is also used in some parts of the world. While both drugs are effective against *P. falciparum*, they are relatively ineffective against *P. vivax (Scholer et al.*, 1984).

2.6.2. Mechanism of Action

Sulfa drugs are structural analogues of *p*-aminobenzoic acid (Figure 3) and act as competitive inhibitors of dihydropteroate synthetase, an early step in *de*

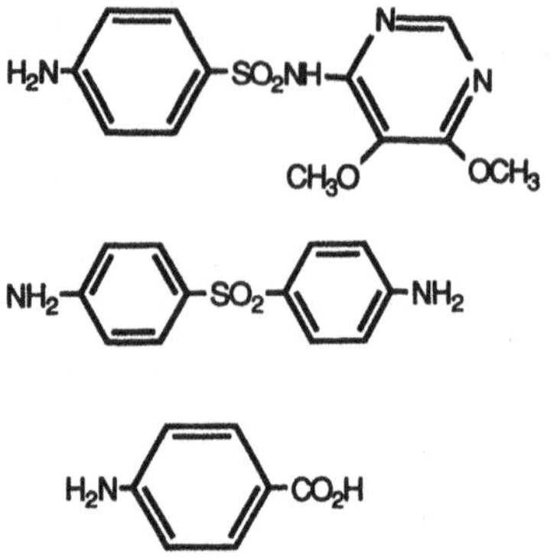

Sulfadoxine

Dapsone

p-Aminobenzoic acid

FIGURE 3. Sulfa drugs.

novo folate biosynthesis (Figure 4). Sulfa drugs are selectively toxic to malarial parasites, since mammals lack this enzyme and only use folate acquired through the diet. However, there is some evidence that the parasites can also take up preformed folates from their host, since folic acid antagonizes the antimalarial effects of sulfa drugs *in vitro* (Watkins *et al.*, 1984). This antagonistic effect is

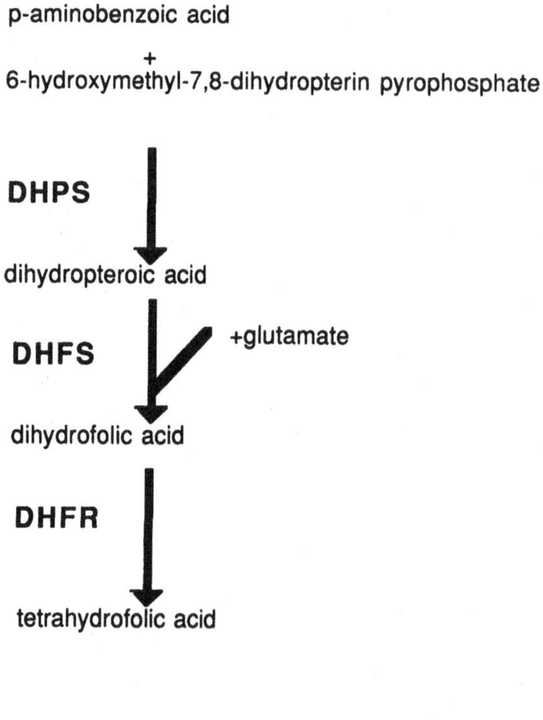

FIGURE 4. A portion of the *de novo* folate biosynthesis pathway. DHPS, dihydropteroate synthetase; DHFS, dihydrofolate synthetase; DHFR, dihydrofolate reductase.

not due to the breakdown of folate to produce *p*-aminobenzoic acid, since *p*-aminobenzoylglutamate, an intermediate breakdown product of folates, does not antagonize sulfonamides. Sulfa drugs are synergistic with drugs such as pyrimethamine and proguanil, which inhibit the later dihydrofolate reductase step in the folate pathway.

Sulfa drugs inhibit the *P. falciparum* dihydropteroate synthetase (DHPS) with inhibitory concentrations ranging from 6–90 μM, but inhibit parasite growth *in vitro* at 100- to 1000-fold lower concentrations (30–500 nM). There is evidence for the active concentration of sulfa drugs by malarial parasites which could explain this difference (Zhang and Meshnick, 1991).

2.6.3. Resistance

Sulfa-resistant strains of *P. falciparum* have been isolated from American travelers returning from West Africa (Milhous *et al.*, 1989).

2.6.4. Pharmacology

There are many thousands of sulfa drugs that have been synthesized and tested, some of which have very short half-lives and some of which have very long ones. Sulfadoxine has a half-life of from 123–195 hr and dapsone has a half-life of 21–30 hr (Scholer *et al.*, 1984).

Sulfonamides must be used cautiously since sulfa allergies are common. Long-term use of sulfa drugs, including their use in malaria prophylaxis, has been associated with Stevens-Johnson syndrome or toxic epidermal necrolysis, a potentially fatal exfoliative disease (Hernborg, 1985). Dapsone, in addition, becomes metabolized to dapsone monohydroxylamine, which can act as an activated oxygen generator and cause hemolytic anemia in G6PDH-deficient persons (Hjelm and DeVerdier, 1965; Glader and Conrad, 1973). In addition, dapsone can cause agranulocytosis and peripheral neuropathy (Scholer *et al.*, 1984).

2.7. Antifolates: Dihydrofolate Reductase Inhibitors

2.7.1. History and Clinical Use

Biguanides, including proguanil (chlorguanide) (Figure 5), were found to have antimalarial activity in the 1940s. Later, it was discovered that proguanil's antimalarial activity was dependent on its metabolic activation to cycloguanil (Figure 5), which is a competitive inhibitor of dihydrofolate reductase (DHFR). Pyrimethamine (Figure 5), which resembles cycloguanil, was introduced in the 1960s (Hitchings, 1962). Pyrimethamine is combined with sulfadoxine and sold as Fansidar and combined with dapsone and sold as Maloprim.

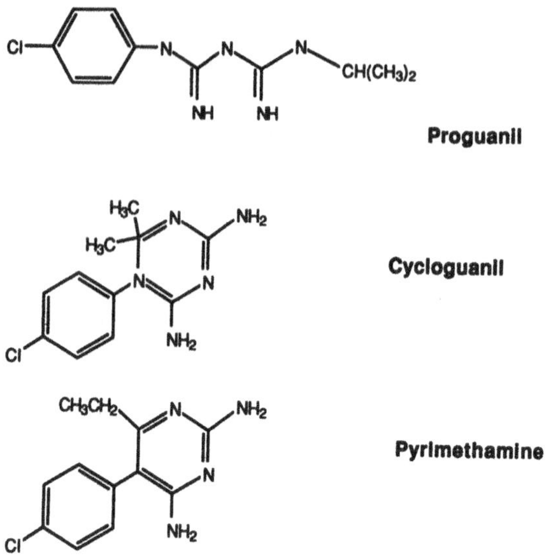

FIGURE 5. Dihydrofolate synthetase inhibitors.

2.7.2. Mechanism of Action

Cycloguanil, like pyrimethamine, is a structural analogue of dihydropteroate, the substrate on DHFR (Figure 4). Both drugs are competitive inhibitors of DHFR and inhibit the parasite DHFR at much lower concentrations than those at which they inhibit mamalian enzymes (Ferone *et al.*, 1969). Dihydrofolate reductase inhibitors are synergistic with the sulfa drugs because both classes of drug inhibit the same pathway, *de novo* folate biosynthesis (Figure 4).

2.8. Artemisinin (Qinghaosu) and Derivatives

2.8.1. History and Clinical Use

Artemisia annua ("qinghao") has been used in China for the treatment of fever since at least 340 A.D. The active antimalarial ingredient of this herb—artemisinin or qinghaosu—was isolated in 1972 (Klayman, 1985) and found to be an unusual sesquiterpene endoperoxide (Figure 6). A variety of artemisinin derivatives (Figure 6) have also been synthesized and have undergone clinical studies. Water-soluble derivatives, such as artesunate, have been tested as oral agents. A new, promising water-soluble derivative, artelinic acid, has demonstrated good activity both *in vitro* and *in vivo* (Lin *et al.*, 1987). In early studies, therapy with both artemisinin and artesunate was associated with a high incidence of recrudescence. Artemether, given intramuscularly in oil, was found to

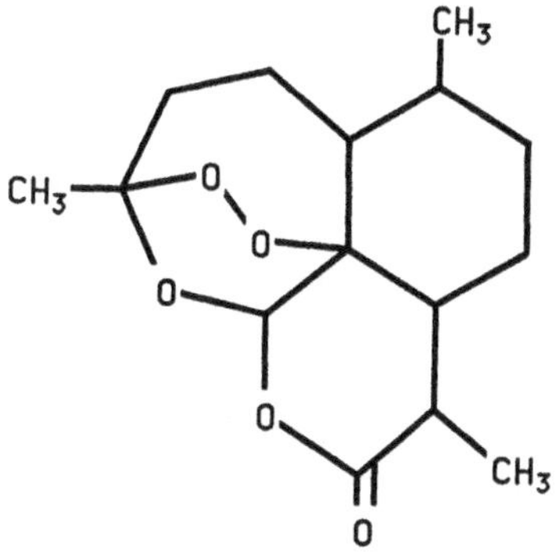

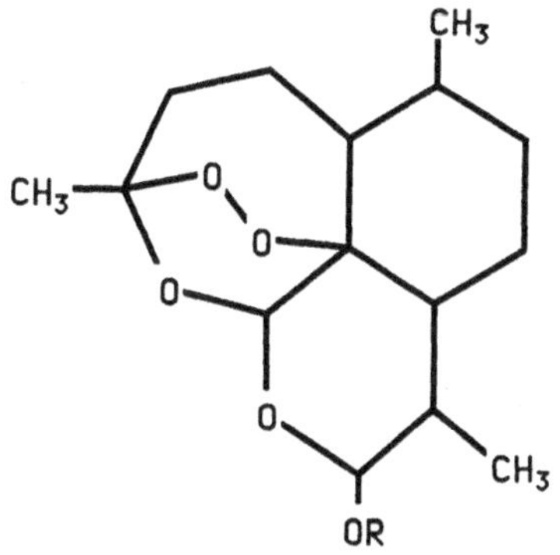

FIGURE 6. Artemisinin and derivatives. R = H, dihydroartemisinin; R = $CO(CH_2)_2CO_2H$, artesunate; R = CH_3, artemether; R = CH_2CH_3, arteether.

be much more effective, with a substantially lower incidence of recrudescence than artemisinin (4% versus 25%, Wang and Xu, 1985). Over two million doses of artemether have been given in China (UNDP/World Bank/WHO, 1991). Artesunate in combination with mefloquine has proven to be particularly effective against multidrug-resistant falciparum malaria in Thailand (Looareesuwan *et al.*, 1992).

2.8.2. Mechanism of Action

Artemisinin contains a potentially labile endoperoxide bridge. There are several pieces of evidence suggesting that the antimalarial activity of artemisinin and its derivatives are mediated by the breakdown of this bridge to generate activated oxygen (i.e., superoxide, hydrogen peroxide, hydroxyl radicals). First, artemisinin derivatives without the endoperoxide bridge are inactive (Brossi *et al.*, 1988). Second, free radical scavengers, such as vitamin E (Krungkrai and Yuthavong, 1987; Levander *et al.*, 1989) and ascorbic acid (Meshnick *et al.*,

1989), antagonize the antimalarial activity of artemesinin *in vitro* and *in vivo*. Third, evidence of lipid peroxidation was found in artemisinin-treated infected red cells *in vitro* (Meshnick *et al.*, 1989).

But why is artemisinin selectively toxic to plasmodia? One reason appears to be the presence of hemozoin—a storage form of hemin—in the parasite. *In vitro,* artemisinin reacts with free hemin to generate activated oxygen and a covalent drug-hemin adduct (Meshnick *et al.*, 1991). This adduct was found in parasites treated with a radiolabelled drug, but not in uninfected red cells (Meshnick *et al.*, 1991), suggesting that the same reaction occurs *in situ* as well. The importance of hemozoin is also indicated by the observation that a chloroquine-resistant strain of *P. berghei* lacking hemozoin was found to be highly resistant to artemisinin (Peters *et al.*, 1986).

Another reason for the selectivity might be selective uptake of the drug by the parasites. *In vitro,* artemisinin is taken up and concentrated by the parasites up to 300-fold compared to the media concentration (Gu *et al.*, 1984).

2.8.3. Resistance

Clinically important artemisinin resistance has not been reported.

2.8.4. Pharmacology

Artemisinin and its derivatives are extremely fast-acting and easily penetrate the blood–brain barrier and are thus particularly useful in the treatment of cerebral malaria. These compounds act more rapidly than either quinine or chloroquine. For example, Wang and Xu (1985) found that artemether cleared parasitemia in about 24 hr, while it took twice as long for parasites to clear after chloroquine treatment (Wang and Xu, 1985).

Artemisinin and its derivatives have short half-lives. They are rapidly metabolized by the liver (Luo and Shen, 1987), and after intravenous injection have half-lives of 30–40 min (White, 1985).

2.9. Miscellaneous Agents

2.9.1. Halofantrine

Halofantrine (Figure 7) is a phenanthrene methanol that was developed by the U.S. Army during World War II, but only underwent clinical trials in 1984. It has demonstrated good efficacy in clinical trials against *P. falciparum* and *P. vivax,* including chloroquine-resistant strains of *P. falciparum*. Its mechanism of action is unknown. The drug is orally active and has a half-life of 1–2 days (Editorial, 1989).

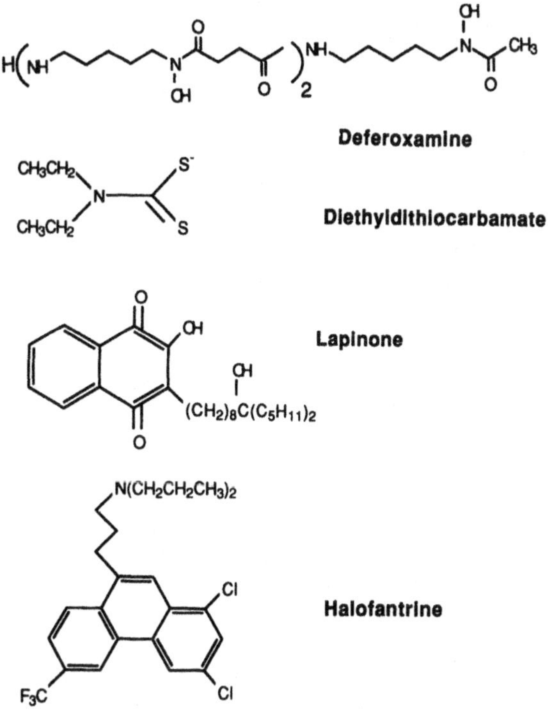

FIGURE 7. Miscellaneous agents.

2.9.2. Chelators

A variety of chelators (compounds which bind metals) have been found to have antimalarial activity *in vitro*. There are two categories of antimalarial chelators: (1) those whose activities are antagonized by added exogenous metal ions, that is, deferoxamine, and (2) those whose activities are potentiated by added metal, that is, diethyldithiocarbamate. The first group of chelators probably act by the removal of a metal ion from a necessary enzyme. The second group of chelators probably act by forming a toxic chelator–metal complex.

Deferoxamine (desferrioxamine) (Figure 7) is a potent iron chelator that is used for the treatment of iron overload. It has demonstrated curative activity against *P. falciparum in vitro* (Raventos-Suarez *et al.*, 1982). Deferoxamine does not act by depriving the parasites of iron (Hershko and Peto, 1988). Instead, its activity appears to depend on its ability to enter *P. falciparum*-infected red cells (Fritsch and Jung, 19986). Macromolecular derivatives of deferoxamine, which cannot pass through membranes, lack antimalarial activity (Scott *et al.*, 1990).

The prospects for clinical development of deferoxamine are limited since the drug can cause many adverse effects and is not orally absorbable (Klaasen, 1985). Several orally absorbable iron chelators, however, have also been shown to have antimalarial activity *in vitro* (Heppner *et al.*, 1988).

Diethyldithiocarbamate (Figure 7) is the breakdown product of disulfiram (Antabuse), which is used to treat alcoholism. It is a potent copper chelator whose antimalarial activity was first demonstrated by Scheibel *et al.* (1979). The antimalarial activity of this compound is potentiated by both extracellular and intraerythrocytic copper (Meshnick *et al.*, 1990a).

Scheibel and Adler (1980, 1982) have identified a number of other chelators with antimalarial activity, including 8-hydroxyquinoline derivatives. The activities of these chelators also appear to be dependent on the formation of toxic metal chelates.

2.9.3. Naphthoquinones

Research on the antimalarial activity of naphthoquinones began during World War II. Lapinone (Figure 7) was effective in patients with *P. vivax* in 1951, but only at high intravenous doses. Menoctone, which is orally absorbable, was tested in 1973 in humans, but was only slightly active (Hudson, 1984). Very good results have been obtained recently with BW566C80, a menoctone derivative, which is currently undergoing clinical trials (Davies *et al.*, 1989).

Naphthoquinones may work either as respiration inhibitors or as oxidant drugs. In mammalian mitochondria, naphthoquinones have been shown to inhibit electron transport, possibly at the ubiquinone locus (Hudson, 1984). In *P. falciparum,* one naphthoquinone, BW58C, has been shown to inhibit pyrimidine biosynthesis, probably by inhibiting the electron-transport-dependent enzyme, dihydroorotate oxidase (Hammond *et al.*, 1985). The antimalarial activity of naphthoquinones might also be mediated by activated oxygen generation, since the naphthoquinone menadione has been shown to exert oxidant stress on *P. falciparum in vitro* (Friedman, 1979).

3. LEISHMANIASIS

3.1. Background

Leishmaniasis is a group of protozoan zoonoses caused by various species of the genus *Leishmania.* These infections occur in the tropical and subtropical areas of the globe. It is estimated that more than 400,000 new cases occur each year with a total of 10–50 million persons affected. The natural cycle is transmitted between phlebotomine sandflies and wild or domestic animals. Humans are infected when they are close enough to the natural reservoir to serve as a substitute host for the biting sandfly. The diseases are chronic and either cutaneous or visceral in presentation. The former is manifest as a persisting ulcer on the exposed parts of the body, which may persist for months or even years. In the Old World, where cutaneous disease is usually caused by *L. major* or *L. aethiopica*, the disease is self-limited and eventually undergoes spontaneous cure. In the New World, where cutaneous disease is caused by strains of the *L. mexicana* or *L. braziliensis* taxonomic groups, a percentage of persons will develop mutilating metastatic infection of the mucosal portions of the nose, mouth, and pharynx. Visceral disease occurs in both hemispheres but is far more common in Africa and India. It is generally caused by *L. donovani* and is principally an infection of the reticuloendothelial system. The leishmaniases are diseases of great public health importance and have been designated as among the six most important human health problems by the World Health Organization.

3.2. Microbiology

Leishmania are dimorphic protozoa. In the sandfly vector, they are motile promastigotes about 15–20 μm in length with an anterior flagellum. They multiply in the midgut of the sandfly and migrate forward to the pharynx and buccal cavity. When the sandfly takes a blood meal from a mammalian host, the promastigotes are deposited into the wound where they enter the tissue macrophages and also pass into the bloodstream. Depending on the species of organism and its ability to withstand body temperature, they may persist in the reticuloendothelial system (visceral disease) or other tissue macrophages (mucocutaneous disease) or the macrophages in the skin at the site of the bite wound (cutaneous disease). The parasites within the macrophages are about one fourth the size of the promastigotes and are termed amastigotes since they have no flagellum visible by light microscopy. These forms replicate within the phagolysosomes of macrophages, probably protected by the unusual mannose polymer that comprises the major portion of their surface membranes. The amastigotes replicate, destroy the macrophage, and reinvade other macrophages in the vicinity. Sandflies complete the cycle when they feed and take up the amastigotes from blood or the skin.

3.3. Pathogenesis

The type of clinical disease is determined to a considerable degree by the species of the parasite. Although the taxonomy of these organisms is changing somewhat are a result of the application of monoclonal antibody, isoenzyme electrophoresis, and DNA hybridization techniques, the principal causes of the syndromes are still those organisms mentioned above. Resistance to the disease and spontaneous cure are probably due to cell-mediated immunity. The leishmaniases are not conceptually different form other infectious processes in which the cell-mediated immune response plays the dominant role; tuberculosis and leprosy are two other examples. In each of these, the organisms can be contained by a host with good nutrition and a vigorous immune response. Leishmaniasis may be self-healing, although parasites may remain in the healed cutaneous lesions. When these lesions are active they have relatively few parasites, a brisk lymphocytic response, and there is a hypersensitivity to leishmanin. This hypersensitivity, along with the vigorous tissue damage that accompanies it, appears to be responsible for the ulcers in the cutaneous form and for the destruction and ulceration in the nasal septum and the soft cartilages of the palate and larynx in the mucocutaneous form. In visceral disease there are more parasites in the tissue, fewer lymphocytes, and an absence of skin reaction to leishmanin. In some instances this progresses to a disseminated diffuse cutaneous leishmaniasis analogous to the diffuse cutaneous disease that occurs in leprosy when the cell-mediated immune response is inadequate.

The manifestations in a given infected person appear also to be a function of genetics. In certain parts of South America, for example, the same species of *L. braziliensis* may cause only cutaneous disease in persons of Indian descent and mucocutaneous disease in others who are descended from African blacks (Walton, 1979).

3.4. Current Treatment

The drug of choice for visceral disease is pentavalent antimony (Pentostam). In adults the dose is 20 mg/kg per day, intravenously or intramuscularly, with a maximum dose of 850 mg/day. The usual course of treatment is 20–30 days and it may need to be repeated several times. The duration of therapy and the clinical response vary from region to region and general statements often are misleading. In the Old World, the ulcer that occurs at the site of in the insect bite tends to be self-contained. The lesions, although disfiguring, heal spontaneously and generally are not treated. In the New-World form of the disease, cutaneous disease is usually self-limiting, but can progress to involve areas of the nasal, oral, and laryngeal mucosa. Unlike its progenitor, it does not heal. In the Western hemisphere, the more common preparation is meglumine antimoniate (Glucantime). This preparation has 85 mg/ml of antimony rather than 100 mg/ml as in Pentostam. For greater detail on the clinical manifestations and management of this disease, see Marr (1992a).

The mechanism of action of pentavalent antimonials is unknown. Although the drug is given in the pentavalent form, it is reduced to the trivalent form in the body before it takes effect (Goodwin and Page, 1943). Chen (1948) proposed that trivalent antimonials inactivate sulfhydryl-containing enzymes. Bueding and Mansour (1957) determined that phosphofructokinase is inhibited by these drugs in the trematode, *Schistosoma mansoni*. Treatment of *L. tropica* promastigotes with trivalent antimony decreases the flow of glucose into the tricarboxylic acid (TCA) cycle and causes accumulation of glycolytic products, suggesting inhibition of enzymes of the TCA cycle. Thus, in *Leishmania*, antimonials appear to inhibit initial steps in glycolysis and certain enzymes of the TCA cycle to decrease energy production by the parasite (Gutteridge and Coombs, 1977).

Although the diamidines (e.g., pentamidine) and amphotericin B have also been used for the treatment of leishmaniasis, they are second-line agents and are not commonly employed (Marr, 1984).

3.5. Experimental Chemotherapy

3.5.1. Purine Metabolism and Purine Analogues in Chemotherapy

Recent studies of purine metabolism in *Leishmania* have demonstrated that their pathways and enzymes have characteristics that differentiate them from their counterparts in humans. Moreover, exploration of various purine analogues, in particular the pyrazolopyrimidines, has led to the development of two compounds that have been successful not only in laboratory models but also in clinical studies.

The metabolism of purines in protozoan parasites is significantly different from that in humans (Figure 8) while pyrimidine metabolism is substantially the same (Marr and Berens, 1985). In humans, purines are synthesized *de novo* from nonnucleotide precursors (amino acids, carbon dioxide, and ammonia) as well as from preformed purine bases via the salvage pathways. The general route for *de novo* biosynthesis is the same in all of the species of mammals, birds, yeasts, and bacteria that have been studied. The pathway is a sequence of reactions by which the purine ring is assembled on carbon-1 of ribose-5′-phosphate, leading to the formation of inosinic acid (IMP). Neither free purines nor nucleosides appear as intermediates in this sequence. Inosinic acid is then converted to AMP or to GMP, with subsequent conversion to nucleoside di- and triphosphates and nucleic acids.

Nucleotides also can be formed from free purines and purine nucleosides. These routes are regarded as salvage pathways that permit the reutilization of purines from the degradation of nucleic acid or nucleotides. These pathways are present in both humans and pathogenic protozoans. In the latter they are the only means of synthesizing purine nucleotides. Free purines react with phosphoribosyl-1-pyrophosphate (PRPP) to yield nucleoside-5-monophosphate. These reversible reactions are catalyzed by distinct phosoribosyltransferases (PRTases) for the activation of a particular purine base. Since the pyrophosphate released in these reactions is hydrolyzed rapidly, the synthesis of purine nucleosides proceeds irreversibly. Other salvage pathways involve the conversion of free purines

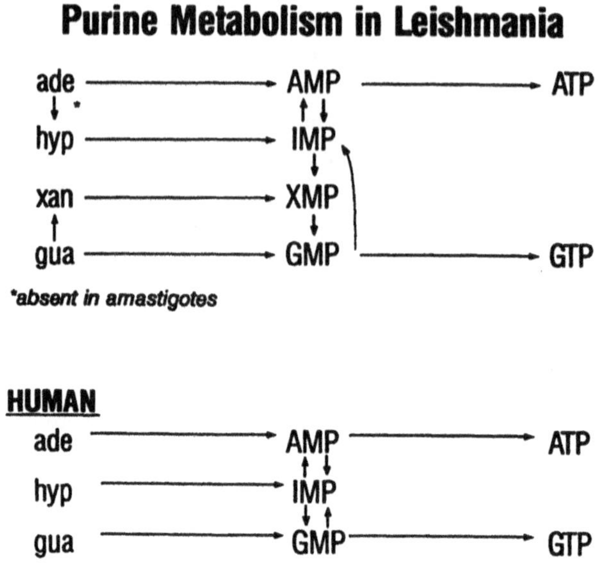

FIGURE 8. Summary of purine metabolism in humans and *Leishmania*.

to nucleosides and then to nucleotides. The former reactions are catalyzed by purine nucleoside phosphorylase, in which ribose-1-phosphate serves as the ribose donor and inorganic phosphate is released along with the purine nucleoside. Conversion of the nucleoside to the nucleotide occurs through the addition of a phosphate group to the 5-position on the ribose moiety by a nucleoside kinase or a nucleoside phosphotransferase. The former is usually relatively specific for a given purine nucleoside while the latter is relatively nonspecific with respect to both the donor and recipient of the phosphate group.

In humans, the first step in the *de novo* synthetic pathway and many steps in the salvage pathways are under metabolic regulation. Unfortunately, there is little presently known about regulation of purine and pyrimidine biosynthesis and degradation in the parasitic protozoans. There is also little known about the biosynthesis or degradation of nucleic acids in these organisms, except that mRNA degradation is accelerated at 38 °C (Marr and Berens, 1985) and by the exposure of promastigotes to pyrazolopyrimidines (Marr and Berens, 1983). Purine nucleotides that are released from either RNA or DNA are degraded by nonspecific phosphatases or nucleotidases to yield nucleosides and inorganic phosphate. This is true for both parasitic protozoans and humans. The nucleosides are acted on by nucleoside-cleaving enzymes, which can be either nucleoside phosphorylases (which require phosphate as a participant in the reaction to yield a purine and ribose-1-phosphate) or nucleoside hydrolases (which use water to produce the base and free ribose) (Tuttle and Krenitsky, 1980). Both types of enzymes exist in humans and pathogenic protozoans. The purine base is not known to be degraded further in protozoans and probably is reutilized in the salvage pathways. In humans, the purine ring usually is salvaged as described, but small amounts are partially degraded to urea and allantoin. Most of the bases that are not salvaged are deaminated and then oxidized by xanthine oxidase to uric acid and then excreted. There appears to be no xanthine oxidase in protozoans.

In summary, purine metabolism in the pathogenic protozoans appears to offer several opportunities for chemotherapy: (1) Since there is no *de novo* synthesis of these compounds, interdiction of the salvage pathway has far greater implications than interruption of similar pathways in humans; (2) some of the enzyme systems in the hemoflagellates are capable of accepting purine analogues and metabolizing them to nucleotide analogues, which serve as metabolic inhibitors.

The biochemical differences between these organisms and humans are significant enough that one can predict that purine analogues should be effective inhibitors of these organisms. Indeed, this prediction has been borne out already with respect to the hemoflagellates *Leishmania donovani, L.b.panamensis,* and *Trypanosome cruzi.*

3.5.2. Purine Metabolism in *Leishmania*

3.5.2.1. Promastigotes. Purine metabolism in *Leishmania* is similar to that found in mammalian cells except for the means of entry into the metabolic scheme (Figure 8). The primary means of ingress is through the PRTases. *L. donovani* has three distinct enzymes: HGPRTase, XPRTase, and APRTase. The finding of an XPRTase is unusual (Marr and Berens, 1985; Tuttle and Krenitsky, 1980). Mammalian tissues do not contain this enzyme. The major substrates for the PRTases appear to be hypoxanthine and xanthine. There is a very active adenase, which deaminates adenine to hypoxanthine. The investigations using radiolabeled substrates shows that the data for adenine and hypoxanthine are identical. This indicates that the former is deaminated to the latter before entry into the salvage pathways. The same is true for guanine and xanthine. Guanine is deaminated rapidly by a guanase to xanthine. The latter is then activated through the PRTase and enters the salvage pathway. The inosine and guanosine nucleosides can interchange. There is also an active GMP reductase and an AMP deaminase (Figure 8).

There are active nucleoside-cleaving activities in *Leishmania*. There is an adenosine phosphorylase (Konigk and Rasoul, 1978) and an inosine nucleosidase (Konigk, 1978). *Leishmannia tropica* contains a nucleotidase that has a broad substrate specificity (Pereira and Konigk, 1981). A study by Kozalka and Krenitsky (1979) has provided detailed information on the nucleosidases of *L. donovani*. One is a novel purine 2-deoxyribonucleosidase; another is specific for the purine nucleosides inosine and guanisine; and a third cleaves both purine and pyrimidine ribonucleosides with a clear preference for the latter.

3.5.2.2. Amastigotes. Studies by Looker and co-workers (1983), using isolated and purified amastigotes from hamster spleen, described the purine pathways in these forms. Purine metabolism in the amastigote is similar to that in the promastigote except for adenine and adenosine metabolism. In these forms, there is no adenase. Adenine is activated to AMP by an APRTase. This is a significant departure from purine metabolism in the promastigote. Less than 20% of adenine label can be found in guanine nucleotides compared to 45% when hypoxanthine is the precursor. In addition, the relative flow is 55% to AMP and 45% to GMP; this differs from the flow in promastigotes where IMP is converted 75% to AMP and 25% to GMP. The metabolism of guanine, xanthine, and their respeçtive nucleosides is similar to that described for the promastigote.

3.5.2.3. Therapeutic Implications. The therapeutic implications of purine metabolism in these organisms are quite substantial (Berens *et al.*, 1980; Looker *et al.*, 1983). Their HGPRTase is relatively nonspecific: it will accept pyrazolopyrimidines (Figure 9) as though they were purines.

The best studied of these is allopurinol (Figure 9). This hypoxanthine analogue is activated to the nucleotide analogue of IMP. The nucleotide is an excellent inhibitor of succinyl-AMP synthase. The net effect is to strongly inhibit this enzyme and prevent the formation of AMP from purine precursors. In addition, this nucleotide analogue inhibits the GMP reductase. This prevents the conversion of GMP to AMP, which might help to overcome the metabolic block. Thus, allopurinol brings about virtually complete inhibition of purine biosynthesis in these organisms and halts growth. In addition, the IMP analogue is converted to an AMP analogue, which is then phosphorylated to the ATP analogue. This aminopyrazolopyrimidine nucleotide analogue is incorporated into the RNA of the parasite (Figure 10). Once incorporated into RNA it brings about the net efflux of mRNA from the polymer to individual nucleotides. This net breakdown of mRNA probably accounts, to a large degree, for the observed inhibition of protein synthesis. The aminated compounds appear to account for the antiparasitic action (Looker *et al.*, 1986).

These biochemical findings have been translated into clinical benefits. There have been several clinical reports of the benefits of allopurinol in visceral leishmaniasis and one study using allopurinol riboside in cutaneous leishmaniasis (Saenz *et al.*, 1989). In the latter study, allopurinol riboside was shown to be as effective as glucantime when the total cure rate (cure plus relapse) was considered. In a study of cutaneous leishmaniasis in Colombia, allopurinol was shown to be more than twice as effective as glucantime (Martinez and Marr, 1992).

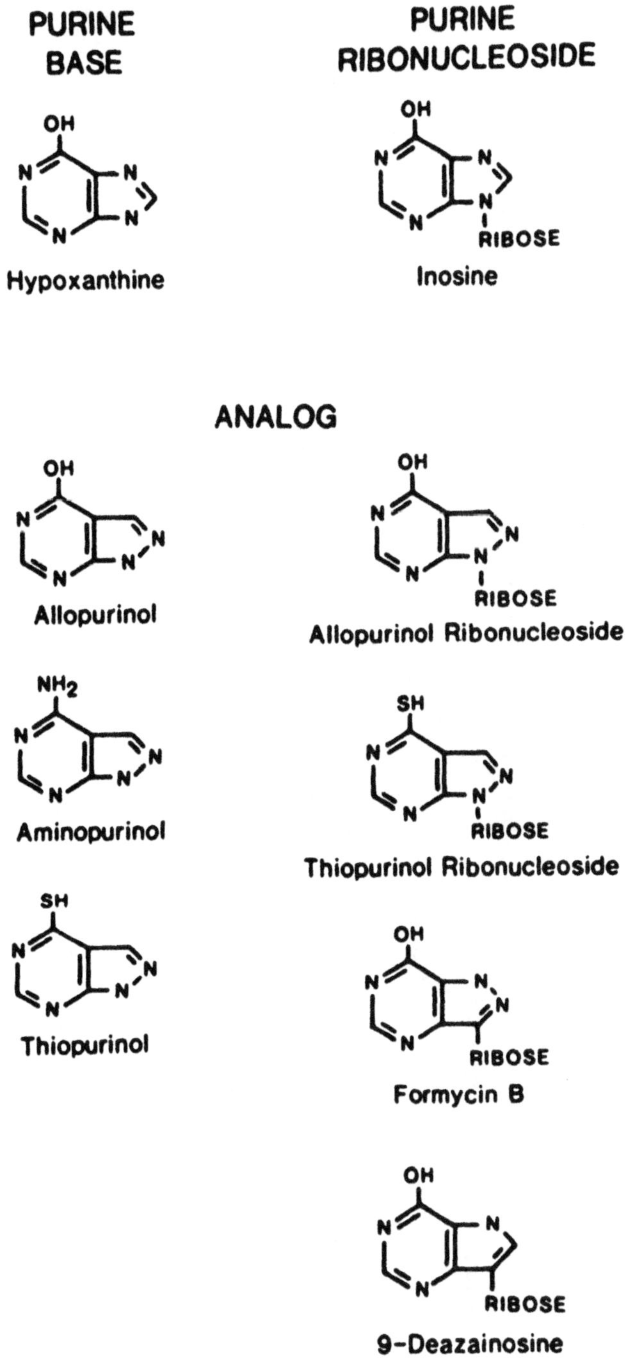

FIGURE 9. Structures of purines and purine analogues.

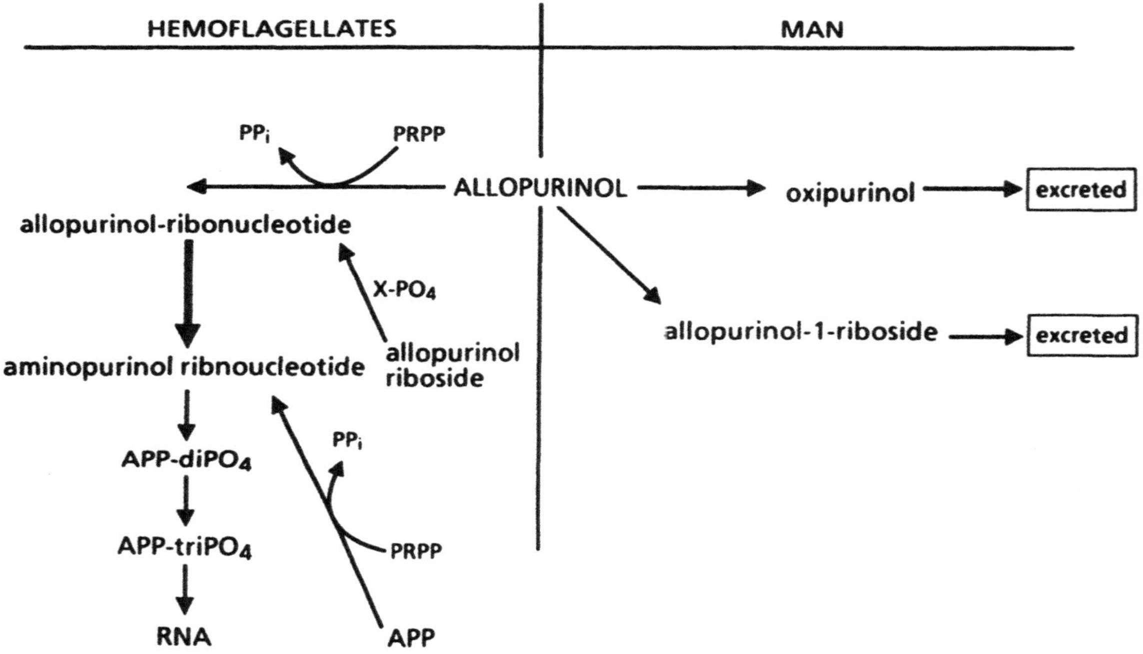

FIGURE 10. Metabolism of pyrazolopyrimidines in hemoflagellates and humans. PP_i, inorganic pyrophosphate; PRPP, phosphoribosylpyrophosphate; X-PO_4, phosphate donor; APP-diPO_4, 4-aminopyrazolo-[3,4-*d*]-pyrimidine ribonucleotide triphosphate; APP, 4-aminopyrazolo-[3,4-*d*]-pyrimidine.

4. AMERICAN TRYPANOSOMIASIS (CHAGAS' DISEASE)

Chagas' disease is a public health problem primarily in rural areas; the incidence of new infections is not known, since the majority of cases go undiagnosed. The mean age at which infection occurs in areas where transmission is common is quite low, probably around four to six years. In one survey it was found that 85% of acute cases occurred in children under the age of 10 (Laranja *et al.*, 1956). The case fatality rate associated with new infections is probably less than 8%, but in this particular study the case fatality rate was 12%. Although the prevalence of chronic infections due to *Trypanosoma cruzi* is not known with certainty, it has been estimated that 10–12 million persons are infected. More than one half of those infected live in Brazil (Brener, 1982). A recent study from that country showed that of 1.5 million people from rural areas, approximately 10% had serologic evidence of infection (Silveira, 1982).

4.1. Background

Trypomastigotes are capable of infecting a wide rage of cells. Once they are within the intracellular milieu, they transform into amastigotes, which reside in the cytoplasm. The amastigotes replicate by binary fission and then differentiate into trypomastigotes, which are released from the cells. These circulate in the bloodstream and then enter a new tissue. The acute phase of the infection lasts one to two months in humans and is characterized by extensive tissue parasitism and by a prominent parasitemia. In the chronic infection that follows the acute phase, the detection of parasites, either in the bloodstream or in tissues, is difficult. The disease process appears to be held in check, presumably by the action of the cell-mediated immune response.

Since this pathogenic hemoflagellate is biochemically and biologically similar to the host in many respects, it is not surprising that compounds toxic to *T. cruzi* are likewise toxic to the host cell. The high therapeutic index of agents used in the treatment of bacterial diseases is absent in the treatment of Chagas' disease. Some excellent molecular biological and biochemical studies are being directed to these functions, and our understanding of the organism is expanding. Certain aspects of the biology of *T. cruzi* have relevance for chemotherapy.

4.2. Compounds of Current Clinical and Experimental Interest

4.2.1. Agents Capable of a Reductive Metabolism: Nitro Compounds

Reductive metabolism is the means by which the trypanocidal action of many biologically active compounds is expressed (Docampo and Moreno, 1984a and b; Marr and Docampo, 1986). The selectivity of this action depends on several factors, including cellular uptake, subcellular localization of the agent,

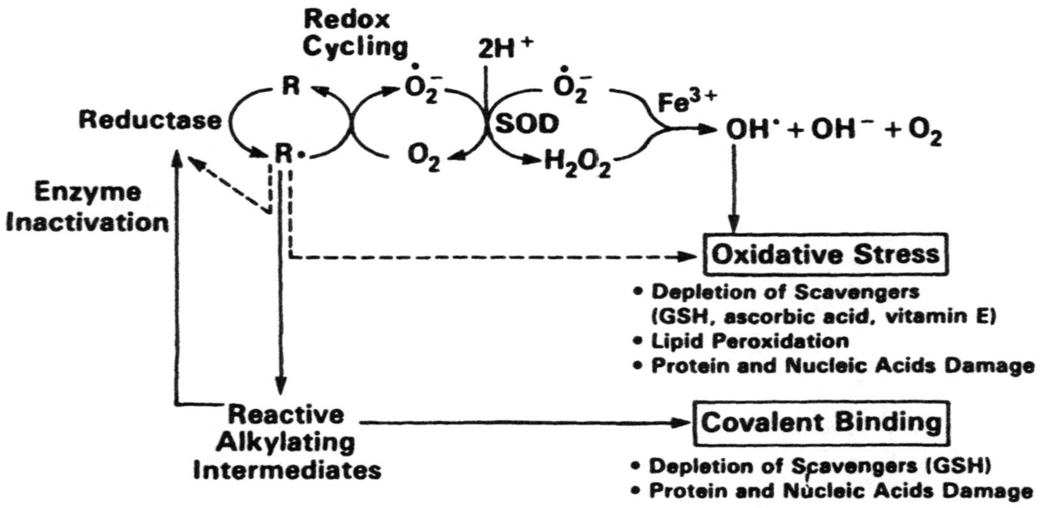

FIGURE 11. Scheme of redox cycling of reactive alkylating intermediates of different pharmacologic agents leading to either oxidative stress or covalent binding and cellular damage. SOD, superoxide dismutase; GSH, glutathione. From Marr and Do Campo (1986). Reproduced with permission.

oxidant defenses (e.g., catalase, superoxide dismutase, glutathione), amplifying reagents (e.g., oxygen), and ultimately, the availability of enzymes competent to initiate the reductive process. Reductive metabolism appears frequently as an adventitious process—the results of a chance encounter between a drug possessing an oxidizing functional group and an enzyme capable of transferring electrons to that group. When the reductases are available and cell toxicity occurs, it is typically the consequence of one of two possible mechanisms; the production of potential alkylating intermediates or redox cycling at the expense of molecular oxygen (Figure 11).

A wide variety of chemicals with vastly different structures may undergo reduction in *T. cruzi* (Marr and Docampo, 1986). One-electron reduction is particularly important for the trypanocidal action of several compounds because it results in the formation of free radicals (compounds with an odd or spin-unpaired electron in their outer orbital; Figure 11). Because of the thermodynamic potential of the unpaired electron to form an electron pair, free-radical intermediates are extremely reactive. They can add across double bonds, abstract hydrogen from other molecules, or combine with themselves to form dimers (Pryor, 1976). In addition, free-radical intermediates can activate molecular oxygen by univalent reduction (redox cycling) to the superoxide anion (O_2-), which in turn can produce H_2O_2 either spontaneously or by the action of the enzyme superoxide dismutase (SOD). Redox-active metals such as iron then catalyze the reaction of O_2- and H_2O_2 to produce hydroxyl radicals ($OH\cdot$), which is one of the most damaging radical species (see Section 2.2.2) (Halliwell and Gutteridge, 1989). The formation of free radicals (drug-derived and oxygen-derived radicals) may lead to extensive cellular damage through the production of secondary and tertiary free radicals from lipids, amino acids, and components of nucleic acids. In addition, reduction of some drugs (e.g., some nitro compounds) may result in the formation of other reactive intermediates—not necessarily free radicals—capable of binding and alkylating proteins or nucleic acids and interfering with different cellular functions.

4.2.1.1. Nifurtimox. Packchanian (1957) demonstrated that nitrofuran compounds were active against *T. cruzi*. The following year Mintzer *et al.* (1953) reported on the chemotherapeutic effects of nitrofurantoin. Since that time a number of nitrofuran derivatives have been tested and reviews of the structure–activity relationships have been written by Steck (1981) and Grunberg and Titsworth (1973). The clinical use of these compounds has been reviewed by Brener (1979, 1982). The only one of these agents that has achieved any measure of success in the treatment of Chagas' disease is nifurtimox (Figure 12). This agent is in use for the treatment of Chagas' disease in humans, particularly in Argentina, Brazil, and Chile. The results (parasitologic cure) of therapy have varied from 80 to 86% and vary with the population studied. The efficacy of *nifurtimox* in chronic Chagas' disease has not been established. Adverse effects

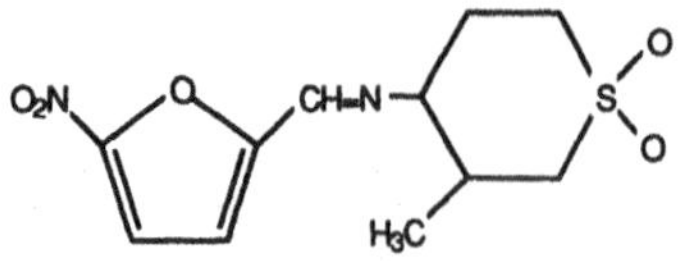

Nifurtimox

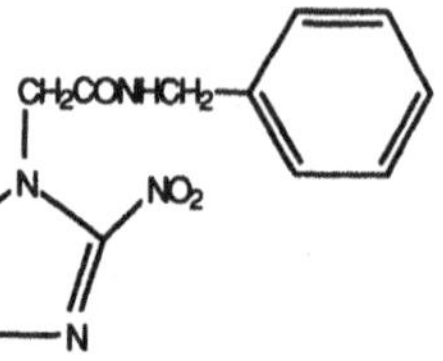

Benznidazole

FIGURE 12. Antitrypanosomal nitro compounds.

of this compound occur in 40–70% of patients. Most of the adverse reactions are related to the central nervous system and the gastrointestinal tract. Nausea and vomiting are common, as are weight loss and significant myalgias with stiffness and weakness. Convulsions, parathesias, polyneuritis, and psychotic symptoms may occur.

There is general agreement that metabolic reduction of the nitro group is of central importance in the toxicity of nitro compounds. The first step in the reduction of nitro compounds is the formation of the nitro anion radical ($R\text{-}NO_2$). The intermediate ($R\text{-}NO_2$) is the first in a sequence of reduction products any of which, except the amine product, may be responsible for the cytotoxicity. The reaction proceeds as follows:

$$R\text{-}NO_2 \xrightarrow{e-} R\text{-}NO_2^- \xrightarrow{2H+} R\text{-}NO \xrightarrow[2H+]{2e-} R\text{-}NHOH \xrightarrow[2H+]{2e-} R\text{-}NH_2$$

It seems likely that some of the cytotoxic effects of nifurtimox on *T. cruzi* are mediated by the generation of activated oxygen. It also seems likely that oxidative stress is responsible for some of the nifurtimox-induced toxic effects that have been observed in animals. It is not clear what contribution the metabolic activation of nifurtimox makes to the drug's mutagenic and possible carcinogenic effects in intact cells or tissues. So far no evidence is available on covalent binding of nifurtimox metabolites to cell macromolecules (Docampo and Moreno, 1985).

4.2.1.2. Benznidazole. The only other agent used with any frequency for the treatment of Chagas' disease is benznidazole (Figure 12). As with nifurtimox, it remains controversial whether benznidazole is curative. In addition, adverse effects such as progressive "purpuric" dermatitis, polyneuropathy, digestive symptoms, and psychic disturbances are usually also observed after prolonged treatment with benznidazole. In spite of many similarities in the mode of action of nitrofurans and nitroimidazoles in several biologic systems (Mason, 1982; Docampo and Moreno, 1984a and b), nifurtimox and benznidazole seem to differ significantly in their mechanisms of toxicity to *T. cruzi* and mammalian cells.

Addition of high concentrations of benznidazole to NADH-containing homogenates of *T. cruzi* epimastigotes results in the appearance of a weak electron spin resonance (ESR) signal corresponding to the nitro anion radical. However, no nitro anion radical formation is observed on incubation of benznidazole with NAD(P)H and subcellular fractions of *T. cruzi,* and the generation of O_2- and H_2O_2 is not stimulated under the same experimental conditions used to detect their formation by nifurtimox (Moreno *et al.*, 1982). In addition, benznidazole inhibits *T. cruzi* respiration, in contrast to the stimulation produced by nifurtimox. This rules out any major role for redox cycling of benznidazole under physiologic conditions. Moreover, since benznidazole inhibits growth of *T. cruzi* at concentrations that do not stimulate generation of O_2- and H_2O_2, it is unlikely that the trypanocidal action of the drug depends on the effect of oxygen radicals (Moreno *et al.*, 1982).

In summary, nitro compounds are toxic both to *T. cruzi* and to the mammalian host. The particular mechanism whereby a nitro compound is toxic can vary somewhat from one compound to another, but the ultimate result is the same. It is presumably for this reason that these compounds, as active as they are against trypanosomes, have such poor therapeutic indices. Unless highly specific compounds that are active only with enzyme systems in the trypanosome can be developed, it will be difficult to separate efficacy from toxicity. The toxicity of the niitro compounds is apparently related to the presence of the nitro group, which is also responsible for the antiparasitic activity.

4.2.2. Purine Analogues

4.2.2.1. Purine Metabolism: Epimastigotes. Studies in which purine-defined culture conditions and incubation with radiolabeled precursors were used have shown that epimastigotes of *T. cruzi* lack the pathways of *de novo* purine synthesis and must rely on salvage of exogenous purines (Berens *et al.*, 1981b; Gutteridge and Gaborak, 1979; Gutteridge *et al.*, 1979; Hammond and Gutteridge, 1984). Gutteridge and Gaborak (1979) showed that epimastigotes incorporated both purine bases and ribonucleosides, and that adenine and guanine nucleotides were readily interconvertible. The results of the study by Ceron *et al.* (1979) were essentially identical. Berens *et al.* (1981b), using more sensitive methods, evaluated the relative significance of the various purine salvage pathways and found that all four bases served as purine sources. Quantitatively, adenine and hypoxanthine were used more efficiently than either guanine or xanthine. The former two compounds were incorporated preferentially into ade-

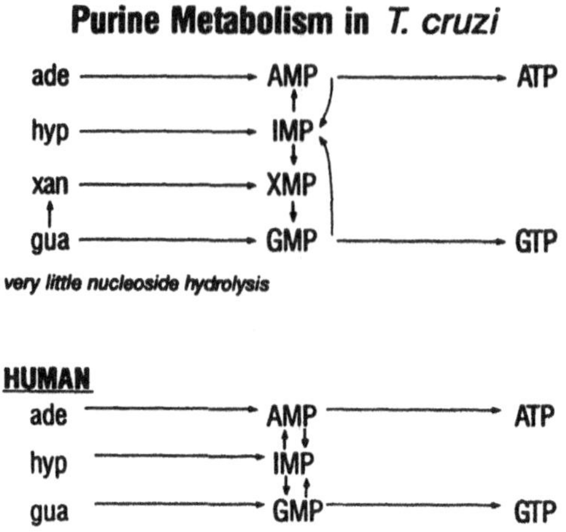

FIGURE 13. Summary of purine metabolism in *T. cruzi.*

nine nucleotides, whereas the latter were used equally for both guanine and adenine nucleotide synthesis.

A number of individual enzymes of purine salvage have been investigated (Figure 13). Gutteridge and Davies (1981, 1982) reported the presence of phosphoribosyl-transferase (PRTase) activities for adenine, hypoxanthine, and guanine. A second protein which was active with both hypoxanthine and guanine, was found to be associated with the glycosome. The finding that *T. cruzi* is capable of converting allopurinol to its ribonucleoside (Marr *et al.*, 1978) indicates that it contains a PRTase that is similar to that found in *L. donovani* in its ability to react with pyrazolopyrimidines.

4.2.2.2. Purine Metabolism: Trypomatigote and Amastigote Forms. Much less work has been done with the trypomastigote and amastigote forms because of the difficulties involved in obtaining them. Gutteridge and Gaborak (1979) reported that both trypomastigotes and amastigotes are unable to synthesize the purine ring *de novo* and are dependent on salvage pathways. A more detailed investigation of the individual salvage enzymes (Gutteridge and Davies, 1981) showed that the trypomastigote, amastigote, and epimastigote are qualitatively identical.

Berens *et al.* (1981b) reported that both trypomastigotes and amastigotes metabolize allopurinol in a manner identical to that of the epimastigotes: this suggests that the substrate specificities of hypoxanthine phosphorbosyl transferase (HPRTase), adenylosuccinate synthetase, and adenylosuccinate lyase are similar in these three forms of *T. cruzi*. From these investigations it appears that there are no major differences in purine metabolism among the morphologic forms of the parasite.

The ability of pyrazolopyrimidines and other purine analogues to inhibit the growth of *T. cruzi* has implications for both basic biochemistry and chemotherapy. Since *T. cruzi* cannot synthesize purines *de novo,* it is dependent on preformed bases and nucleosides for nucleic acid synthesis. There are other important differences between *T. cruzi* and the mammalian host with respect to transport of the purine analogues across the cell membrane and their subsequent phosphorylation, amination, and incorporation into RNA (Figure 10). The mammalian cell does not concentrate these analogues to the same degree as does *T. cruzi,* perhaps because the mammalian cell does not phosphorylate them to inosine 5′-monophosphate (IMP) analogues and, therefore, does not maintain the necessary concentration gradient.

The important purine analogues that are active against *T. cruzi* fall into two groups: (1) those that are related to the purine base hypoxanthine and (2) those that are related to the purine ribonucleoside inosine (Figure 9). Of the former group, allopurinol was the first described and it remains the most potentially important (Marr *et al.*, 1978; Berens *et al.*, 1981a). All of the morphologic forms of *T. cruzi* metabolize allopurinol as described above (Figure 10). It undergoes this same sequence of conversions in forms of *T. cruzi* infecting vertebrates. Infection with this organism in both mice and tissue cultures can be eliminated with the compound (Marr *et al.*, 1978; Avila *et al.*, 1980; Avila and Avila, 1981; Berens *et al.*, 1981a). The aminated metabolites, which appear to be more toxic, are formed only by the parasite and not by the mammalian host. When studied in a tissue culture model, allopurinol at a concentration of 25 μg/ml was able to cure a cell culture infected with *T. cruzi* (Berens *et al.*, 1982). This indicates that this compound can eradicate the infection even in the absence of an immune response.

Avila and Avila (1981) showed that the administration of allopurinol intraperitoneally protected mice from the acute form of Chagas' disease when the drug was given two to six days after infection in C57B1-6J mice. Parasitemia was undetectable by direct methods for 310 days. Inoculation of blood from these mice into susceptible mice, however, showed that there were small numbers of circulating trypanosomes. Mice that received a second schedule of treatment were cured of infection after 275 days. These data provide evidence for the activity of this pyrazolopyrimidine against *T. cruzi.* Although the early parasitemia can be eliminated, the fact that almost a year is required for the parasite to be completely eliminated from the hosts suggests that cure in this animal model was due to the action of allopurinol in concert with the immune system.

Is this activity of allopurinol uniform across all species or strains and is it active against all morphologic forms of the parasite? Studies published by Avila and co-workers (Avila *et al.*, 1981, 1984) and Cruz *et al.* (1980) have shown that the trypomastigote is insensitive to this compound. This finding suggests that pyrazolopyyrimidine nucleotides act only on dividing cells and probably are active only on the intracellular amastigote forms or during the transformation of trypomastigotes to amastigotes.

Avila and co-workers (1981) have shown that the incorporation of allopurinol may vary from one strain of *T. cruzi* to another, and this variation is reflected in their sensitivities to the compounds. These differences in sensitivity may be related to transport into the cell or to the rates of phosphorylation or amination of the drug. Studies by Avila *et al.* (1984) and by Spector *et al.* (1982) have shown that there is some correlation between sensitivity of *T. cruzi* to allopurinol and the ability of the organism to aminate the compound. At this time, the mechanism(s) of partial resistance to this agent remains unclear.

4.2.2.3. Therapeutic Implications. The therapeutic implications for *T. cruzi* are similar to those for *Leishmania.* The metabolic pathways, with respect to pyrazolopyrimidines, are essentially the same. Inhibition of growth by allopurinol is rapid and can be shown not only *in vitro,* but also in tissue culture and in animals (Looker *et al.*, 1986; Avila and Avila, 1981).

A recent clinical study has shown that allopurinol at doses of 600 and 900 mg per day is equal in efficacy to both nifurtimox and benznidazole. All patients studied had chronic, asymptomatic Chagas' disease. Allopurinol was as efficacious as the nitro drugs as measured by three serologic tests and by xenodiagnosis. Allopurinol had the additional advantage of having little or no toxicity; the nitro drugs showed evidence of significant toxicity in approximately 30% of patients (Gallerano *et al.*, 1990). Two large clinical studies of allopurinol have been undertaken by the World Health Organization to verify these findings.

5. TOXOPLASMOSIS

5.1. Background

Toxoplasmosis is a protozoan infection that is widespread in nature. The organism, *Toxoplasma gondii,* can infect all orders of mammals, and the disease is actually a zoonosis for which humans are incidental hosts. This obligate intracellular parasite usually causes asymptomatic infection and the disease, when it occurs in normal hosts, is of short duration and mild in its manifestations. After infection, the cysts persist in the tissues of the person, but this chronic infection does not cause disease unless the individual becomes immunocompromised. Toxoplasmosis is conceptually similar to other intracellular infections that are contained by cell-mediated immunity and that manifest as disease only when such immunity is suppressed.

5.2. Microbiology

Toxoplasma gondii is an obligate intracellular protozoan parasite that is classified among the *Sporozoa*. It exists in three forms: the tachyzoite, the cyst (bradyzoite), and the oocyst. The tachyzoite, formerly termed the trophozoite, is a crescent-shaped organism that can invade virtually any mammalian cell except the erythrocyte. It resides within a vacuole in the cell and divides until the host cell ruptures to release the new organisms. These enter the intestinal epithelium and reinvade to repeat the cycle or they can disseminate throughout the body to invade other cells and form cysts therein. This period of replication and dissemination is associated with acute disease. The cyst, which can develop within any tissue, can contain thouusands of organisms. Cysts probably are viable for the life of the host and are the source of organisms when the infection reactivates to produce disease in an immunocompromised person. These forms are generally found in muscle, including cardiac muscle, and in brain. The oocyst, released in the intestine, is of major importance in propagating the organism. It is found only in felines, with the domestic cat being the major animal to disseminate the organism.

5.3. Treatment

The treatment of this disease varies with the clinical circumstances. Immunologically normal patients generally do not require treatment unless there is specific organ dysfunction such as myocarditis, pneumonitis, or encephalitis. The immunocompromised patient usually responds to specific treatment with pyrimethamine and sulfadiazine. The usual course of treatment is pyrimethamine, 25 mg/day, annd sulfadiazine, 6–8 gm/day, both until 4–6 weeks after resolution of all signs and symptoms. The above agents are active against tachyzoites and are synergistic. The tissue cyst form is resistant to therapy. For further description of the manifestations and management of toxoplasmosis, see Marr (1992b).

The mechanism of action of the dihydrofolate reductase inhibitors and sulfonamides are discussed in the Sections 2.6 and 2.7. The *T. gondii* dihydropteroate synthetase and dihydrofolate reductase have recently been characterized (Allegra *et al.*, 1990; Kovacs *et al.*, 1990).

Other agents are available to treat this disease but none is as effective as the two mentioned above. Spiromycin is less toxic than pyrimethamine plus sulfadiazine and has been used in Europe for treatment in pregnant women. It may be effective in the treatment of infants with congenital infection. It is not available in the United States except by request of the Food and Drug Administration. The combination of trimethoprim and sulfamethoxazole is less active than the

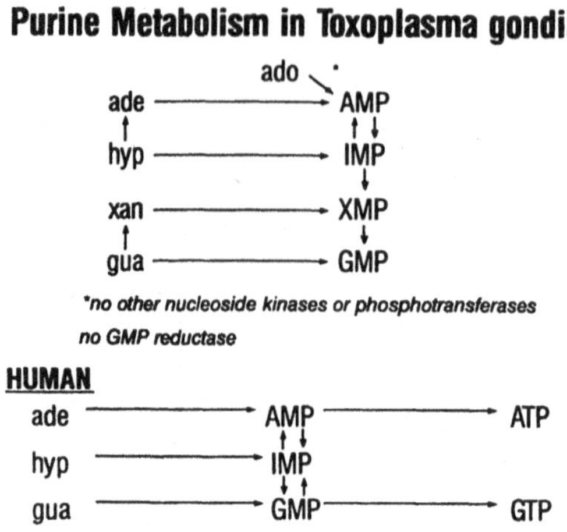

FIGURE 14. Summary of purine metabolism in *T. gondii*.

combination of pyrimethamine and sulfadiazine. Clindamycin has been shown to be effective against the disease in a rabbit model and in some human infections. Its mechanism of action is presumed to be inhibition of protein synthesis, as in bacteria.

Recent studies in purine metabolism in these organisms have demonstrated that, like other intracellular protozoan parasites, they have no *de novo* purine biosynthesis. They do have salvage pathways and all of the PRTase activities (Figure 14). The adenosine kinase is a particularly active enzyme. Of particular interest with respect to chemotherapy is the fact that these organisms cannot cross from the guanosine to the adenosine nucleoside pathways (Krug *et al.*, 1989). The exact relationship of purine metabolism to chemotherapy is unclear. The inosine analogues, which have proven effective in leishmaniasis and Chagas' disease, are not effective against *T. gondii in vitro*. This aspect of chemotherapy remains to be explored.

It has been shown that artemisinin (qinghaosu) (see Section 2.8 and Figure 6) is effective against these organisms *in vitro* and in a tissue culture model (Ke *et al.*, 1990). The mechanism of action of this compound has been discussed in Section 2. At present this remains an interesting finding but has not been confirmed in animal tests.

6. REFERENCES

Aikawa, M., 1972, High-resolution autoradiography of malarial parasites treated with ^{3}H-chloroquine, *Am. J. Pathol.* **67**:277–280.
Allahyari, R., Strother, A., Fraser, I. M., and Verbiscar, A. J., 1984, Synthesis of certain hydroxy

analogs of the antimalarial drug priamquine and their *in vitro* methemoglobin-producing and glutathione-depleting activity in human erythrocytes, *J. Med. Chem.* **27**:407–410.

Allegra, C., Boarman, J. D., Kovacs, J. A., Morrison, P., Beaver, J., Chabner, B. A., and Masur, H., 1990, Interaction of sulfonamide and sulfone compounds with Toxoplasma gondii dihydropteroate synthase, *J. Clin. Invest.* **85**:371–379.

Avila, J. L., and Avila, A., 1981, *Trypanosoma cruzi:* Allopurinol in the treatment of mice with experimental acute Chagas disease, *Exp Parasitol.* **51**:204–28.

Avila, J. L., Avila, A., and Monzon, H., 1984, Differences in allopurinol and 4-aminopyrazolo (3,4-d)pyrimidine metabolism in drug sensitive and insensitive strains of *Trypanosoma cruzi, Mol. Biochem. Parasitol.* **11**:51–60.

Avila, J. L., Avila, A., and Munoz, E., 1981, Effect of allopurinol on different strains of *Trypanosama cruzi, Am. J. Trop. Med. Hyg.* **30**:769–774.

Bass, G. E., Hudson, D. R., Parker, J. E., and Purcell, W. P., 1971, Mechanism of antimalarial activity of chloroquine analogs from quantitative structure–activity studies. Free enery related model, *J. Med. Chem.* **14**:275–283.

Bates, M. D., Meshnick, S. R., Sigler, C. I., Leland, P., and Hollingdale, M. R., 1990, *In vitro* effects of primaquine and primaquine metabolites on exoerythrocytic stages of *Plasmodium berghei, Am. J. Trop. Med. Hyg.* **42**:532–537.

Berens, R. L., and Nelson, D. J., 1981, Purine metabolism in *T. cruzi, Mol. Biochem. Parasitol.* **3**:187–196.

Berens, R. L., Marr, J. J., Nelson, D. J., and LaFon, S. W., 1980, Antileishmanial effect of allopurinol and allopurinol ribonucleoside on the intracellular forms of *Leishmani donovani:* Biology and biochemical action, *Biochem. Pharmacol.* **29**:2397–2402.

Berens, R. L., Marr, J. J., da Cruz, F. S., and Nelson, D. J., 1989, Effect of allopurinol on *Trypanosoma cruzi:* Metabolism and biological activity in intracellular and blood stream forms, *Antimicrob. Agents Chemother.* **22**:657–661.

Berens, R. L., Marr, J. J., LaFon, S. W., and Nelson, D. J., 1981b, Purine metabolism in *Trypanosoma cruzi, Mol. Biochem. Parasitol.* **3**:187–197.

Berens, R. L., Marr, J. J., Cruz, F. S., and Nelson, D. J., 1982, Metabolism of allopurinol by intracellular and bloodstream forms *Trypanosoma cruzi,* Antimicrob. Agents Chemother. **22**:657–661.

Bitonti, A. J., Sjoerdsma A., McCann, P. P., Kyle, D. E., Oduola, A. M. J., Rossan, R. N., Milhous, W. K., and Davidson, D. E., 1988, Reversal of chloroquine resistance in malaria parasite *Plasmodium falciparum* by desipramine, *Science* **242**:1301–1303.

Boudreau, E. F., Webster, H. K., Pavanand, K., and Thosingha, L., 1982, Type II mefloquine-resistance in Thailand, *Lancet* **2**:1335.

Breckenridge, A., Back, D. J., Edwards, I. G., Mihaly, G., Orme, M., Purba, H., and Ward, S., 1984, The clinical and biochemical pharmacology of primaquine, in *Primaquine: Pharmacokinetics, Metabolism, Toxicity, and Activity* (W. H. Wernsdorfer and P. I. Trigg, eds.), pp. 65–76, Wiley, New York.

Brener, Z., 1982, Recent developments in the field of Chagas' disease, *Bull. WHO* **60**:463–473.

Brossi, A., Venugopaplan, B., Gerpe, L. D., Yeh, H. J. C., Flippen-Anderson, J. L., Luo, X. D., Milhous, W., and Peters, W., 1988, Arteether, a new antimalarial drug: Synthesis and antimalarial properties, *J. Med. Chem.* **31**:645–650.

Bueding, E., and Mansour, J. M., 1957, The relationship between inhibition of PFK activity and the mode of action of trivalent organic antimonials on *Schistosoma mansoni, Br. J. Pharmacol.* **12**:159–163.

Carson, P. E., 1984, 8-Aminoquinolines, in *Antimalarial Drugs II* (W. Peters and W. H. G. Richards, eds.), pp. 83–121, Springerr Verlag, Berlin.

Ceron, C. R., Caldas, R. D. A., Felix, C. R., Mundim, M. H., and Roitman, I., 1979, Purine metabolism in trypanosomatids, *J. Protozool.* **26:**479–483.

Chen, G., 1948, Effects of arsenicals and antimonials on the activity of glycolytic enzymes in lysed preparations of *Trypanosoma equiperdium, J. Inf. Dis.* **82:**226–231.

Chevli, R., and Fitch, C. D., 1982. The antimalarial drug mefloquine binds to membrane phospholipids, *Antimicrob. Agents Chemother.* **21:**581–586.

Chou, A. C., Chevli, R., and Fitch, C. D., 1980, Ferriprotoporphyrin IX fulfills the criteria for identification as the chloroquine receptor of malaria parasites, *Biochemistry* **19:**1543–1549.

Clark, I. A., Chaudri, G., and Cowden, W. B., 1989, Some roles of free radicals in malaria, *Free Radicals Biol. Med.* **6:**315–321.

Cruz, F. S., Marr, J. J., and Berens, R. L., 1980, Prevention of transfusion-induced Chagas' disease by amphotericin B, *Am. J. Trop. Med. Hyg.* **29:**761–765.

Davidson, M. W., Griggs, B. G., Boykin, D. W., and Wilson, W. D., 1977, Molecular structure effects involved in the interaction of quinolinemethanols with DNA. Implications for antimalarial action, *J. Med. Chem.* **20:**1117–1122.

Davies, C. S., Pudney, M., Matthews, P. J., and Sinden, R. E., 1989, The causal prophylactic activity of the novel hydroxynaphthaquinone 566C80 against Plasmodium berghei infections in rats, *Acta Leiden.* **58:**115–128.

Docampo, R., and Moreno, S. N. J., 1984a, Free radical metabolites in the mode of action of chemotherapeutic agents and phagocytic cells on *Trypanosoma cruzi, Rev. Infect. Dis.* **6:**223–238.

Docampo, R., and Moreno, S. N. J., 1984b, Free radical intermediates in the trypanocidal action of drugs and phagocytes, in *Oxygen Radicals in Chemistry and Biology* (M. Saran and D. Tait, eds.), pp. 749–751, Walter de Gruyter, Berlin.

Docampo, R., and Moreno, S. N. J., 1985, Biochemical toxicology of antiparasitic compounds used in the chemotherapy and chemoprophylaxis of American trypanosomiasis (Chagas' disease), *Rev. Biochem. Toxicol.* **7:**159–204.

Editorial, 1989, Halofantrine in the treatment of malaria, *Lancet* **2:**537–538.

Estensen, R. D., Krey, A. K., and Hahn, F. E., 1969, Studies on a deoxyribonucleic acid–quinine complex, *Mol. Pharmacol.* **5:**532–541.

Fairfield, A. S., Abosch, A. Ranz, A., Eaton, J. W., Meshnick, S. R., 1986a, Oxidative defense enzymes of *Plasmodium falciparum, Mol. Biochem. Parasitol.* **30:**77–82.

Fairfield, A., Eaton, J. W., and Meshnick, S. R., 1986b, Superoxide dismutase and catalase in the murine malaria *Plasmodium berghei:* Content and subcellular distribution, *Arch. Biochem. Biophys.* **250:**526–629.

Ferone, R., 1984, Dihydrofolate reductase inhibitors, in *Antimalarial Drugs II* (W. Peters and W. H. G. Richards, eds.), pp. 207–221, Springer Verlag, Berlin.

Ferone, R., Burchall, J. J., and Hitchings, G. H., 1969, *Plasmodium berghei* dihydrofolate reductase. Isolation, properties, and inhibition by antifolates, *Mol. Pharmacol.* **5:**49–59.

Fitch, C. D., 1972, Chloroquine resistance in malaria: Drug binding and cross resistance patterns, *J. Parasitol.* **39:**265–271.

Fitch, C. D., 1986, Antimalarial schizonticide: Ferriprotoporphyrin IX interaction hypothesis, *Parasitol. Today* **2:**330–331.

Fitch, C. D., Yunis, N. G., Chevli, R., and Gonzales, Y., 1974, High-affinity accumulation of chlroquine by mouse erythrocytes infected with *Plasmodium berghei, J. Clin. Invest.* **54:**24–43.

Foote, S. J., Kyle, D. E., Martin, R. K., Oduola, A. M., Forsyth, K., Kemp, D. J., and Cowman, A. F., 1990, Several alleles of the multidrug-resistance gene are closely linked to chloroquine resistance in *Plasmodium falciparum, Nature* **345:**255–258.

Friedman, M. J., 1979, Oxidant damage mediates variant red cell resistance to malaria, *Nature* **280:**245–247.

Fritsch, G., and Jung, A., 1986, ^{14}C-Desferrioxamine B: Uptake into erythrocytes infected with *Plasmodium falciparum*, *Z. Parasitenkd.* **72:**709–713.

Gallerano, R. H., Sosa, R. R., and Marr, J. J., 1990, Therapeutic efficacy of allopurinol in patients with chronic Chagas' disease, *Am. J. Trop. Med. Hyg.* **43:**(2):146–158.

Ginsburg, H., Kutner, S., Krugliak, M., and Cabantchik, Z. I., 1985, Characterizattion of permeation pathways appearing in the host membrane of *Plasmodium falciparum* infected red blood cells, *Mol. Biochem. Parasitol.* **14:**312–322.

Glader, B. E., and Conrad, M. E., 1973, Hemolysis by diphenylsulphones: Comparative effects of DDS and hydroxylamine-DDS, *J. Lab. Clin. Med.* **81:**267–272.

Goldberg, D. E., Slater, A. F. G., Cerami, A., and Henderson, G. B., 1990, Hemoglobin degradation in the malaria parasite *Plasmodium falciparum:* An ordered process in a unique organelle, *Proc. Natl. Acad. Sci. USA* **87:**2931–2953.

Goodwin, L. G., and Page, J. E., 1943, A study of the eccretin of certain antimonials using a polarographic procedure, *Biochem. J.* **37:**198–205.

Grunberg, E., and Titsworth, E. H., 1973, Chemotherapeutic properties of heterocyclic compounds: Monocyclic compounds with five-membered rings, *Annu. Rev. Microbiol.* **27:**317–346.

Gu, H. M., Warhurst, D. C., and Peters, W., 1984, Uptake of [^{3}H]dihydroartemisinin by erythrocytes infected with *Plasmodium falciparum in vitro, Trans. R. Soc. Trop. Med. Hyg.* **78:**265–270.

Gutteridge, W. E.., and Coombs, G. H., 1977, *Biochemistry of Parasitic Protozoa,* University Park Press, Baltimore.

Gutteridge, W. E., and Davies, M. J., 1981, Enzymes of purine salvage in *Trypanosoma cruzi, FEBS Lett.* **127:**211–214.

Gutteridge, W. E., and Davies, M. J., 1982, Properties of the purine phosphoribosyltransferases of *Trypanosoma cruzi, FEMS Microbiol. Lett.* **13:**207–212.

Gutteridge, W. E., and Gaborak, M., 1979, A re-examination of purine and pyrimidine synthesis in the three main forms of *Trypanosoma cruzi, Int. J. Biochem.* **10:**415–422.

Gutteridge, W. E., Dave, D., Richards, W. H. G., 1979, Conversion of dihydroorotate to orotate in parasitic protozoa, *Biochim. Biophys. Acta* **582:**390–401.

Halliwell, B., and Gutteridge, J. M. C., 1989, *Free Radicals in Biology and Medicine,* Clarendon Press, Oxford.

Hammond, D. J., and Gutteridge, W. E., 1984, Purine and pyrimidine metabolism in trypanosomatidae, *Mol. Biochem. Parasitol.* **13:**243–261.

Hammond, D. J., Burchell, J. R., and Pudney, M., 1985, Inhibition of Pyrimidine biosynthesis de novo in Plasmodium Falciparum by 2-(4-■+-butycycloyhexyl)-3-hydroxy-1, 4-naphthoquinone *in vitro, Mol. Biochem. Parasitol.* **14:**97–109.

Hatton, C. S., Peto, T. E., Bunch, C., Pasvol, G., Rusell, S. J., Singer, C. R., Edwards, G., and Winstanley, P., 1986, Frequency of severe neutropenia associated with amodiaquine prophylaxis against malaria, *Lancet* **1:**411–414.

Heppner, D. G., Hallaway, P. E., Kontoghiorghes, G. J., and Eaton, J. W., 1988, Antimalarial properties of orally active iron chelators, *Blood* **72:**358–361.

Hernborg, A., 1985, Stevens-Johnson syndrome after mass prophylaxis with sulfadoxine for cholera in Mozambique, *Lancet* **2:**1072–1073.

Hershko, C., and Peto, T. E., 1988, Deferoxamine inhibition of malaria is independent of host iron status, *J. Exp. Med.* **168:**375–387.

Hitchings, G. H., 1962, Daraprim as an antagonist of folic and folinic acids, *Trans. R. Soc. Trop. Med. Hyg.* **46:**467–473.

Hjelm, M., and DeVerdier, C. H., 1965, Biochemical effects of aromatic amines I. Methemoglobinemia, hemolysis, and Heinz body formation induced by 4,4′-diaminodiphenylsulfone, *Biochem. Pharmacol.* **14:**1119–1128.

Hoffheinz, W., and Merkli, B., 1984, Quinine and quinine analogs, in *Antimalarial Drugs II* (W. Peters and W. H. G. Richards, eds.), pp. 61–81, Springer Verlag, Berlin.

Homewood, C. A., Warhurst, D. C., Peters, W., and Baggaley, V. C., 1972, Lysosomes, pH, and the antimalaria action of chloroquine, *Nature* **235**:50–52.

Hudson, A. T., 1984, Lapinone, menoctone, hydroxyquinolinequinones and similar structures, in *Antimalarial Drugs II* (W. Peters and W. H. G. Richards, eds.), pp. 343–361, Springer Verlag, Berlin.

Irvin, J. L., and Irvin, E. M., 1947, Spectrophotometric and potentiometric evaluation of apparent acid dissociation exponents of various 4-aminoquinolines, *J. Am. Chem. Soc.* **69**:1091–1099.

Irvin, J. L., Irvin, E. M., and Parker, F. S., 1949, The interaction of antimalarials with nucleic acids, *Science* **110**:426–428.

Ke, O.-Y., Krug, E. C., Marr, J. J., and Berens, R. L., 1990, Inhibition of *Toxoplasma gondii* growth by qinghaosu and derivatives, *Antimicrob. Agents Chemother.* **34**:1961–1965.

Keystone, J. S., 1990, Prevention of malaria, *Drugs* **39**:337–354.

Klaasen, C. D., 1985, Heavy metals and heavy metal antagonists, in *The Pharmacological Basis of Therapeutics*, 7th ed. (A. G. Gilman, L. S. Goodman, T. W. Rall, and F. Murad, eds.), pp. 1605–1607, Macmillan, New York.

Klayman, D. L., 1985, Quinghaosu (artemisinin): An antimalarial drug from China, *Science* **228**:1049–1055.

Konigk, E., 1978, Purine nucleotide metabolism in promastigotes of *Leishmania tropica:* Inhibitory effect on allopurinol and analogues of purine nucleosides, *Tropenmed. Parasitol.* **29**:435–438.

Konigk, E., and Rasoul, S. A., 1978, Catabolism of adenosine-5′-monophosphate in promastigotes of *Leishmania tropica, Tropenmed. Parasitol.* **28**:319–322.

Kovacs, J. A., Allegra, C. J., and Masur, H., 1990, Characterization of dihydrofolate reductase of Pneumocystis carinii and Toxoplasma gandii, *Exp. Parasitol.* **71**;60–68.

Kozalka, G. W., and Krenitsky, T. A., 1979, Nucleosidases from *Leishmania donoani, J. Biol. Chem.* **254**:8185–8193.

Krogstad, D. J., and Schlesinger, P. H., 1987, The basis of antimalarial action: Non-weak base effects of chloroquine on acid vesicle pH, *Am. J. Trop. Med. Hyg.* **36**:213–220.

Krogstad, D. J., Gluzman, I. Y., Kyle, D. E., Odvola, A. M. J., Martin, S. K., Milhjouse, W. K., and Schlesinger, P. H., 1987, Efflux of chloroquine from *Plasmodium falciparum:* Mechanism of chloroquine resistance, *Science* **238**:1283–1285.

Krug, E., Marr. J. J., and Berens, R. L., 1989, Purine metabolism in *Toxoplasma gondii, J. Biol. Chem.* **264**:10601–10607.

Krungkrai, S. R., and Yuthavong, Y., 1987, The antimalarial action on *Plasmodium falciparum* of qinghaosu and artesunate in combination with agents which modulate oxidant stress, *Trans. R. Soc. Trop. Med. Hyg.* **81**:710–714.

Kwakye-Berko, F., and Meshnick, S. R., 1989, Binding of chloroquine to DNA, *Mol. Biochem. Parasitol.* **35**:51–56.

Kwakye-Berko, F., and Meshnick, S. R., 1990, Sequence preference of chloroquine binding to DNA and prevention of Z-DNA formation, *Mol. Biochem. Parasitol.* **39**:275–278.

Laranja, F. S., Diaz, E., Nobrega, G., and Miranda, A., 1956, Chagas' disease: Clinical, epidemiologic, and pathologic study, *Circulation* **14**:1035–1060.

Levander, O. A., Ager, A. L., Morris, V. C., and May, R. G., 1989, Qinghaosu, dietary vitamin E, selenium, and cod liver oil: Effect on the susceptibility of mice to the malarial parasite *Plasmodium yoelii, Am. J. Clin. Nutr.* **50**:346–352.

Lin, A. J., Klayman, D. L., and Milhaus, W. K., 1987, Antimalarial activity of new water-soluble dihydroartemisinin derivatives, *J. Med. Chem.* **30**:2147–2150.

Looareesuwan, S., Viravan, C., Varijanonta, S., Wilairatana, P., Suntharasamai, P., Charoenlarp, P., Arnold, K., Kyle, D., Canfield, C., and Webster, K., 1992, Randomized trial of artensunate and mefloquine alone and in sequence for acute uncomplicated falciparum malaria, *Lancet* **i**:821–824.

Looker, D. L., Berens, R. L., and Marr, J. J., 1983, Purine metabolism in *Leishmania donovani* in amastigotes and promastigotes, *Mol. Biochem. Parasitol.* **9**:15–28.

Looker, D. L., Martinez, S., Horton, J. M., and Marr, J. J., 1986, Growth of *Leishmania donovani* amastigotes in a continuous human macrophage culture, *J. Inf. Dis.* **154**:323–327.

Luo, X.-D., and Shen, C.-C., 1987, The chemistry, pharmacology, and clinical applications of qinghaosu (artemisinin) and its derivatives, *Med. Res. Rev.* **7**:29–52.

Malin, A. S., and Hall, A. P., 1990, Falciparum malaria resistant to quinine and pyrimethamine-sulfadoxine successfully treated with mefloquine, *Br. Med. J.* **300**:1175.

Marr, J. J., 1984, Chemotherapy of leishmaniasis, in *Parasitic Diseases,* Vol. 2, *The Chemotherapy* (J. M. Mansfield, ed.), pp. 201–229, Marcel Dekker, New York and Basel.

Marr, J. J., and Berens, R. L., 1983, Pyrazolopyrimidine metabolism in the pathogenic trypanosomatidae, *Mol. Biochem. Parasitol.* **7**(4):339–356.

Marr, J. J., and Berens, R. L., 1985, Purine metabolism in *Leishmania,* in *Parasitic Diseases,* Vol. 1 (Leishmaniasis) (K. P. Chang and R. S. Bray, eds.), pp. 69–78, Elsevier, Amsterdam.

Marr, J. J., and Docampo, R., 1986, Chemotherapy for Chagas' disease: A perspective of current therapy and considerations for future research, *Rev. Infect. Dis.* **8**:893–904.

Marr, J. J., Berens, R. L., and Nelson, D. J., 1978, Antitrypanosomal effect of allopurinol: Conversion *in vivo* to aminopyrazolopyrimidine nucleotides by *Trypanosoma cruzi, Science* **201**:1018–1020.

Marr, J. J., 1992a, Leishmaniasis, in *Textbook of Internal Medicine,* (W. N. Kelley, ed.), pp. 1525–1527, J. B. Lippincott Company, Philadelphia.

Marr, J. J., 1992b, Toxoplasmosis, in *Textbook of Internal Medicine,* (W. N. Kelley, ed.), pp. 1535–1539, J. B. Lippincott Company, Philadelphia.

Martin, S. K., Oduola, A. M., and Milhous, W. K., 1987, Reversal of chloroquine resistance in *Plasmodium falciparum* by verapamil, *Science* **235**:899–901.

Martinez, S., and Marr, J. J., 1992, Allopurinol in the treatment of cutaneous leishmaniasis, *N. Engl. J. Med.* **326**:741–744.

Mason, R. P., 1982, Free-radical intermediates in the metabolism of toxic chemicals, in *Free radicals in biology,* Vol. 5 (W. A. Pryor, ed.), pp. 161–222, Academic Press, New York.

McChesney, W. W., and Fitch, C. D., 1984, 4-Aminoquinolines, in *Antimalaral Drugs II* (W. Peters and W. H. G. Richards, eds.), pp. 4–59, Springer Verlag, Berlin.

Meshnick, S. R., 1990, Chloroquine as intercalator: A hypothesis revived, *Parasitol. Today* **6**:77–79.

Meshnick, S. R., Tsang, T. W., Lin, F. B., Pan, H. Z., Chang, C. N., Kuypers, F., Chiu, D., and Lubin, B., 1989, Activated oxygen mediates the antimalarial activity of qinghaosu, *Prog. Clin. Biol. Res.* **313**:95–104.

Meshnick, S. R., Scott, M. D., Lubin, B. H., Ranz, A., and Eaton, J. W., 1990a, The antimalarial activity of diethyldithiocarbamate: Potentiation by copper, *Biochem. Pharmacol.* **40**:213–216.

Meshnick, S. R., Thomas, A., Ranz, O., Xu, C. M., and Pon, H. Z., 1991, Artemisinin (qinghaosu): The role of intracellular hemin in its mechanism of antimalarial action, *Mol. Biochem. Parasitol.* **49**:181–190.

Milhous, W. K., Gerena, L., Kyle, D. E., and Oduola, A. M. J., 1989, *In vitro* strategies for circumventing antimalarial drug resistance, *Prog. Clin. Biol. Res.* **313**:61–72.

Mintzer, S., Kadison, E. L., Shales, W. M., and Felsenfeld, O., 1953, Treatment of urinary tract infections with a new antibacterial nitrofuran, *Antibiot. Chemother.* **3**:151–157.

Moreno, S. N. J., Docampo, R., Mason, R. P., Leon, W., and Stoppani, A. O. M., 1982, Different behaviors of benznidazole as free radical generator with mammalian and *Trypanosoma cruzi* microsomal preparations, *Arch. Biochem. Biophys.* **218**:585–591.

O'Brien, R. L., and Hahn, F. E., 1965, Chloroquine structural requirements for binding to deoxynucleic acid and antimalarial activity, *Antimicrob. Agents Chemother.* 315–320.

Packchanian, A., 1952, Chemotherapy of experimental Chagas' disease with nitrofuran compounds, *Antibiot. Chemother.* **7:**13–23.

Panisko, D. M., and Keystone, J. S., 1990, Treatment of Malaria—1990, *Drugs* **39:**160–189.

Pereira, N. M., and Konigk, E., 1981, A nucleotidase from *Leishmania tropica* promastigotes: Partial purification and properties, *Tropenmed. Parasitol.* **32:**209–214.

Peters, W., Lin, L., Robinson, B. L., and Warhurst, D. C., 1986, The chemotherapy of rodent malaria XL: The action of artemisinin and related sequiterpenes, *Ann. Trop. Med. Hyg.* **80:**483–489.

Peters, W., 1987, *Chemotherapy and Drug Resistance in Malaria,* vols. I and II, Academic Press, London.

Peters, W., and Richards, W. H. G., 1984, *Antimalarial Drugs,* vols. I and II, Springer Verlag, Berlin.

Peterson, D. S., Milhous, W. K., and Wellems, T. E., 1990, Molecular basis of differential resistance to cycloguanil and pyrimethamine in *Plasmodium falciparum malaria, Proc. Natl. Acad. Sci. USA* **87:**3018n–3022.

Polet, H., and Barr, C., 1968, Chloroquine and dihydroquinine. *In vitro* studies of their antimalarial effect upon *Plasmodium knowlesii, J. Pharmacol. Exp. Ther.* **164:**380–386.

Pollack, S., 1989, *P. falciparum* iron metabolism, *Prog. Clin. Biol. Res.* **313:**151–161.

Pryor, W. A., 1976, The role of free radical reactions in biological systems, in *Free Radicals in Biology,* Vol. 1 (W. A. Pryor, ed.), pp. 1–49, Academic Press, New York.

Ranz, A., and Meshnick, S. R., 1989, *Plasmodium falciparum:* Inhibitor sensitivity of the endogenous superoxide dismutase, *Exp. Parasitol.* **69:**125–128.

Raventos-Suarez, C., Pollack, S., and Nagel, R. L., 1982, *Plasmodium falciparum:* Inhibition of *in vitro* growth by desferrioxamine, *Am. J. Trop. Med. Hyg.* **31:**319–922.

Rodriguez, M. H., and Jungery, M., 1986, A protein on *Plasmodium falciparum* infected erythrocytes functions as a transferrin receptor, *Nature* **324:**388–391.

Roth, E. F., Raventos-Suarez, C., Ribaldi, A., and Nagel, R., 1983, Glucose-6-phosphate dehydrogenase deficiency inhibits growth of *Plasmodium falciparum, Proc. Natl. Acad. Sci. USA* **80:**298–299.

Saenz, R. E., Paz, H. M., Johnson, C. M., Marr, J. J., Nelson, D. J., Pattishall, K. H., and Rogers, M. D., 1989, Treatment of cutaneous leishmaniasis with orally administered allopurinol riboside, *J. Infect. Dis.* **160:**153–158.

Scheibel, L. W., and Adler, A., 1980, Antimalarial activity of selected aromatic chelators, *Mol. Pharmacol.* **18:**320–325.

Scheibel, L. W., and Adler, A., 1982, Antimalarial activity of selected aromatic chelators. III. 8-hydroxyquinoline (oxine) substituted in positions 5 and 7 and oxines annelated in position 5,6 by an aromatic ring, *Mol. Pharmacol.* **22:**140–144.

Scheibel, L. W., Adler, A., and Trager, W., 1979, Tetraethylthiuram disulfide (Antabuse) inhibits the human malaria parasite *Plasmodium falciparum, Proc. Natl. Acad. Sci. USA* **76:**5303–5307.

Scholer, H. J., Leimer, R., and Richle, R., 1984, Sulphonamides and sulphones, in *Antimalarial Drugs II* (W. Peters and W. H. G. Richards, eds.), pp. 123–206, Springer Verlag, Berlin.

Scott, M. D., Ranz, A., Kuypers, F. A., Lubin, B. H., and Meshnick, S. R., 1990, Parasite uptake of deferoxamine: A prerequisitte for antimalarial activity, *Br. J. Hematol.* **75:**598–602.

Sherman, I. W., 1984, Metabolism, in *Antimalarial Drugs I* (W. Peters and W. H. G. Richards, eds.), pp. 21–81, Springer Verlag, Berlin.

Silveira, A. C., 1982, Epidemiolgia da doenca da Chagas, in *Epidemiologic das Doencas Endemicas* [A. Prata (Sucam Ministerial da Saude Brazil), ed.] Editoria Universidad de Brasilia, Brazil.

Slater, A., and Cerami, A., 1992, Inhibition by chloroquine of a novel haem polymerase enzyme activity in malaria trophozoites, *Nature* **355**:167–169.

Spector, T., Berens, D. L., and Marr, J. J., 1982, Adenylosuccinate synthetase and adenylosuccinate lyase from *Trypanosoma cruzi;* Specificity studies with potential chemotherapeutic agents, *Biochem. Pharmacol.* **31**:225–230.

Steck, E. A., 1971, *The Chemotherapy of Protozoan Diseases,* Walter Reed Army Institute of Medicine, Washington.

Steck, E. A., 1981, The chemotherapy of protozoal infections: Whither man? *J. Protozool.* **28**:30–35.

Strother, A., Allahyari, R., Buchholz, X., Fraser, I. M., and Tilton, B. E., 1984, *In vitro* metabolism of the antimalarial agent primaquine by mouse liver enzymes and identification of a methemoglobin-forming metabolite, *Drug Metabl. Disposit.* **12**:35–44.

Sturchler, D., 1989, How much malaria is there world-wide? *Parasitol. Today* **5**:39.

Sturchler, D., Handschin, J., Kaiser, D., Kerr, L. Mittelholzer, M. L., Reber, R., and Fernex, M., 1990, Neuropsychiatric side effects of mefloquine, *N. Engl. J. Med.* **322**:1752–1753.

Sweeney, T. R., 1984, Drugs with quinine-like action, in *Antimalarial Drugs II* (W. Peters and W. H. G. Richards, eds.), pp. 267–324, Springer Verlag, Berlin.

Sweeney, T. R., and Strube, R. E., 1980, Antimalarials, in *Burger's Medicinal Chemistry,* Fourth ed. (M. Wolff, ed.), pp. 333–413, Wiley, New York.

Tuttle, J. B., and Krenitsky, T. A., 1980, Purine phosphoribosyltransferases from *Leishmania donovani, J. Biol. Chem.* **255**:909–916.

UNDP/World Bank/WHO Special Programme for Research and Training in Tropical Diseases, 1991, *Tropical Diseases: Progress in Research, 1989–1990,* pp. 29–40, World Health Organization, Geneva, Switzerland.

Walton, B. C., 1979, Racial differences in espundia, *Ann. Trop. Med.. Parasitol.* **73**:23–28.

Wang, T. Y., and Xu, R. C., 1985, Clinical studies of the treatment of falciparum malaria with artemether, a derivative of qinghaosu, *J. Trad. Chin. Med.* **5**:240–242.

Warhurst, D. C., 1981, The quinine-hemin interaction and its relationship to antimalarial activity, *Biochem. Pharmacol.* **30**:3323–3327.

Warhurst, D. C., 1986, *Parasitol. Today* **2**:331–334.

Warhurst, D. C., 1987, Antimalarial drugs. An update, *Drugs* **33**:50–65.

Watkins, W. M., Oloo, J. A., Lury, J. D., Mosoba, M., Kariuki, D., Mjumba, M., Koech, D. K., and Gilles, H. M., 1988, Effects of multiple-dose halofantrine in treatment of chloroquine-resistant malaria in children in Kenya, *Lancet* **2**:247–251.

Watkins, W. M., Sixsmith, D. G., Chulay, J. D., and Spencer, H. C., 1984, Antagonism of sulfadoxine and pyrimethamine antimalarial activity *in vitro* by *p*-aminobenzoic acid, *p*-aminobenzoylglutamic acid, and folic acid, *Mol. Biochem. Parasitol.* **14**:55–61.

Wellems, T. E., Panton, L. J., Gluzman, I. Y., do Rosario, V. E., Gwadz, R. W., Walker-Jonah, A., and Krogstad, D. J., 1990, Chloroquine resistance not linked to *mdr*-like genes in a *Plasmodium falciparum* cross, *Nature* **345**:253–255.

White, N. J., 1985, Clinical pharmacokinetics of antimalarial drugs, *Clin. Pharmacokinet.* **10**:187–215.

Zhang, Y., and Hempelmann, E., 1987, Lysis of malarial parasites and erythrocytes by ferriprotoporphyrin IX-chloroquine and the inhibition of this effect by proteins, *Biochem. Pharmacol.* **36**:1267–1273.

Zhang, Y., and Meshnick, S. R., 1991, Inhibition of the *P. falciparum* dihydropteroate synthetase and of parasite growth *in vitro, Antimic. Agents Chemother.* **35**:267–271.

Index

GPSR Compliance

The European Union's (EU) General Product Safety Regulation (GPSR) is a set of rules that requires consumer products to be safe and our obligations to ensure this.

If you have any concerns about our products, you can contact us on ProductSafety@springernature.com

In case Publisher is established outside the EU, the EU authorized representative is:

Springer Nature Customer Service Center GmbH
Europaplatz 3
69115 Heidelberg, Germany

Batch number: 10761599

Printed by Printforce, the Netherlands